Table of Atomic Weights (Based on Carbon-12)

Element	Symbol	Atomic Number	Atomic Weight	Element	Symbol	Atomic Number	Atomic Weight
Actinium	Ac	89	(227)*	Mercury	Hg	80	200.6
Aluminum	Al	13	27.0	Molybdenum	Mo	42	95.9
Americium	Am	95	(243)	Neodymium	Nd	60	144.2
Antimony	Sb	51	121.8	Neon	Ne	10	20.2
Argon	Ar	18	39.9	Neptunium	Np	93	237.0
Arsenic	As	33	74.9	Nickel	Ni	28	58.7
Astatine	At	85	(210)	Niobium	Nb	41	92.9
Barium	Ba	56	137.3	Nitrogen	N	7	14.0
Berkelium	Bk	97	(245)	Nobelium	No	102	(254)
Beryllium	Be	4	9.01	Osmium	Os	76	190.2
Bismuth	Bi	83	209.0	Oxygen	O	8	16.0
Boron	B	5	10.8	Palladium	Pd	46	106.4
Bromine	Br	35	79.9	Phosphorus	P	15	31.0
Cadmium	Cd	48	112.4	Platinum	Pt	78	195.1
Calcium	Ca	20	40.1	Plutonium	Pu	94	(242)
Californium	Cf	98	(251)	Polonium	Po	84	(210)
Carbon	C	6	12.0	Potassium	K	19	39.1
Cerium	Ce	58	140.1	Praseodymium	Pr	59	140.9
Cesium	Cs	55	132.9	Promethium	Pm	61	(145)
Chlorine	Cl	17	35.5	Protactinium	Pa	91	231.0
Chromium	Cr	24	52.0	Radium	Ra	88	226.0
Cobalt	Co	27	58.9	Radon	Rn	86	(222)
Copper	Cu	29	63.5	Rhenium	Re	75	186.2
Curium	Cm	96	(245)	Rhodium	Rh	45	102.9
Dysprosium	Dy	66	162.5	Rubidium	Rb	37	85.5
Einsteinium	Es	99	(254)	Ruthenium	Ru	44	101.1
Erbium	Er	68	167.3	Samarium	Sm	62	150.4
Europium	Eu	63	152.0	Scandium	Sc	21	45.0
Fermium	Fm	100	(254)	Selenium	Se	34	79.0
Fluorine	F	9	19.0	Silicon	Si	14	28.1
Francium	Fr	87	(223)	Silver	Ag	47	107.9
Gadolinium	Gd	64	157.3	Sodium	Na	11	23.0
Gallium	Ga	31	69.7	Strontium	Sr	38	87.6
Germanium	Ge	32	72.6	Sulfur	S	16	32.1
Gold	Au	79	197.0	Tantalum	Ta	73	180.9
Hafnium	Hf	72	178.5	Technetium	Tc	43	98.9
Helium	He	2	4.00	Tellurium	Te	52	127.6
Holmium	Ho	67	164.9	Terbium	Tb	65	158.9
Hydrogen	H	1	1.008	Thallium	Tl	81	204.4
Indium	In	49	114.8	Thorium	Th	90	232.0
Iodine	I	53	126.9	Thulium	Tm	69	168.9
Iridium	Ir	77	192.2	Tin	Sn	50	118.7
Iron	Fe	26	55.8	Titanium	Ti	22	47.9
Krypton	Kr	36	83.8	Tungsten	W	74	183.8
Lanthanum	La	57	138.9	Uranium	U	92	238.0
Lawrencium	Lr	103	(257)	Vanadium	V	23	50.9
Lead	Pb	82	207.2	Xenon	Xe	54	131.3
Lithium	Li	3	6.94	Ytterbium	Yb	70	173.0
Lutetium	Lu	71	175.0	Yttrium	Y	39	88.9
Magnesium	Mg	12	24.3	Zinc	Zn	30	65.4
Manganese	Mn	25	54.9	Zirconium	Zr	40	91.2
Mendelevium	Md	101	(256)				

*Parentheses around atomic weight indicate that weight given is that of the most stable known isotope

Editor: Paul Corey
Production Supervisor: Elisabeth Belfer
Production Manager: Pamela Kennedy Oborski
Text and Cover Designer: Natasha Sylvester
Cover Photograph: © Ian Bradshaw/Woodfin Camp & Associates, Inc.
Photo Researcher: Chris Migdol   Except as otherwise indicated, color photographs are from Office of Learning Technologies, Indiana University–Purdue University.

This book was set in ITC Garamond by Polyglot Pte. Ltd.
printed and bound by Von Hoffmann Press
The cover was printed by Von Hoffmann Press

Copyright © 1991 by Macmillan Publishing Company, a division of Macmillan, Inc.
Printed in the United States of America

All rights reserved. No part of this book may be reproduced or transmitted in any form or by any means, electronic or mechanical, including photocopying, recording, or any information storage and retrieval system, without permission in writing from the publisher.

Earlier editions, entitled *General, Organic, and Biological Chemistry: Foundations of Life* copyright © 1986 and 1983 by Burgess Publishing

Macmillan Publishing Company
866 Third Avenue, New York, New York 10022

Collier Macmillan Canada, Inc.
1200 Eglinton Avenue, E.
Don Mills, Ontario, M3C 3N1

Library of Congress, Cataloging-in-Publication Data
Feigl, Dorothy M.
    Foundations of life: an introduction to general, organic, and biological chemistry/Dorothy M. Feigl, John W. Hill, Erwin Boschmann. — 3rd ed.
       p.   cm.
    Rev. ed. of: General, organic, and biological chemistry. 2nd ed. c1986.
    Includes index.
    ISBN 0-02-336737-7
    1. Chemistry. 2. Biochemistry. I. Hill, John William (date).
II. Boschmann, Erwin. III. Feigl, Dorothy M. General, organic, and biological chemistry. IV. Title.
QD31.2.F44 1991
540—dc20   90-6415
                                                                           CIP

Credits for Chapter Opening Photographs
1 © Fabian/Sygma   2 © Michael Gilbert/Photo Researchers, Inc./Science Photo Library   3 © Chuck O'Rear/Woodfin Camp & Associates   4 © Chuck O'Rear/Woodfin Camp & Associates   5 National Aeronautics and Space Administration   6 © Sygma   7 Hokusai, "The Wave," Private Collection, Art Resource   8 © Richard Megna/Fundamental Photographs   9 © Terrence Moore/Woodfin Camp & Associates   10 © Patrick Ward/Stock Boston   11 © Paul Silverman/Fundamental Photographs   12 © John Blaustein/Woodfin Camp & Associates   13 © Keith Gunnar/Photo Researchers, Inc.   14 Courtesy of the Colgate Palmolive Company   15 © Jerry Howard/Stock Boston   16 © Joseph Szabo/Photo Researchers, Inc.   17 © Dr. Stanley Cohen/Photo Researchers, Inc./Science Photo Library   18 © Michal Heron/Woodfin Camp & Associates

Printing: 1 2 3 4 5 6 7 8   Year: 1 2 3 4 5 6 7 8 9 0

THIRD EDITION

FOUNDATIONS OF LIFE

An Introduction to General, Organic, and Biological Chemistry

DOROTHY M. FEIGL
Saint Mary's College
Notre Dame, Indiana

JOHN W. HILL
University of Wisconsin–River Falls
River Falls, Wisconsin

ERWIN BOSCHMANN
Indiana University–Purdue University at Indianapolis
Indianapolis, Indiana

MACMILLAN PUBLISHING COMPANY New York

COLLIER MACMILLAN CANADA Toronto

PREFACE

Our world has been transformed by science and technology. The impact of science on the quality of human life is profound. Yet, to beginning students, the sciences that daily influence their lives often seem mysterious and incomprehensible. Those of us who enjoy the study of a science, however, find it a fascinating and rewarding experience precisely because it *can* provide reasonable explanations for seemingly mysterious phenomena.

Foundations of Life has been written in that spirit. Seemingly obscure phenomena are explained in an informal, readable style. We assume that the student has little or no chemistry background and clearly explain each new concept as it is introduced. Chemical principles and biological applications are carefully integrated throughout the text, with liberal use of drawings, diagrams, and photographs.

Our selection of topics and choice of examples make the text especially appropriate for students in health and life sciences, but it is also suitable for anyone seeking to become a better-informed citizen of our technological society. The text is appropriate for a one-quarter, one-semester, or two-quarter course.

Changes in the Third Edition

In response to the suggestions of users of the previous editions, our own experience, and the comments made by reviewers, we have made numerous modifications.

- ▶ The first eight chapters contain expanded treatments of
  measurements
  the periodic table
  electronic structure
  molecular geometry
  ideal gas law
  the mole
  solution preparation
  water
- ▶ The number of problems, exercises, and illustrations has been increased.

- Marginal comments and definitions have been added throughout.
- Each chapter opens with a series of interest-arousing "Did You Know That ..." questions.
- A self-quiz has been added at the end of each chapter.
- The problems at the ends of the chapters have been divided into groups that parallel the sections in the text.
- An extensive Glossary has been added.

Supplements

STUDY GUIDE. The Study Guide by Charles Baker and Claire Baker presents study skills and strategies, questions, paired additional problems (one of each pair is worked out), and a practice test for each chapter. The text itself includes many learning aids, but the Study Guide should help your students learn the material more effectively and more efficiently.

LABORATORY MANUAL. *Chemistry and Life in the Laboratory: Experiments in General, Organic, and Biological Chemistry* by Victor L. Heasley, Val J. Christensen, and Gene E. Heasley, complements the text very well. It also uses examples from students' lives to communicate the science of chemistry. The manual covers the same general topics as the textbook, instructions are clear and thorough, and the experiments are well written and imaginative. There are 36 experiments to choose from, and all have been thoroughly tested with students in the laboratory. Special care has been taken to minimize the cost of chemicals required in the laboratory and to make the experiments safe. An Instructor's Manual is available from the publisher.

INSTRUCTOR'S RESOURCE MANUAL. This manual contains lecture outlines, answers to all problems in the text, solutions to selected problems, and a test bank.

Acknowledgments

We are grateful for the advice and constructive criticism provided by those who reviewed the manuscript: Warren Bosch, Elgin Community College; Don Glover, Bradley University; Henry Harris, Armstrong State University; David Macaulay, William Rainey Harper College; Milica Nedelson, Oakton Community College; Justine Walhout, Rockford College; Lavern Weidler, Black Hawk College; and Normal Wells.

No book—or other educational device—can replace a good teacher; thus we have designed this book as an aid to the classroom teacher. The only valid test of this or any text is in a classroom. We would greatly appreciate receiving comments and suggestions based on your experience with this book.

D. M. F.
J. W. H.
E. B.

Contents

To the Student — xv

1 Matter and Measurement — 1

- 1.1 Matter and How to Classify It — 2
- 1.2 Energy and How to Classify It — 7
- 1.3 Electric Forces — 8
- 1.4 Measurement: The Modern Metric System — 10
- 1.5 Problem Solving: Dimensional Analysis — 13
- 1.6 Conversions Within the Metric System — 15
- 1.7 Conversions Between Systems — 20
- 1.8 Measuring Energy: Temperature and Heat — 22
- 1.9 Density — 26

Chapter Summary 31   Chapter Quiz 31   Problems 32

2 Atoms — 35

- 2.1 Dalton's Atomic Theory — 36
- 2.2 The Nuclear Atom — 38
- 2.3 The Bohr Model of the Atom — 43
- 2.4 Electron Arrangements — 46
- 2.5 The Periodic Table — 51
- 2.6 Some Chemistry of the Elements — 54
- 2.7 With So Many Models to Choose from … — 64

Chapter Summary 64   Chapter Quiz 65   Problems 65

3 Chemical Bonding — 69

- 3.1 The Stability of the Noble Gases — 70
- 3.2 The Ionic Bond — 71

3.3 Electron-Dot Formulas 74
3.4 Ionic Compounds 76
3.5 Names of Simple Ions and Ionic Compounds 78
3.6 Covalent Bonds 83
3.7 Polar Covalent bonds 85
3.8 Shapes of Molecules 89
3.9 Polyatomic Ions 94
3.10 Bonding Forces and the States of Matter 95
3.11 Dipole Forces 97
3.12 Hydrogen Bonds 98
3.13 Dispersion Forces 99
3.14 Forces in Solutions 100
3.15 Some Common Ions of Physiological Importance 102
Chapter Summary 103   Chapter Quiz 104   Problems 104

4 Nuclear Processes 107

4.1 Nuclear Arithmetic 108
4.2 Discovery of Radioactivity 111
4.3 Types of Radioactivity 113
4.4 Penetrating Power 116
4.5 Radiation Measurement 118
4.6 Half-life 119
4.7 Artificial Transmutation and Induced Radioactivity 121
4.8 Fission and Fusion 124
4.9 Radiation and Living Things 126
4.10 Nuclear Medicine 127
4.11 Medical Imaging 129
4.12 Other Applications 132
4.13 The Nuclear Age Revisited 133
Chapter Summary 134   Chapter Quiz 135   Problems 135

5 Gases 139

5.1 Air: A Mixture of Gases 141
5.2 The Kinetic Molecular Theory 142
5.3 Atmospheric Pressure 143
5.4 Boyle's Law: Pressure–Volume Relationship 144
5.5 Charles's Law: Volume–Temperature Relationship 150
5.6 Gay-Lussac's Law: Pressure–Temperature Relationship 153
5.7 Putting It All Together: The Combined Gas Law 156
5.8 Dalton's Law of Partial Pressures 158

5.9 Henry's Law: Pressure–Solubility Relationship 161
5.10 Partial Pressure and Respiration 162

Chapter Summary 165   Chapter Quiz 166   Problems 167

6 Chemical Reactions   171

6.1 The Chemical Equation 172
6.2 Chemical Arithmetic and the Mole 176
6.3 Gases Revisited 186
6.4 Energy Changes in Chemical Reactions 188
6.5 Reaction Rates 190
6.6 Equilibrium in Chemical Reactions: Le Châtelier's Principle 193
6.7 Oxidation–Reduction Reactions 196
6.8 Oxidation and Antiseptics 204

Chapter Summary 205   Chapter Quiz 206   Problems 207

7 Systems of Solvents, Solutes, and Solutions   211

7.1 Solutions: A Review of Some Definitions 213
7.2 Water: So Common, Yet So Unusual 213
7.3 Some Properties of Solutions 214
7.4 Qualitative Aspects of Solubility 215
7.5 Dynamic Equilibria and Saturated Solutions 218
7.6 Quantitative Aspects of Solubility 219
7.7 Strengths of Drugs 233
7.8 Dilution of Stock Solutions 234
7.9 Colligative Properties of Solutions 236
7.10 Osmosis 238
7.11 Colloids: The In-Between State 242
7.12 Dialysis 244
7.13 Solutions of the Body 246

Chapter Summary 248   Chapter Quiz 249   Problems 250

8 Acids and Bases   253

8.1 Acids and Bases: Some Definitions 254
8.2 Acids and Bases in Aqueous Solutions 256
8.3 Neutralization in Aqueous Solution 259

    8.4 Salts in Water: Acidic, Basic, or Neutral 264
    8.5 Reactions of Acids with Carbonates and Bicarbonates 268
    8.6 Antacids 271
    8.7 The pH Scale 273
    8.8 Buffers: The Control of pH 278
    8.9 Buffers in the Blood 280
    8.10 Acidosis and Alkalosis 281
Chapter Summary 283   Chapter Quiz 284   Problems 284

9 Hydrocarbons 287

    9.1 A Comparison of Organic and Inorganic Compounds 288
    9.2 The Hydrocarbons 289
    9.3 The Alkanes 290
    9.4 Isomerism 293
    9.5 IUPAC Nomenclature 295
    9.6 Properties of Alkanes 300
    9.7 Alkenes: Structure and Nomenclature 302
    9.8 Geometric Isomerism in Alkenes 305
    9.9 Properties of Alkenes 309
    9.10 Alkynes 312
    9.11 Cyclic Hydrocarbons 314
    9.12 Benzene 315
    9.13 Aromatic Hydrocarbons 316
    9.14 Hydrocarbons and Health 319
Chapter Summary 320   Chapter Quiz 321   Problems 321

10 Some Oxygen-Containing Organic Compounds 325

    10.1 Functional Groups 326
    10.2 The Alcohols: Nomenclature 328
    10.3 Physical Properties of Alcohols 332
    10.4 Chemical Properties of Alcohols 333
    10.5 Physiological Properties of Alcohols 335
    10.6 Multifunctional Alcohols 339
    10.7 The Phenols 340
    10.8 The Ethers 342
    10.9 General Anesthetics 343
    10.10 The Carbonyl Group: A Carbon–Oxygen Double Bond 345
    10.11 Naming Common Aldehydes 347

10.12 Naming Common Ketones 348
10.13 Physical Properties of Aldehydes and Ketones 350
10.14 Chemical Properties of Aldehydes and Ketones 351
Chapter Summary 355 Chapter Quiz 356 Problems 356

11 Organic Acids and Derivatives 359

11.1 Acids and Their Derivatives: The Functional Groups 360
11.2 Some Common Acids: Structures and Names 361
11.3 Physical Properties of Carboxylic Acids 364
11.4 Chemical Properties of Carboxylic Acids 365
11.5 Esters: Structures and Nomenclature 367
11.6 Physical Properties of Esters 369
11.7 Chemical Properties of Esters: Hydrolysis 370
11.8 Physiological Properties of Selected Esters 371
11.9 Amides: Structures and Nomenclature 374
11.10 Physical Properties of Amides 376
11.11 Chemical Properties of Amides: Hydrolysis 376
11.12 Physiological Properties of Selected Amides 377
Chapter Summary 379 Chapter Quiz 380 Problems 381

12 Amines 385

12.1 Structure and Classification of Amines 386
12.2 Simple Amines 387
12.3 Heterocyclic Amines 389
12.4 Physical Properties of Amines 391
12.5 Chemical Properties of Amines 392
12.6 Physiological Properties of Selected Amines and Derivatives 398
Chapter Summary 406 Chapter Quiz 407 Problems 407

13 Carbohydrates: Structure and Metabolism 411

13.1 Carbohydrates: Definitions and Classifications 412
13.2 Monosaccharides: Further Classifications 413
13.3 Mirror-Image Isomerism 413
13.4 Some Common Monosaccharides 415

13.5 More on the Structure of Monosaccharides 416
13.6 Some Reactions of Monosaccharides 419
13.7 Some Important Disaccharides 422
13.8 Polysaccharides 424
13.9 Digestion of Carbohydrates 426
13.10 Metabolism of Carbohydrates 427
13.11 Insulin and Diabetes Mellitus 431
Chapter Summary 432   Chapter Quiz 433   Problems 434

14 Lipids: Structure and Metabolism 437

14.1 What Is a Lipid? 438
14.2 Fatty Acids 439
14.3 Fats: The Triglycerides 440
14.4 Soaps and the Saponification of Fats 443
14.5 Phosphatides 446
14.6 Nonsaponifiable Lipids: Steroids and Prostaglandins 447
14.7 Digestion and Metabolism of Fats 450
14.8 The Ketone Bodies 454
14.9 The Chemistry of Starvation 456
Chapter Summary 457   Chapter Quiz 458   Problems 458

15 Proteins Structure and Metabolism 461

15.1 Amino Acids: Structure and Physical Properties 462
15.2 The Essential Amino Acids 466
15.3 Amino Acids as Buffers 467
15.4 Formation of Proteins: The Peptide Bond 467
15.5 The Sequence of Amino Acids in Proteins 469
15.6 More on the Structure of Proteins 472
15.7 Denaturation of Proteins 479
15.8 Classification of Proteins 480
15.9 Enzymes 481
15.10 Digestion and Metabolism of Proteins 483
15.11 Endorphins and Enkephalins 488
Chapter Summary 489   Chapter Quiz 490   Problems 490

Contents

16 Energy and Life 493

16.1 Life and the Laws of Thermodynamics 494
16.2 Predicting Spontaneous Change: Free Energy 495
16.3 Coupled Reactions 496
16.4 ATP: Universal Energy Currency 498
16.5 Synthesis of ATP 500
16.6 Oxidative Phosphorylation 502
16.7 Mitochondria 506
16.8 Efficiency of Cellular Energy Transformations 507
16.9 Muscle Power 508

Chapter Summary 512   Chapter Quiz 512   Problems

17 Nucleic Acids 515

17.1 The Structure of Nucleic Acids 516
17.2 The Secondary Structure of Nucleic Acids: Base Pairing 521
17.3 DNA: Self-replication 524
17.4 RNA: Protein Synthesis and the Genetic Code 527
17.5 Genetic Engineering 531

Chapter Summary 534   Chapter Quiz 534   Problems 535

18 Vitamins, Minerals, and Hormones 539

18.1 Minerals 540
18.2 Vitamins 542
18.3 Hormones 549
18.4 Cyclic AMP: An Opportunity to See How Far We Have Come 555

Chapter Summary 558   Chapter Quiz 558   Problems 559

Appendixes

A The International System of Measurement 561

B Significant Figures 565

C Exponential Notation 569

Glossary	575
Answers to Chapter Quizzes	585
Answers to Selected Problems	586
Index	597

TO THE STUDENT

What is chemistry?

Chemistry is such a broad, all-encompassing area of study that people almost despair in trying to define it. Indeed, some have given up and simply state that chemistry is "what chemists do." But that won't do; it's much too narrow a view.

Chemistry is what we all do. We bathe, clean, and cook. We put chemicals on our faces, hands, and hair. Collectively, we use more than 10 000 consumer chemical products in our homes. Professionals in the health and life sciences use thousands of additional chemicals as drugs, antiseptics, or reagents for diagnostic tests.

Your body itself is a remarkable chemical factory. You eat and breathe, taking in raw materials for the factory. You convert these supplies into an unbelievable array of products, some incredibly complex. This chemical factory—your body—also generates its own energy. It detects its own malfunctions and can regenerate and repair some of its component parts. It senses changes in its environment and adapts to these changes. With the aid of a neighboring facility, this fabulous factory can create other factories much like itself.

Everything you do involves chemistry. You read this sentence; light energy is converted to chemical energy. You think; protein molecules are synthesized and stored in your brain. All of us are chemists.

Chemistry affects society as well as individuals. Chemistry is the language—and the principal tool—of the biological sciences, the health sciences, and the agricultural and earth sciences.

Chemistry has illuminated all of the natural world, from the tiny atomic nucleus to the immense cosmos. We believe that a knowledge of chemistry can help you. We have written this book in the firm belief that beginning chemistry can be related immediately to problems and opportunities in the life and health sciences. And we believe that this can make the study of chemistry interesting and exciting, especially to nonchemists.

For example, an "ion" is more than a chemical abstraction. Mercury ions in the wrong place can kill you, but calcium ions in the right place can keep you from bleeding to death. "$P \times V = c$" is an equation, but it is also the basis for the respiratory therapy that has saved untold lives in hospitals. "Hydrogen bonding" is a chemical phenomenon, but it also accounts for the fact that a dog has puppies, whereas a cat has kittens and a human has human offspring. Hundreds of similar fundamental and interesting applications of chemistry to life can be cited.

A knowledge of chemistry has already had a profound effect on the quality of life. Its impact on the future will be even more dramatic. At present we can control diabetes, cure some forms of cancer, and prevent some forms of mental retardation because of our understanding of the chemistry of the body. We can't *cure* diabetes or cure *all* forms of cancer or *all* mental retardation because our knowledge is still limited. So learn as much as you can. Your work will be enhanced and your life enriched by your greater understanding.

1

MATTER AND MEASUREMENT

DID YOU KNOW THAT...

- all life is sustained by chemical processes?
- there are over 100 known elements?
- mass and weight are not synonymous?
- water is not an atom, an element, or a mixture?
- you have been using the metric system most of your life?
- heat and temperature are totally different concepts?
- a Calorie is actually 1000 calories?

The title of this book is *Foundations of Life*. What does chemistry have to do with life? What is chemistry? For that matter, what is life? The last question is more than rhetorical. The definition of human life has been debated before committees of Congress. Progress in science, technology, and medicine has blurred the distinction between life and death. Is someone whose heart has stopped beating dead? Is someone whose vital functions are being maintained by machine alive? We will not even attempt in this book to supply a definitive answer to the question "What is life?" We will simply note the critical significance of the question for our society.

How about the first question. "What does chemistry have to do with life?" A chemist would say, "Just about everything." The human body, for example, is the most extraordinarily complicated, most elegantly designed, and most efficiently operated chemical laboratory there is. Our attempt to answer the first question will fill most of this text.

Chemistry—the science of matter

That leaves the middle question, "What is chemistry?" **Chemistry** is defined as a study of matter and the changes it undergoes. Since changes in matter are accompanied by changes in energy, chemistry must also concern itself with energy transformations. The American Chemical Society says

> The chemical sciences deal with the composition, structure, and properties of substances and with the transformations they undergo.

Matter and energy make up the physical universe. Therefore the study of chemistry takes us from atoms to stars, from rocks to living organisms.

Scientific advances are in part the result of either curiosity or a perceived need. Thus, Roy Plunkett was curious about the residue of a reaction and discovered Teflon, whereas Fritz Haber saw a need to make ammonia and discovered a process that is now used to make over 30,000 tons of ammonia per day. Advances are most likely to occur when scientists observe with total lack of bias and use only reproducible data. Scientific theories eventually result when the hard facts allow for a generalized explanation. The process of asking, observing, and generalizing is referred to as the **scientific process**.

To begin our study of chemistry it is necessary to define some terms that we will encounter again and again and to familarize ourselves with the system of measurement used by chemists and other scientists.

1.1 Matter and How to Classify It

Matter occupies space and has mass.

Matter and energy are such fundamental concepts that definitions are difficult. **Matter** is the stuff that makes up all material things. It occupies space and has mass. Wood, sand, water, and people have mass and occupy space. So does air, although one usually needs a flat tire or a 30-mph wind to emphasize the point.

Mass is a measure of the quantity of matter. **Weight**, on the other hand, measures a force. On Earth, it is a measure of the force of attraction (the gravitational force) between the Earth and the mass in question. For most of its history, the human race was restricted to the surface of the Earth and found it convenient to use the terms *mass* and *weight* interchangeably. If something had twice the mass of something else, it also weighed twice as much. So, in most cases, it did not really matter whether mass or weight was discussed. When the exploration of space began, however, astronauts and cosmonauts began to demonstrate dramatically that mass and weight are not the same thing. The mass of an astronaut on the moon is the same as his or her mass on Earth. The matter that makes up the astronaut does not change. His or her weight on the moon, however, is one-sixth that on Earth because the moon's pull is only one-sixth as strong as the Earth's (Figure 1.1). Weight varies with gravity; mass does not.

Matter is characterized by its *properties*. This means that we know we have water and not gasoline because water has certain characteristics or properties that distinguish it from gasoline. For example, it is hard to start a fire by dousing something with water and then lighting it with a match. **Chemical properties** describe how one substance reacts with other substances. To demonstrate a chemical property, a

Mass—a measure of the amount of matter

Weight—a measure of the force of attraction

Chemical properties are observed in chemical reactions.

FIGURE 1.1
Astronaut John W. Young leaps from the lunar surface, where gravity pulls at him with only one sixth the force on Earth. [NASA photograph.]

Physical properties are observed without changing the composition.

substance must undergo a change in composition. In exhibiting its **physical properties**, a substance undergoes no change in composition. Characteristics such as color, hardness, density, and melting point are physical properties. Sulfur is a brittle yellow solid that is more dense than water. Each of these physical properties of sulfur can be evaluated without changing the composition of the sulfur. We can also cite some chemical properties of sulfur: it reacts with oxygen, with carbon, and with iron. These reactions yield, in turn, sulfur dioxide, carbon disulfide, and iron sulfide—all new substances.

Matter is also characterized by its *changes*. The burning of a candle demonstrates both a physical change and a chemical change. After the candle is lighted, the solid wax near the burning wick melts (a physical change because in both solid form and liquid form the composition of the wax is the same). Some of the melted wax is drawn into the burning wick, where a chemical change occurs. The wax combines with oxygen in the air to form carbon dioxide gas and water vapor. As the candle burns and the wax undergoes this chemical change, the candle becomes smaller and smaller. However, the apparent disappearance of something is not necessarily a sign that we are observing a chemical change. If a glass of water is allowed to stand for many days, the water seems to disappear. In this instance, however, we are observing a physical change. The water is evaporating, changing from a liquid to a gas, but in both forms it is still water. Thus, it is not always easy to identify a physical or a chemical change. The critical question is: Has the fundamental composition of the substance been changed? In a chemical change it has; in a physical change it has not.

Some changes are **reversible**, and some are **irreversible**. Thus melted wax returns to solid when it cools, and water vapor (steam) can become liquid again. However, making flour from wheat is an irreversible physical process. Mercuric oxide can be decomposed to mercury and oxygen, and these can reverse the process to form mercuric oxide. The burning wax, on the other hand, undergoes an irreversible chemical change to carbon dioxide, water, and energy.

Matter comes in three familiar **states**: solid, liquid, and gas. **Solid** objects ordinarily maintain their shape and volume regardless of their location. **Liquids** assume the shape of their container (except for a generally flat surface at the top). Like solids, however, liquids maintain a fairly constant volume. If you have a 12-oz soft drink, you have 12 oz whether the soft drink is in a can, a bottle, or, through some slight mishap, on the floor—which demonstrates another property of liquids. Unlike solids, liquids flow readily. **Gases** maintain neither shape nor volume. They expand to fill completely whatever container one puts them in. Gases can be easily compressed. For example, enough air for many minutes of breathing can be compressed into a steel tank for underwater diving. We shall take up the topic of the states of matter in more detail in later chapters.

1.1 Matter and How to Classify It

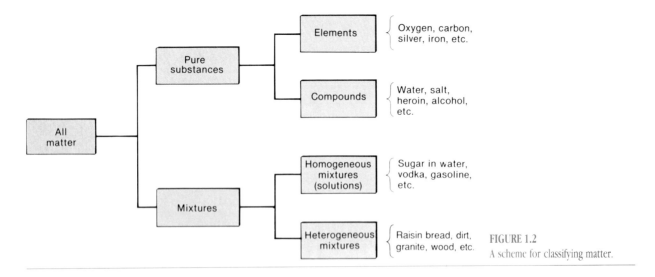

FIGURE 1.2
A scheme for classifying matter.

There are other ways of classifying matter. As you can see from Figure 1.2, we can subdivide matter into pure substances and mixtures. **Pure substances** have a definite or fixed composition. The composition of **mixtures** may vary. Water is a pure substance; it always contains 11% hydrogen and 89% oxygen. Pure gold is pure gold, that is, 100% gold. Lemonade, on the other hand, is a mixture. The proportions of lemon juice and water change depending on who is preparing the lemonade. Mixed nuts are a mixture; the ratio of peanuts to pecans depends on how much you are willing to pay per pound.

Homogeneous mixtures have an even distribution of all mixed parts. So, in sugar water a drop of the solution has the same composition whether taken from the top, the middle, or the bottom of the beaker. On the other hand, a **heterogeneous mixture** has its component parts separated as in milk of magnesia or dirt in water.

Pure substances may be either elements or compounds. **Elements** are those fundamental substances from which all material things are constructed. **Compounds** are pure substances that are made up of two or more elements combined in fixed proportions. Our concepts of elements have changed over time. Fire was once considered an element but is now regarded as nonmaterial, that is, as a form of energy. Water was once considered an element, but we now know it to be a compound composed of two elements, hydrogen and oxygen. We currently regard as elements more than 100 pure substances that cannot be broken down by chemical means into simpler substances. Sulfur, oxygen, carbon, and iron are elements. Sulfur dioxide, carbon disulfide, and iron sulfide are compounds.

Because elements are so fundamental to our study of chemistry, we shall find it useful to refer to them in a shorthand form. Each element can be represented by a symbol of one or two letters derived from the name of the element (or, sometimes, the Latin name of the element).

Elements cannot be chemically broken down into anything simpler.

Compounds are elements combined in definite proportions.

TABLE 1.1 Names, Symbols, and Physical Characteristics of Some Common Elements

Name (Latin name)	Symbol	Selected Properties
Aluminum	Al	Silvery lightweight metal
Argon	Ar	Colorless gas
Arsenic	As	Grayish white solid
Barium	Ba	Silvery white metal
Beryllium	Be	Steel gray, hard, lightweight metal
Boron	B	Black or brown powder; several crystal forms
Bromine	Br	Reddish brown liquid (Br_2)
Cadmium	Cd	White ductile metal
Calcium	Ca	Silvery white metal
Carbon	C	Soft black solid (graphite) or hard brilliant crystal (diamond)
Cerium	Ce	Steel gray metal
Cesium	Cs	Soft silvery metal
Chlorine	Cl	Greenish yellow gas (Cl_2)
Chromium	Cr	Hard silvery metal
Cobalt	Co	Hard silvery metal
Copper (cuprum)	Cu	Light reddish brown metal
Fluorine	F	Pale yellow gas (F_2)
Gallium	Ga	Steel gray metal
Germanium	Ge	Grayish white solid
Gold (aurum)	Au	Yellow malleable metal
Helium	He	Colorless gas
Hydrogen	H	Colorless gas (H_2)
Iodine	I	Lustrous black solid (I_2)
Iron (ferrum)	Fe	Silvery white, ductile, malleable metal
Krypton	Kr	Colorless gas
Lead (plumbum)	Pb	Bluish white, soft, heavy metal
Lithium	Li	Silvery white, soft, lightweight metal

The first letter of the symbol is always capitalized; the second is always lowercase. (It does make a difference. For example, Hf is the symbol for hafnium, an element; HF is the formula for hydrogen fluoride, a compound. Similarly, Co is cobalt, an element; CO is carbon monoxide, a compound.)

Symbols for some of the more important elements are given in Table 1.1. It is well worth your time to memorize the names and symbols of as many of the elements shown in this table as you can. The structure of compounds will be examined in detail in Chapter 3. You will find that discussion much easier to follow if you are familiar with the common elemental symbols.

Seven elements exist in nature as bonded pairs: H_2, O_2, N_2, F_2, Cl_2, Br_2, I_2. The subscript 2 means "two bonded atoms." For consistency,

TABLE 1.1 (*Continued*)

Name (Latin name)	Symbol	Selected Properties
Magnesium	Mg	Silvery white, ductile, lightweight metal
Manganese	Mn	Grayish white, hard, brittle metal
Mercury (hydrargyrum)	Hg	Silvery white, liquid, heavy metal
Molybdenum	Mo	Silvery white metal
Neon	Ne	Colorless gas
Nickel	Ni	Silvery white, ductile, malleable metal
Nitrogen	N	Colorless gas (N_2)
Oxygen	O	Colorless gas (O_2)
Palladium	Pd	Silvery white metal
Phosphorus	P	Yellowish white waxy solid or red powder (P_4)
Platinum	Pt	Grayish white heavy metal
Plutonium	Pu	Silvery white radioactive metal
Potassium (kalium)	K	Silvery white soft metal
Radon	Rn	Colorless gas
Scandium	Sc	Silvery metal
Selenium	Se	Bluish gray solid
Silicon	Si	Lustrous gray solid
Silver (argentum)	Ag	Silvery white metal
Sodium (natrium)	Na	Silvery white soft metal
Strontium	Sr	Silvery white metal
Sulfur	S	Yellow solid (S_8)
Technetium	Tc	Synthetic radioactive element
Tin (stannum)	Sn	Silvery white soft metal
Titanium	Ti	Silvery or dark gray hard metal
Uranium	U	Silvery radioactive metal
Vanadium	V	Pale gray metal
Xenon	Xe	Colorless gas
Zinc	Zn	Bluish white metal

the periodic table lists them just as H, O, N, etc.; however, when used as they appear in nature, they must be shown as diatomic species.

1.2 ENERGY AND HOW TO CLASSIFY IT

Energy is defined as the capacity for doing work. (By this definition, play that involves exercise is considered work.) But energy is more than that; it is the basis for change in the material world. When something moves or breaks or cools or shines or grows or decays, energy is involved. Energy can be divided into two general categories, potential and kinetic.

An object has **potential energy** in a particular arrangement or place. Its potential energy is derived from its position relative to another

Energy is the ability to cause change.

Potential energy is stored energy.

FIGURE 1.3
A boulder on a cliff represents potential energy. A falling boulder represents kinetic energy. The falling boulder can do work (smash a house, for example). The boulder on the cliff has a potential for doing the same thing.

object. The usual example given for such a system is a boulder on a cliff (Figure 1.3). Because it is at the top of the cliff and not at the bottom, the boulder has potential energy. The boulder can fall from the cliff and destroy a house at the base (thereby doing work), whereas a similar boulder resting on the ground at the base of the cliff cannot do the same thing. The larger the boulder and the higher its position relative to the ground, the greater is its potential energy. One extremely important form of potential energy is the energy stored in chemical compounds. Sugar contains energy (calories). Life depends on the controlled release of this store of potential energy, and biological systems have developed marvelously intricate mechanisms for utilizing the energy.

Kinetic energy is active energy.

Kinetic energy is the energy an object possesses because of its motion. A boulder at the base of the cliff can destroy the house if it is rolling along the ground at a good clip. Its ability to do work in this case depends on its motion. If you were really determined to destroy that house, you would use a large boulder rather than a small stone. Given the choice, you should also arrange to have the boulder moving rapidly rather than slowly. In other words, kinetic energy *depends on mass and velocity*. The bigger an object is and the faster it is moving, the more kinetic energy it has and the more work it can do.

1.3 ELECTRIC FORCES

Chemistry often borrows concepts from its neighboring discipline, physics. One such concept is force. A **force** is a push or a pull that sets an object in motion, or stops a moving object, or holds an object in

1.3 Electric Forces

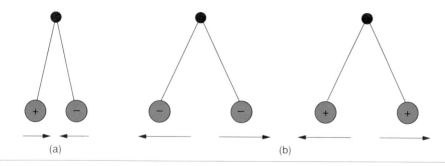

FIGURE 1.4
(a) Particles with unlike charges attract one another. (b) Those with like charges repel one another.

place. As we noted earlier, gravity is a force, and weight is one measure of that force.

Electric forces are extremely important in chemistry. Some particles of matter are neutral, but some bear an electric charge, either positive (+) or negative (−). No one can tell you exactly what an electric charge is. We simply accept the fact that a particle with a "charge" can exert a force, that is, can push or pull another particle that also has a "charge" on it. The particles do not have to be touching to attract or repel each other. For this reason, we say that charged particles have force fields around them. Even at a distance they attract and repel one another, although these forces get weaker as the particles get farther apart. Particles with like charges (both positive or both negative) repel one another. Those with unlike charges (one positive and one negative) attract one another (Figure 1.4).

Opposite charges attract. Like charges repel.

This phenomenon of charged substances is not unfamiliar to you. Anyone who has pulled clothes from an automatic dryer on a cold winter day has probably seen what commercials like to call "static cling," pieces of clothing sticking to one another (Figure 1.5). The

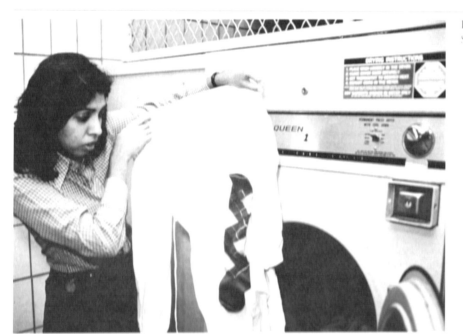

FIGURE 1.5
Static cling. [Ken Karp/MGH.]

"cling" is due to the attraction of *unlike* charges. If, on the other hand, you brush your hair vigorously on a cold day, your hair may become "unmanageable" (another term used often in advertisements). Instead of lying flat against your head, the hair sticks out in all directions, each strand seemingly trying to get way from all the other strands. That is exactly what is happening; the strands have *like* charges on them and repel one another.

Most matter, as we encounter it, is electrically neutral. Positively charged particles attract negatively charged particles. Because of this attractive force, particles of opposite charge tend to combine to produce samples of matter that carry no *net* charge.

1.4 Measurement: The Modern Metric System

Accurate measurements of such quantities as mass (weight), volume, time, and temperature are essential to the compilation of dependable scientific "facts." A chemist interested in basic research may use such facts, but similar information is of critical importance in every science-related field. Certainly we are all aware that measurements of both temperature and blood pressure are routinely made in medicine. It is also true that modern medical diagnosis depends on a whole battery of other measurements, including careful chemical analysis of blood and urine.

A variety of systems of measurement have been used around the world. The familiar English system of feet, pounds, and quarts is now being phased out in many quarters in the United States, one of the last nations to do so. Scientists around the world have long used the *metric system,* and hospitals and the health professions are well along in their conversion to this system. More recently, scientists adopted an updated metric plan called the International System of Units, or SI (from the French *Système International*).

The metric system is a decimal system, and that is its beauty. Converting from one unit in the system to another involves shifting a decimal point, as is the case with our monetary system, where one dollar contains 100 cents and a dime is $0.1. No multiplication or division by some odd factor is necessary. In the metric system, the basic unit of length is the **metre** (spelled **meter** in the United States); a unit of mass is the **gram**; and a unit of volume* is the **litre** (**liter**).

---

*Volume can be expressed as length cubed. The SI unit of volume is the cubic meter (m^3). A more convenient unit for us is the cubic centimeter (cm^3). The symbol cm^3 is also read (and written) sometimes as "cc." Note that 1 mL = 1 cm^3 = 1 cc.

1.4 Measurement: The Modern Metric System

All other units can be derived from these by the use of prefixes. For example,

$$1 \text{ millimeter (mm)} = 0.001 \text{ meter (m)}$$
$$1 \text{ milligram (mg)} = 0.001 \text{ gram (g)}$$
$$1 \text{ milliliter (mL)} = 0.001 \text{ liter (L)}$$
$$1 \text{ kilogram (kg)} = 1000 \text{ grams (g)}$$
$$1 \text{ kilometer (km)} = 1000 \text{ meters (m)}$$

Table 1.2 contains a more complete list of the metric prefixes and Table 1.3 shows metric conversions. Refer to Appendixes A, B, and C for more details. If you are not familiar with metric units, you might find the following English equivalents helpful for establishing in your mind the "size" of the metric units (see also Figure 1.6).

$$1.0 \text{ m} = 39 \text{ in. or } 1.1 \text{ yd}$$
$$1.0 \text{ L} = 1.1 \text{ qt}$$
$$1.0 \text{ kg} = 2.2 \text{ lb}$$

TABLE 1.2 Approved Numerical Prefixes

Exponential Expression	Decimal Equivalent	Prefix	Phonetic	Symbol
10^{12}	1 000 000 000 000	tera-	ter' a	T
10^{9}	1 000 000 000	giga-	ji' ga	G
10^{6}	1 000 000	mega-	meg' a	M
10^{3}	1 000	kilo-	kil' o	k
10^{2}	100	hecto-	hek' to	h
10	10	deka-	dek' a	da
10^{0}	1 (unit)			
10^{-1}	0.1	deci-	des' i	d
10^{-2}	0.01	centi-	sen' ti	c
10^{-3}	0.001	milli-	mil' i	m
10^{-6}	0.000 001	micro-	mi' kro	μ
10^{-9}	0.000 000 001	nano-	nan' o	n
10^{-12}	0.000 000 000 001	pico-	pe' ko	p
10^{-15}	0.000 000 000 000 001	femto-	fem' to	f
10^{-18}	0.000 000 000 000 000 001	atto-	at' to	a

The most commonly used units are in color.

TABLE 1.3 Metric Conversions

Factors	Length	Mass	Volume
kilo means × 1000	1 m × 1000 = 1 km	1 g × 1000 = 1 kg	1 L × 1000 = 1 kL
hecta means × 100	1 m × 100 = 1 hm	1 g × 100 = 1 hg	1 L × 100 = 1 hL
deca means × 10	1 m × 10 = 1 dm	1 g × 10 = 1 dag	1 L × 10 = 1 daL
Unit × 1	1 m × 1 = 1 meter	1 g × 1 = 1 gram	1 L × 1 = 1 liter
deci means × 0.1	1 m × 0.1 = 1 dm	1 g × 0.1 = 1 dg	1 L × 0.1 = 1 dL
centi means × 0.01	1 m × 0.01 = 1 cm	1 g × 0.01 = 1 cg	1 × 0.01 = 1 cL
milli means × 0.001	1 m × 0.001 = 1 mm	1 g × 0.001 = 1 mg	1 L × 0.001 = 1 mL
1 kilo = 10 hecta	1 km = 10 hm	1 kg = 10 hg	1 kL = 10 hL
= 100 deca	= 100 dam	= 100 dag	= 100 daL
= 1000 units	= 1000 m	= 1000 g	= 1000 L
= 10,000 deci	= 10,000 dm	= 10,000 dg	= 10,000 dL
= 100,000 centi	= 100,000 cm	= 100,000 cg	= 100,000 cL
= 1,000,000 milli	= 1,000,000 mm	= 1,000,000 mg	= 1,000,000 mL
1 milli = 0.1 centi	1 mm = 0.1 cm	1 mg = 0.1 cg	1 mL = 0.1 cL
= 0.01 deci	= 0.01 dm	= 0.01 dg	= 0.01 dL
= 0.001 units	= 0.001 m	= 0.001 g	= 0.001 L
= 0.000 1 deca	= 0.000 1 dam	= 0.000 1 dag	= 0.000 1 daL
= 0.000 01 hecta	= 0.000 01 hm	= 0.000 01 hg	= 0.000 01 hL
= 0.000 001 kilo	= 0.000 001 km	= 0.000 001 kg	= 0.000 001 kL

Try to develop a "sense" for the size of metric units. As you work problems in this text, and, more important, when your professional responsibilities include working with numerical data, you will want to be able to assess the reasonableness of an answer. If you were told that a room measures 2 in. by 4 in. or that you should take a 2-gal dose of cough medicine or that the average human heart weighs 300 lb, you would know that a mistake had been made. But if you were instructed to draw a 12-L sample of blood for a clinical test, would you recognize that by the time half of that amount was collected the patient literally would have been drained dry? If not, then now is the time to develop the ability to detect obvious errors with metric units.

Physical measurement, as contrasted to counting, is never exact. One can count exactly 42 people in a classroom, but measurement of the width of the room would give differing results, depending on the care taken and the accuracy of the measuring device. Measurement is discussed further in Appendix A, where more extensive conversion tables are provided. Significant figures are treated in Appendix B, and exponential notation is reviewed in Appendix C.

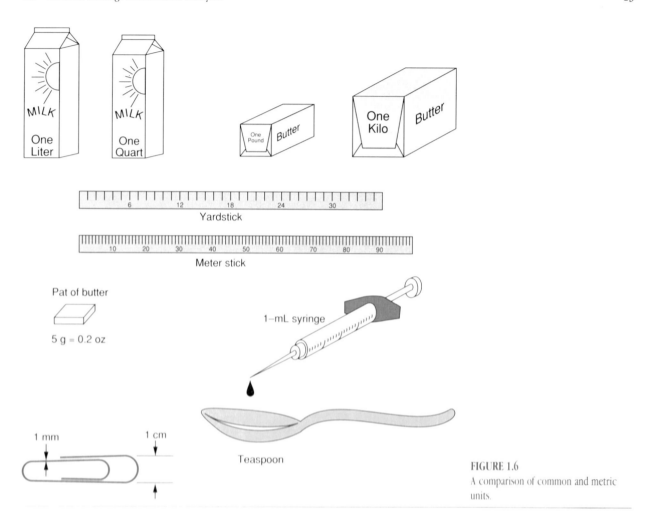

FIGURE 1.6
A comparison of common and metric units.

1.5 PROBLEM SOLVING: DIMENSIONAL ANALYSIS

One of the advantages of the metric system is the ease with which one can convert between units in the system. Before we demonstrate that, we would like to consider the broader topic of problem solving, and specifically solving numerical problems. Chemistry is not all math, but a number of key concepts are best expressed in the language of mathematics. An even more compelling reason for you to develop an ability to carry out mathematical calculations is that many professions, particularly the health professions, require it. Medication dosages, clinical tests, even a patient's chart involve the manipulation of numerical data. Medication that can save a life if used in the proper concentration can kill if misused.

A widely useful approach to numerical problems is called **dimensional analysis** or sometimes the **factor-label method** or **unit-conversion method**. The method employs units such as L, cm^3, or

Dimensional analysis uses units to set up problems

mph as aids in setting up and solving problems. The general approach is to multiply the known quantity (and its units!) by one or more conversion factors so that the answer is obtained in the desired units.

$$\begin{matrix}\text{Known quantity}\\ \text{and unit(s)}\end{matrix} \times \begin{matrix}\text{conversion}\\ \text{factor(s)}\end{matrix} = \begin{matrix}\text{answer in}\\ \text{desired units}\end{matrix}$$

Quantities can be expressed in a variety of units. For example, you can buy beverages by the 12-oz can or by the pint, quart, gallon, or liter. If you wish to compare prices, you must be able to convert from one unit to another (Figure 1.7). Such a conversion changes the numbers and units, but it does not change the quantity. Your actual height, for example, remains unchanged whether it is expressed in feet or centimeters.

You know that multiplying a number by 1 does not change its value. Multiplying by a fraction equal to 1 also leaves the value unchanged. A fraction is equal to 1 when the numerator is equal to the denominator. For example, we know that

$$1 \text{ ft} = 12 \text{ in.}$$

Therefore, the fraction

$$\frac{1 \text{ ft}}{12 \text{ in.}} = 1$$

Similarly,

$$\frac{12 \text{ in.}}{1 \text{ ft}} = 1$$

If we wish to convert from inches to feet, we can do so by choosing one of the fractions as a *conversion factor*. *Which* one do we choose? The one that gives us an answer with the right units! Let us illustrate by an example. (Appendix B discusses significant figures in numerical calculations. We urge you to review that material before proceeding.)

FIGURE 1.7
Using conversions to find the best buy.
[Meredith Davenport.]

EXAMPLE 1.1

My bed is 72 in. long. What is its length in feet?

You know the answer, of course, but let us proceed to show *how* the answer is obtained by using dimensional analysis. We need to multiply 72 in. by one of the above fractions. Which one? The known quantity and unit is 72 in.

$$72 \text{ in.} \times \text{conversion factor} = ? \text{ ft}$$

For the conversion factor, choose the fraction that, when inserted

1.6 Conversions Within the Metric System

in the equation, cancels the unit *in.* and becomes the unit *ft.*

$$72 \text{ in.} \times \frac{1 \text{ ft}}{12 \text{ in.}} = 6.0 \text{ ft}$$

Just for kicks, let us try the other conversion factor.

$$72 \text{ in.} \times \frac{12 \text{ in.}}{1 \text{ ft}} = \frac{860 \text{ in.}^2}{\text{ft}}$$

Absurd! How can a bed be 860 in.²/ft? You should have no difficulty in choosing between the two possible answers.

1.6 Conversions Within the Metric System

We stated that one of the advantages of the metric system was the ease of conversion between units. Let us try to demonstrate that with a few examples. Remember that a list of equivalent values, such as those on page 11, is actually a list of conversion factors. Thus, the equality

$$1 \text{ kilogram} = 1000 \text{ grams}$$

can be rearranged into two useful conversion factors:

$$\frac{1 \text{ kilogram}}{1000 \text{ grams}} \quad \text{and} \quad \frac{1000 \text{ grams}}{1 \text{ kilogram}}$$

EXAMPLE 1.2

Convert 0.742 kg to grams.

$$0.742 \text{ kg} \times \frac{1000 \text{ g}}{1 \text{ kg}} = 742 \text{ g}$$

EXAMPLE 1.3

Convert 1247 cm to meters.

$$1247 \text{ cm} \times \frac{1 \text{ m}}{100 \text{ cm}} = 12.47 \text{ m}$$

The process is exactly the same as it is with conversion within the English system. Consider the next two examples.

EXAMPLE 1.4

Convert 0.742 lb to ounces.

$$0.742 \text{ lb} \times \frac{16 \text{ oz}}{1 \text{ lb}} = 11.9 \text{ oz}$$

EXAMPLE 1.5

Convert 1247 in. to yards.

$$1247 \text{ in.} = \frac{1 \text{ yd}}{36 \text{ in.}} = 34.64 \text{ yd}$$

In conversions using English units, you multiply and divide by factors such as 16 or 36. In metric conversions, you multiply and divide by 100 or 1000 and so on. You need only shift the decimal point when doing metric conversions.

Conversion factors are not usually given in a problem. They can be obtained from listings such as those on page 11 or the more extensive listings in Appendix A. However, you would be wise to learn to convert within the metric system without the need of tables. Also, you should remember that the equality

$$1 \text{ centimeter} = 0.01 \text{ meter}$$

is equivalent to

$$100 \text{ centimeters} = 1 \text{ meter}$$

All these fractions

$$\frac{1 \text{ cm}}{0.01 \text{ m}} \qquad \frac{0.01 \text{ m}}{1 \text{ cm}} \qquad \frac{100 \text{ cm}}{1 \text{ m}} \qquad \frac{1 \text{ m}}{100 \text{ cm}}$$

are valid conversion factors.

1.6 Conversions Within the Metric System

EXAMPLE 1.6

Knowing that 1 mL = 0.001 L, write four conversion factors relating milliliters and liters.

The first two conversion factors can be formed by arranging the two sides of the equality in the form of a fraction.

$$\frac{1 \text{ mL}}{0.001 \text{ L}} \quad \text{or} \quad \frac{0.001 \text{ L}}{1 \text{ mL}}$$

To derive the other two conversion factors, first multiply both sides of the equality by 1000 (in order to obtain the equality in terms of 1 L).

$$1000 \times 1 \text{ mL} = 1000 \times 0.001 \text{ L}$$

$$1000 \text{ mL} = 1 \text{ L}$$

Now just arrange this last equality in fractional form.

$$\frac{1000 \text{ mL}}{1 \text{ L}} \quad \text{or} \quad \frac{1 \text{ L}}{1000 \text{ mL}}$$

The conversion factors 1 mL/0.001 L and 1000 mL/1 L would give exactly the same answer if used in a problem. The only difference is convenience. Some people would rather multiply by 1000 than divide by 0.001. In this age of the electronic calculator, perhaps even this difference is no longer significant.

Let us try some more conversions within the metric system.

EXAMPLE 1.7

How many milliliters are there in a 2-L bottle of soda pop?
From memory or from the tables in Appendix A you find that

$$1 \text{ L} = 1000 \text{ mL}$$

$$2 \text{ L} \times \frac{1000 \text{ mL}}{1 \text{ L}} = 2000 \text{ mL}$$

Notice that we picked the conversion factor that allowed us to cancel liters and obtain an answer in the desired units, milliliters.

Sometimes it is necessary to carry out more than one conversion in a problem.

EXAMPLE 1.8

In the United States, the usual soda pop can holds 360 mL. How many such cans could be filled from one 2.0-L bottle?
 The problem tells us that

$$1 \text{ can} = 360 \text{ mL}$$

Using that equivalence, we can calculate the answer.

$$2.0 \text{ L} \times \frac{1000 \text{ mL}}{1 \text{ L}} \times \frac{1 \text{ can}}{360 \text{ mL}} = 5.6 \text{ cans}$$

EXAMPLE 1.9

How many 325-mg aspirin tablets can be made from 875 g of aspirin?
 The problem asks us to convert the *given* value of 875 g to tablets. The problem also includes a necessary conversion factor:

$$1 \text{ tablet} = 325 \text{ mg}$$

From memory or the tables, we have another required conversion factor:

$$1 \text{ g} = 1000 \text{ mg}$$

By multiplying the given value by the appropriately arranged conversion factors, we arrive at the answer.

$$875 \text{ g} \times \frac{1000 \text{ mg}}{1 \text{ g}} \times \frac{1 \text{ tablet}}{325 \text{ mg}} = 2690 \text{ tablets}$$

EXAMPLE 1.10

Treatment of a bacterial infection calls for 2 g of ampicillin per day. The label on the bottle indicates that each tablet contains

1.6 Conversions Within the Metric System

250 mg of ampicillin. How many tablets per day are required for the prescribed dosage?

This is essentially the same problem as the preceding one; only some details have been changed. We know that

$$1 \text{ tablet} = 250 \text{ mg}$$

We need to convert the prescribed 2 g of ampicillin to a number of tablets.

$$2 \text{ g} \times \frac{1000 \text{ mg}}{1 \text{ g}} \times \frac{1 \text{ tablet}}{250 \text{ mg}} = 8 \text{ tablets}$$

EXAMPLE 1.11

A sprinter runs the 100-m dash in 11 sec. What is her speed in kilometers per hour?

The given speed is 100 m per 11 sec.

$$\frac{100 \text{ m}}{11 \text{ sec}}$$

The conversion factors that we need are found in the tables or recalled from memory.

$$\frac{100 \text{ m}}{11 \text{ sec}} \times \frac{1 \text{ km}}{1000 \text{ m}} \times \frac{60 \text{ sec}}{1 \text{ min}} \times \frac{60 \text{ min}}{1 \text{ hr}} = 33 \text{ km/hr}$$

Note that the first conversion factor changes *m* to *km*. It takes two factors to change *sec* to *hr*. Note also that we could have first applied the factors that convert *sec* to *hr* and then converted *m* to *km*. The answer would have been the same.

EXAMPLE 1.12

If your heart beats at a rate of 72 times per minute, and your lifetime will be 70 years, how many times will your heart beat during your lifetime?

Two equivalences are given in the problem.

$$72 \text{ beats} = 1 \text{ min}$$
$$1 \text{ lifetime} = 70 \text{ yr}$$

Three others that you will need you can recall from memory.

$$1 \text{ yr} = 365 \text{ days}$$
$$1 \text{ day} = 24 \text{ hr}$$
$$1 \text{ hr} = 60 \text{ min}$$

Start with the factor 72 beats/1 min (the known quantities and units) and apply the conversion factors as needed to get an answer in beats/lifetime (the desired units).

$$\frac{72 \text{ beats}}{1 \text{ min}} \times \frac{60 \text{ min}}{1 \text{ hr}} \times \frac{24 \text{ hr}}{1 \text{ day}} \times \frac{365 \text{ days}}{1 \text{ yr}} \times \frac{70 \text{ yr}}{1 \text{ lifetime}}$$
$$= 2\,600\,000\,000 \text{ beats/lifetime}$$

1.7 Conversions Between Systems

To convert from one system of measurement to another, you need a list of equivalents such as the one on page 11 or the more extensive lists in Appendix A. Let us plunge right in and work some examples.

EXAMPLES 1.13

How many kilograms are there in 33 lb?

$$33 \text{ lb} \times \frac{1.0 \text{ kg}}{2.2 \text{ lb}} = 15 \text{ kg}$$

EXAMPLE 1.14

You know that your weight is 142 lb, but the job application form asks for your weight in kilograms. What is it?

From the table we find

$$1.00 \text{ lb} = 0.454 \text{ kg}$$

The solution is simple.

1.7 Conversions Between Systems

$$142 \text{ lb} \times \frac{0.454 \text{ kg}}{1.00 \text{ lb}} = 64.5 \text{ kg}$$

EXAMPLE 1.15

A recipe calls for 750 mL of milk, but your measuring cup is calibrated in fluid ounces. How many ounces of milk will you need?

$$750 \text{ mL} \times \frac{1.00 \text{ fl oz}}{29.6 \text{ mL}} = 25 \text{ fl oz}$$

EXAMPLE 1.16

How many meters are there in 764 ft?

$$764 \text{ ft} \times \frac{12 \text{ in.}}{1 \text{ ft}} \times \frac{1.00 \text{ m}}{39.4 \text{ in.}} = 233 \text{ m}$$

EXAMPLE 1.17

How would you describe a young man who is 1.6 m tall and weighs 91 kg?

$$1.6 \text{ m} \times \frac{39 \text{ in.}}{1.0 \text{ m}} \times \frac{1 \text{ ft}}{12 \text{ in.}} = 5.2 \text{ ft}$$

$$0.2 \text{ ft} \times \frac{12 \text{ in.}}{1 \text{ ft}} = 2.4 \text{ in}$$

$$91 \text{ kg} \times \frac{2.2 \text{ lb}}{1.0 \text{ kg}} = 200 \text{ lb}$$

The young man is about 5 ft 2 in. tall and weighs 200 lb. Let us be generous and say that he is well muscled.

It is possible (and frequently necessary) to manipulate units in the denominator as well as in the numerator of a problem. Just remember to use conversion factors in such a way that the unwanted units cancel.

1.8 Measuring Energy: Temperature and Heat

In the United States, weather reports and cooking recipes still use the Fahrenheit temperature scale for the most part. This scale defines the freezing temperature of water as 32 °F and the boiling point as 212 °F. Between these two points, the scale is divided into 180 (that is, 212 − 32) equal units, each a Fahrenheit degree. Most people in the rest of the world—and scientists everywhere, including the United States—use the Celsius scale, which defines the freezing temperature of water as 0 °C and the boiling point as 100 °C. The scale between these two points is divided into 100 equal Celsius degrees. Thus, the Celsius degree is larger than the Fahrenheit degree. Only 100 of the larger Celsius degrees are required to cover the same temperature range as 180 Fahrenheit degrees (Figure 1.8).

Now assume that we would like to relate a reading of 20 °C to the Fahrenheit scale. While the Celsius scale begins at 0 °C, the Fahrenheit scale has a head start and begins at 32 °F. Thus °F = 32 + ⋯. Now, every Celsius degree unit is larger than the Fahrenheit degree unit by 1.8 or 9/5. So, the complete formula is

$$°F = 32 + \frac{9}{5}(20) = 68 \, °F$$

To rearrange the relationship for Celsius, simply solve for °C to obtain

$$°C = \frac{5}{9}(°F - 32)$$

FIGURE 1.8
A comparison of the Fahrenheit and Celsius temperature scales.

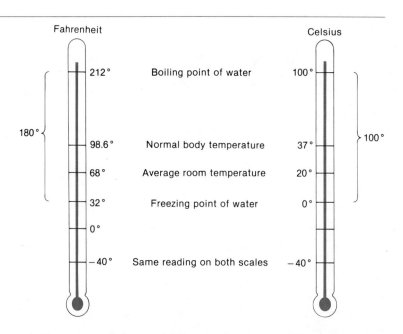

1.8 Measuring Energy: Temperature and Heat

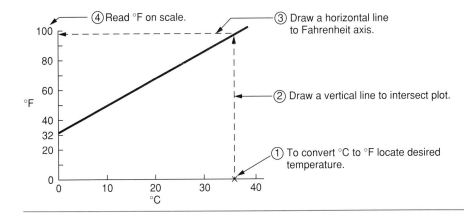

FIGURE 1.9
Relating Fahrenheit to Celsius scales.

Figure 1.9 illustrates the relationship between Fahrenheit and Celsius temperatures. Note that when the Celsius scale is at 0, the Fahrenheit scale is at 32.

EXAMPLE 1.18

The temperature in Tucson reached 113 °F on a summer day. What temperature would that be on the Celsius scale?

$$\frac{5}{9}(113 - 32) = \frac{5}{9}(81) = 45\,°C$$

EXAMPLE 1.19

If a woman has a temperature of 40 °C, how sick is she?

$$32 + \frac{9}{5}(40) = 32 + 72 = 104\,°F$$

Pretty sick.

EXAMPLE 1.20

The temperature at St. Louis dropped to −10 °C on a winter night. What was the temperature on the Fahrenheit scale?

$$32 + \frac{9}{5}(-10) = 32 - 18 = 14\,°F$$

> **EXAMPLE 1.21**
>
> The temperature of a food freezer should be maintained at 0 °F. What is the temperature on the Celsius scale?
>
> $$\frac{5}{9}(0-32) = \frac{5}{9}(-32) = -18 \text{ °C}$$

The preceding problems should have given you a "feel" for the Celsius temperature scale.

The SI unit for temperature is the kelvin (K). The kelvin and the Celsius degree are the same size, but the zero point on the Kelvin scale equals -273 °C. The Kelvin scale is called an *absolute scale* because its zero point is the coldest temperature possible, called absolute zero. This fact was determined by theoretical considerations and has since been confirmed by experiment. The absolute scale has no negative temperatures. To convert from degrees Celsius to kelvins, add 273 to the Celsius temperature.

$$K = °C + 273$$

Note that K carries no degree sign.

0 K = −273 °C

> **EXAMPLE 1.22**
>
> What is the boiling point of water in kelvins?
>   The boiling point of water is 100 °C.
>
> $$100 + 273 = 373 \text{ K}$$

> **EXAMPLE 1.23**
>
> The surface temperature of Titan, Saturn's largest moon, was found by Voyager I to be 94 K. What is the temperature in °C?
>   If K = °C + 273, then °C = K − 273.
>
> $$94 - 273 = -179 \text{ °C}$$

Scientists often need to measure amounts of heat energy. You should not confuse heat with temperature. **Heat** is a measure of quantity, that is, of *how much energy* a sample contains. **Temperature** is a measure

Heat measures the amount of energy in matter; calorie is a unit of energy.

1.8 Measuring Energy: Temperature and Heat

of intensity, that is, of *how energetic* each individual particle of the sample is. A glass of water at 70 °C contains less heat than a bathtub of water at 60 °C. The particles of water in the glass are more energetic, on the average, than those in the tub, but there is far more water in the tub and its total heat content is greater. A familiar unit of heat energy is the calorie (cal). A **calorie** is the amount of heat required to raise the temperature of 1 g of water 1 °C. There is a more precise definition than this, but for our purposes this simpler version will do.

Temperature measures how energetic individual particles are.

If we were to heat 2 g of water to increase the temperature by 1 °C, it would take 2 cal, and if the 2 g was heated up by 3 °C, it would take 6 cal. Generally, then,

Calories = mass × temperature change × heat capacity

The **heat capacity** is a measure of the amount of heat needed to effect a temperature change. For water it is 1 cal per gram per °C. Heat conductors such as metals heat up fast, and insulators such as water heat up slowly. The specific heat of iron is 0.12 cal/[(g)(°C)], which means it takes only 0.12 cal to increase the temperature of 1 g of iron by 1 °C.

EXAMPLE 1.24

How many calories would it take to raise the temperature of 250 g of water from 25 °C to 100 °C?

The definition of *calorie* indicates that 1 cal is required for every degree of change for every gram of water. Thus, the **heat capacity** of water is

$$\frac{1 \text{ cal}}{(1 \text{ g})(1 \text{ °C})}$$

The temperature change is 75 °C (that is, 100 °C − 25 °C), and the amount of water is 250 g.

$$250 \text{ g} \times 75 \text{ °C} \times \frac{1 \text{ cal}}{(1 \text{ g})(1 \text{ °C})} = 19\,000 \text{ cal}$$

The **large Calorie** (note the capital C) is used for the measurement of the energy content of foods. The large Calorie is more correctly called a *kilocalorie*. A dieter may be aware that a banana split contains 1500 Cal (kcal). If the same dieter realized that meant 1 500 000 calories, giving up the banana split might be easier (Figure 1.10).

FIGURE 1.10
A section from a Calorie chart. The units here are large Calories or kilocalories. Multiply by 1000 to get values in calories.

Desserts	Portion	Calories
Cake		
Angel Food Cake	$\frac{1}{12}$ of tube cake	135
Devils Food Cake, with chocolate icing	$\frac{1}{16}$ of 2-layer cake	235
Coffee Cake	$\frac{1}{6}$ cake, $2\frac{1}{2}$ oz.	230
Gingerbread	$\frac{1}{9}$ of 8″ cake	175
Fruit Cake, dark	$\frac{1}{2}$ oz.	55
Pound, loaf	$\frac{1}{7}$ of a loaf	195
Brownies, with nuts	$1\frac{3}{4}″ \times 1\frac{3}{4}″$	95
Cookies		
Chocolate Chip, commercial	1	50
Oatmeal, with raisins	1	60
Vanilla Wafers	1	18
Gingersnap, 2″ diameter	1	22

EXAMPLE 1.25

If a hot-water bag contains 1.0 kg of water at 65 °C, how much heat will it have supplied to someone's aching muscles by the time it has cooled to 20 °C?

$$1.0 \text{ kg} = 1000 \text{ g}$$

$$1000 \text{ g} \times (65 - 20) \text{ °C} \times \frac{1 \text{ cal}}{(1 \text{ g})(1 \text{ °C})}$$

$$= 1000 \text{ g} \times 45 \text{ °C} \times \frac{1 \text{ cal}}{(1 \text{ g})(1 \text{ °C})}$$

$$= 45\,000 \text{ cal or } 45 \text{ kcal}$$

1.9 Density

An important property of matter, particularly in scientific work, is density (a measure of compactness). When we speak of lead as "heavy" or aluminum as "light," we are referring to the density of these metals. **Density** is defined as the amount of mass per unit volume.

$d = \dfrac{m}{V}$

$$\text{Density} = \frac{\text{mass}}{\text{volume}}$$

1.9 Density

Density values are usually reported in grams per milliliter (g/mL) or grams per cubic centimeter (g/cm^3 or g/cc).

Remember:
$1 \text{ mL} = 1 \text{ cm}^3 = 1 \text{ cc}$

EXAMPLE 1.26

A bottle filled to the 25-mL mark contains 75 g of bromine. What is the density of bromine?

If you know that density is ordinarily reported in grams per milliliter, then you know how to set up the problem. Arrange the data so that the answer will come out in grams per milliliter. The density of bromine is

$$\frac{75 \text{ g}}{25 \text{ mL}} = 3.0 \text{ g/mL}$$

EXAMPLE 1.27

What is the density of iron if 94 g of iron occupies a volume of 12 cm^3?

$$\frac{94 \text{ g}}{12 \text{ cm}^3} = 7.8 \frac{\text{g}}{\text{cm}^3}$$

EXAMPLE 1.28

The density of alcohol is 0.79 g/mL. What is the mass of 150 mL of alcohol?

The problem asks that we convert a volume (150 mL) to the equivalent mass. The density serves as a conversion factor that relates volume and mass.

$$150 \text{ mL} \times \frac{0.79 \text{ g}}{1.0 \text{ mL}} = 120 \text{ g}$$

EXAMPLE 1.29

What is the mass of 1.0 L of gasoline if the density of gasoline is 0.66 g/mL?

We are given the volume in liters, but our conversion factor (the density) uses milliliters as the volume unit. Therefore, the given volume must be converted from liters to milliliters before the density is used to convert from volume to mass.

$$1.0 \,\cancel{L} \times \frac{1000 \,\cancel{mL}}{1 \,\cancel{L}} \times 0.66 \,\frac{g}{\cancel{mL}} = 660 \text{ g}$$

EXAMPLE 1.30

What volume is occupied by 451 g of mercury? The density of mercury is 13.6 g/mL.

This time we want to convert from mass to volume. Density will still serve as the conversion factor, but it must be inverted to be used correctly.

$$461 \,\cancel{g} \times \frac{1 \text{ mL}}{13.6 \,\cancel{g}} = 33.9 \text{ mL}$$

The density of water is 1 g/mL, or 1 g/cm³, or 1 g/cc (remember, 1 mL = 1 cm³).* This nice round number for the density of water is not an accident. Metric units were frequently defined to give "neat" values for many physical properties of water (melting point, boiling point, density, heat capacity).

A term related to density is **specific gravity**. Specific gravity is the ratio of the mass of any substance to the mass of an equal volume of water. Another way of saying this is

$$\text{sp gr}_{subst.} = \frac{d_{subst.}}{d_{water}}$$

$$\text{Specific gravity of a substance} = \frac{\text{density of the substance}}{\text{density of water}}$$

The specific gravity of water itself, therefore, is always 1.

Because it is the ratio of two values with equivalent units, specific gravity is a number without units. If we work in the metric system, in which the density of water is 1 g/mL, the specific gravity of a substance is numerically the same as its density.

---

*Density actually varies with temperature, but for liquids and solids the variation is small. The density of water is greatest at 4 °C and is defined as 1 g/mL at that temperature. At 0 °C, water has a density of 0.999 87 g/mL, and at 25 °C, 0.997 07 g/mL.

1.9 Density

EXAMPLE 1.31

The density of chloroform is 1.5 g/mL. What is its specific gravity?

$$\frac{1.5 \text{ g/mL}}{1.0 \text{ g/mL}} = 1.5$$

Mercury (the silvery liquid in thermometers, not the planet) has a specific gravity of 13.6, that is, it has a density 13.6 times that of water. If a mixture of water and mercury were placed in a container, the mercury would sink to the bottom of the container while the water layer would float on top of the mercury layer. The specific gravity of human fat is 0.903. If mixed with water, the fat would float on top. The higher the proportion of body fat in a person, the more buoyant that person is in water (Figure 1.11).

Potential blood donors are often screened for anemia through a simple test that determines the specific gravity of their blood. A drop of blood is released into a solution of known specific gravity. The blood droplet either rises (because its specific gravity is less than that of the solution) or sinks (because its specific gravity is greater) or remains stationary (because the specific gravities are equal). A low specific gravity suggests that the concentration of iron-rich hemoglobin in the

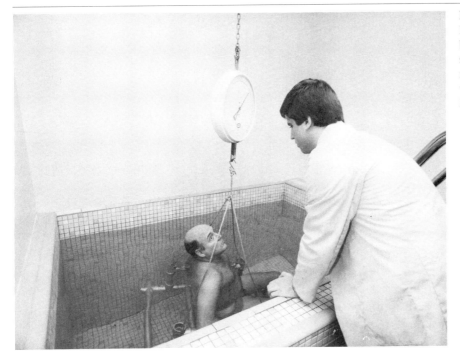

FIGURE 1.11
Percent of body fat can be determined by weighing a person who is completely submerged in water. The calculation must include a correction for the volume of air in the lungs. [Will & Deni McIntyre/Photo Researchers, Inc.]

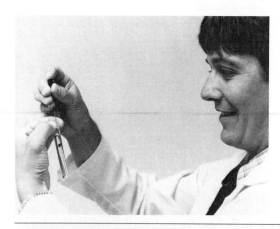

FIGURE 1.12
Tests for anemia can be conducted using a simple test tube or a complex machine. *Left:* A technician determines the relative density of blood in the test tube. *Below:* A machine counts the number of blood cells. [Photographs by Raymond C. Carballada, Department of Medical Photography, Geisinger Medical Center.]

FIGURE 1.13
A hydrometer. The one shown here measures specific gravities over a range of 0.700–0.770.

blood is too low. Normally the specific gravity of blood ranges between about 1.05 and 1.07 (Figure 1.12).

A device called a hydrometer measures the specific gravity of a liquid. The hydrometer is placed in a solution whose specific gravity is to be measured. How far the hydrometer dips down into the liquid is determined by the density of the liquid (Figure 1.13). The stem of the hydrometer is calibrated in such a way that the specific gravity can be read directly at the surface of the liquid. Hydrometers can be used to measure the strength of the "battery acid" in your car, the sugar content in maple syrup, dissolved solids in urine, antifreeze in the radiator, and many other properties related to the specific gravity of solutions.

CHAPTER SUMMARY

Chemistry is the study of matter and its changes. Matter is anything that has mass and occupies space. Mass is a measure of the amount of matter, but weight is sometimes used interchangeably with mass, even though weight actually is a measure of the force of attraction between two masses.

Matter is characterized by its chemical and physical properties. Chemical properties describe the way in which the composition of substances changes, whereas physical properties demonstrate qualities associated with no change in composition of a substance.

There are three states of matter. In the solid state, matter maintains a fixed shape and volume. Liquids maintain their volume but flow much more readily than solids. Gases maintain neither their volume nor their shape but can expand or contract to fill any container they occupy.

Matter may be in the form of a pure substance of fixed composition, either an element or a compound. An element is one of the slightly more than 100 fundamental substances that cannot be converted by chemical means into simpler substances. Compounds are substances that consist of fixed proportions of more than one element. In chemistry, the elements are represented by shorthand symbols involving one or two letters of the name of the element (Table 1.1). In addition to pure substances, matter exists in the form of mixtures, whose composition may vary.

Changes in matter are accompanied by changes in energy, defined as the capacity for doing work. The two general classifications are potential energy, or energy due to position, and kinetic energy, or energy due to motion.

Particles of matter may bear two different types of electric charges, positive or negative. Those particles with like charges repel one another, and those with unlike charges attract one another.

Measurement in chemistry and all other sciences involves the use of the metric or SI system. In the metric system the meter is a unit of length, the gram is a unit of mass, and the liter is a unit of volume. Other units within the system can be derived from these through the use of prefixes (Table 1.2).

While the Fahrenheit temperature scale is still widely used in the United States, scientists employ either the Celsius scale or an absolute temperature scale whose unit is the kelvin. Conversion among these scales is easily accomplished through the use of simple equations. Temperature is a measure of heat *intensity*, that is, the average energy of individual particles of matter. The *quantity* of heat energy is reported in a different unit, the calorie. A calorie is the amount of heat required to raise the temperature of 1 g of water 1 °C. The energy content of food is frequently reported in terms of *kilocalories*, although many listings use the term Calorie (capital *C*) in place of kilocalorie.

The amount of matter in a given volume is reported as density. In the metric system, density is usually given in g/mL or g/cm^3. Specific gravity gives a related measure of the concentration of matter. Specific gravity is the ratio of the density of any substance to the density of water.

Science often requires mathematical calculations. An excellent approach to numerical problem solving involves the use of conversion factors. Conversion factors are fractions relating two different units. In the method known as dimensional analysis, the units serve as aids in setting up and solving problems. In this text, dimensional analysis will be used whenever possible to solve problems.

CHAPTER QUIZ

1. The symbol Na stands for which element?
   a. nitrogen   b. sodium
   c. potassium   d. gold
2. The height of a tall man is
   a. 2 cm   b. 2 m   c. 2 km   d. 2 mm
3. A glass of milk contains
   a. 0.3 μL   b. 0.3 mL
   c. 0.3 L   d. 0.3 cm^3
4. The weight of a newborn baby might be
   a. 3 mg   b. 3 g   c. 3 kg   d. 3 cg
5. If the outside temperature is 45 °C, in the midwest it is
   a. summer   b. winter
   c. fall   d. spring

6. A "mild fever" would correctly describe a body temperature of
   a. 35 °C   b. 38 °C   c. 42 °C   d. 101 °C

7. The density of water is
   a. 1.0 g/cm^3
   b. 1.2 g/cm^3
   c. depends on volume
   d. depends on mass

8. To cool 10.0 g of water from 45 °C to 40 °C requires how many calories?
   a. 5   b. 10   c. 50   d. 400

9. Ten inches approximately equals
   a. 25 cm   b. 9 cm   c. 1 m   d. 0.5 km

10. Ten m is the same as
    a. 100 cm   b. 1 mm   c. 0.010 km   d. 1000 hm

PROBLEMS

1.1 Matter and How to Classify It

1. Define the following terms.
   a. chemistry
   b. matter
   c. element
   d. compound

2. Distinguish between chemical and physical properties.

3. Which describes a physical change, and which describes a chemical change?
   a. Sheep are sheared and the wool is spun into yarn.
   b. Silkworms feed on mulberry leaves and produce silk.

4. Which describes a physical change, and which describes a chemical change?
   a. Because a lawn is watered and fertilized it grows thicker.
   b. An overgrown lawn is manicured by mowing it with a lawn mower.

5. Which describes a physical change, and which describes a chemical change?
   a. Ice cubes form when a tray filled with water is placed in a freezer.
   b. Milk left outside a refrigerator for many hours turns sour.

6. How do pure substances and mixtures differ?

7. All samples of the sugar glucose consist of 8 parts (by weight) oxygen, 6 parts carbon, and 1 part hydrogen. Is glucose a pure substance or a mixture?

8. Identify each of the following as a pure substance or a mixture.
   a. carbon dioxide   b. oxygen
   c. smog

9. Identify each of the following as a pure substance or a mixture.
   a. gasoline   b. mercury
   c. soup

10. Identify each of the following as a pure substance or a mixture.
    a. a carrot   b. 24-karat gold

11. How do gases, liquids, and solids differ in their properties?

12. Which is the most compressible form of water: steam, ice, or liquid water?

13. Explain the difference between mass and weight.

14. Two samples are weighed under identical conditions in a laboratory. Sample A weighs 1 lb, and sample B weighs 2 lb. Does B have twice the mass of A?

15. What has changed when a person completes a successful diet: the person's weight or the person's mass?

16. Sample A, which is on the moon, has exactly the same mass as sample B, which is on Earth. Do the two samples weigh the same?

17. Which of the following represent elements, and which represent compounds?
    a. H   b. He   c. HF

18. Which of the following represent elements, and which represent compounds?
    a. C   b. CO   c. Ca

19. Which of the following represent elements, and which represent compounds?
    a. CO_2 b. Cl_2 c. $CaCl_2$

1.2 Energy and How to Classify It

20. What is the difference between kinetic energy and potential energy?

21. Which has the greater kinetic energy: a sprinter or a long-distance runner (assume the two weigh the same)?

22. Which has the greater kinetic energy: a cannonball or a bullet fired at the same speed?

23. Which has the greater kinetic energy: a bicyclist traveling at 15 mph or an automobile traveling at 50 mph?

24. Which has the greater kinetic energy: a 110-kg football tackle moving slowly across the field or an 80-kg halfback racing quickly down the field?
25. Which has the greater potential energy: a diver on the 1-m board or that diver on the 10-m platform?
26. Which has the greater potential energy: an elevator stopped at the twentieth floor or one stopped at the twelfth floor?
27. Which has the greater potential energy: a roller coaster as it starts to climb the first hill or as it reaches the top of the hill?

1.3 Electric Forces

28. Describe what happens to two particles with like charges when they are brought close together. What happens to particles with unlike charges when they are brought close together?

1.4 Measurement: The Modern Metric System

29. For each of the following, indicate which is the larger unit.
    a. mm or cm   b. kg or g
    c. dL or μL
30. For each of the following, indicate which is the larger unit.
    a. L or cc   b. cm^3 or mL
31. For each of the following, indicate which is the larger unit.
    a. in. or m   b. lb or g
    c. L or gal

1.5 Problem Solving: Dimensional Analysis

32. How many milliliters are there in 1 cm^3? In 15 cc?
33. Convert 0.10 km/sec to miles per minute.
34. Calculate your weight (mass) in kilograms and your height in centimeters.

1.6 Conversions Within the Metric System

35. How many meters are there in each of the following?
    a. 50 km   b. 25 cm   c. 76 mm

36. How many millimeters are there in each of the following?
    a. 1.5 m   b. 16 cm   c. 2.5 km
37. How many deciliters are there in each of the following?
    a. 1.0 L   b. 20 mL   c. 15 cm^3
38. How many liters are there in each of the following?
    a. 2056 mL   b. 47 kL   c. 5.2 dL
39. Make the following conversions.
    a. 15 000 mg to grams
    b. 0.086 g to milligrams
40. Make the following conversions.
    a. 0.149 L to milliliters
    b. 47 mL to liters

1.7 Conversions Between Systems

41. Make the following conversions.
    a. 1.5 qt to milliliters
    b. 18 L to quarts

1.8 Measuring Energy: Temperature and Heat

42. For each of the following, indicate which is the larger unit.
    a. °C or °F   b. cal or Cal
43. Convert the following to degrees Celsius.
    a. 68 °F   b. 25 °F   c. −10 °F
44. Convert the following to degrees Fahrenheit.
    a. 15 °C   b. −31 °C   c. 212 °C
45. Convert the following to kelvins.
    a. 37 °C   b. −100 °C   c. 273 °C
46. Which is hotter: 100 °C or 100 °F?
47. Order the temperatures from coldest to hottest: 0 K, 0 °C, 0 °F.
48. How many calories are there in each of the following?
    a. 2.75 kcal   b. 0.74 Cal
49. How many calories of heat would be required to raise the temperature of 50 g of water from 20 °C to 50 °C?
50. How much heat is released if 2.0 kg of water cools from 90° C to 20 °C?
51. One gram of fat provides 9 kcal of energy. A nutrition article says margarine, a fat, provides 100 kcal per tablespoon (Tbs). How many grams of margarine are there in 1.0 Tbs of margarine?

1.9 Density

52. What is the density of a salt solution if 50 cm^3 of the solution has a mass of 57 g?
53. A 1.0-L container of carbon tetrachloride has a mass of 1.6 kg. What is the density of carbon tetrachloride?
54. What is the mass of 50 mL of mercury? The density of mercury is 13.6 g/mL.
55. In Problem 50, you calculated the mass of a tablespoon (Tbs) of margarine. If 1.0 Tbs has a volume of 15 mL, what is the density of the margarine?
56. If the density of a normal urine sample is 1.02 g/mL, what is its specific gravity?
57. A tall, layered drink called a pousse-café is made of four different liqueurs. The liqueurs (20.0 mL of each) are carefully poured into a single container so that they do not mix. The names, colors, and masses (per 20.0 mL) of the liqueurs follow.

Liqueur	Color	Mass (g)
Cassis (black currant)	Dark red	23.2
Creme de menthe	Green	21.8
Irish Mist	Amber	20.4
Triple Sec	White	21.2

Calculate the density of each liqueur, and make a diagram of the drink, indicating the color of each layer.

2
ATOMS

DID YOU KNOW THAT...

- the idea of atoms has been around for 2000 years?
- many of John Dalton's proposals about atoms are wrong?
- most of the mass of an atom resides in a negligible volume?
- all chemistry depends on electrons, yet electrons have essentially no mass and occupy no space?
- electrons are both matter and energy?
- colored flames are caused by electron motions?
- an electron is here, there, and everywhere—yet nowhere?
- the arrangement of the periodic table was first conceived of on the basis of chemical properties and has since been reconfirmed on the basis of electron structure?
- the periodic table is the only true language of the universe?

An atom is the smallest chemically obtainable portion of an element.

An **atom** is defined as the smallest characteristic particle of an element. This concept has been around for thousands of years. For most of that time, however, it was not a popular idea. The development of chemistry as a science was largely a matter of accepting the existence of atoms and then working to define the properties of the atom more and more precisely.

As originally proposed by both ancient Hindu and Greek scholars, the concept of the atom was as much philosophical as practical. Both groups spoke of earth, water, fire, and air as elements composed of eternal and unchanging atoms in perpetual motion. For the Hindus, it was Brahma, the soul of the universe, who set the atoms in motion. Some Greek philosophers believed that atoms were in motion because they were falling toward the center of the universe. The atomistic view of matter fell into disfavor when the greatest of the Greek philosophers, Aristotle, elected to side with those who believed that matter was continuous, that is, infinitely divisible. According to Aristotle and his followers, there was no particle of matter so small that it could not be subdivided into still smaller pieces. It is a tribute to Aristotle that his view prevailed for two millennia.

A gradual accumulation of data that could not be explained within Aristotle's philosophy finally forced a reevaluation of his views. The reassessment and ultimate rejection of Aristotle's concept of the nature of matter was a major victory for experimental science. From that point onward, scientific theories were accepted or rejected on the basis of their ability to explain experimental data and not on their beauty or elegance or even their appeal to common sense.

2.1 Dalton's Atomic Theory

The man whose scientific theory did explain all the available experimental data was John Dalton (1766–1844) an English schoolteacher (Figure 2.1). Dalton accepted the atomistic view of matter and presented the modern atomic theory. The theory is called "modern" because it was based on the best evidence available in Dalton's time (the early nineteenth century) rather than on ideas formulated in Aristotle's time (fourth century B.C.). The most important points of Dalton's atomic theory are

1. All elements are made up of small, indestructible, and indivisible particles called atoms.
2. All atoms of a given element are identical, but the atoms of one element differ from the atoms of any other element.
3. Atoms of different elements can combine to form compounds.
4. A chemical reaction involves a change not in the atoms themselves, but in the way atoms are combined to form compounds.

2.1 Dalton's Atomic Theory

FIGURE 2.1
John Dalton. [Courtesy of The Smithsonian Institution, Washington, DC.]

What did Dalton mean when he said that all atoms of a given element are the same? Were the atoms of an element all the same color or the same shape? Rather—and this was Dalton's most important insight—he proposed that the mass of an atom identifies it uniquely. Atoms are extremely minute. In the nineteenth century, it was impossible to determine actual masses of atoms. Indirect measurements, however, could indicate their relative masses. These relative masses are now expressed in terms of **atomic mass units (amu)**. This unit of mass was invented for atoms. According to Dalton, hydrogen atoms have a mass of 1 amu, and oxygen atoms have a mass of six times as much as hydrogen atoms, or 6 amu.

Atomic mass units (amu) are used to express relative masses.

Dalton was wrong. He was wrong about the mass of oxygen atoms. He was wrong about all atoms of the same element being identical. He was even wrong about atoms being indivisible and indestructible. The most amazing thing about Dalton is that he was wrong about so many things, yet he is still regarded, and rightly so, as one of the giants of modern chemistry. Other scientists took up, and then modified, his ideas, and a new era in chemistry began. Science is not simply an accumulation of correct bits of information; it is the gradual development of a model for our universe, a model that enables us to understand the workings of nature. Ideas are important if they help us understand. Some ideas have to be corrected or modified as our understanding increases, but they are nonetheless valuable. Without them it might not be possible to advance to a higher level of understanding. This situation is not unique to chemistry. Accepted medical treatment of 50 (to say nothing of 100) years ago is likely to be unacceptable today. Yet such now outdated treatment did once keep people alive and helped medical researchers develop better, safer, more effective procedures.

2.2 THE NUCLEAR ATOM

Although Dalton may have erred in some details, he started us along the right path. If one begins with the premise that atoms exist, then it is possible to devise experiments to find out more about these atoms. In the century following the work of Dalton, many such experiments were conducted. Data continued to accumulate, and many of the results could not be explained by Dalton's atomic theory. An international array of scientists (see Figure 2.2) were discovering that the atom was not indivisible. It could still be regarded as the smallest characteristic unit of an element, but it certainly was not the smallest particle of matter.

(a)

(b)

(c)

(d)

(e)

FIGURE 2.2
(a) Sir William Crookes, a British scientist who invented the cathode-ray tube. (b) Antoine Henri Becquerel, a French physicist, whose experiments showed that certain atoms emit subatomic particles, some of which are positive and some negative. (c) Marie Curie, the Polish scientist who named the phenomenon of radioactivity. (d) Ernest Rutherford, a British scientist, who found that atoms are made up mostly of empty space. (e) James Chadwick, a student of Rutherford, who discovered the neutron. [(a) Reprinted with permission from Mary E. Weeks, *Discovery of the Elements*, Chemical Education Publishing Company, Easton, PA, 1968. Copyright © 1968 by Chemical Education Publishing Company. (b) reproduced by courtesy of the Trustees of the British Museum. (c)–(e) The Bettman Archive.]

2.2 The Nuclear Atom

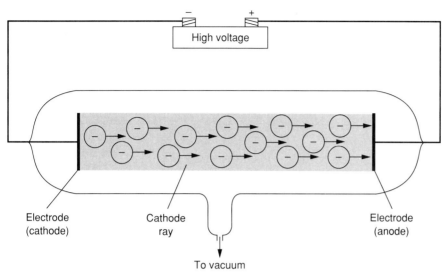

FIGURE 2.3
A simple discharge tube.

The studies of a German physicist, Eugen Goldstein, and an English physicist, William Crookes, demonstrated that atoms could be torn apart. In an apparatus called a discharge tube (Figure 2.3), electric energy was used to kick small pieces from atoms of the metal making up one electrode. The tiny *subatomic* particles were shown to be negatively charged and could be detected as a beam of particles (called a **cathode ray**) that crossed from one electrode in the apparatus to the other. No matter what the elemental constitution of the electrode, identical negatively charged subatomic particles were kicked out to form the cathode ray. These particles were named **electrons**.

When a small amount of gas was admitted into the tube, the cathode ray particles struck the atoms of the gas, knocking more electrons loose from these atoms (Figure 2.4). Those atoms that lost negatively charged

An electron is a subatomic particle with a negative charge and essentially no mass.

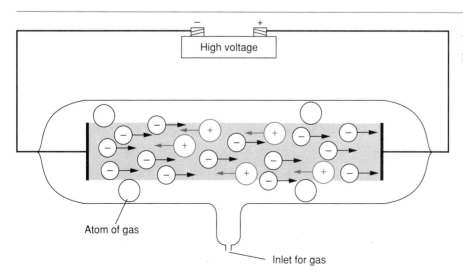

FIGURE 2.4
When a gas is in the discharge tube, positive particles are formed.

electrons were left with a positive charge. Although all the negatively charged particles were identical, the positively charged particles differed depending on what gas was in the tube. For example, helium produced more massive positively charged particles than did hydrogen. The lightest positive particles found were those from hydrogen and were given a special name, **protons**. The charges on the proton and the electron are the same size (although opposite in sign), but the proton was found to be 1837 times as massive as the electron. Thus, the smallest positive particle isolated was many times heavier than the negatively charged electron.

A proton is a subatomic particle with a positive charge and is 1837 times as massive as an electron.

A French physicist, Antoine Henri Becquerel, discovered that some atoms fall apart all by themselves. Atoms of the element uranium, for example, emitted "rays," some of which proved to be subatomic particles that were a fraction of the size of the original atoms (Figure 2.5). Some of the emitted particles were negatively charged and were shown to be identical to the electrons formed in a cathode ray tube. Uranium also emitted positively charged particles much smaller than the uranium atom itself but much more massive than the electrons. A Polish-born chemist, Marie Sklodowska Curie, named these phenomena **radioactivity**.

Certain atoms naturally decay and form subatomic particles.

Radioactivity—the emission of subatomic particles

A British scientist, Ernest Rutherford, used radioactive atoms as "atomic guns," with the tiny emitted particles playing the role of bullets (Figure 2.6). By shooting such bullets at other atoms, he discovered that the target atoms were mostly empty space. Most of the bullets passed right through the target atoms. Some of the bullets were deflected, however, suggesting that there was a concentrated bit of

FIGURE 2.5
Uranium was the first element shown to be radioactive. The radiation emitted by a sample of uranium includes both negatively and positively charged subatomic particles. The phenomenon of radioactivity is described in more detail in Chapter 4.

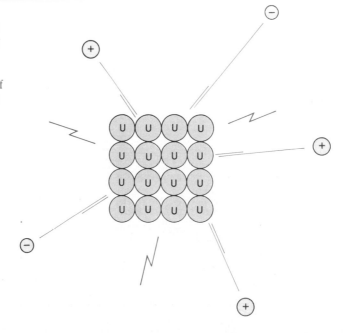

2.2 The Nuclear Atom

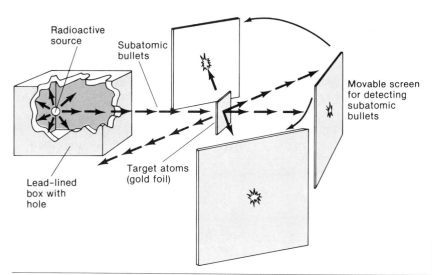

FIGURE 2.6
Rutherford's gold foil experiment.

"solid" material in the atom. A student of Rutherford, James Chadwick, discovered a subatomic particle unlike those previously found. This one, called a **neutron**, was neutral, being neither positively nor negatively charged (Figure 2.7).

Early in this century a new model of the atom was proposed to account for all these new data. According to this model there are three types of subatomic particles: protons, neutrons, and electrons. A proton and a neutron have virtually the same mass, 1.007 276 amu and 1.008 665 amu, respectively. That is equivalent to saying that two different people weigh 100.7 lb and 100.9 lb; the difference is so small that it can be ignored. Thus, for many purposes, we assume that the masses of the proton and the neutron are 1 amu each. The proton has a charge equal in magnitude but opposite in sign to that of an electron.

Neutrons are similar in mass to protons but have no charge.

FIGURE 2.7
Model explaining the results of Rutherford's experiment.

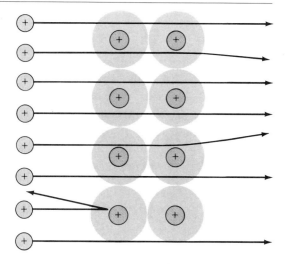

TABLE 2.1 Subatomic Particles

Particle	Symbol[a]	Approximate Mass (amu)	Charge	Location in Atom
Proton	p^+	1	1+	Nucleus
Neutron	n^0	1	0	Nucleus
Electron	e^-	0	1−	Outside nucleus

[a] The superscripts are often omitted. We will keep them to remind us of the charge of each particle.

This charge on a proton is written as 1+. The electron has a charge of 1− and a mass of 0.000 549 amu. The electrons in an atom contribute so little to its total mass that the mass of the electrons is usually disregarded. Electrons are treated as if their mass were zero.* These properties of the subatomic particles are summarized in Table 2.1.

When combined to form an atom, the more massive particles (protons and neutrons) are crowded into the **nucleus**, a tiny volume of space at the very center of the atom (Figure 2.8). The electrons are scattered throughout the remaining volume of the atom. To visualize how "empty" an atom is, picture a balloon that stands 10 stories high. If the balloon were an atom, then its nucleus would be the size of a BB. The rest of the space in the balloon-atom is the domain of the electrons, whose mass is smaller than that of the nucleus by a factor of thousands.

Atomic number = number of protons

The number of protons in the nucleus of an atom of any element is equal to the **atomic number** of that element. This number determines the kind of atom, that is, the identity of the element. Dalton had said that the mass of an atom determines the element. We now say it is not the mass but the number of protons that determines the element. For example, an atom with 26 protons (one whose atomic number is 26) is an atom of iron (Fe). An atom with 92 protons is an atom of uranium (U).

The number of neutrons in the nuclei of atoms of a given element may vary. For example, most hydrogen (H) atoms have a nucleus consisting of a single proton and no neutrons (and therefore a mass of 1 amu). About 1 hydrogen atom in 5000, however, does have a neutron as well as a proton in the nucleus. This heavier hydrogen is called **deuterium** and has a mass of 2 amu. Both kinds are hydrogen atoms (any atom with atomic number 1, that is, with one proton, is a hydrogen atom). Atoms that have this sort of relationship—the same

FIGURE 2.8
The Li atom.

---

*To emphasize how small subatomic particles are, we list here their masses in grams: proton = 1.673×10^{-24} g (0.000 000 000 000 000 000 000 001 673 g); neutron = 1.675×10^{-24} g; electron = 9.107×10^{-28} g. (Appendix C reviews exponential notation.)

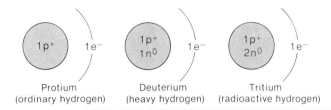

FIGURE 2.9
The isotopes of hydrogen.

number of protons but differing numbers of neutrons—are called **isotopes** (Figure 2.9). A third, very rare isotope of hydrogen is **tritium**, which has two neutrons and one proton in the nucleus (and thus a mass of 3 amu). Most, but not all, elements exist in nature in isotopic forms. This fact also requires a major modification of Dalton's original theory. He said that all atoms of the same element have the same mass. We now say that most elements have several isotopes, that is, atoms with different numbers of neutrons and therefore different masses.

An element's isotopes have the same number of protons but different numbers of neutrons.

The number of electrons in an atom equals the number of protons. Therefore, positively and negatively charged particles balance in an atom. The atom as a whole is neutral.

2.3 THE BOHR MODEL OF THE ATOM

Have you ever watched colored fireworks (Figure 2.10) or thrown chemicals into a fireplace to color the flames? If you have, then you have seen a phenomenon that puzzled scientists and triggered another modification of our concept of the atom. Light is pure energy. Light of

FIGURE 2.10
The colors of a fireworks display result from changes in electronic energy levels in atoms. [Photo by D. Gorton/The New York *Times*.]

FIGURE 2.11
Niels Bohr. [The Bettmann Archive.]

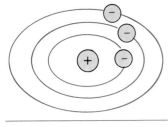

FIGURE 2.12
Bohr visualized the atom as planetary electrons circling a nuclear sun.

different color is light of different energy. For example, blue light packs more energy than red light. If you throw compounds containing calcium into a fire, you see the flames colored red. With sodium compounds (like ordinary table salt), the light is yellow-orange; with barium, green; with potassium, lavender. Why? That is what early-twentieth-century scientists asked themselves. A Danish physicist, Niels Bohr (Figure 2.11), provided the answer. He said that electrons cannot be located just any place outside the nucleus. They must move around only in well-defined paths or orbits (Figure 2.12). An electron in one of these orbits has a characteristic amount of energy (a certain amount of kinetic energy because of its motion around the orbit, a certain amount of potential energy because it is located a certain distance from the nucleus). If an electron changes from one orbit to another, its energy

FIGURE 2.13
The processes of (a) absorption and (b) emission of energy by atoms.

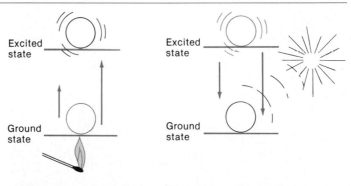

(a) Absorption of energy by an atom

(b) Emission of energy by an atom

2.3 The Bohr Model of the Atom

also changes. If it moves from a higher energy orbit to a lower energy orbit, it loses energy, and that lost energy appears as light (Figure 2.13). The color of the light depends on the difference in energy between the two orbits. Atoms in which all electrons are in their lowest possible **energy levels** are said to be in the *ground state*. If one or more electrons occupy higher energy levels than in the ground state, the atom is in an *excited state*.

Energy levels—regions around the atomic nucleus in which electrons have different energies

Bohr also said that different orbits within an atom can handle only a certain number of electrons (see Figure 2.14). We shall simply state Bohr's findings in this regard. The maximum number of electrons that can be in a given energy level is given by the formula $2n^2$, where n is equal to the energy level being considered. For the first energy level ($n = 1$), the maximum population is $2(1^2)$, or 2. For the second energy level ($n = 2$), the maximum number of electrons is $2(2^2)$, or 8. For the third level, the maximum is $2(3^2)$, or 18.

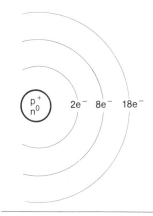

Imagine building up atoms by adding one electron to the proper energy level as protons are added to the nucleus. Electrons will go to the lowest energy level available. For hydrogen (H), with a nucleus of only one proton, the one electron goes into the first energy level. For helium (He), with a nucleus of two protons (and two neutrons), two electrons go into the first energy level. According to Bohr, the maximum population of the first energy level is two electrons; thus, that level is filled in the helium atom. With lithium (Li), which has three electrons, the extra electron goes into the second energy level. This process of adding electrons is continued until the second energy level is filled with eight electrons, as in the neon (Ne) atom (which has a total of 10 electrons, two in the first level and eight in the second). This buildup is diagrammed in Figure 2.15. One could continue to build atoms (in one's imagination) in this manner and, with a few modifications, build up the entire collection of known elements.

FIGURE 2.14
A Bohr diagram for the atoms of an element pictures specified numbers of electrons in distinct energy levels.

FIGURE 2.15
Bohr diagrams of the first 18 elements.

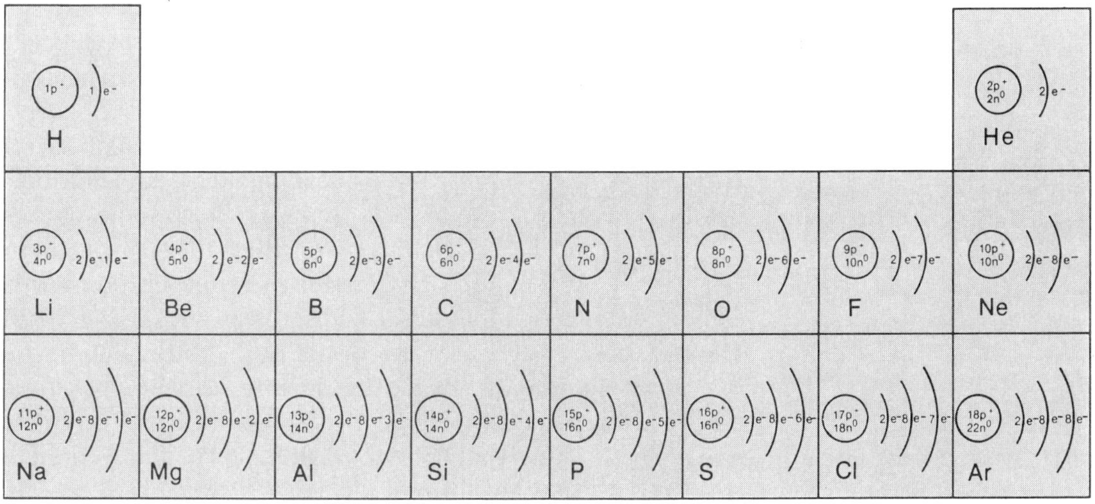

EXAMPLE 2.1

Draw a Bohr diagram for fluorine (F).

Fluorine is element number 9; it has nine protons and nine electrons. Two of the electrons go into the first energy level, the remaining seven into the second.

EXAMPLE 2.2

Draw a Bohr diagram for sodium (Na).

Sodium is element number 11. It has 11 protons and 11 electrons. Two of the electrons can go into the first energy level. Of the nine remaining electrons, eight go into the second energy level. That leaves one electron for the third energy level.

2.4 Electron Arrangements

The Bohr model of the atom has been replaced for many purposes by more sophisticated models that explain more data in greater detail than does the simpler planetary model of Bohr. This presents us with something of a quandary. We can more accurately interpret the nature of matter only by using a model that is more difficult to understand. Fortunately, however, we need not understand the laborious mathematics that generate these models to make use of some of the results.

Quantum mechanics is a highly mathematical discipline used in the late 1920s by the Austrian physicist Erwin Schrödinger to develop elaborate equations that describe the properties of electrons in atoms. The results provide a measure of the probability of finding an electron in a given volume of space. The definite planetary orbits of the Bohr model are replaced by shaped volumes of space in which electrons move. The term **orbital** (replacing Bohr's *orbits*) was used for this new description of the location of electrons.

Orbital—the defined volume of an atom where an electron exists

2.4 Electron Arrangements

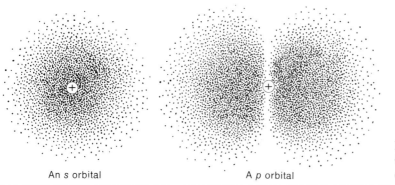

FIGURE 2.16
Charge cloud representations of atomic orbitals.

Suppose you had a camera that could photograph electrons (there is no such thing, but we are just supposing) and you left the shutter open while an electron zipped about the nucleus. When you developed the picture, you would have a record of where the electron had been. Doing the same thing with an electric fan that was turned on would give you a picture in which the blades of the fan look like a disk of material. The blades move so rapidly that their photographic image is blurred. Similarly, electrons in the first energy level of an atom would appear in our imaginary photograph as a fuzzy ball (Figure 2.16). The fuzzy ball (frequently referred to as a *charge cloud* or *electron cloud*) is the rough equivalent of an orbital.

Like Bohr, Schrödinger concluded that only two electrons could occupy the first energy level in an atom. In the quantum mechanical atom, this level is referred to as the 1s orbital (which is spherical, that is, shaped like a ball). Also like Bohr, Schrödinger stated that the second energy level of an atom could contain a maximum of eight electrons. However, Schrödinger also concluded that these eight electrons must be located in four different orbitals. Each individual orbital could contain a maximum of two electrons. One of the orbitals of the second energy level is spherical in shape and is referred to as the 2s orbital. The remaining three orbitals of the second level have identical dumbbell shapes (Figure 2.16). They differ in their orientation in space (i.e., the direction in which they point; see Figure 2.17). As a group these orbitals are called the 2p orbitals, and to distinguish them from one another they are individually referred to as the $2p_x$, $2p_y$, and $2p_z$ orbitals. Electrons in these three orbitals all lie at the same energy level and possess slightly more energy than the electrons in the 2s orbital. Thus, the quantum mechanical atom distinguishes between main energy levels (1, 2, 3, and so on) and sublevels (2s and 2p, for example). See Tables 2.2 and 2.3.

The third energy level of Bohr's model, and of Schrödinger's, could hold 18 electrons. In the quantum mechanical atom, however, the third main energy level is divided into three sublevels totaling nine orbitals:

An orbital can hold no more than two electrons.

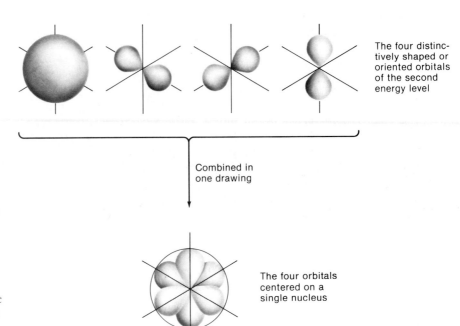

FIGURE 2.17
Electron orbitals. In these drawings the nucleus of the atom is located at the intersection of the axes. The eight electrons that would be placed in the second energy level of Bohr's model are distributed among these four orbitals in the current model of the atom, two electrons per orbital.

one $3s$ orbital, three $3p$ orbitals, and five $3d$ orbitals. Each higher main energy level adds another sublevel. The orbitals in each new sublevel have their own special shapes.

Just as one could build up atoms for all the elements by fitting electrons into the orbits of the Bohr model of the atom, so it is possible to construct a table of the elements by placing electrons into the orbitals of the quantum mechanical model. The energy-level diagram for nitrogen (which has seven electrons) is shown in Figure 2.18. The quantum mechanical description of the nitrogen atom focuses on a more detailed description of its **electron configuration** (arrangement): $1s^2 2s^2 2p^3$. The superscripts indicate the total number of electrons contained in a particular energy sublevel. For example, in nitrogen there are a total of three electrons in the $2p$ orbitals. Table 2.4

Electron configuration—the specific ordering of electrons in an atom

TABLE 2.2 Energy Levels and Orbitals

Energy Level	Orbital(s)
1	s
2	s,p
3	s,p,d

TABLE 2.3 Electrons in Orbitals

Orbital Type	Orientation	Number of Orbitals	Electrons per Orbital	Total Number of Electrons
s	Spherical	1	2	2
p	Perpendicular	3	2	6
d	Perpendicular	5	2	10

2.4 Electron Arrangements

TABLE 2.4 Electron Structures for Atoms of the First 20 Elements

Name	Atomic Number	Electron Structure
Hydrogen	1	$1s^1$
Helium	2	$1s^2$
Lithium	3	$1s^2 2s^1$
Beryllium	4	$1s^2 2s^2$
Boron	5	$1s^2 2s^2 2p^1$
Carbon	6	$1s^2 2s^2 2p^2$
Nitrogen	7	$1s^2 2s^2 2p^3$
Oxygen	8	$1s^2 2s^2 2p^4$
Fluorine	9	$1s^2 2s^2 2p^5$
Neon	10	$1s^2 2s^2 2p^6$
Sodium	11	$1s^2 2s^2 2p^6 3s^1$
Magnesium	12	$1s^2 2s^2 2p^6 3s^2$
Aluminum	13	$1s^2 2s^2 2p^6 3s^2 3p^1$
Silicon	14	$1s^2 2s^2 2p^6 3s^2 3p^2$
Phosphorus	15	$1s^2 2s^2 2p^6 3s^2 3p^3$
Sulfur	16	$1s^2 2s^2 2p^6 3s^2 3p^4$
Chlorine	17	$1s^2 2s^2 2p^6 3s^2 3p^5$
Argon	18	$1s^2 2s^2 2p^6 3s^2 3p^6$
Potassium	19	$1s^2 2s^2 2p^6 3s^2 3p^6 4s^1$
Calcium	20	$1s^2 2s^2 2p^6 3s^2 3p^6 4s^2$

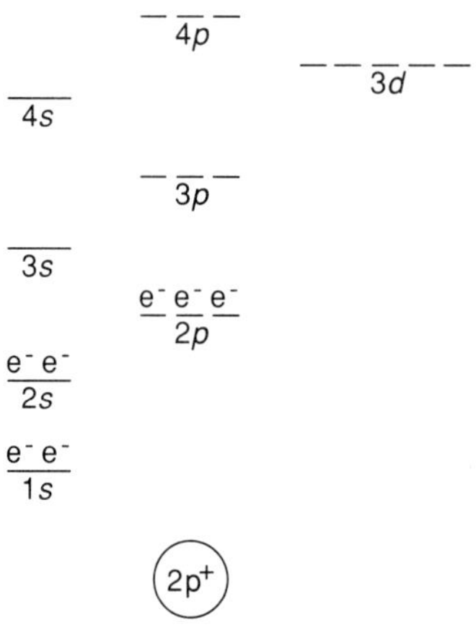

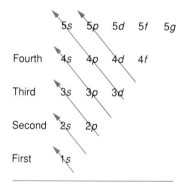

FIGURE 2.18
Energy level diagram for nitrogen.

lists the electron configurations of the first 20 elements. Notice that the orbitals fill in order of increasing energy: first the 1s, then the 2s, 2p, 3s, and so on. The d orbitals introduce a minor complication because the 3d orbitals turn out to be at a higher energy level than the 4s orbitals. Figure 2.19 illustrates the order in which orbitals are filled.

To check whether you understand the fundamentals of orbitals and electron configurations, work through the following examples.

FIGURE 2.19
An order-of-filling chart for determining the electron configuration of atoms. The first energy level has only s orbitals, the second has s and p orbitals, the third s, p, and d orbitals, and so on.

EXAMPLE 2.3

Write the electron configuration for carbon, atomic number 6.

The atomic number indicates that there are six electrons in the carbon atom. The first energy level can accommodate only two of these electrons in its 1s orbital.

$$1s^2 \ldots$$

The remaining four electrons must occupy orbitals on the second main energy level. The lowest energy orbital on the second level is the 2s orbital, which can accommodate two electrons.

$$1s^2 2s^2 \ldots$$

The remaining two electrons must go to the 2p orbitals, which can handle up to six electrons if necessary.

$$1s^2 2s^2 2p^2$$

EXAMPLE 2.4

What is the electron configuration of hydrogen (atomic number 1)?

The single electron in hydrogen resides in the lowest energy orbital. Therefore, the electron configuration for hydrogen is $1s^1$.

EXAMPLE 2.5

What is the electron configuration for phosphorus (atomic number 15)?

We have 15 electrons to distribute. Two go into the first main energy level ($1s^2$), eight go into the second ($2s^2 2p^6$), and five go into the third ($3s^2 3p^3$). In summary, the configuration of phosphorus is

$$1s^2 2s^2 2p^6 3s^2 3p^3$$

When an atom gains or loses electrons, it becomes charged and is referred to as an **ion**. Thus, if Li were to lose an electron, its electron configuration would be $1s^2 2s^0$. It would then be Li^+. In the same way, if F were to gain an electron, it would become F^- with an electron configuration $1s^2 2s^2 2p^6$. More on ions in Chapter 3.

Quantum mechanics offers to scientists detailed descriptions of the electron structure of atoms. However, we shall make most use of the picture it paints of electrons as clouds of negative charge. Sometimes the shape of the cloud (that is, the orbital) is presented simply in outline. (Figure 2.17 uses shaded drawings to present the combined orbitals of the second energy level.) Later we shall see how the shape and orientation of the electron cloud can determine the shape of what we will call molecules.

2.5 THE PERIODIC TABLE

As our picture of the atom becomes more detailed, we find ourselves in a dilemma. With more than 100 elements to deal with, how can we keep all this information straight? One way is by using the **periodic table of the elements**. (See the inside front cover of this book.) The periodic table neatly tabulates information about atoms. It records how many protons and electrons the atoms of a particular element contain. It permits us to calculate the number of neutrons in the most common isotope for most elements. It even stores information about how electrons are arranged in the atoms of each element. The most extraordinary thing about the periodic table is that it was largely developed before anyone knew there were protons or neutrons or electrons in atoms.

The periodic table of the elements provides a wealth of information.

Not long after Dalton presented his model for the atom (an indivisible particle whose mass determined its identity), chemists began preparing listings of elements arranged according to their atomic weights. While working out such tables of elements, these scientists observed patterns among the elements. For example, it became clear that elements that occurred at specific intervals shared a similarity in certain properties. Among the approximately 60 elements known at that time, the second and ninth showed similar properties, as did the third and tenth, the fourth and eleventh, the fifth and twelfth, and so on.

In 1869, Dmitri Ivanovich Mendeleev (Figure 2.20), a Russian chemist, published his periodic table of the elements. Mendeleev prepared his table by taking into account both the atomic weights and the periodicity of certain properties of the elements. The elements were

FIGURE 2.20
Dmitri Mendeleev, a Russian chemist, who devised the periodic table of the elements. [Reprinted with permission from Mary E. Weeks, *Discovery of the Elements*, Easton, PA, Chemical Education Publishing Company, 1968. Copyright © 1968 by Chemical Education Publishing Company.]

TABLE 2.5 Mendeleev's Periodic Table, as Published in the *Journal of the Russian Chemical Society* (1869)

			Ti = 50	Zr = 90	? = 180
			V = 51	Nb = 94	Ta = 182
			Cr = 52	Mo = 96	W = 186
			Mn = 55	Rh = 104.4	Pt = 197.4
			Fe = 56	Ru = 104.4	Ir = 198
			Ni = Co = 59	Pl = 106.5	Os = 199
H = 1			Cu = 63.4	Ag = 108	Hg = 200
	Be = 9.4	Mg = 24	Zn = 65.2	Cd = 112	
	B = 11	Al = 27.4	? = 68	Ur = 116	Au = 197?
	C = 12	Si = 28	? = 70	Sn = 118	
	N = 14	P = 31	As = 75	Sb = 122	Bi = 210
	O = 16	S = 32	Se = 79.4	Te = 128?	
	F = 19	Cl = 35.5	Br = 80	I = 127	
Li = 7	Na = 23	K = 39	Rb = 85.4	Cs = 138	Tl = 204
		Ca = 40	Sr = 87.4	Ba = 137	Pb = 207?
		? = 45	Ce = 92		
		?Er = 56	La = 94		
		?Yt = 60	Di = 95		
		?In = 75.6	Th = 118?		

arranged primarily in order of increasing atomic weight. In a few cases, Mendeleev placed a slightly heavier element before a lighter one. He did this only when it was necessary in order to keep elements with similar chemical properties in the same row. For example, he placed tellurium (atomic weight = 128) ahead of iodine (atomic weight = 127) because tellurium resembled sulfur and selenium in its properties, whereas iodine was similar to chlorine and bromine.

Mendeleev left a number of gaps in his table (Table 2.5). Instead of looking upon those blank spaces as defects, he boldly predicted the existence of elements as yet undiscovered. Furthermore, he even predicted the properties of some of these missing elements. In succeeding years, many of the gaps were filled in by the discovery of new elements. The properties were often quite close to those Mendeleev had predicted. The predictive value of this great innovation led to the wide acceptance of Mendeleev's table.

It is now known that properties of an element depend mainly on the number of electrons in the outermost energy level of the atoms of the element (Chapter 3). Sodium atoms have one electron in their outermost energy level (the third). Lithium atoms have a single electron in their outermost level (the second). The chemical properties of sodium and lithium are similar. The atoms of helium and neon have filled outer electron energy levels, and both elements are inert, that is, they do not undergo chemical reactions readily. Apparently, not only are similar

2.5 The Periodic Table

chemical properties shared by elements whose atoms have similar electron **configurations** (arrangements) but also certain configurations appear to be more stable (less reactive) than others. This is a point we shall explore in Chapter 3.

In Mendeleev's table, the elements were arranged by atomic weights for the most part, and this arrangement revealed the periodicity of chemical properties. Because the number of electrons determines the element's chemical properties, that number should (and now does) determine the order of the periodic table. In the modern periodic table, the elements are arranged according to atomic number. Remember, this number indicates both how many protons and how many electrons there are in a neutral atom of the element. The modern table, arranged in order of increasing atomic number, and Mendeleev's table, arranged in order of increasing atomic weight, parallel one another because an increase in atomic number is generally accompanied by an increase in atomic weight. In only a few cases (noted by Mendeleev) do the weights fall out of order. Atomic weights do not increase in precisely the same order as atomic numbers because *both protons and neutrons* contribute to the mass of an atom. It is possible for an atom of lower atomic number to have more neutrons than one with a higher atomic number. Thus, it is possible for an atom with a lower atomic number to have a greater mass than an atom with a higher atomic number. Thus the atomic mass of Ar (no. 18) is more than that of K (no. 19), and Te (no. 52) has a mass greater than that of I (no. 53); see the periodic table.

The modern periodic table (inside front cover) has vertical columns called **groups** or **families**. Each group includes elements with the same number of electrons in their outermost energy levels and, therefore, with similar chemical properties (Figure 2.21). The horizontal rows of the table are called **periods**. Each new period indicates the opening of the next main electron energy level. For example, sodium starts row three, and the outermost electron in sodium is the first electron to be placed in the third energy level. Because each row begins a new energy level, we can predict that the size of atoms increases from top to bottom. And since electrons are easier to remove when farther from the nucleus, we can also predict that the larger the atom the lower its *ionization energy*, the energy needed to remove an electron.

Columns in the periodic table are called groups or families.

Rows in the periodic table are called periods.

Certain groupings of elements in the periodic table are designated by special names. The heavy, stepped, diagonal line on the table divides the elements into two major classes. Those to the left of the line are called **metals**, and those to the right, **nonmetals**. Group IA elements are known as **alkali metals**; Group IIA are **alkaline earth metals**; Group VIIA, **halogens**. The group at the extreme right of the table contains the **noble gases**. All the Group B elements are called **transition metals**.

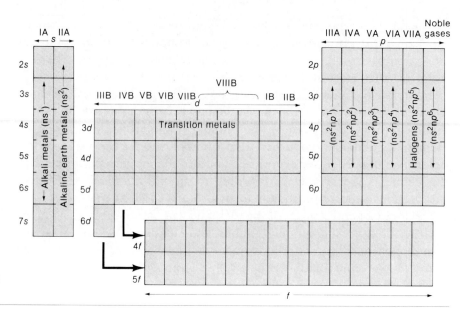

FIGURE 2.21
Electron configurations and the periodic table.

We think you will find the periodic table an invaluable aid in your study of chemistry.

2.6 Some Chemistry of the Elements

In this section we examine some of the properties of the various groups of elements and relate the properties to their electron structure. We start with the noble gases. Then we contrast the reactivity of the elements in other groups to the inactivity of the noble ones.

The easier it is to remove outermost electrons, the more metallic the element. Metallic character decreases from left to right in parallel with ionization energies. It also increases from top to bottom with increasing atomic size. The alkali metals are all quite metallic and the halogens are all distinctly nonmetallic. In the middle of the table we see intermediate behavior. Group IVA has carbon, a nonmetal, at the top, and two metals, tin and lead, at the bottom. In between are the "metalloids—metal-like elements—silicon and germanium. The following brief description of the groups of elements will emphasize periodic trends.

Group VIIIA: The Noble Gases

In the last decade of the nineteenth century a series of elements was discovered that made up an entirely new family, one completely unexpected by Mendeleev and his contemporaries. Nonetheless, this new

2.6 Some Chemistry of the Elements

group, called the noble gases, fit neatly between the very reactive alkali metals (Group IA) and the highly active nonmetals of Group VIIA. In the usual flat form of the periodic table the noble gases are placed to the far right in the column immediately following Group VIIA.

The six noble gases are helium, neon, argon, krypton, xenon, and radon. All are components of the atmosphere. Argon is rather abundant, making up nearly 1% by volume of the atmosphere. Xenon, on the other hand, is quite rare, making up only 91 parts per billion (by volume) of the atmosphere. Radioactive radon seeps from the ground. It makes up only a tiny portion of the atmosphere. It can be a problem, however, when it concentrates in a building with little ventilation.

The noble gases are exceptionally resistant to chemical reaction. This lack of reactivity is a reflection of their electron structure. Helium has the configuration $1s^2$, a filled first energy level. Neon has the structure $1s^2 2p^6$, a filled second energy level. The others all have the valence configuration $ns^2 np^6$, an especially stable configuration. Other elements undergo reactions in order to achieve the more stable electronic configurations of the noble gases. All the noble gases exist in elemental form as monatomic species.

Because of their stability, the noble gases were once called "inert gases." But in 1962 it was discovered that a few compounds of krypton and xenon could be formed, necessitating the change from "inert" to "noble." As yet, despite many attempts, no compounds have been made of the lighter elements helium, neon, and argon. While this family of elements is no longer called inert, its nobility is still unquestioned. More than any other family, the noble gases disdain interactions with the masses of other elements.

Helium is found in natural gas deposits, where it was formed from elements within the Earth. Helium is used to fill balloons and blimps. Its lifting power is more than 90% that of hydrogen, the lightest of all the gases, and helium has the advantage of being nonflammable. Helium is also used to provide an inert atmosphere for the welding of metals that otherwise might be attacked by oxygen in the air. Liquid helium is used to achieve extremely low temperatures. It boils at only 4.2 K.

Neon is used in lighted signs for advertising. A tube with electrodes is shaped into letters or symbols and is filled with neon at low pressure. An electric current is passed through the tube, causing the atoms to emit their characteristic orange glow.

Argon is the most plentiful of the noble gases. It is used to fill incandescent light bulbs. Unlike nitrogen and oxygen, it does not react with the tungsten filament. It also decreases the tendency of the filament to vaporize, thus extending the filament's life. Fluorescent lights are filled with a mixture of argon and mercury vapor.

Krypton and xenon are too expensive to have many important commercial applications, although krypton has found some use in light bulbs. Radon, although exceedingly rare in the atmosphere, can be collected from the radioactive decay of radium. This radon may be sealed in small vials and used for radiation therapy of certain malignancies. Radon accumulates in well-insulated homes; it is thought to cause about 15% of all cases of lung cancer.

Group IA: The Alkali Metals

The Group IA elements, the alkali metals, all have the outer electron configuration ns^1. There are six elements in the group: lithium, sodium, potassium, rubidium, cesium, and francium.

In the elemental form, the alkali metals are soft solids with low melting points. Indeed, on a hot day cesium would be a liquid, for it melts at 29 °C. These metals, when freshly cut, are bright and shiny, but they tarnish readily as they become oxidized by the atmosphere. They are the most reactive of the metals. The atoms all have a great tendency to give up one electron each and form +1 ions. Using symbols, we write the process for sodium as

$$Na \longrightarrow Na^+ + e^-$$

In compounds, the alkali metals occur as the +1 ions. All form oxides of the general formula M_2O. Nearly all compounds of the alkali metals are soluble in water.

Lithium salts are found in certain naturally occurring brines. Lithium carbonate (Li_2CO_3)* is used in medicine to level out the dangerous "manic" highs that occur in manic-depressive psychoses. Some practitioners also recommend lithium carbonate for the depression stage of the cycle. It appears to act by affecting the transport of chemical substances across cell membranes in the brain.

Sodium salts are quite common. Ordinary table salt (NaCl) supplies the body with chloride ions, necessary for the production of hydrochloric acid by our stomachs. Living tissues also require a balance of sodium ions and potassium ions.

Potassium ion is an essential nutrient for plants. It is generally abundant and is readily available to plants except in soil depleted by high-yield agriculture. The usual form of potassium in commercial fertilizers is potassium chloride (KCl).

In animals, potassium ions are the principal positive ions inside cells, and sodium ions are the principal positive ions in the extracellular fluid.

> Although it is a member of Group IA, hydrogen is **not** an alkali metal; it isn't even a metal. It exists as H_2 molecules and is a gas at room temperature.

---

* This is not the time to concern yourself with *how* chemical formulas are determined. You will learn about that in Chapter 3. Here the formulas are given simply as examples.

2.6 Some Chemistry of the Elements

Group IIA: The Alkaline Earth Elements

The six alkaline earth metals are beryllium, magnesium, calcium, strontium, barium, and radium. All have the outer electron configuration ns^2. In the elemental form these metals are fairly soft and reactive. The atoms show a tendency to give up two electrons each and form +2 ions. For example,

$$Mg \longrightarrow Mg^{2+} + 2\,e^-$$

In compounds these metals occur almost exclusively as the +2 ions. All form oxides of the general types MO.

Beryllium is something of an oddball member of the family. Unlike the others, it does not react with water. The metal itself is rather hard, rigid, and strong. Its lightness makes it quite valuable in structural alloys. The element is poisonous in all its forms.

Magnesium ions are essential to both plants and animals. In plants magnesium ions are incorporated in chlorophyll molecules. Thus, this ion is essential to photosynthesis. Both calcium and magnesium are essential for proper functioning of the nerves that control muscles.

Calcium ions are necessary for the proper development of bones and teeth. For this reason growing children are usually encouraged to drink milk, a rich source of calcium. Adults also require calcium because it is necessary for clotting of blood and maintenance of a regular heartbeat.

Calcium carbonate (limestone) and other rocks containing calcium ions (Ca^{2+}), magnesium ions (Mg^{2+}), or iron ions (Fe^{2+} or Fe^{3+}) are widely distributed in nature. Water containing calcium, magnesium, or iron is what is known as hard water. The ions react with the soaps to form curdy precipitates sometimes called bathtub ring.

> Mendeleev based his periodic table to a large degree on the fact that elements within a group formed oxides and hydrides with the same general formula.

Group IIIA

Group IIIA consists of the metalloid boron and the metals aluminum, gallium, indium, and thallium. All have the electron configuration ns^2np^1. Boric acid (H_3BO_3) is a familiar ingredient of mild antiseptic eye rinses. The other elements in Group IIIA are typical metals and tend to form +3 ions.

$$Al \longrightarrow Al^{3+} + 3\,e^-$$

Aluminum is the most abundant metal in the Earth's crust, but it is tightly bound in compounds in nature. Much energy, mainly electricity, is required to extract aluminum from its ores.

Aluminum is light and strong. It is widely used in airplanes. As we try to decrease the weight of automobiles in order to increase gas mileage, aluminum has replaced steel in many automobile parts. Aluminum corrodes much more slowly than iron.

Group IVA: Some Compounds of Carbon

Group IVA is made up of the nonmetal carbon, the metalloids silicon and germanium, and the metals tin and lead. All have the electron configuration ns^2np^2. Of these, carbon is easily the most important. Carbon forms thousands—perhaps millions—of compounds with hydrogen. These compounds, called hydrocarbons, and their derivatives are called organic chemicals, and their study is called organic chemistry. There are a few simple compounds of carbon, though, that are more like inorganic than organic chemicals. Among these are carbon monoxide (CO), carbon dioxide (CO_2), and such minerals as limestone and marble (calcium carbonate, $CaCO_3$).

Carbon exists in three allotropic forms. Allotropes are modifications of an element that can exist in more than one form in the same physical state. The solid element can be found in the form of graphite (the "lead" of pencils), carbon black (the soot that forms on the bottom of a casserole warmed over a candle flame), and diamond (the precious jewel). Coal is composed of varying amounts of elemental carbon, from about 6% in peat up to 88% or more in anthracite. When burned, the carbon in coal combines with oxygen to form carbon dioxide. We can use symbols and formulas to write this as a chemical equation.*

$$C + O_2 \longrightarrow CO_2$$

Hydrocarbons also are burned as fuels. Methane, the simplest hydrocarbon, burns with a hot flame. If sufficient oxygen is present, the main products are relatively innocuous carbon dioxide and water.

$$CH_4 + 2\,O_2 \longrightarrow CO_2 + 2\,H_2O$$

With less oxygen available, poisonous carbon monoxide may form. Carbon monoxide is an invisible, odorless, tasteless gas. It exerts its insidious effect by tying up the hemoglobin in the blood. The normal function of hemoglobin is to transport oxygen. Carbon monoxide binds tenaciously to hemoglobin—once on, it refuses to get off. The hemoglobin is thus prevented from transporting oxygen.

---

* This is not the time to concern yourself with *how* to write chemical equations. You will learn about that in Chapter 6. Equations are given here simply as illustrations. You can read them by translating the formulas into names. The plus sign is read as "and," and the arrow is read as "react to form" or "yields."

2.6 Some Chemistry of the Elements

Group VA: Some Nitrogen Compounds

Group VA includes the nonmetals nitrogen and phosphorus, the metalloids arsenic and antimony, and the metal bismuth. They have the electron configuration ns^2np^3. We limit our discussion to the chemistry of nitrogen.

Although nitrogen makes up 78% of the atmosphere, the molecules of nitrogen gas (N_2) cannot be used directly by higher plants or by animals. They first have to be "fixed," that is, converted to a more readily used form. Certain types of bacteria convert atmospheric nitrogen to nitrates. Lightning also serves to "fix" nitrogen by causing it to combine with oxygen. Nitric oxide (NO) and nitrogen dioxide (NO_2) are formed. The equations are

$$N_2 + O_2 \xrightarrow{\text{lightning}} 2\,NO$$

$$2\,NO + O_2 \longrightarrow 2\,NO_2$$

The nitrogen dioxide reacts with water to form nitric acid (HNO_3).

$$3\,NO_2 + H_2O \longrightarrow 2\,HNO_3 + NO$$

The nitric acid falls in rainwater, adding to the supply of available nitrates in sea and soil. Scientists have undertaken substantial intervention in the nitrogen cycle by industrial fixation—the manufacture of nitrogen fertilizers. This intervention has greatly increased our food supply, for the availability of fixed nitrogen is often the limiting factor in the production of food.

The combustion process in an automobile engine leads to the formation of nitric oxide (NO).

$$N_2 + O_2 \longrightarrow 2\,NO$$

Nitric oxide is oxidized slowly by oxygen to nitrogen dioxide (NO_2).

$$2\,NO + O_2 \longrightarrow 2\,NO_2$$

By forming nitric acid, nitrogen oxides contribute to acid rain.

Group VIA: Compounds of Oxygen and Sulfur

Group VIA includes the nonmetals oxygen, sulfur, and selenium, the metalloid tellurium, and the radioactive metal polonium. They have the electron configuration ns^2np^4. We limit our discussion to the chemistry of oxygen and sulfur.

The element oxygen occurs in the atmosphere mainly as the diatomic molecule O_2. Oxygen is involved in oxidation processes such as combustion (rapid burning) and rusting and other forms of corrosion, and in respiration. Oxygen reacts rapidly with more active metals. Magnesium, for example, burns with a brilliant white flame when ignited in air.

$$2\ Mg + O_2 \longrightarrow 2\ MgO + heat + light$$

At room temperature a reaction occurs on the surface of freshly prepared magnesium metal. This magnesium oxide forms a thin, transparent coating that is impervious to air, preventing further oxidation. Such oxide coatings are common on metals, making it possible for us to use those otherwise quite reactive metals in utensils and machines. Magnesium, aluminum, and titanium are familiar examples of highly reactive metals that can be used as structural materials because of their ability to form protective oxide coatings.

Generally, oxygen reacts with metals by acquiring electrons and forming oxide ions, the ions that make solutions basic (Chapter 8).

$$O_2(g) + 4\ e^- \longrightarrow 2\ O^{2-}$$

The oxide ions react with water to form hydroxide ions.

$$O^{2-} + H_2O \longrightarrow 2\ OH^-$$

Oxides of metals, then, are generally basic oxides.

Oxygen also reacts with many nonmetals. Sulfur, for example, burns in air to form sulfur dioxide.

$$S(s) + O_2(g) \longrightarrow SO_2(g)$$

When dissolved in water, sulfur dioxide reacts to form sulfurous acid.

$$SO_2(g) + H_2O \longrightarrow H_2SO_3(aq)$$

Generally, the oxides of nonmetals form acidic solutions (Chapter 8); thus, we call them acidic oxides.

Small amounts of oxygen occur as ozone (O_3), an allotropic form of oxygen. Ozone is quite unstable. At room temperature, it breaks down to ordinary oxygen.

$$2\ O_3(g) \longrightarrow 3\ O_2(g) + heat$$

2.6 Some Chemistry of the Elements

Ozone is formed by electrical discharges through oxygen and by ultraviolet lamps. The pungent odor around electrical equipment is due to ozone.

The ozone shield in the upper atmosphere protects us from harmful ultraviolet radiation.

Sulfur occurs in nature in both the combined and elemental forms. Free sulfur occurs as S_8, a ring of eight atoms. For simplicity, however, sulfur is often represented in equations only by the letter S, just as if it were monatomic.

Sulfur atoms can accept two electrons to form sulfide ions (S^{2-}).

$$S + 2\,e^- \longrightarrow S^{2-}$$

Sulfur gains electrons from the more active metals to form sulfides.

$$Ca(s) + S(s) \longrightarrow CaS(s)$$

Sulfur burns in air to form sulfur dioxide.

$$S(s) + O_2(g) \longrightarrow SO_2(g)$$

The sulfur dioxide may react further with oxygen to form sulfur trioxide.

$$2\,SO_2(g) + O_2(g) \longrightarrow 2\,SO_3(g)$$

The sulfur trioxide can react in turn with water to form sulfuric acid.

$$SO_3(g) + H_2O \longrightarrow H_2SO_4$$

The oxides of sulfur from the burning of sulfur-containing coal and from other industrial processes contribute greatly to acid rain. Acidic rainwater can corrode metals, eat holes in nylon stockings, decompose stone buildings and statuary, and render lakes so acidic that all life is destroyed.

Group VIIA: The Halogens

Fluorine, chlorine, bromine, iodine, and astatine make up Group VIIA, the halogen family. All are nonmetals, and they have the electron configuration ns^2np^4. The Group IA metal sodium reacts with chlorine to form sodium chloride, the familiar table salt. Indeed, it is characteristic of the entire group that they react with metals to form salts. The word halogen is derived from Greek words meaning "salt former". We will limit our discussion to the first four halogens, for astatine is rare and highly radioactive.

In the elemental form all the halogens exist as diatomic molecules. The atoms can achieve a noble gas configuration by gaining an electron to form a negative ion.

$$F_2 + 2\ e^- \longrightarrow 2\ F^-$$

$$Cl_2 + 2\ e^- \longrightarrow 2\ Cl^-$$

This tendency, combined with that of many metals to give up electrons readily, is responsible for the ability of the halogens to form so many salts.

Fluoride salts, in moderate to high concentrations, are acute poisons. Small amounts of fluoride ion, however, are essential for our well-being. Concentrations of 0.7 to 1.0 parts per million, by mass, of fluoride have been added to the drinking water of many communities. Evidence indicates that such fluoridation results in a reduction in the incidence of dental caries (cavities).

Chlorine in the elemental form (Cl_2) is used to kill bacteria in water-treatment plants. Sodium hypochlorite (NaOCl) is an ingredient of common household bleaching solutions. Chlorine is present in our bodies as chloride ion (Cl^-). This ion is essential to life.

Several compounds of bromine are of importance. Silver bromide (AgBr) is sensitive to light and is used in photographic film. Sodium bromide (NaBr) and potassium bromide (KBr) have been used medicinally as sedatives. Bromide ions depress the central nervous system. Unfortunately, prolonged intake can cause mental deterioration and other problems. Bromides have been largely replaced by other, presumably safer, pain relievers, but the word *bromide* remains in our language as a synonym for platitude.

Iodine is another essential nutrient. Compounds of iodine are necessary for the proper action of the thyroid gland. Iodine is also used as a topical antiseptic. Tincture of iodine is a solution of elemental iodine (I_2) in a mixture of alcohol and water. Iodine-releasing compounds, called iodophors, are often used as antiseptics in hospitals.

The halogens also react with hydrogen to form hydrogen halides.

$$H_2(g) + Cl_2(g) \longrightarrow 2\ HCl(g)$$

$$H_2(g) + Br_2(g) \longrightarrow 2\ HBr(g)$$

In water solutions these compounds form acids. Hydrochloric acid is a familiar example. This acid is present in the stomach and is involved in the digestive process.

The B Groups: Some Representative Transition Elements

Most of our attention so far has been focused on the A groups of the periodic table. The chemistry of the B groups is also important, if

2.6 Some Chemistry of the Elements

perhaps more complicated. The B groups collectively often are called the transition elements. All are metals in the elemental form. They conduct electricity and have characteristic metallic luster.

The transition elements, in general, are those in which inner electron energy levels are being filled. (Group IB and IIB elements are exceptions, yet they share many properties with the other B groups and are often included as transition elements.) Recall that the third period ends with argon, which has eight electrons in its outermost energy level. The fourth period begins with potassium, which has one electron in its fourth energy level. The next element, calcium, has two electrons in its fourth energy level. Recall, however, that the $2n^2$ rule predicts a maximum of 18, not 8, electrons for the third energy level ($2n^2 = 2 \times 3^2 = 2 \times 9 = 18$). The fourth energy level has begun to fill before the third one is full. Things return to "normal" with the next element, scandium. This element skips back and continues to fill the third energy level. The first transition series (from scandium to zinc) corresponds to a filling of the $3d$ sublevel. With gallium ($Z = 31$), we return to an A group element. All inner energy levels are filled, and the next electron enters the outer energy level.

Physical properties vary widely in the transition series. For example, mercury (Hg) is a liquid at room temperature (its melting point is −38 °C). Tungsten (W), however, melts at 3410 °C. Chemically, the elements also exhibit a variety of properties. Most form more than one kind of simple ion. Iron is a familiar example. It readily forms both iron(II) and iron(III) ions (Fe^{2+} and Fe^{3+}, respectively).

Many compounds of the transition metals are colored, often brilliantly so. For a given element, the color is different for different ions. Aqueous solutions of Fe^{2+} are often pale green; those of Fe^{3+} are often yellow. With polyatomic ions, color variation often is greater still. An aqueous solution containing manganese(II) ions (Mn^{2+}) is a faint pink. Solutions of MnO_4^{2-} are an intense green, and those of MnO_4^- are deep purple.

Transition metals known to be essential to life are iron, copper, zinc, cobalt, manganese, vanadium, chromium, and molybdenum. Nickel is sometimes found in body tissues but has not been shown to be essential. Other transition metals, most notably cadmium and mercury, are dangerous poisons.

Iron is essential for the proper functioning of hemoglobin, the red protein molecule involved in oxygen transport. This iron must be in the form of the +2 ion. If it is changed to the +3 form, the resulting compound (called methemoglobin) is incapable of carrying oxygen. The oxygen-deficiency disease that results is called methemoglobinemia. In infants this condition is called the blue-baby syndrome.

Cobalt is a component of vitamin B_{12}. Deficiency leads to pernicious anemia. Vitamin B_{12} is found only in meat products. One hazard of a strict vegetarian diet is vitamin B_{12} deficiency.

2.7 With So Many Models to Choose from...

For some purposes in this text, we will use Bohr diagrams of atoms to picture the distribution of electrons among the main energy levels. At other times we will find that the electron clouds of the quantum mechanical model are more useful. Even Dalton's model will sometimes prove to be the best way to describe certain phenomena (the behavior of gases, for example). Our choice of model will always be based on which one is most helpful for understanding a particular concept. That, after all, is the whole purpose of scientific models.

CHAPTER SUMMARY

The continuous view of matter held that matter was infinitely divisible, but experimental science has confirmed the atomistic view of matter. The atom is defined as the smallest characteristic particle of an element. According to John Dalton's atomic theory, the atom is indestructible and indivisible, and an element is uniquely identified by the mass of its atoms. These masses are measured in atomic mass units (amu).

Later experiments required modifications in the model of the atom. Work with cathode-ray tubes and with radioactive elements demonstrated that the atom is neither indivisible nor indestructible. The discovery of isotopes, atoms of the same element with different masses, established that an element was not uniquely identified by the mass of its atoms. The gold foil experiment of Rutherford led to the nuclear model for the atom. In this model, three fundamental subatomic particles are recognized: the proton, the electron, and the neutron. The positively charged proton and the uncharged neutron both have a mass of 1 amu and are confined to a small volume in the center of the atom called the nucleus. The negatively charged electron, with a mass close to zero, is located outside the nucleus.

The number of protons in an atom, called the atomic number, uniquely identifies the element. The atomic number also gives the number of electrons in a neutral atom.

Niels Bohr first provided a more detailed picture of the arrangement of electrons (the electron configuration) that was later refined by quantum mechanical calculations. Bohr placed the electrons in orbits set at definite distances from the nucleus. The orbits represent different energy levels for the electrons. The maximum number of electrons that can occupy a particular energy level is given by the formula $2n^2$, where n is the number of the energy level ($n = 1$ is the lowest level, $n = 2$ is the next higher level, and so on).

In the quantum mechanical atom, electrons are described as occupying shaped volumes of space referred to as orbitals. An orbital may hold a maximum of two electrons. Bohr's main energy levels were subdivided by quantum mechanics into sublevels of slightly different energies. A specific set of orbitals is associated with each of these sublevels. The first energy level consists of a single spherical 1s orbital. The second main energy level includes a spherical 2s orbital and three dumbbell-shaped 2p orbitals. Higher main energy levels involve more sublevels.

The periodic table is a device for ordering elements in a way that reveals their atomic structure and their chemical reactivity. The columns of the table are called groups or families, and the horizontal rows are called periods. Some families of elements have been given special names, including the alkali metals (Group IA), the alkaline earth metals (Group IIA), the halogens (Group VIIA), the noble gases, and the transition metals (Group B). Noble gases resist chemical reaction, whereas the alkali and alkaline earth metals—except for beryllium—are quite reactive. Aluminum is the most abundant metal on Earth, and carbon is one of the most important elements. Nitrogen and oxygen are the main constituent elements of the atmosphere. The halogen family includes fluorine, chlorine, bromine, and iodine. The transition elements are metals whose d sublevels are being filled.

CHAPTER QUIZ

1. Dalton's atomic theory proposed
   a. indivisible atoms   b. nuclear protons
   c. negative electrons   d. isotopes
2. The symbol, approximate amu, and charge for the electron are, in that order,
   a. e^-, 1, 1−   b. e^-, 0, 1−
   c. e^-, 1, 1+   d. e^-, 1, 0
3. Which can exhibit radioactivity?
   a. protons   b. electrons
   c. neutrons   d. atoms
4. Atoms in an excited state, compared to the ground state, have
   a. more electrons
   b. fewer electrons
   c. promoted electrons
   d. promoted protons
5. The total number of electrons possible in the first energy level is
   a. 0   b. 2   c. 6   d. 8
6. The number of orbitals in the second energy level is
   a. 1   b. 2   c. 3   d. 4
7. The electron configuration $1s^2 2s^2 2p^1$ refers to
   a. Li   b. Be   c. B   d. C
8. Which is a halogen?
   a. C   b. Cl   c. Ca   d. Cu
9. Which has the smallest radius?
   a. V   b. Ca   c. Mg   d. Rb
10. Which is a nonmetal?
    a. P   b. Mn   c. Sc   d. Al

PROBLEMS

1. What is the distinction between the atomistic view and the continuous view of matter?

2. If foods were described as atomistic or continuous, which designation would you use for each of the following?
   a. peas
   b. mashed potatoes
   c. milk
   d. hot dogs
   e. hard-boiled eggs
   f. scrambled eggs

2.1 Dalton's Atomic Theory

3. Outline the main points of Dalton's atomic theory.
4. An atom of calcium has a mass of 40 amu, and an atom of cobalt has a mass of 59 amu. Are these findings in agreement with Dalton's atomic theory?
5. An atom of calcium has a mass of 40 amu, and an atom of potassium has a mass of 40 amu. Are these findings in agreement with Dalton's atomic theory?
6. An atom of calcium has a mass of 40 amu, and another atom of calcium has a mass of 44 amu. Are these findings in agreement with Dalton's atomic theory?

2.2 The Nuclear Atom

7. How did the discovery of radioactivity contradict Dalton's atomic theory?
8. Give the distinguishing characteristics of the proton, the neutron, and the electron.
9. Should the proton and electron attract or repel one another?
10. Should the neutron and proton attract or repel one another?
11. Which subatomic particles are found in the nuclei of atoms?
12. What are the extranuclear (outside the nucleus) subatomic particles?
13. Compare Dalton's model of the atom with the nuclear model of the atom.

14. If the nucleus of an atom contains 10 protons,
    a. how many electrons are there in the neutral atoms?
    b. how many neutrons are there in the nucleus of the atom?
15. What are isotopes?
16. The table below describes four atoms.

	Atom A	Atom B	Atom C	Atom D
Number of protons	10	11	11	10
Number of neutrons	11	10	11	10
Number of electrons	10	11	11	10

Are atoms A and B isotopes? Are atoms A and C isotopes? A and D? B and C?

17. What are the masses of the atoms in Problem 16?
18. Use the table inside the front cover to determine the number of protons in atoms of the following elements.
    a. helium (He)
    b. sodium (Na)
    c. chlorine (Cl)
    d. oxygen (O)
    e. magnesium (Mg)
    f. sulfur (S)
19. How many electrons are there in the neutral atoms of the elements listed in Problem 18?

2.3 The Bohr Model of the Atom

20. How did Bohr refine the nuclear model of the atom?
21. What subatomic particles travel in the "orbits" of the Bohr model of the atom?
22. According to Bohr, what is the maximum number of electrons in the fourth energy level ($n = 4$)?
23. If the third energy level of electrons contains two electrons, what is the total number of electrons in the atom?
24. Define the terms
    a. ground state
    b. excited state
25. When light is emitted by an atom, what change has occurred within the atom?
26. Which atom absorbed more energy: one in which an electron moved from the second energy level to the third energy level, or an otherwise identical atom in which an electron moved from the first energy level to the third?
27. Draw Bohr diagrams for the elements listed in Problem 18.
28. The following Bohr diagram is supposed to represent the neutral atoms of an element. The diagram is incorrectly drawn. Identify the error.

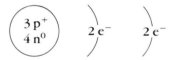

2.4 Electron Arrangements

29. In the quantum mechanical notation $2p^6$, how many electrons are described? What is the general shape of the orbitals described in the notation? How many orbitals are included in the notation?
30. Use quantum mechanical notation to describe the electron configuration of the atom represented in the Bohr diagram.

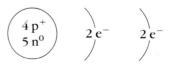

31. Give the electron configurations (using quantum mechanical notation) for the elements in Problem 18.

32. Identify the elements from their electron configurations.
    a. $1s^2 2s^2$
    b. $1s^2 2s^2 2p^3$
    c. $1s^2 2s^2 2p^6 3s^2 3p^1$
33. Give the atomic numbers of the elements described in Problem 32.
34. None of the following electron configurations is reasonable. In each case explain why.
    a. $1s^2 2s^2 3s^2$
    b. $1s^2 2s^2 2p^2 3s^1$
    c. $1s^2 2s^2 2p^6 2d^5$
35. What is the difference in the electron configurations of oxygen (O) and fluorine (F)?
36. What is the difference in the electron configurations of fluorine (F) and sulfur (S)?
37. If three electrons were added to the outermost energy level of a phosphorus atom, the new electron configuration would resemble that of what element?
38. If two electrons were removed from the outermost energy level of a magnesium atom, the new electron configuration would resemble that of what element?

2.5 The Periodic Table

39. Elements are defined on a theoretical basis as being composed of atoms that share the same atomic number. On the basis of this theory, would you think it possible for someone to discover a new element that would fit between magnesium (atomic number 12) and aluminum (atomic number 13)?
40. Referring only to the periodic table, indicate what similarity in electron structure is shared by fluorine (F) and chlorine (Cl). What is the difference in their electron structures?
41. Where are the metals and nonmetals located in the periodic table?
42. Identify the following elements as metals or nonmetals. You may refer to the periodic table. The numbers in parentheses are the atomic numbers of the elements.
    a. sulfur (16)
    b. chromium (24)
    c. iodine (53)
43. Indicate the group number or numbers of the following families.
    a. alkali metals
    b. transition metals
    c. halogens
    d. alkaline earth metals
44. Identify the period of the following elements. You may refer to the periodic table. The numbers in parentheses are the atomic numbers of the elements.
    a. chlorine (17)
    b. osmium (76)
    c. hydrogen (1)
45. Which of the following elements are halogens?
    a. Ag
    b. At
    c. As
46. Which of the following elements are alkali metals?
    a. K
    b. Y
    c. W
47. Which of the following elements are noble gases?
    a. Fe
    b. Ne
    c. Ge
    d. He
    e. Xe
48. Which of the following elements are transition metals?
    a. Ti (22)
    b. Tc (43)
    c. Te (52)
49. Which of the following elements are alkaline earth metals?
    a. Bi
    b. Ba
    c. Be
    d. Br
50. How many electrons are in the outermost energy level of the halogens?

51. How many electrons are in the outermost energy level of Group IIA elements?

52. In what period of elements are electrons first introduced into the fourth energy level?

53. What element is found at the intersection of
    a. the fourth row and Group VA?
    b. the third row and the halogens?
    c. the period representing the second energy level and the alkaline earth metal group?

54. Which is larger?
    a. K or Rb
    b. Ne or Na
    c. Zn or Zr

55. Which will ionize more easily?
    a. Li or Cs
    b. N or Sb
    c. Ar or Ac

3
CHEMICAL BONDING

DID YOU KNOW THAT...

- only the outermost electrons determine the chemistry of an atom?
- there are two major types of bonding?
- the naming of most compounds is quite simple?
- like any other physical object, molecules have distinctive shapes?
- H_2O is not a linear molecule?
- most salts are very hard to melt, but many dissolve readily in water?
- under certain conditions, adding heat to a substance does not cause its temperature to rise?
- hydrogen bonds are among the most important life-sustaining forces?
- like dissolves like?

Chemical bonds: the forces that keep elements together in compounds

There are approximately 100 chemical elements. There are millions of chemical compounds, and about 600 000 new compounds are prepared every year. To form these compounds, atoms of different elements must be held together in specific combinations. **Chemical bonds** are the forces that maintain these arrangements. Chemical bonding also plays a role in determining the state of matter. At room temperature, water is a liquid, carbon dioxide is a gas, and table salt is a solid because of differences in chemical bonding.

As scientists developed an understanding of the nature of chemical bonding, they gained the ability to manipulate the structure of compounds. Dynamite, birth control pills, synthetic fibers, and a thousand other products were fashioned in chemical laboratories and have dramatically changed the way we live. We are now entering an era that promises (some would say forebodes) even greater change.

The DNA molecule—the chemical basis of heredity—carries its genetic message in its bonds. Whether an organism is fish, fowl, hippopotamus, or human is determined by the arrangement of bonds in DNA. Scientists already have the ability to rearrange these bonds, and this ability has given them limited control over the structure of living matter. As techniques of genetic engineering improve, scientists may literally be able to custom-tailor genes.

Let us begin our consideration of chemical bonding so that we, too, can understand the forces that control the structure of matter, living and nonliving.

3.1 THE STABILITY OF THE NOBLE GASES

As we have seen in the previous chapter, the elements in the extreme right-hand column of the periodic table are called the *noble gases* (Figure 3.1). They were once known as the inert gases. Both names

FIGURE 3.1
The noble gases.

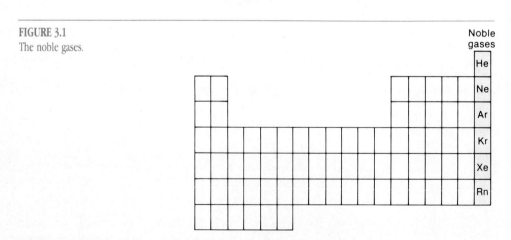

3.2 The Ionic Bond

refer to the fact that these elements are reluctant to undergo chemical reactions, that is, to form compounds with other elements. It is as if the atoms of the noble gases are so perfectly constructed that they resist being altered in any way by interaction with the atoms of other elements. This is, in essence, what has been proposed as an explanation for their inertness.

What the noble gases have in common is a feature of their electron structures. Except for helium, each has eight electrons in its outermost main energy level. In quantum mechanical terms, each has its outermost s and p orbitals filled. Apparently this is an unusually stable electron arrangement, one might even say an especially desirable one.

The outermost energy level of the helium atom is the first energy level, consisting of a single s orbital. Thus, the outermost electron level in helium can hold only two electrons. Since helium has all the inertness of a noble gas, its filled outermost energy level must also represent an especially stable arrangement of electrons.

We now ask our favorite question: So what? If you can accept the fact that even atoms have their own standards of beauty, then we can explain why sodium atoms react with chlorine atoms to form a compound called sodium chloride, common table salt; why hydrogen reacts with oxygen to form water; even why salt dissolves in water.

For the moment, then, hold this thought: to an atom, an **octet** (eight outer electrons), or in special cases a duet (two outer electrons), is a desirable goal.

As a rule of thumb, an atom strives to obtain an octet of electrons.

3.2 THE IONIC BOND

Let us look at an atom of the element sodium (Na). It has 11 electrons, of which two are in the first energy level, eight in the second, and one in the third. If the sodium atom could get rid of an electron, then the product, called a *sodium ion*, would have the same electron structure as an atom of the noble gas neon (Ne).

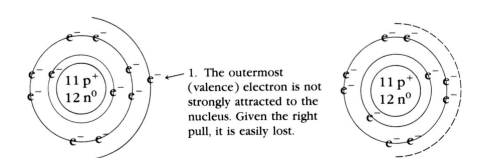

1. The outermost (valence) electron is not strongly attracted to the nucleus. Given the right pull, it is easily lost.

2. The electron has been lost, exposing an inner octet.

This can be represented algebraically by

$$\begin{pmatrix} 11\,p^+ \\ 12\,n^0 \end{pmatrix} 2\,e^-\ 8\,e^-\ 1\,e^- \longrightarrow \begin{pmatrix} 11\,p^+ \\ 12\,n^0 \end{pmatrix} 2\,e^-\ 8\,e^- + 1\,e^-$$

$$\text{Na} \hspace{5em} \text{Na}^+$$

Recall that neon has the structure

$$\begin{pmatrix} 10\,p^+ \\ 10\,n^0 \end{pmatrix} 2\,e^-\ 8\,e^-$$

$$\text{Ne}$$

Let us immediately emphasize that the sodium ion (Na^+) and the neon atom (Ne) are not identical. The electron arrangement is the same, but the nuclei—and resulting charges—are not. As long as sodium keeps its 11 protons, it is still a form of sodium, but it is the sodium *ion*, not the sodium *atom*. **Ions** are charged particles, particles in which the number of electrons does not equal the number of protons. Positively charged ions are called **cations** (pronounced "cat-ions"). The sodium ion is a cation.

If a chlorine atom (Cl) could gain an electron, it would have the same electron structure as the noble gas argon (Ar).

$$\begin{pmatrix} 17\,p^+ \\ 18\,n^0 \end{pmatrix} 2\,e^-\ 8\,e^-\ 7\,e^- + 1\,e^- \longrightarrow \begin{pmatrix} 17\,p^+ \\ 18\,n^0 \end{pmatrix} 2\,e^-\ 8\,e^-\ 8\,e^-$$

$$\text{Cl} \hspace{5em} \text{Cl}^-$$

The structure of the argon atom is

$$\begin{pmatrix} 18\,p^+ \\ 22\,n^0 \end{pmatrix} 2\,e^-\ 8\,e^-\ 8\,e^-$$

$$\text{Ar}$$

The chlorine atom, having gained an electron, becomes negatively charged. It has 17 protons (17+) and 18 electrons (18−). It is written Cl^- and is called a *chloride ion*. (We shall discuss the rules for naming ions in Section 3.5.) Negatively charged ions are called **anions** (pronounced "ann-ions"). The chloride ion is an anion.

A sodium atom forms a less reactive species, a sodium ion, by *losing* an electron. A chlorine atom becomes a less reactive chloride ion by

---

Ion: a particle that contains unequal numbers of protons and electrons and carries a charge

A positive ion is a cation.

A negative ion is an anion.

3.2 The Ionic Bond

gaining an electron. A chlorine atom cannot just pluck an electron from empty space, nor can a sodium atom kick out an electron unless something else is willing to take it on. What happens when sodium comes into contact with chlorine? The obvious. A chlorine atom removes an electron from a sodium atom.

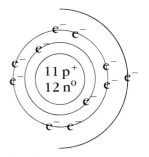 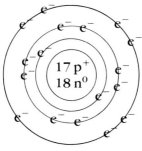

1. The outermost electron is easily removed.

2. The outermost shell in Cl is lacking only one electron. It seeks to complete its octet.

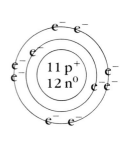 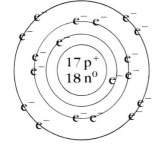

3. Na has lost an electron and exposed an octet. Cl has gained an electron and created an octet.

Seen differently,

$$\begin{pmatrix} 11\,p^+ \\ 12\,n^0 \end{pmatrix} \; 2\,e^- \; 8\,e^- \; 1\,e^- \;+\; \begin{pmatrix} 17\,p^+ \\ 18\,n^0 \end{pmatrix} \; 2\,e^- \; 8\,e^- \; 7\,e^-$$

$$\downarrow$$

$$\begin{pmatrix} 11\,p^+ \\ 12\,n^0 \end{pmatrix} \; 2\,e^- \; 8\,e^- \;\;\;\;+\; \begin{pmatrix} 17\,p^+ \\ 18\,n^0 \end{pmatrix} \; 2\,e^- \; 8\,e^- \; 8\,e^-$$

The sodium *ion* and the chloride *ion* have electron arrangements (electron configurations) like those of two noble gases (neon and argon, respectively). Not only do the ions have stable octets of electrons, they also have opposite charges. Everyone knows that opposites

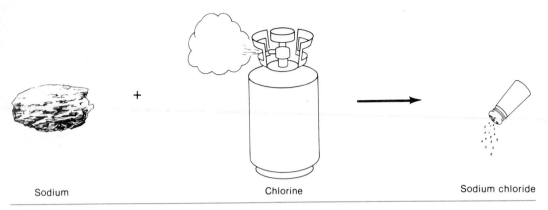

FIGURE 3.2
Sodium, a soft silvery metal, reacts with chlorine, a greenish gas, to form sodium chloride (ordinary table salt).

attract. While this rule of thumb may not always work when applied to people, it works quite well for cations and anions. The attractive force between oppositely charged ions is called an **ionic bond**, and the combination of sodium ions and chloride ions is the compound sodium chloride or table salt (Figure 3.2).

3.3 Electron-Dot Formulas

Assume that the room you are sitting in is an atom. The nucleus would be suspended as a small dot in the middle of the room. Surrounding that small nucleus is vast open space in which the electrons that surround the nucleus move, each in its own orbital. During a chemical reaction this atom would come in contact with another atom. The first point of contact between the two atoms would naturally be the outermost, or valence, electrons.

In forming ions, the nuclei and the inner (i.e., lower) energy levels of electrons do not change. Only the outermost or **valence** electron energy level is altered when ionic compounds are formed from atoms. To illustrate these changes for sodium and chlorine we can draw modified Bohr diagrams for the atoms and ions involved. However, we can simplify matters greatly if we let the nucleus and inner energy levels of electrons be represented by the atomic symbol alone (Na for sodium and Cl for chlorine). The **valence electrons**, that is, the electrons in the outermost main energy level, can then be represented by dots surrounding the symbol. Thus, Na has one, and Cl seven, valence electrons.

Valence electron: electron in the outermost main energy level

$(11\,p^+,\ 12\,n^0)\ \ 2\,e^-\ \ 8\,e^-\ \ 1\,e^-$ becomes Na· $(17\,p^+,\ 18\,n^0)\ \ 2\,e^-\ \ 8\,e^-\ \ 7\,e^-$ becomes :Cl·

3.3 Electron-Dot Formulas

Thus, the ionization of sodium is reduced to

$$\text{Na} \cdot \longrightarrow \text{Na}^+ + 1\,e^-$$

and that of chlorine is

$$:\!\ddot{\text{Cl}}\!\cdot + 1\,e^- \longrightarrow :\!\ddot{\text{Cl}}\!:^-$$

Representations of this sort are called **electron-dot formulas**. In electron-dot form, the reaction of a sodium atom with a chlorine atom to form the compound sodium chloride is written

$$\text{Na} \cdot + :\!\ddot{\text{Cl}}\!\cdot \longrightarrow \text{Na}^+ \;:\!\ddot{\text{Cl}}\!:^-$$

The chemical formula for the compound sodium chloride is NaCl. This formula indicates that in the compound there is one sodium ion for each chloride ion.

Writing electron-dot formulas for elements in the first three periods (horizontal rows) of the periodic table is easy. The number of valence electrons (those in the outermost energy level) is equal to the group number. Aluminum (Al) is in Group IIIA; therefore, it has three outer electrons. Sulfur (S) is in Group VIA; thus, it has six outer electrons. This generalization works well for elements in A groups even beyond the first three periods, and, for the moment, we shall limit our discussion to elements in the A groups. Table 3.1 gives the electron-dot formulas for the first 20 elements. Notice that there is a pattern to the way in which the dots are drawn. First the s orbital fills, followed by the three p orbitals. Since the latter are all of the same energy, they each acquire one electron before going to the less desirable condition of pairing up. Notice that Mg has the two dots as a pair, but Si has one pair (the s orbital electrons) and two separate dots (in p orbitals). On the other hand, in O we find two pairs, one for the s electrons and one for a p orbital that had to be paired with four p electrons present.

TABLE 3.1 Electron-Dot Formulas for the First 20 Elements

IA	IIA	IIIA	IVA	VA	VIA	VIIA	Noble Gases
H·							He:
Li·	Be:	·B·	·C:	:N·	:Ö·	:F:	:Ne:
Na·	Mg:	·Al·	·Si:	:P·	:S·	:Cl:	:Ar:
K·	Ca:						

It does not matter whether you draw lithium in any of the following ways.

$$\dot{\text{Li}} \quad \text{Li}\cdot \quad \underset{\cdot}{\text{Li}} \quad \cdot\text{Li}$$

And magnesium can be drawn

$$\text{Mg}: \quad \underset{\cdot\cdot}{\text{Mg}} \quad :\text{Mg} \quad \overset{\cdot\cdot}{\text{Mg}}$$

Similarly, nitrogen can be represented by

$$\cdot\overset{\cdot\cdot}{\underset{\cdot}{\text{N}}}\cdot \quad \cdot\underset{\cdot}{\dot{\text{N}}}\cdot \quad \cdot\overset{\cdot}{\underset{\cdot}{\text{N}}}\cdot \quad :\dot{\text{N}}\cdot$$

As long as you show five outer electrons, you will be right. As we get into a discussion of compounds, you will see that occasionally sticking to one choice of structures will simplify writing symbols for compounds.

3.4 Ionic Compounds

Generally speaking, elements on the left side of the periodic table (Groups IA, IIA, and IIIA) tend to give up their outer electrons to form positively charged ions with noble gas configurations. Groups VA, VIA, and VIIA try to pick up electrons to form negative ions with complete octets. The elements on the left (called metals) react with elements on the right (nonmetals) to form compounds held together by strong ionic bonds. The following examples illustrate this general chemical principle.

EXAMPLE 3.1

Draw the electron-dot formula for the ionic compound formed from potassium and chlorine.

Potassium is in Group IA.

$$\text{K}\cdot \longrightarrow \text{K}^+ + 1\ e^-$$

Chlorine is in Group VIIA.

$$:\overset{\cdot\cdot}{\underset{\cdot\cdot}{\text{Cl}}}\cdot + 1\ e^- \longrightarrow :\overset{\cdot\cdot}{\underset{\cdot\cdot}{\text{Cl}}}:^-$$

The electron-dot formula for the compound formed from potassium and chlorine is

$$\text{K}\cdot + \cdot\overset{\cdot\cdot}{\underset{\cdot\cdot}{\text{Cl}}}: \longrightarrow \text{K}^+ :\overset{\cdot\cdot}{\underset{\cdot\cdot}{\text{Cl}}}:^-$$

3.4 Ionic Compounds

EXAMPLE 3.2

Show the formation of an ionic compound from magnesium atoms and oxygen atoms.

Magnesium is in Group IIA.

$$\overset{..}{Mg} \longrightarrow Mg^{2+} + 2\ e^-$$

Oxygen is in Group VIA.

$$\cdot \overset{..}{\underset{..}{O}} : +\ 2\ e^- \longrightarrow\ : \overset{..}{\underset{..}{O}} :^{2-}$$

Notice that the magnesium ion has a double positive charge because it must give up two electrons to achieve a noble gas configuration. The oxide ion has a double negative charge because it acquires two electrons to fill in its octet. In the compound formed from magnesium and oxygen, one magnesium atom supplies the two electrons required by one oxygen atom:

$$\overset{..}{Mg} + \cdot \overset{..}{\underset{..}{O}} : \longrightarrow Mg^{2+} : \overset{..}{\underset{..}{O}} :^{2-}$$

EXAMPLE 3.3

What is the formula for the compound formed from potassium and oxygen?

Potassium is in Group IA.

$$K\cdot \longrightarrow K^+ + 1\ e^-$$

Oxygen is in Group VIA.

$$\cdot \overset{..}{\underset{..}{O}} : +\ 2\ e^- \longrightarrow\ : \overset{..}{\underset{..}{O}} :^{2-}$$

Oxygen requires two electrons, but a potassium atom has only one to give. The problem can be solved by having each oxygen atom react with two potassium atoms.

$$\begin{matrix} K\cdot \\ \ \ \ \ \ + \cdot \overset{..}{O} \cdot \longrightarrow K^+ : \overset{..}{\underset{..}{O}} :^{2-} K^+ \\ K\cdot \end{matrix}$$

The formula is K_2O.

EXAMPLE 3.4

What is the formula for the compound formed from magnesium and nitrogen?

Magnesium is in Group IIA.

$$\ddot{Mg} \longrightarrow Mg^{2+} + 2\,e^-$$

Nitrogen is in Group VA.

$$\cdot\ddot{N}\cdot + 3\,e^- \longrightarrow :\!\ddot{\underset{..}{N}}\!:^{3-}$$

To proceed, first note the number of electrons involved in the formation of each ion: two for magnesium and three for nitrogen. Then determine the smallest number into which each of these numbers can be divided evenly (called the **least common multiple**). In this case, that number is 6; this indicates the *smallest* number of electrons that can be *evenly* exchanged between the two elements. Now—how many magnesium atoms are required to supply six electrons? The answer is three. How many nitrogen atoms are required to accept six electrons? The answer is two.

$$\begin{array}{c}\ddot{Mg}\\ \ddot{Mg} + \begin{array}{c}\cdot\ddot{N}\cdot\\ \cdot\ddot{N}\cdot\end{array} \\ \ddot{Mg}\end{array} \longrightarrow Mg^{2+}:\!\ddot{\underset{..}{N}}\!:^{3-} Mg^{2+} :\!\ddot{\underset{..}{N}}\!:^{3-} Mg^{2+}$$

The formula is Mg_3N_2.

3.5 Names of Simple Ions and Ionic Compounds

Simple positive ions are named by adding the word ion to the name of the element.

Simple negative ions are named by changing the usual ending to -ide and adding the word ion.

Names of simple positive ions are derived from those of the parent elements by addition of the word *ion*. A potassium atom (K), upon losing an electron, becomes *potassium ion* (K^+). A magnesium atom (Mg), upon losing two electrons, becomes a *magnesium ion* (Mg^{2+}). Names of simple negative ions are derived from those of the parent elements by changing the usual ending to *-ide* and adding the word *ion*. A chlor*ine* atom (Cl), upon gaining an electron, becomes a chlor*ide* ion (Cl^-). A sulf*ur* atom (S), upon gaining two electrons, becomes a sulf*ide* ion (S^{2-}).

Names and formulas for several important simple ions are given in

3.5 Names of Simple Ions and Ionic Compounds

TABLE 3.2 Formulas and Names for Some Simple Ions

Group	Element	Name of Ion	Formula for Ion
IA	Lithium	Lithium ion	Li^+
	Sodium	Sodium ion	Na^+
	Potassium	Potassium ion	K^+
IIA	Magnesium	Magnesium ion	Mg^{2+}
	Calcium	Calcium ion	Ca^{2+}
IIIA	Aluminum	Aluminum ion	Al^{3+}
VA	Nitrogen	Nitride ion	N^{3-}
VIA	Oxygen	Oxide ion	O^{2-}
	Sulfur	Sulfide ion	S^{2-}
VIIA	Chlorine	Chloride ion	Cl^-
	Bromine	Bromide ion	Br^-
	Iodine	Iodide ion	I^-
IB	Copper	Copper(I) ion (cuprous ion)	Cu^+
		Copper(II) ion (cupric ion)	Cu^{2+}
	Silver	Silver ion	Ag^+
IIB	Zinc	Zinc ion	Zn^{2+}
VIIIB	Iron	Iron(II) ion (ferrous ion)	Fe^{2+}
		Iron(III) ion (ferric ion)	Fe^{3+}

Table 3.2. Note that the charge on an ion of a Group IA element is +1 (usually written simply as +). The charge on an ion of a Group IIA element is 2+, and that on an ion of a Group IIIA element is 3+. You can calculate the charge on the negative ions in the table by subtracting eight from the group number. For example, the charge on the oxide ion (oxygen is in Group VIA) is $6 - 8 = -2$. The charge on a nitride ion (nitrogen is in Group VA) is $5 - 8 = -3$. The periodic relationship of these simple ions is shown in Figure 3.3.

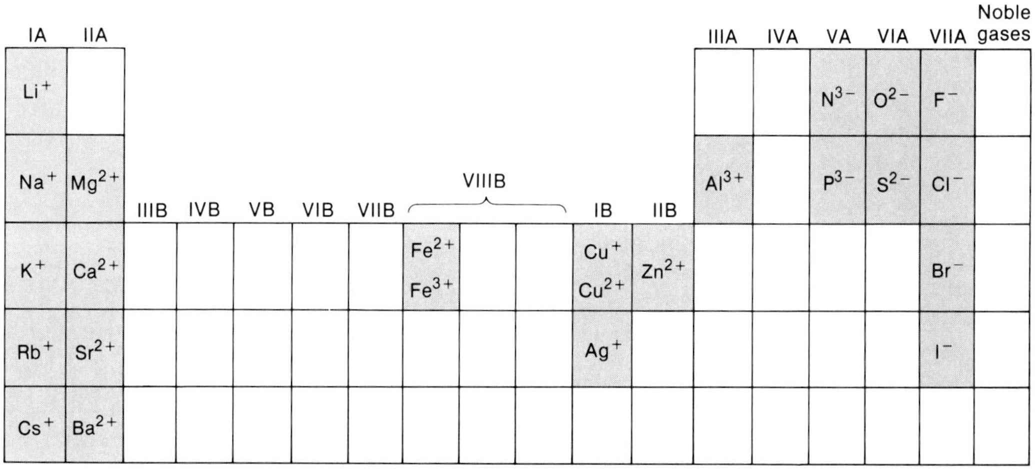

FIGURE 3.3
The periodic relationship of some simple ions.

IA	IIA	IIIB	IVB	VB	VIB	VIIB	VIIIB			IB	IIB	IIIA	IVA	VA	VIA	VIIA	Noble gases
Li^+														N^{3-}	O^{2-}	F^-	
Na^+	Mg^{2+}											Al^{3+}		P^{3-}	S^{2-}	Cl^-	
K^+	Ca^{2+}						Fe^{2+} Fe^{3+}			Cu^+ Cu^{2+}	Zn^{2+}					Br^-	
Rb^+	Sr^{2+}									Ag^+						I^-	
Cs^+	Ba^{2+}																

There is no simple way to determine the most likely charge on ions formed from Group VIIIB elements and from those in the B subgroups. Indeed, you may have noticed that these can form ions with different charges. In such cases, chemists use Roman numerals with the names to indicate the charge. Thus, Fe^{2+} is the iron(II) ion, and Fe^{3+} is the iron(III) ion. An older terminology calls Fe^{2+} ferr*ous* ion and Fe^{3+} ferr*ic* ion. See Table 3.2 for a similar situation involving copper ions.

To name an ionic compound, first name its cation and then its anion. The term *ion* is not included in the compound name. Thus, the compound consisting of sodium ions and chloride ions is called sodium chloride. The compounds described in Examples 3.1 to 3.4 are potassium chloride, magnesium oxide, potassium oxide, and magnesium nitride, respectively.

The constituent units of ionic compounds are charged particles—ions. Yet the compound as a whole is electrically neutral. Instead of working with electron-dot formulas (as we did in Examples 3.1 to 3.4), we can use the principle of electrical neutrality to determine the combining ratios for sets of ions. Let us use this principle to determine the formula for the compound lithium chloride. Lithium ions (Li^+) must combine with chloride ions (Cl^-) in a one-to-one ratio in order to maintain electrical neutrality in the compound. The formula LiCl expresses this ratio. The combining ratio (1:1) and the ionic charges are understood.

Let us try another example. To maintain electrical neutrality, one calcium ion (Ca^{2+}) combines with *two* chloride ions (Cl^-). This ratio is expressed in the formula $CaCl_2$. In this formula, the ionic charges are understood. As with a coefficient of 1 in algebra, a subscript of 1 is understood where no other number appears; therefore, in the formula $CaCl_2$, the subscript 1 for calcium ion is understood, but the 2 for chloride is explicitly written (not Ca_1Cl_2 but $CaCl_2$). Thus, $CaCl_2$ not only gives us the combining ratio (1:2) but also stands for the compound calcium chloride. It is a shorthand way of writing (1 Ca^{2+} and 2 Cl^-).

You can use the charges on the ions in Table 3.2 to determine formulas for compounds of these elements. You can also use a periodic table to predict the charge on ions formed from Subgroup A elements, with Group IVA being the dividing line between positive and negative ions.

EXAMPLE 3.5

What is the formula for sodium sulfide?

First, write the symbols for the ions (positive ion first). Sodium is in Group IA; therefore, its charge is $+1$. Sulfur is in Group VIA, and its charge is -2. The symbols are Na^+ and S^{2-}. The smallest number into which both charges can be evenly divided, that is, the least common multiple (LCM), is 2. The subscript for each symbol can be determined by dividing its charge (without the plus or minus) into the least common multiple. This step determines how many atoms of each element are needed to supply or accept the smallest common number of electrons. For Na^+,

$$\frac{2\,(LCM)}{1\,(charge)} = 2$$

For S^{2-},

$$\frac{2\,(LCM)}{2\,(charge)} = 1$$

Thus, we have the formula Na_2S_1 (2 Na^+ and 1 S^{2-}), or Na_2S.

EXAMPLE 3.6

What is the formula for aluminum oxide?

The symbols are Al^{3+} and O^{2-} (Al is in Group IIIA and O is in Group VIA). The LCM is 6. For Al^{3+},

$$\frac{6}{3} = 2$$

For O^{2-},

$$\frac{6}{2} = 3$$

The formula is therefore Al_2O_3 (2 Al^{3+} and 3 O^{2-}).

Another way to look at this is the cross-multiplication method: the charge number on one ion becomes the subscript on the other. Thus, in Example 3.6, the charge on the aluminum ion is 3, which becomes the subscript for oxygen in aluminum oxide; and the charge on oxygen is 2, which becomes the subscript for the aluminum ion.

$$Al^{3+} \quad O^{2-} \qquad Al_2O_3$$

Thus, two aluminum ions have six positive charges, and three oxygen ions have six negative charges. The resulting compound, Al_2O_3, is therefore neutral—as all compounds are.

EXAMPLE 3.7

What is the formula for calcium sulfide?
The symbols are Ca^{2+} and S^{2-}. The LCM is 2. For Ca^{2+},

$$\frac{2}{2} = 1$$

For S^{2-},

$$\frac{2}{2} = 1$$

The formula is therefore CaS (Ca^{2+} and S^{2-}).

Let us also practice naming simple ionic compounds.

EXAMPLE 3.8

What is the name for the compound Na_2S?
Find the constituent ions in Table 3.2. They are sodium ion (Na^+) and sulfide ion (S^{2-}). The compound is sodium sulfide.

EXAMPLE 3.9

What is the name for the compound FeS?
There are two kinds of iron ions. Since sulfur exists as the S^{2-} ion and one iron ion is combined with it, the iron ion in this compound must be Fe^{2+}. The name of FeS is iron(II) sulfide.

EXAMPLE 3.10

What is the name of the compound $FeCl_3$?
Since the charge on the chloride ion is 1− and three of these ions are combined with one iron ion, the iron ion must be Fe^{3+}. The name of $FeCl_3$ is iron(III) chloride.

3.6 Covalent Bonds

One might expect a hydrogen atom, with its one electron, to acquire another electron and assume the helium configuration. Indeed, hydrogen atoms do just that in the presence of atoms of a reactive metal such as lithium, that is, a metal that finds it easy to give up an electron.

$$Li \cdot \longrightarrow H \qquad Li^+ \quad :H^- \qquad Li^+ :H^-$$

1. In an attempt to complete its duet, H pulls the lone Li electron.

2. The result is a Li^+ ion (Li minus an electron) and an H^- (hydride) ion (H plus one electron).

3. Since opposites attract, lithium hydride is formed.

The process can be summarized as

$$Li \cdot + H \cdot \longrightarrow Li^+ \ H:^-$$

But what if there are no other kinds of atoms around? What if there are only hydrogen atoms (as in a sample of the pure element)? One hydrogen atom can scarcely grab an electron from another, for among hydrogen atoms all have equal attraction for electrons. (Even more important, perhaps, hydrogen atoms do not have a tendency to lose electrons at all, for the result would be a highly reactive bare proton— the hydrogen nucleus.) Still—hydrogen wants a duet of electrons like helium's. If one hydrogen cannot capture another's electron, the two atoms can compromise by *sharing* their electrons.

$$H \longleftrightarrow H \qquad\qquad H:H$$

1. Since both atoms are the same element, they have the same pull for electrons. The compromise is to share.

2. Both hydrogens have given *and* taken an electron. Therefore, they truly share the pair of electrons.

This is summarized chemically as

$$H \cdot + \cdot H \longrightarrow H:H$$

It is as if the two hydrogen atoms, in approaching one another, get their electron clouds or orbitals so thoroughly enmeshed that they cannot easily pull them apart again.

Molecule—a group of atoms held together by shared pairs of electrons. This binding force is called a covalent bond.

Most of the time the electrons are located between the two nuclei. The electron-dot formula usually used, H:H, is therefore a fairly good picture. (If we were to attribute human qualities to hydrogen atoms, we would suggest that they are a bit nearsighted. Each one looks around, sees two electrons, and decides that these electrons are its very own and that therefore it has an arrangement like that of helium, one of the noble gases.) This combination of hydrogen atoms is called a *hydrogen molecule.* **Molecules** are discrete groups of atoms held together by shared pairs of electrons. The bond formed by a shared pair of electrons is called a **covalent bond**.

A chlorine atom will pick up an extra electron from anything willing to give one up. But, again, what if the only thing around is another chlorine atom? Chlorine atoms, too, can attain a more stable arrangement by sharing a pair of electrons.

$$:\ddot{\text{Cl}}\cdot\ +\ \cdot\ddot{\text{Cl}}:\ \longrightarrow\ :\ddot{\text{Cl}}:\ddot{\text{Cl}}:$$

Each chlorine atom in the chlorine molecule counts eight electrons around itself and concludes that it has an arrangement like that of the noble gas argon. The shared pair of electrons in the chlorine molecule also creates a covalent bond.

For simplicity, the hydrogen molecule is often represented as H_2 and the chlorine molecule as Cl_2. The subscripts indicate two atoms *per* molecule. In each case, the covalent bond between the atoms is understood. Sometimes the covalent bond is indicated by a dash, H—H and Cl—Cl.

Let us be sure we understand the meaning of numbers in formulas. Take a moment to establish in your mind the differences among the following: H, H_2, 2 H, 2 H_2, H_2O, 2 H_2O. Is it clear to you that although H represents a single atom of hydrogen, H_2 implies two atoms of H bonded together, whereas 2 H represents two separate, free, and independent atoms of H? On the other hand, the meaning of H_2 in H_2O is totally different from that of H_2 as a molecule. In H_2O it means that two atoms of H are individually attached to O (not to themselves!) to form a molecule of water. Finally, 2 H_2O simply refers to two individual molecules of water.

Covalent bonds are not limited to the sharing of one pair of electrons. Consider, for example, the nitrogen atom. Its electron-dot symbol is

$$:\dot{\text{N}}\cdot$$

Now, after all we have learned about the octet rule, we know that this electron arrangement is not complete. It has only five electrons in its outermost energy level. It could share a pair of electrons with another nitrogen atom and would then look like this:

$$:\!\overset{..}{\underset{..}{N}}\!:\!\overset{..}{\underset{..}{N}}\!: \quad \text{(incorrect structure)}$$

The situation has not improved a great deal. Each nitrogen atom in this arrangement has only six electrons surrounding it (not eight). Each nitrogen atom has two electrons hanging out there without partners, so, to solve the dilemma, each nitrogen atom shares two additional pairs of electrons, for a total of three pairs.

$$:\!N\!::\!:\!N\!: \quad (\text{or} \quad :\!N\!\equiv\!N\!: \quad \text{or} \quad N\!\equiv\!N)$$

In drawing the nitrogen molecule (N_2), we have placed all the electrons being shared by the two atoms in the space between the two atoms. Each nitrogen atom has now satisfied the octet rule. A molecule in which *three pairs* of electrons (a total of *six individual* electrons) are being shared is said to contain a **triple bond**. Each nitrogen atom also has an *unshared* pair of electrons. Note that we could have drawn the unshared pair of electrons above or below the atomic symbol. Such a drawing would represent the same molecule.

3.7 POLAR COVALENT BONDS

So far we have seen that atoms combine in two different ways. Some that are quite different in electron structure (from opposite ends of the periodic table) react by the complete transfer of one or more electrons from one atom to another (ionic bond formation). Atoms that are identical combine by sharing one or more pairs of electrons (covalent bond formation). Now let us look at some "in-betweeners."

Hydrogen and chlorine react to form a colorless, toxic gas called hydrogen chloride. This reaction can be represented schematically by

$$H\cdot + \cdot\overset{..}{\underset{..}{Cl}}\!: \longrightarrow H\!:\!\overset{..}{\underset{..}{Cl}}\!: \quad (\text{or} \quad H\!-\!Cl)$$

Both the hydrogen atom and the chlorine atom want an electron, so they compromise by sharing and form a covalent bond. Since the substances hydrogen and chlorine actually consist of diatomic molecules rather than single atoms, the reaction is more accurately represented by the scheme

$$H\!:\!H + :\!\overset{..}{\underset{..}{Cl}}\!:\!\overset{..}{\underset{..}{Cl}}\!: \longrightarrow 2\ H\!:\!\overset{..}{\underset{..}{Cl}}\!:$$

This can be more simply written as

$$H_2 + Cl_2 \longrightarrow 2\ HCl$$

One might reasonably ask why the hydrogen molecule and the chlorine molecule react at all. Have we not just explained that they themselves were formed to provide a more stable arrangement of electrons? Yes, indeed, we did say that. But there is stable and there is more stable. The chlorine molecule represents a more stable arrangement than separate chlorine atoms, but, given the opportunity, a chlorine atom would rather bond to hydrogen than to another chlorine atom.

1. Hydrogen gas and chlorine gas by themselves are stable, with complete electron clouds.

   H:H   :Cl:Cl:
   stable   stable

2. If brought into contact with each other (and providing the right encouragement exists), both H_2 and Cl_2 can make *even more* stable arrangements through a mutual exchange.

   H:H   :Cl:Cl:

3. The net result is the same electron environment for each atom but with increased bond strength.

   H:Cl: H:Cl:
   more stable

In a molecule of hydrogen chloride, a chlorine atom shares a pair of electrons with a hydrogen atom. In this case, and in others we shall consider, sharing does not mean sharing equally. Some atoms within molecules attract electrons more strongly than do other atoms. The term **electronegativity** is used to describe the affinity of an element in a molecule for electrons. The higher the electronegativity, the more strongly the atoms of an element attract electrons to themselves. The most electronegative element is fluorine, which is located in the upper right corner of the periodic table. The electronegativity of elements decreases as one moves away from fluorine in the periodic table. Thus, electronegativity decreases as one moves down a group (column) or left across a period (row). Figure 3.4 lists some values for the relative electronegativities of some common elements.

The greater an element's attraction for electrons, the higher its electronegativity.

Chlorine is more electronegative than hydrogen. In a hydrogen chloride molecule, the chlorine atom has a much greater attraction than the hydrogen atom for the shared electron pair. Because the shared electrons are held more tightly by the chlorine atom, the chlorine end of the molecule is more negative than the hydrogen end. If you think of an orbital as a fuzzy-looking cloud, then the cloud is denser near the chlorine atom. When the electrons in a covalent bond are not equally shared, the bond is said to be **polar**. Thus, the bonding in hydrogen chloride is described as **polar covalent**, whereas the bonding in the hydrogen molecule or in the chlorine molecule is **nonpolar covalent**. The polar covalent bond is not an ionic bond. In an ionic

In a polar covalent bond, one atom attracts the electrons more strongly than the other atom.

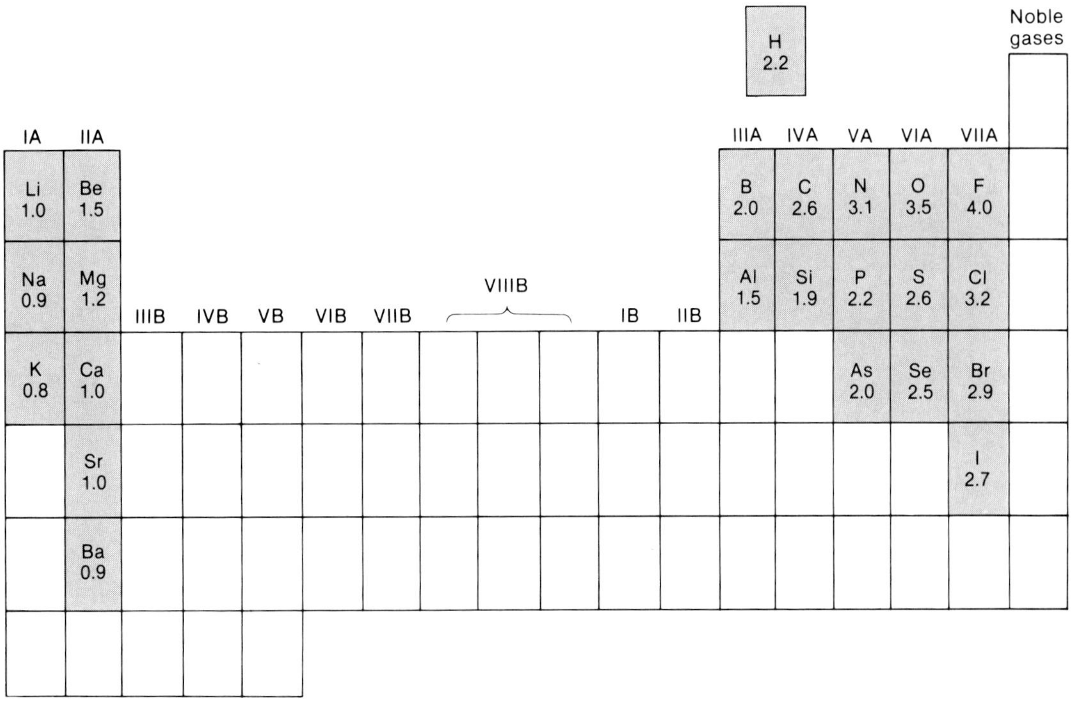
Noble gases
H 2.2
IA
IIA
IIIA
IVA
VA
VIA
VIIA
Li 1.0
Be 1.5
B 2.0
C 2.6
N 3.1
O 3.5
F 4.0
Na 0.9
Mg 1.2
IIIB
IVB
VB
VIB
VIIB
VIIIB
IB
IIB
Al 1.5
Si 1.9
P 2.2
S 2.6
Cl 3.2
K 0.8
Ca 1.0
As 2.0
Se 2.5
Br 2.9
Sr 1.0
I 2.7
Ba 0.9

FIGURE 3.4
Relative electronegativities of some common elements.

bond, one atom completely loses an electron. In a polar covalent bond, the atom at the positive end of the bond (hydrogen in HCl) still has some share in the bonding pair of electrons (Figure 3.5). To distinguish this arrangement from that in an ionic bond, the following notation is used.

$$\overset{\delta+}{\mathrm{H}}—\overset{\delta-}{\mathrm{Cl}}$$

The line between the atoms represents the covalent bond, a pair of shared electrons. The $\delta+$ and $\delta-$ (read "delta plus" and "delta minus") signify which end is partially positive and which is partially negative (the word *partially* is used to distinguish this charge from the full charge on an ion).

Another polar covalent molecule, unquestionably the most important polar covalent molecule on Earth, is water, H_2O. Oxygen has six outer electrons and needs two more to complete its octet. Hydrogen atoms need only one electron each to complete their duets. Therefore, an oxygen atom bonds with two hydrogen atoms.

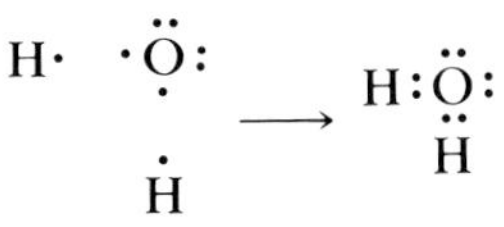

This arrangement completes the outer shell of oxygen, which now has the neon structure, and that of the hydrogen atoms, which now have the helium structure.

FIGURE 3.5
Representation of the polar hydrogen chloride molecule. (a) The electron-dot formula with the shared electron pair shown nearer the chlorine atom. The symbols $\delta+$ and $\delta-$ indicate partial positive and negative charges, respectively. (b) A diagram depicting the unequal distribution of electron density in the hydrogen chloride molecule.

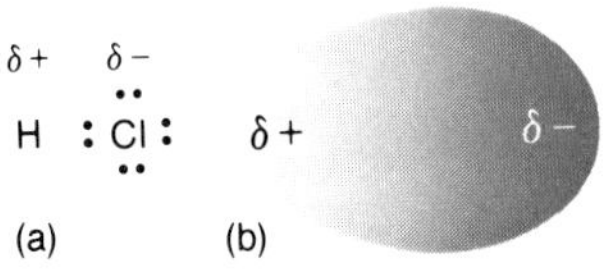
δ+
δ−
H
Cl
δ+
δ−
(a)
(b)

We should expect the bonds in water to be polar because oxygen is more electronegative than hydrogen. (Like chlorine, oxygen is to the right in the periodic table, and electronegativity increases as one moves to the right in the table.) Just because a molecule contains polar bonds, however, does not mean that the molecule as a whole is polar. If the atoms in the water molecule were in a straight row (that is, in a linear arrangement), the two polar bonds would cancel one another out.

$$\overset{\delta+}{H}—\overset{\delta-}{O}—\overset{\delta+}{H} \quad \text{(incorrect structure)}$$

Instead of having one end of the molecule positive and the other end negative, the electrons would be pulled toward the right in one bond and toward the left in the other. Overall there would be no net dipole. By a molecular **dipole**, we mean a molecule that has a positive end and a negative end.

In a molecular dipole, the molecule has a negative end and a positive end.

But water *does* act like a dipole. If you place a sample of water between two electrically charged plates, the water molecules align themselves, one end attracted toward the positive plate and the other end toward the negative plate. For water to act like a dipole, the bonds in the water molecule must be in a bent arrangement so that they do not cancel each other out.

$$^{\delta+}H:\overset{..}{\underset{..}{O}}:^{\delta-} \quad \text{or} \quad \overset{\delta+H}{\underset{H^{\delta+}}{O^{\delta-}}}$$
$$\phantom{^{\delta+}H:}H^{\delta+}$$

Such molecules would align themselves between charged plates in the following way.

Some molecules with polar bonds do not act like dipoles. Carbon dioxide (CO_2) is one of these. Its structure is

$$:\overset{..}{O}::C::\overset{..}{O}: \quad \text{or} \quad O{=}C{=}O$$

Since oxygen is more electronegative than carbon, the bonds in this molecule should be polar. Carbon dioxide is a linear molecule, however, and the polar covalent bonds cancel one another to give a nonpolar molecule.

Many of the properties of compounds, like melting point, boiling point, and solubility, depend on the polarity of the molecules of the compound. Later we will consider this point, but now let us look at the reason for the different shapes.

3.8 SHAPES OF MOLECULES

We said that electrons are essentially without mass and therefore occupy no space. However, we must keep three things in mind.

1. Electrons are not fixed in place but move about in orbitals, and for their motion they demand a certain amount of space. This is true whether the electrons are part of a bond or not.
2. Electrons are negatively charged, which means that the negative cloud created by electrons in one orbital will repel that of another orbital. So, if we had an atom with only two orbitals occupied by electrons, they would try to get away from each other as much as possible and find themselves on opposite sides of the atom.
3. The geometry of a molecule will be determined by the space requirements of the valence electrons only. Since they are the outermost electrons, they are the only ones not shielded by other electrons and therefore their motion outlines a certain space.

With these facts in mind, let us establish some basic geometries. We said that two groups, pairs, of electrons surrounding an atom would result in a linear structure. An example might be $BeCl_2$. Similarly, if an atom is surrounded by three groups of electrons, a flat trigonal (triangular) arrangement results. An example is $AlCl_3$. (Think of three balloons tied together; they point out in triangular fashion.) Now, for four groups of electrons surrounding a central atom, the best arrangement is not a flat square geometry, but rather a three-dimensional arrangement known as a tetrahedron. CH_4 is one of many examples. (Again, it is most helpful to take four balloons filled to the same level and tie them together. Try it!) Table 3.3 summarizes these observations for up to six surrounding groups of electrons.

The examples given above all have only bonded electrons (those that bond atoms together) surrounding the central atom. Many molecules have both bonded and nonbonded electrons surrounding the central atom. Water and ammonia are such examples. The oxygen in water is surrounded by four groups (two bonded and two nonbonded). Regard-

TABLE 3.3 Shapes of Covalent Molecules

Surrounding Electron Groups		Example	Shape	Structural Formula
Bonded	Nonbonded			
1		H_2	Linear 180°	H—H
2		$BeCl_2$	Linear 180°	Cl—Be—Cl
3		BCl_3	Flat trigonal 120°	
4		CH_4	Tetrahedral 109.5°	
5		PCl_5	Trigonal bipyramid 90°, 120°	
6		SF_6	Octahedral 90°	
2	2	H_2O	Bent 104.5°	
3	1	NH_3	Tripod 109.5°	

3.8 Shapes of Molecules 91

less of their function, they all occupy space and therefore participate in determining the structure. So, in water the four groups point away from oxygen in a tetrahedral shape. Two of those electron groups are attached to hydrogen. So the structure of water is not linear, but rather bent in an angle close to 104.5°. By the same token, ammonia has its three hydrogens pointing away from nitrogen in tripod fashion.

EXAMPLE 3.11

Draw the electron-dot structure for H_2S.

The electron-dot symbol for hydrogen is H·, and for sulfur, ·S̈:.
We have two hydrogen atoms and one sulfur to combine.

$$H \cdot \quad H \cdot \quad \cdot \ddot{S} :$$

If the two hydrogen atoms share electrons, then they will be satisfied, but the sulfur atom will be left out. If the sulfur atom shares one of its lone electrons with one hydrogen atom and its other single electron with the other hydrogen atom, however, everyone ends up happy.

If we were asked to determine its structure, we would see that there are four pairs. Therefore, as in water, the electron groups describe a tetrahedral geometry with the H—S—H in bent position.

H:S̈:   or
H

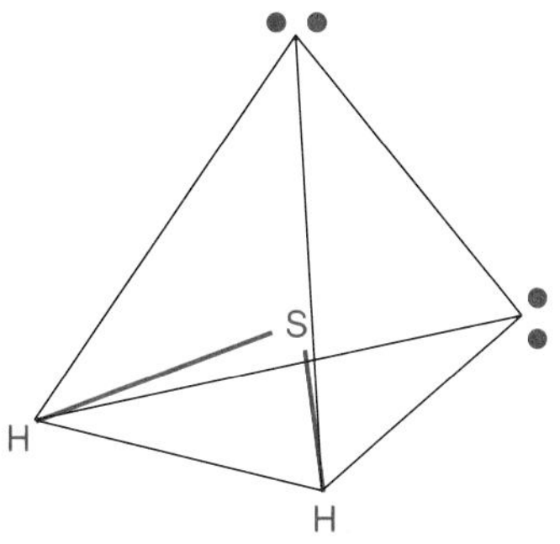

EXAMPLE 3.12

Draw the electron-dot structure for methane, CH_4.
Here are the parts:

$$\cdot \dot{C} \cdot \quad H \cdot \quad H \cdot \quad H \cdot \quad H \cdot$$

We have four hydrogen atoms, each wanting to share its single electron, and we have a carbon atom, which has four single electrons to be shared. The only way to combine these is

$$\begin{array}{c} \text{H} \\ \text{H}:\overset{..}{\underset{..}{\text{C}}}:\text{H} \\ \text{H} \end{array}$$

Four electron groups lead to a tetrahedral arrangement.

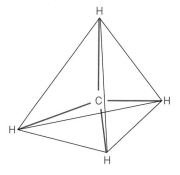

EXAMPLE 3.13

The compound phosgene has the formula $COCl_2$. What is its structure?

Once again we should list the parts:

$$\cdot\overset{\cdot}{\underset{\cdot}{\text{C}}}\cdot \quad \cdot\overset{..}{\text{O}}: \quad :\overset{..}{\underset{..}{\text{Cl}}}\cdot \quad :\overset{..}{\underset{..}{\text{Cl}}}\cdot$$

The carbon atom should form four bonds; oxygen, two; the chlorine atoms, one each. The answer:

$$\begin{array}{c} :\overset{..}{\underset{..}{\text{Cl}}}:\text{C}:\overset{..}{\underset{..}{\text{Cl}}}: \\ :\overset{..}{\text{O}}: \end{array}$$

The total number of electron groups around C pointing in different directions is three, and hence the structure is flat trigonal.

$$\underset{\underset{\text{O}}{\parallel}}{\overset{\text{Cl}\diagdown\diagup\text{Cl}}{\text{C}}}\,120°$$

Notice that we have here an example of a **double bond**, the sharing of two pairs of electrons between the carbon and oxygen atoms.

3.8 Shapes of Molecules

EXAMPLE 3.14

The formula for hydrogen cyanide is HCN. From its electron-dot structure, determine the geometry of this molecule.

The parts are

$$H\cdot \quad \cdot \overset{\cdot}{\underset{\cdot}{C}}\cdot \quad \cdot \overset{\cdot\cdot}{\underset{\cdot}{N}}\cdot$$

The hydrogen atom wants to form one covalent bond (because it has a single unpaired electron), the carbon atom wants to form four (it has four unpaired electrons), and the nitrogen atom wants to form three (of its five electrons, three are unpaired). The only combination that satisfies everyone is one that incorporates a triple bond.

$$H:C:::N: \quad \text{or} \quad H-C\equiv N$$

Notice that the **triple bond** has three pairs of electrons. But since they all point in the same direction (because they bond the same two atoms) the six electrons can be considered as a single group for purposes of determining the geometry. The resulting structure is linear.

Many covalent compounds have common and widely used names. Examples are water (H_2O), methane (CH_4), and ammonia (NH_3). For other molecular compounds, the Greek prefixes *mono-*, *di-*, *tri-*, and *tetra-* are used to indicate the number of atoms of each element in the molecule. For example, the compound N_2O_4 is called *dinitrogen tetroxide*. (The *a* often is dropped from tetra- when it precedes another vowel.) We often leave off the mono- prefix (NO_2 is nitrogen dioxide) but do include it to distinguish between two compounds of the same pair of elements (CO is carbon monoxide, CO_2 is carbon dioxide).

EXAMPLE 3.15

What is the name for SCl_2? For BF_3?

SCl_2 has one sulfur and two chlorine atoms; it is sulfur dichloride. BF_3 is boron trifluoride.

EXAMPLE 3.16

Write the formula for carbon tetrachloride.

The name indicates one carbon atom and four chlorine atoms. The formula is CCl_4.

3.9 Polyatomic Ions

Many compounds contain both ionic and covalent bonds. Sodium hydroxide, commonly known as lye, consists of sodium ions (Na^+) and hydroxide ions (OH^-). The hydroxide ion contains an oxygen atom covalently bonded to a hydrogen atom plus an "extra" electron. Whereas the sodium atom becomes a cation by giving up an electron, the hydroxide group becomes an anion by gaining an electron.

$$e^- + \cdot \ddot{\underset{\cdot\cdot}{O}} \cdot + \cdot H \longrightarrow [:\ddot{\underset{\cdot\cdot}{O}}:H]^-$$

The formula for sodium hydroxide is NaOH. For each sodium ion there is one hydroxide ion.

Many groups like the hydroxide ion exist in nature. They remain intact through most chemical reactions. **Polyatomic ions** are charged particles containing two or more covalently bonded atoms. A list of the common polyatomic ions is given in Table 3.4.

EXAMPLE 3.17

Using Table 3.4, determine how many sodium ions will combine with a nitrate ion. With a sulfate ion. With a phosphate ion.

The sodium ion has a single positive charge, Na^+. The nitrate ion has a single negative charge, NO_3^-. To produce a compound in which the positive and negative charges balance, each Na^+ combines with one NO_3^- to form the neutral compound $NaNO_3$, sodium nitrate.

The sulfate ion has a double negative charge, SO_4^{2-}. Therefore, it requires two Na^+ to balance the charge on one SO_4^{2-}. The formula for sodium sulfate is Na_2SO_4.

The phosphate ion is PO_4^{3-}. To balance the triple negative charge, three Na^+ are needed. Sodium phosphate is Na_3PO_4.

EXAMPLE 3.18

What is the formula for ammonium carbonate?

From Table 3.4 we note that the ammonium ion is NH_4^+ and the carbonate ion is CO_3^{2-}. We need two ammonium ions to balance the charge on the carbonate ion. The formula for ammonium carbonate is therefore $(NH_4)_2CO_3$. Notice that we have enclosed the ammonium ion in parentheses. The subscript 2 indicates that we need two of everything enclosed in the parentheses, that is, two polyatomic ammonium ions.

3.10 Bonding Forces and the States of Matter

TABLE 3.4 Some Common Polyatomic Ions

Charge	Name	Formula
1+	Ammonium ion	NH_4^+
1−	Hydrogen carbonate (bicarbonate) ion	HCO_3^-
	Hydrogen sulfate (bisulfate) ion	HSO_4^-
	Acetate ion	$CH_3CO_2^-$ (or $C_2H_3O_2^-$)
	Nitrite ion	NO_2^-
	Nitrate ion	NO_3^-
	Cyanide ion	CN^-
	Hydroxide ion	OH^-
	Dihydrogen phosphate ion	$H_2PO_4^-$
2−	Carbonate ion	CO_3^{2-}
	Sulfite ion	SO_3^{2-}
	Sulfate ion	SO_4^{2-}
	Monohydrogen phosphate ion	HPO_4^{2-}
	Oxalate ion	$C_2O_4^{2-}$
3−	Phosphate ion	PO_4^{3-}

EXAMPLE 3.19

What is the formula for calcium phosphate?

Calcium is a Group IIA element and forms the Ca^{2+} ion. From Table 3.4 we know that phosphate is PO_4^{3-}. Again we can use the least-common-multiple approach to solve this problem. The calcium ion has a charge of +2 and the phosphate ion has a charge of −3. The least common multiple is 6. We need 3 Ca^{2+} (a total of six positive charges) to balance 2 PO_4^{3-} (six negative charges). The formula for calcium phosphate is $Ca_3(PO_4)_2$.

3.10 BONDING FORCES AND THE STATES OF MATTER

The states of matter—solid, liquid, and gas—are obviously different from one another. Chemists offer a model to explain these differences. The model is referred to as the **kinetic molecular theory**. Right now we need only consider how the model pictures solids, liquids, and gases. **Solids** are viewed as highly ordered assemblies of particles in close contact with one another. **Liquids** are pictured as much more

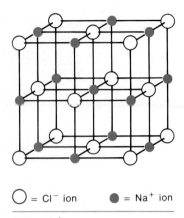

○ = Cl⁻ ion   ● = Na⁺ ion

FIGURE 3.6
Ball-and-stick model of a sodium chloride crystal.

Melting point: the temperature at which

solid → liquid

Vaporization: the process

liquid → gas

loosely organized collections of particles. In a liquid, the particles are still in close contact with one another, but they are much more free to move about. Finally, in **gases** the particles are no longer in close contact with one another but are separated by relatively great distances and are moving about at random.

Table salt (sodium chloride) is a typical crystalline solid. In even a small amount of salt there are billions and billions of sodium ions and chloride ions, always in a one-to-one ratio. In a crystal of salt (Figures 3.6 and 3.7), each sodium ion is surrounded (and attracted) by six chloride ions (the ones to the front and back, the top and the bottom, and both sides). Each chloride ion is similarly surrounded and attracted by six sodium ions. In this solid, ionic bonds hold the ions in position and maintain the orderly arrangement. Not all solids are held together by ionic bonds, but some attractive force is necessary to maintain the characteristic orderly array of particles in a solid.

To get a better image of a liquid at the molecular level, think of a box of marbles that is being shaken continuously. The marbles move back and forth, rolling over one another. The particles of a liquid (like the marbles) are not so rigidly held in place as are particles in a solid. The particles in a liquid are being held close together, however, and that means there must be some force attracting them.

Gas particles do not experience significant attractive forces. They move about at great distances from one another and interact only during occasional collisions.

Solids can be changed to liquids; that is, they can be **melted**. The solid is heated, and the heat energy is absorbed by the particles of the solid. The energy causes the particles to vibrate in place with more and more vigor until finally the forces holding the particles in a particular arrangement are overcome. The solid has become a liquid. The temperature at which this happens is called the **melting point** of that solid. A high melting point is one indication that the forces holding a solid together are very strong.

A liquid can change to a gas or vapor in a process called **vaporization**. Again, one need only supply sufficient heat to achieve this change. Energy is absorbed by the liquid particles, which move faster and faster as a result. Finally the attractive forces holding the liquid particles in

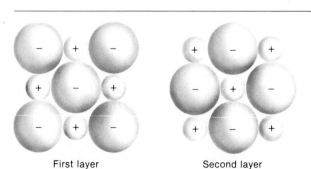

First layer   Second layer

FIGURE 3.7
The arrangement of ions in a sodium chloride crystal.

3.11 Dipole Forces

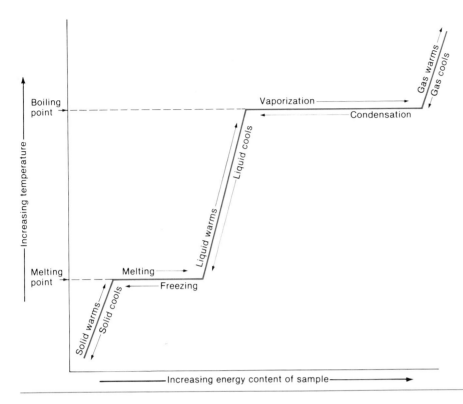

FIGURE 3.8
Diagram of change in state of matter on heating or cooling.

contact are overcome by this increasingly violent motion, and the particles fly away from one another. The liquid has become a gas.

The entire sequence of changes can be reversed by removing energy from the sample and slowing down the particles. Vapor changes to liquid in a process referred to as **condensation**; liquid changes to solid in a process called **freezing**. Figure 3.8 is a diagram of the changes in state that occur as energy is added to or removed from a sample.

The amount of energy required to accomplish these changes depends on the type of forces responsible for maintaining the solid or the liquid state. The ionic bonds in salt crystals are very strong. Sodium chloride must be heated to about 800 °C before it melts. Generally, interionic forces are the strongest of all the forces that hold solids and liquids together. We will now consider some other interactions that hold the particles of solids and liquids together.

Condensation—the reverse of vaporization:

$$gas \rightarrow liquid$$

Freezing—the reverse of melting:

$$liquid \rightarrow solid$$

3.11 Dipole Forces

Hydrogen chloride melts at −112 °C and boils at −85 °C (it is a gas at room temperature). The attractive forces between molecules are not nearly as strong as the interionic forces in salt crystals. We know that covalent bonds hold the hydrogen and chlorine *atoms* together to form

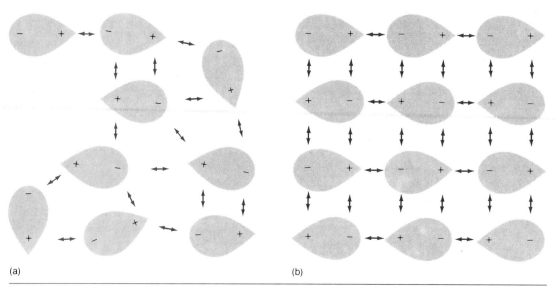

FIGURE 3.9
An idealized representation of dipole forces in (a) a liquid and (b) a solid. In a real liquid or solid interactions are more complex.

Dipole interactions—attractive forces between the positive end of one molecular dipole and the negative end of another

Hydrogen bond: the attraction between a hydrogen bonded to N, O, or F in one molecule and the lone pair of electrons on an atom of another molecule

the hydrogen chloride *molecule*, but what makes one molecule interact with another in the solid or liquid states?

Remember that the hydrogen chloride molecule is a dipole. It has a positive and a negative end. Two dipoles brought close enough will attract one another. The positive end of one molecule attracts the negative end of another. Such forces may exist throughout the structure of a liquid or solid (Figure 3.9). In general, attractive forces between dipoles are much weaker than the attractive forces between ions. **Dipole interactions** are, however, stronger than the forces between nonpolar molecules of comparable size.

3.12 Hydrogen Bonds

Certain polar molecules exhibit stronger attractive forces than expected on the basis of ordinary dipolar interactions. These forces are strong enough to be given a special name, the **hydrogen bond**. Note that "*hydrogen* bond" is a somewhat misleading name, since it emphasizes only one component of the interaction. Not all compounds containing hydrogen exhibit this strong attractive force; the hydrogen *must* be attached to fluorine, oxygen, or nitrogen. It is the presence of these atoms that permits us to offer an explanation for the extra strength of hydrogen bonds as compared with other dipolar forces. Fluorine, oxygen, and nitrogen all have a high electron-attracting power (they are very electronegative), and they are small (they are at the top of the periodic table). A hydrogen–fluorine bond, for example, is very strongly polarized, with a negative fluorine end and a positive hydrogen end. Both hydrogen and fluorine are small atoms, so the

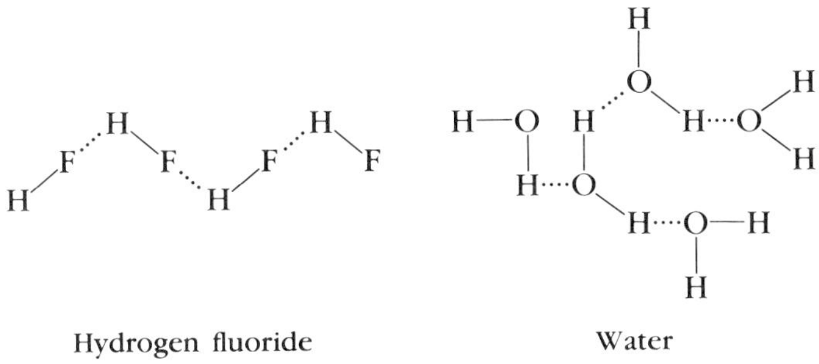

Hydrogen fluoride | Water

FIGURE 3.10
Hydrogen bonding in hydrogen fluoride and in water.

negative end of one dipole can approach very closely the positive end of a second dipole. This results in an unusually strong interaction between the two molecules—the hydrogen bond. Hydrogen bonds are often explicitly represented by *dotted* lines to emphasize their unusual strength compared to ordinary dipolar interactions. A *dotted* line is used to distinguish a hydrogen bond from the much stronger covalent bond, which is represented by a *solid* line (Figure 3.10).

Water has both an unusually high melting point and an unusually high boiling point for a compound with molecules of such small size. These abnormal values are attributed to water's ability to form hydrogen bonds. While the hydrogen bond, at this point, may seem an interesting piece of chemical theory, its importance to life and health is immense. The structure of proteins, chemicals essential to life, is determined in part by hydrogen bonding. The heredity that one generation passes on to the next is dependent on an elegant application of hydrogen bonding.

3.13 Dispersion Forces

If one understands that positive attracts negative, then it is easy enough to understand how ions or polar molecules maintain contact with one another. But how can we explain the fact that nonpolar compounds can exist in the liquid and solid states? Even hydrogen can exist as a liquid or a solid if the temperature is low enough (its melting point is $-259\ °C$). Some force must be holding these molecules in contact with one another in the liquid and solid states.

Up to this point we have pictured the electrons in a covalent bond as being held in place between the two atoms sharing the bond. But the electrons are *not* really static; they actually move about in the bonds. On the average, the two electrons in the hydrogen molecule (or any nonpolar bond) are between and equidistant from the two nuclei. At any given instant, however, the electrons may be at one end of the molecules. At some other time, a moment later, the electrons may be at the other end of the molecule.

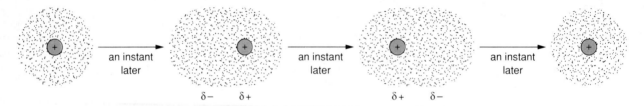

At the instant the electrons of one molecule are at one end, the electrons in the next molecule will move away from its adjacent end. Thus, at this instant there will be an attractive force between the electron-rich end of one molecule and the electron-poor end of the next.

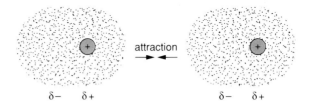

Dispersion forces—attractions between molecules with temporary dipoles

These momentary, usually weak, attractive forces between molecules are called **dispersion forces**. To a large extent, dispersion forces determine the physical properties of nonpolar compounds.

3.14 Forces In Solutions

To complete our look at chemical bonding, we shall briefly examine the interactions that occur in solutions. A **solution** is an intimate, homogeneous mixture of two or more substances. By "**intimate**," we mean that the mixing occurs down to the level of individual ions and molecules. In a salt-water solution, for example, there are not clumps of ions floating around, but single, randomly distributed ions among the water molecules. "**Homogeneous**" means that the mixing is thorough.

Homogeneous solutions have the same composition throughout the sample.

All parts of the solution have the same distribution of components. For the salt solution, the saltiness is the same at the top, bottom, and middle of the solution. The substance being dissolved and usually present in lesser amount (salt in a salt solution) is called the **solute**. The substance doing the dissolving and usually present in greater amount is the **solvent** (water in the salt-water solution) (Figure 3.11).

Heterogeneous solutions have variable composition throughout the sample.

In a *heterogeneous* solution, the solute is not completely dissolved into the solution. In other words, the mixture has two phases.

Ordinarily, solutions form most readily when the substances involved have *similar* bonding characteristics. An old chemical rule is "Like dissolves like." Nonpolar solutes dissolve best in nonpolar solvents. For example, oil and gasoline, both nonpolar, mix (Figure 3.12a),

3.14 Forces in Solutions

but oil and water do not (Figure 3.12b). Alcohol is a liquid held together by hydrogen bonds. So is water. Alcohol readily dissolves in water because the two substances can hydrogen bond with one another (Figure 3.12c). In general, a solute dissolves when attractive forces between it and the solvent overcome the attractive forces operating in the pure solute and in the pure solvent.

Why, then, does salt dissolve in water? Ionic solids are held together by strong ionic bonds. We have already indicated that very high temperatures are required to melt ionic solids and break these bonds. Yet, simply by placing sodium chloride in water at room temperature we can dissolve the salt (or, rather, the water can). And when such a solid dissolves, its bonds *are* broken. The difference between the two processes is the difference between brute force and persuasion. In the melting process, we are simply pouring in enough energy (as heat) to pull the crystal apart. In the dissolving process, we offer the ions an attractive alternative to the ionic interactions in the crystal.

It works this way. Water molecules surround the crystal. Those that approach a negative ion align themselves so that the positive ends of their dipoles point toward the ion. With a positive ion the process is reversed, and the negative end of the water dipole points toward the ion. Still, the attraction between a dipole and an ion is not as strong as that between two ions. To compensate for their weaker attractive power, several molecules surround each ion, and in this way the many **ion–dipole** interactions overcome the **ion–ion** interactions (Figure 3.13).

In an ionic solid, the positive and negative ions are strongly bonded together in an orderly crystalline arrangement. In solution, cations and anions move about more or less independently, each surrounded by a cage of solvent molecules. Water—including the water in our bodies—is an excellent solvent for many ionic compounds. Water also dissolves molecules that are polar covalent like itself.

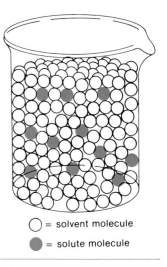

○ = solvent molecule
● = solute molecule

FIGURE 3.11
In a solution the solute molecules are randomly distributed among the solvent molecules.

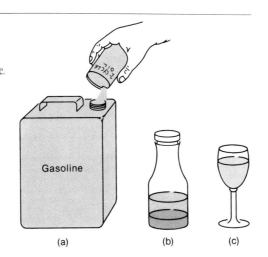

FIGURE 3.12
(a) Lawn mowers with two-cycle engines are fueled and lubricated with a solution of nonpolar lubricating oil in nonpolar gasoline. (b) In Italian dressing polar vinegar and nonpolar olive oil are mixed. However, the two liquids do not form a solution and will separate on standing. (c) Wine is a solution of polar alcohol in polar water.

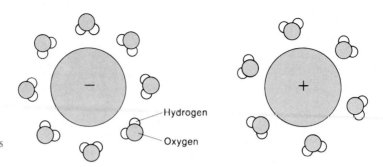

FIGURE 3.13
The interaction of polar water molecules with ions.

3.15 Some Common Ions of Physiological Importance

The most abundant ions in our body fluids are sodium (Na^+), potassium (K^+), and chloride (Cl^-) ions. Cell membranes effectively separate these ions, with most of the potassium ions being found in the intracellular (within the cell) fluid and most of the sodium and chloride ions in the extracellular (outside of the cell) fluid. These dissolved ions play a major role in maintaining balance in the distribution of water in our bodies. They are also essential for proper neuromuscular function and for control of acidity in body fluids. The bicarbonate ion (HCO_3^-) also plays a role in controlling the acidity of body fluids. This function will be discussed in detail in Chapter 8.

Smaller but significant amounts of calcium (Ca^{2+}) and various forms of phosphate ions ($H_2PO_4^-$, HPO_4^{2-}, PO_4^{3-}) are found in body fluids. Calcium ion is important in blood coagulation and in proper neuromuscular function. The various ionic forms of phosphorus are involved in a wide variety of metabolic reactions (body chemical functions), some of them central to the storage and release of energy in the body. Far greater amounts of calcium (more than 98% of the total) and phosphorus (about 85% of the total) are not in solution but in bone and teeth. The mineral part of bone is primarily hydroxyapatite, a complex material incorporating calcium phosphate [$Ca_3(PO_4)_2$] and calcium hydroxide [$Ca(OH)_2$].

The amount of magnesium in the body is much lower than that of calcium, but a major portion (60%) of this ion is also tied up in bone tissue. The remainder is involved in a variety of processes, perhaps the most important being enzyme regulation. Enzymes are special molecules that control metabolic reactions.

We have mentioned only the most abundant ions found in body fluids. Some ions, such as iron(II) ion, are found primarily bound to large molecules (as in hemoglobin). Others, although present in only trace quantities, are nonetheless essential for health. We shall consider a number of these ions later in this text.

CHAPTER SUMMARY

Chemical bonds are the forces that maintain the structure of chemical compounds. Two general classes of bond are ionic bonds and covalent bonds. For both types, the formation of compounds can be explained as reflecting tendency of atoms to achieve noble gas electron configuration. The common feature shared by the noble gases (except helium) is an *octet* of electrons in the outermost energy level (an outer electron configuration of ns^2np^6). This arrangement is associated with particular chemical stability. Helium's filled outermost energy level ($1s^2$) is referred to as a *duet* and shows a similar chemical stability.

An atom may give up or gain electrons to achieve the stable octet (or duet) arrangement. The charged particles formed in this process are called ions. A positively charged cation is formed by elements in Group IA, IIA, and IIIA, and the size of the charge is equal to the group number. Atoms of elements in Group VA, VIA, and VIIA achieve a stable octet (or duet) by forming negatively charged anions. The size of the charge can be calculated by subtracting 8 from the group number.

Compounds formed by the interaction of positive and negative ions are called ionic compounds, and the attraction between the constituent positive and negative ions is called an ionic bond. In forming compounds, positive and negative ions must be combined in ratios that result in overall electrical neutrality. Compounds can be represented by simple formulas that indicate the correct combining ratio (e.g., NaCl) or by electron-dot formulas that convey the electron count in the valence (outermost) energy level and the charge (e.g., $Na^+ : \ddot{C}l : ^-$).

Simple cations are named by adding the term *ion* to the name of the element. Simple anions are named by changing the ending of the element name to *-ide* and adding the term *ion*. Ionic compounds are named by giving first the name of the cation and then the name of the anion. The term *ion* is not included in the compound name.

Compounds may be formed through covalent bonding, in which electrons are shared between two atoms in order to achieve noble gas configurations. Molecules are discrete groups of atoms held together by such shared pairs of electrons. A covalent bond may involve the sharing of one, two, or three pairs of electrons; such bonds are called single, double, and triple bonds, respectively.

Electronegativity can be defined as the tendency of an atom in a molecule to attract electrons. Electrons shared by two atoms of unequal electronegativity are not shared equally. The bond between these two atoms is termed polar covalent; the electron cloud representing the shared electrons is more dense near the electronegative atom. If this unequal distribution of electrons is not canceled by other polar bonds in the molecule, then the molecule is a dipole (that is, displays a positive end and a negative end). Water is the most familiar polar molecule.

Some ions (called polyatomic ions) contain covalent bonds as well as full electric charges.

The conversion of solid to liquid is called melting and occurs for a given substance at a specific temperature, called the melting point. The reverse process is called freezing. The conversion of liquid to gas is called vaporization, and the reverse process is called condensation. The energy changes accompanying these conversions are dependent on the strength of the forces maintaining the particular state.

Forces that hold the particles of the solid and liquid state in contact can be divided into three major categories. Ionic bonds are responsible for maintaining the structure of ionic compounds in the solid and liquid states. Dipole forces (the attraction of the positive end of a dipole for the negative end of another dipole) are responsible for maintaining polar covalent compounds in the solid and liquid states. Hydrogen bonding is an especially strong form of dipole interaction that involves a hydrogen atom bonded to a small electronegative element (N, O, or F). Dispersion forces are transient (short-lived) attractive forces resulting from the movement of electrons within molecules. For nonpolar molecules, dispersion forces are the only attractive forces that function to keep molecules in contact in the liquid and solid phases.

A solution is an intimate, homogeneous mixture of two or more substances. The substance that is taken into solution is called the solute. The substance in which the solute dissolves is the solvent. Solutions are best prepared using solutes and solvent of similar bonding characteristics. Polar solutes dissolve well in polar solvents, and nonpolar solutes dissolve best in nonpolar solvents. Ionic compounds dissolve most readily in polar solvents.

Many simple ions are found in body fluids. Sodium, potassium, and chloride ions are the most abundant dissolved ions, but calcium, magnesium, bicarbonate, and phosphate ions, among others, are also present in significant amounts.

CHAPTER QUIZ

1. The electron dot formula for fluoride is
   a. $:\ddot{F}:^+$;   b. $:\ddot{F}:^-$;   c. $:\ddot{F}:$;   d. $:\ddot{F}:_{2-}$
2. CuO is
   a. copper (I) oxide;   b. cupric oxide;
   c. cuprous oxide;   d. copper oxide.
3. The formula for aluminum sulfide is
   a. Al_2S_3;   b. AlS_3;   c. Al_3S_2;   d. Al_2S.
4. The bonding in O_2 is
   a. ionic;   b. polar;
   c. colavent;   d. triple.
5. Referring only to the periodic table inside the front cover, predict which is most electronegative.
   a. Sr;   b. Al;   c. O;   d. Br.
6. Predict the bond angles in $AlCl_3$.
   a. 90°;   b. 120°;   c. 109.5°;   d. 180°.
7. Predict the bond angles in CH_4.
   a. 90°;   b. 120°;   c. 109.5°;   d. 180°.
8. Sodium nitrate is
   a. Na_2NO_2;   b. $NaNO_2$;
   c. $NaNO_3$;   d. Na_2NO_3.
9. The strongest bonding force is
   a. dipole;   b. vaporization;
   c. dispersion;   d. hydrogen bond.
10. Which would dissolve best in oil?
    a. salt;   b. water;
    c. vinegar;   d. gasoline.

PROBLEMS

3.2 The Ionic Bond

1. What is the structural difference between a sodium atom and a sodium ion?
2. What is the structural difference between a sodium ion and a neon atom? In what way are these two particles similar?
3. What are the structural differences among a chlorine atom, a chlorine molecule, and a chloride ion?
4. What is wrong with the expression "a molecule of sodium chloride"?
5. Indicate the charges on simple ions formed from the following elements.
   a. Group IIIA   b. Group VIA
   c. Group IA   d. Group VIIA

3.3 Electron-Dot Formulas

6. Using electron-dot formulas, draw the formation of an ion from an atom for each of the following.
   a. barium   b. bromine
   c. aluminum   d. sulfur
7. Identify the ions of Problem 6 as cations or anions.

3.5 Names of Simple Ions and Ionic Compounds

8. Draw the formulas (e.g., Cl^-) for the following ions.
   a. lithium ion   b. iodide ion
   c. calcium ion
9. Draw the formulas for the following ions.
   a. ammonium ion
   b. hydrogen carbonate ion
   c. phosphate ion
10. Name the following ions.
    a. K^+ b. O^{2-} c. Al^{3+}
11. Name the following ions.
    a. CO_3^{2-} b. HPO_4^{2-}
    c. NO_3^- d. OH^-
12. Write correct formulas for the following compounds.
    a. lithium fluoride
    b. calcium iodide
    c. aluminum bromide
    d. aluminum sulfide
13. Write correct formulas for the following compounds.
    a. magnesium sulfate
    b. potassium nitrate
    c. sodium cyanide
    d. calcium oxalate

14. Write correct formulas for the following compounds.
    a. ferrous sulfate
    b. ammonium phosphate
    c. magnesium phosphate
    d. calcium monohydrogen phosphate
15. Give correct names for the compounds represented by the following formulas.
    a. NaBr   b. $CaCl_2$ c. Al_2O_3
16. Give correct names for the compounds represented by the following formulas.
    a. KNO_2 b. LiCN
    c. NH_4I d. $NaNO_3$
    e. $CaSO_4$ f. $NaHSO_4$
    g. $KHCO_3$ h. $Al(OH)_3$
    i. Na_2CO_3
17. Give correct names for the compounds represented by the following formulas.
    a. $Mg(CH_3CO_2)_2$ b. $Al(C_2H_3O_2)_3$
    c. $(NH_4)_3PO_4$ d. $(NH_4)_2C_2O_4$
    e. Na_2HPO_4 f. $Ca(H_2PO_4)_2$
    g. $Mg(HCO_3)_2$ h. $Ca(HSO_4)_2$
    i. NH_4NO_2
18. Fill in each cell with the formula for each compound.

	Cl^-	OH^-	S^{2-}	SO_3^{2-}	N^{3-}	HPO_4^{2-}	PO_4^{3-}
Na^+							
NH_4^+							
Ba^{2+}							
Cu(II)							
Fe(III)							
Al^{3+}							

19. Given that titanium chloride is $TiCl_3$ and sodium chromate is Na_2CrO_4, write the formula for titanium chromate.

3.6 Covalent Bonds

20. Fill in this table assuming that elements X, Y, and Z are all in A subgroups in the periodic table.

	Element X	Element Y	Element Z
Group number	IA	——	——
Electron-dot formula	——	$\cdot \dot{Y} \cdot$	——
Charge on ion	——	——	2−

21. Consider the hypothetical elements X, Y, and Z with electron-dot formulas

$$:\ddot{X}\cdot \quad :\dot{Y}\cdot \quad :\dot{Z}:$$

   a. To which group in the periodic table would each belong?
   b. Write the electron-dot formula for the simplest compound of each with hydrogen.
   c. Write electron-dot formulas for the ions formed when X and Y react with sodium.
22. Draw electron-dot formulas for the following covalent molecules.
    a. CH_4O b. NOH_3
    c. CH_5N d. N_2H_4
23. Draw electron-dot formulas for the following covalent molecules.
    a. NF_3 b. C_2H_4
    c. C_2H_2 d. CH_2O
24. Two different covalent molecules are found with the formula C_2H_6O. Draw electron-dot formulas for the two molecules.
25. Shared pairs of electrons can be represented by dashed lines. The electron-dot formula for the hydrogen molecule, H:H, can be translated to H—H. Draw dashed-line formulas for the molecules in Problems 22–24.
26. Name the following covalent compounds.
    a. CS_2 b. CBr_4
    c. N_2S_4 d. PBr_3

3.7 Polar Covalent Bonds

27. If atoms of the two elements in each set below were joined by a covalent bond, which atom would more strongly attract the electrons in the bond? You may refer to Figure 3.4.
    a. N and S
    b. B and Cl
    c. As and F

28. *Without* referring to Figure 3.4 but using the periodic table, indicate which element in each set is more electronegative.
   a. Br or F
   b. Br or Se
   c. Cl or As

29. Solutions of iodine chloride (ICl) are used as disinfectants. Are the molecules of ICl ionic, polar covalent, or nonpolar covalent?

30. For each compound, indicate whether the molecules will act as dipoles. The correct shape of the molecules is presented in the drawings.
   a. F—C≡C—F
   b. Cl—B(—Cl)—Cl (trigonal)
   c. H₂C=N—O—H

3.8 Shapes of Molecules

31. Give an example (not mentioned in the text) for each shape.
   a. linear
   b. trigonal
   c. tetrahedral
   d. bent
   e. octahedral

32. State the approximate bond angles in
   a. PH_3
   b. $AlBr_3$
   c. PCl_5
   d. CCl_4

33. Which molecules are linear? Why
   a. H_2S
   b. CO_2
   c. $BeBr_2$
   d. $SnCl_2$

3.10 Bonding Forces and the States of Matter

34. In what ways are liquids and solids similar? In what ways are they different?

35. List four types of interactions between particles in the liquid state and in the solid state. Give an example of each type.

36. Define:
   a. melting
   b. vaporization
   c. condensation
   d. freezing

37. In which process is energy absorbed by the material undergoing the change of state?
   a. melting or freezing
   b. condensation or vaporization

38. Label each arrow with the term listed in Problem 36 that correctly identifies the process described.

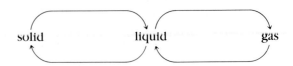

3.12 Hydrogen Bonds

39. In which of the following would hydrogen bonding be an important intermolecular force?
   a. H—S—H
   b. H—C(H)—N(H)—H
   c. H—C(H)(H)—F
   d. H—C(H)(H)—O—H
   e. H—C(H)(H)—C(H)(H)—H

3.14 Forces in Solutions

40. Define:
   a. solution
   b. solute
   c. solvent

41. Alcohol that is used as a disinfectant to clean the skin prior to an injection is actually a solution of 3 parts of water and 7 parts of alcohol. Which component is the solvent, and which is the solute?

42. Explain why a salt dissolves in water.

43. Benzene (C_6H_6) is a nonpolar solvent. Would you expect NaCl to dissolve in benzene? Explain.

44. Motor oil is nonpolar. Would you expect it to dissolve in water? In benzene? Explain.

3.15 Some Common Ions of Physiological Importance

45. Name some ions of physiological significance, and identify their physiological roles.

Nuclear Processes

Did You Know That...

- X-rays have been known longer than radioactivity?
- a neutron can change into an electron and a proton?
- helium is a product of many nuclear reactions?
- women were prominent leaders in the discovery of radioactivity?
- a proton can change into a positron plus a neutron?
- curies, roentgens, rads, and rems are all units of radioactivity?
- this chapter has pictures of a stomach and the face of a fetus?
- radioactive iodine is used in thyroid and prostate cancer treatments?
- your educated judgment will, in part, determine the fate of mankind?

In our consideration of chemical bonding we emphasized the importance of the outermost electrons in atoms. Through most of this text, we shall focus on chemical reactions, and in chemical reactions only these outer regions of atoms are changed. For the moment, however, let us turn our attention to that tiny speck of matter called the nucleus. The volume of a whole atom is about 10 000 times that of its nucleus, yet it is the nucleus that holds the power that has become the symbol of our age.

Nuclear power confronts us with a great paradox. Although nuclear power unleashed in wrath can destroy cities and perhaps civilizations, controlled nuclear power can provide the energy necessary to run our cities and maintain our civilization. Yet even the peaceful uses of nuclear power are not without potential danger. As citizens of the nuclear age, we have difficult decisions to make. Nuclear bombs may kill, but nuclear medicine saves lives. Our knowledge of nuclear science gives us both power and responsibility. How we exercise that responsibility will determine how future generations remember us—even whether there will be future generations to remember us.

4.1 Nuclear Arithmetic

In Chapter 2, we discussed the existence of isotopes. Isotopes are of little importance in ordinary *chemical* reactions. Such reactions involve the outer electrons of an atom, and differences in the number of neutrons buried deep in the heart of the atom would not be expected to have a major effect. In **nuclear reactions**, reactions that do involve the heart of the atom, however, isotopes are of utmost importance, as we shall see.

Before we talk about nuclear reactions, it will be necessary to discuss a special symbolism used in writing such reactions. To represent different isotopes, symbols with subscripted and superscripted numbers are used. In the generalized symbol

$$^{A}_{Z}X$$

Atomic number: Z = number of protons

Mass number: A = number of protons + number of neutrons

the subscript Z represents the **atomic number**, that is, the number of protons. The superscript A represents the **mass number**, that is, the number of protons plus the number of neutrons. From the symbol

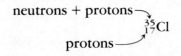

4.1 Nuclear Arithmetic

we know that the number of protons in a chlorine atom is 17. The number of neutrons is 35 − 17 = 18.

The Z value is not really necessary, because the elemental symbol establishes the number of protons in the nucleus being considered. It is covenient to have this information at one's fingertips while writing nuclear equations, however.

EXAMPLE 4.1

Write the symbol for an isotope with a mass number of 58 and an atomic number of 27.

In the general symbol $^A_Z X$, A is the mass number, which is 58 in this case.

$$^{58}_Z X$$

Z is the atomic number, or 27.

$$^{58}_{27} X$$

From the periodic table we can determine that the element with atomic number 27 is cobalt.

$$^{58}_{27} Co$$

EXAMPLE 4.2

Indicate the number of protons, neutrons, and electrons in a neutral atom of the isotope $^{235}_{92}U$.

The atomic number gives the number of protons and electrons in a neutral atom of the isotope.

Atomic number = protons = electrons = 92

The number of neutrons is obtained by subtracting the atomic number from the mass number.

Mass number	235
Atomic number	− 92
Number of neutrons	143

> **EXAMPLE 4.3**
>
> Which of the following are isotopes of the same element? We are using the letter X as the symbol for all so that the symbol will not identify the elements
>
> $$^{16}_{8}X \quad ^{16}_{7}X \quad ^{14}_{7}X \quad ^{14}_{6}X \quad ^{12}_{6}X$$
>
> $^{16}_{7}X$ and $^{14}_{7}X$ are isotopes of the element nitrogen (N). $^{14}_{6}X$ and $^{12}_{6}X$ are isotopes of the element carbon (C).
>
> Which of the original five isotopes have identical mass numbers?
>
> $^{16}_{8}X$ and $^{16}_{7}X$ have the same mass number. The first is an isotope of oxygen, and the second is an isotope of nitrogen. $^{14}_{7}X$ and $^{14}_{6}X$ have the same mass number. The first is an isotope of nitrogen, and the second is an isotope of carbon.
>
> Which of the original five isotopes have the same number of neutrons?
>
> $^{16}_{8}X$ (16 − 8 = 8 neutrons) and $^{14}_{6}X$ (14 − 6 = 8 neutrons) have the same number of neutrons.

An atom of carbon-12 is defined to have a mass of exactly 12 amu. The masses of the other atoms are compared to that of carbon-12.

A specific isotope can be designated by its nuclear symbol (e.g., $^{12}_{6}C$) or by the element name followed by the mass number (e.g., carbon-12).

The atoms of the carbon-12 isotope are defined as having a mass of 12 amu. The relative masses of all other atoms are determined by comparing them to this standard. An atom of the isotope $^{1}_{1}H$ has one-twelfth the mass of an atom of $^{12}_{6}C$; thus, atoms of $^{1}_{1}H$ have an atomic mass of 1 amu. Atoms of deuterium ($^{2}_{1}H$) have one-sixth the mass of atoms of $^{12}_{6}C$; the atomic mass of deuterium atoms is 2 amu.

The weighted average of the masses of the isotopes is the atomic weight of the element.

The atomic weight of an element, as listed in the periodic table, is related to the atomic mass in the following way. The **atomic weight** is the average of the masses of the atoms in a representative sample of an element. Each atom in the sample has a particular atomic mass. For example, in a sample of the element bromine, about half the atoms have a mass number of 79 amu and half have a mass number of 81 amu. The atomic weight of bromine is about 80 amu, the average of 79 amu and 81 amu. (The atomic weight of bromine is listed in the periodic table as 79.9 amu, indicating that there is slightly more of the bromine-79 isotope than of the bromine-81.) Another example: about three-fourths of the atoms of the element chlorine have a mass number of 35 amu, and about one-fourth have a mass number of 37 amu. The atomic weight of chlorine is 35.5 amu; the average mass in this case is much closer to that of the chlorine-35 isotope because there is more of the isotope in the sample (Figure 4.1).

Mass numbers (in amu) of atoms in sample

```
 81
 79
 79
 81
 79
 79
 81
 81
---
640 amu = total mass of sample
```

Average mass = $\frac{640 \text{ amu}}{8 \text{ atoms}}$ = 80 amu = atomic weight

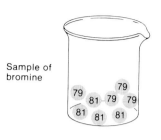

Sample of bromine

Mass numbers (in amu) of atoms in sample

```
 35
 37
 35
 35
 35
 37
 35
 35
---
284 amu = total mass of sample
```

Average mass = $\frac{284 \text{ amu}}{8 \text{ atoms}}$ = 35.5 amu = atomic weight

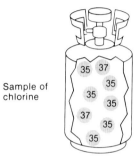

Sample of chlorine

FIGURE 4.1
Atomic weights of elements are averages of isotopic masses.

4.2 DISCOVERY OF RADIOACTIVITY

The discovery of the first nuclear reactions was triggered by the discovery of X-rays, which are not nuclear phenomena. In 1895, a German scientist, Wilhelm Konrad Roentgen, found that he could produce a form of radiation that passed right through solid materials. This mysterious radiation was like visible light in that it was a form of pure energy. Also, like visible light, it resulted from electrons moving from one energy level to another. It was unlike visible light in that you could not see it. For want of a better name, Roentgen called the invisible, penetrating radiation **X-rays**. X-rays are high-energy radiation. They contain more energy than even the most energetic visible light. It is that high energy that gives X-rays such penetrating power.

X-rays are high-energy radiation.

The medical community immediately recognized the significance of the penetrating X-rays. The X-rays picture shown in Figure 4.2 was taken in February 1896, within two months of the publication of Roentgen's discovery. The round black dots are gunshot pellets. This picture was used to establish their position for removal by surgery.

This phenomenon fascinated other scientists, also. One, a Frenchman named Antoine Henri Becquerel, had been studying a different phenomenon called fluorescence. Fluorescent compounds, after exposure to strong sunlight, continue to glow even when taken into a dark room. Becquerel wondred if there were any substances that would give

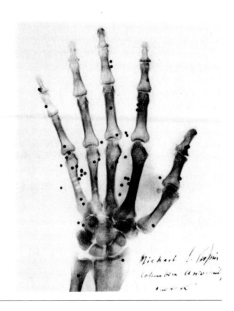

FIGURE 4.2
An early example of the use of X-rays in medicine. Prof. Michael Pupin of Columbia University made this X-ray in 1896 to aid in the removal by surgery of gunshot pellets (dark spots) from the hand of a patient. [Courtesy of Burndy Library, Norwalk, CT.]

off X-rays when they fluoresced. In trying to answer this question, Becquerel tested a large number of materials—among them, uranium compounds. Becquerel found that uranium did emit invisible, penetrating rays. These rays, however, proved to have nothing to do with fluorescence or with X-rays. Becquerel had discovered a totally new phenomenon.

At this point, Marie and Pierre Curie and Ernest Rutherford entered the picture. We first mentioned these scientists in our discussion of models for the atom. Marie Curie (Figure 4.3) named the phenomenon discovered by Becquerel **radioactivity**. The Curies went on to discover a number of new radioactive elements, including radium.

FIGURE 4.3
Marie Curie in her laboratory. [Culver Pictures.]

4.3 Types of Radioactivity

Scientists soon realized that the radiation emanating from uranium and other radioactive elements was of three types. When this radiation was passed through a strong magnetic field, one portion was deflected in one direction, another portion was deflected in the opposite direction, and a third was not deflected at all. Rutherford named these portions **alpha (α) rays**, **beta (β) rays**, and **gamma (γ) rays**, respectively (Figure 4.4).

Alpha (α) particle
$^{4}_{2}$He

Beta (β) particle
$^{0}_{-1}$e

Only the gamma rays are radiation in the same sense that visible light and X-rays are, that is, pure radiant energy. Gamma rays have the highest energy and are the most penetrating form of lightlike radiation yet discovered.

Both alpha and beta rays were found to be streams of tiny particles. An **alpha particle** has a mass four times that of a hydrogen atom and a charge twice that of a proton. These properties make alpha particles identical to helium nuclei. ($^{4}_{2}$He). Beta rays were shown to be streams of electrons. Thus, a **beta particle** ($^{0}_{-1}$e) has a mass only 1/1837 that of a proton and has a negative charge. As a form of radiant energy, gamma rays have no mass and no charge.

We can describe in detail the forms of radioactivity, but we must still answer the question What is It? What is happening to produce radioactivity?

The answer is that some nuclei are unstable as they occur in nature. Radium atoms with a mass number of 226, for example, break down spontaneously, giving off alpha particles. Since alpha particles are identical to helium nuclei, this process can be summarized by the equation

$$^{226}_{88}\text{Ra} \longrightarrow ^{4}_{2}\text{He} + ^{222}_{86}\text{Rn} \quad \text{or} \quad ^{226}_{88}\text{Ra} \longrightarrow \alpha + ^{222}_{86}\text{Rn}$$

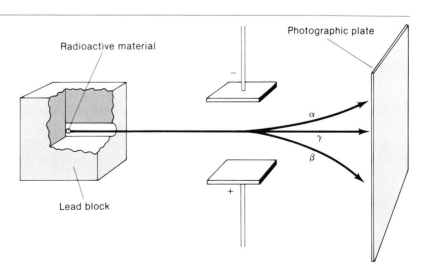

FIGURE 4.4
Behavior of radioactive rays in an electric field.

The new element, with two fewer protons, is identified by its atomic number (86) as radon (Rn). Note that the mass number of the starting material must equal the total of the mass numbers of the products. The same is true for atomic numbers. The symbol $^4_2$He for the alpha particle is preferred to the symbol α, because it allows us to check the balance of mass and atomic numbers more readily.

Tritium nuclei are also unstable. Tritium is one of the heavy isotopes of hydrogen first mentioned in Section 2.2. Like all hydrogen nuclei, the tritium nucleus contains one proton. Unlike the most common isotope of hydrogen, however, the tritium nucleus contains two neutrons, and its mass is therefore 3 amu ($^3_1$H). Tritium decomposes by what is termed **beta decay**. Since a beta particle is identical to an electron, this process can be written

$$^3_1H \longrightarrow\, ^{\,0}_{-1}e + ^3_2He \quad \text{or} \quad ^3_1H \longrightarrow \beta + ^3_2He$$

The product isotope is identified by its atomic number as helium (He). How can the original nucleus, which contains only a proton and two neutrons, emit an electron?

We can envision one of the neutrons in the original nucleus changing into a proton and an electron.

$$^1_0n \longrightarrow\, ^1_1p + ^{\,0}_{-1}e$$

The new proton is retained by the nucleus (therefore, the atomic number of the product increases by one), and the almost massless electron or beta particle is kicked out (the product nucleus has the same mass as the original). A second example of beta decay is pictured in Figure 4.5a. Positron, $^{\,0}_{+1}e$, emission changes a proton into a neutron (Figure 4.5b).

Gamma radiation is pure radiant energy (massless).

The third type of radioactivity is **gamma decay**. No particle is emitted; remember, gamma rays are pure radiant energy. Gamma emission involves no change in atomic number or mass. The process is analogous to the emission of light from an atom (see Figure 2.13). Visible light is emitted when an electron changes from a higher to a lower energy level. In gamma emission, a nucleus in a higher energy state drops to a lower energy state. Thus,

$$^{99m}_{43}Tc \longrightarrow\, ^{99}_{43}Tc + \gamma$$

This type of decay is particularly useful when **radioisotopes** (radioactive isotopes) are used for diagnostic purposes in medicine. We discuss such uses in Section 4.10.

4.3 Types of Radioactivity

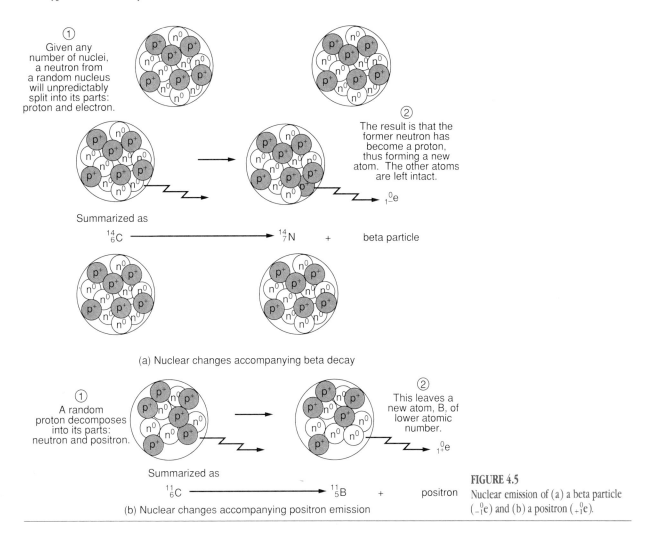

FIGURE 4.5
Nuclear emission of (a) a beta particle ($_{-1}^{0}e$) and (b) a positron ($_{+1}^{0}e$).

The emission of an alpha or beta particle is often accompanied by the emission of a gamma ray.

The major types of radioactive decay and the resulting nuclear changes are summarized in Table 4.1.

TABLE 4.1 Radioactive Decay and Nuclear Change

Type of Radiation	Symbol	Mass Number of Particle	Charge on Particle	Change in Mass Number of Emitting Nucleus	Change in Atomic Number of Emitting Nucleus
Alpha	α	4	2+	Decreases by 4	Decreases by 2
Beta	β	0	1-	No change	Increases by 1
Gamma	γ	0	0	No change	No change

4.4 Penetrating Power

Radioactive materials are dangerous because the radiation emitted by decaying nuclei can damage living tissue (see Section 4.9). The ability of the radiation to inflict damage depends, in part, on its penetrating power.

All other things being equal, the more massive the particle, the less its penetrating power. Of alpha, beta, and gamma rays, alpha rays are the least penetrating. These are streams of helium nuclei, each particle with a mass number of 4. Beta rays are more penetrating than alpha rays. The electrons that make up the stream of beta particles are assigned a mass number of 0. These particles are not really massless, but they are very much lighter than alpha particles. Gamma rays are high-energy radiation and, like light, truly have no mass. This is the most penetrating form of nuclear radiation.

That the biggest particles make the least headway may seem contrary to common sense. Consider that penetrating power reflects the ability of the radiation to make its way through a sample of matter. It is as if you were trying to roll some rocks through a field of boulders. The alpha particle acts as if it were a boulder itself. Because of its size, it cannot get very far before it bumps into and is stopped by other boulders. The beta particle acts like a small stone. It can sneak between boulders and perhaps ricochet off one or another until it has made its way farther into the field (Figure 4.6). The gamma ray can be compared to an insect that can get through the smallest openings: although the

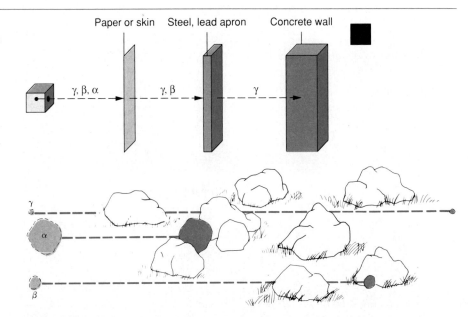

FIGURE 4.6
Shooting radioactive particles through matter is like rolling rocks through a field of boulders—the larger ones are more easily stopped.

4.4 Penetrating Power

insect may brush against some of the boulders, it can, in general, make its way through most of the field without being stopped.

We said that penetrating power is determined by the mass of the particle, all other things being equal. But all other things are not always equal. The faster a particle moves or the more energetic the radiation is, the greater its penetrating power.

If a radioactive source is located outside of the body, an alpha emitter of low penetrating power is least dangerous, and a gamma emitter, most. The alpha particles are stopped by the outer layer of skin, whereas the gamma rays can penetrate to damage vital internal organs. On the other hand, if a radioactive source is within the body, then the nonpenetrating alpha particles can do great damage. All such particles are trapped within the body, which must then absorb all the energy released by the particle. Alpha rays inflict all their damage in a very small area because they do not travel far. Beta rays distribute the damage over a somewhat larger area because they travel farther. Tissue may recover from limited damage spread over a large area; it is less likely to survive very concentrated damage.

Failure to distinguish between damage done by a radiation source external to the body and one that is internal frequently leads to quite different assessments of potential danger. In 1979, radioactive material was released during an emergency shutdown of a nuclear generator at Three Mile Island in Pennsylvania (Figure 4.7). Two different groups disputed the danger associated with this release. One group argued that the people in neighboring communities were exposed to little radiation, not much more than the normal **background radiation** (that

Alpha emitter—least penetrating

Gamma emitter—most penetrating

FIGURE 4.7
Small amounts of radioactive material were released during an emergency shutdown of the nuclear generator at Three Mile Island in Pennsylvania in 1979. [Allied Pix Serivce, Inc.]

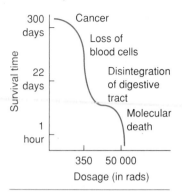

FIGURE 4.8
Causes of radiation death.

which comes from natural sources like radioisotopes in the Earth's crust). On the opposite side of the argument, another group maintained that the released radioactive isotopes would ultimately enter the body through food or inhaled air. The radioisotopes would then concentrate within the body, these people argued, and remain to do long-term damage to various organs. Only the future can tell us which group is correct in its assessment. Figure 4.8 shows some causes of radiation death.

4.5 Radiation Measurement

Several units of measurement are associated with the phenomenon of radioactivity. The *rate* at which nuclear disintegrations occur in a particular sample is measured in **curies** (Ci), named in honor of Marie Curie. A sample whose activity is hundreds or thousands of curies might be used as a source of externally applied radiation for the treatment of cancer. A sample of 10 mCi can be taken internally for diagnostic purposes by an adult, while a sample administered internally to a child might be measured in microcuries.

The *effect of radiation on matter* can be measured in a number of ways. The **roentgen** (R) is a measure of the ability of a source of X-rays or gamma rays to ionize an air sample. The **rad** (*r*adiation *a*bsorbed *d*ose) measures the amount of energy absorbed by any form of matter from any ionizing radiation. Alpha, beta, and gamma rays, as well as X-rays, are forms of **ionizing radiation**, that is, they cause the formation of ions from neutral particles. In the body, ionizing radiation most often interacts with water molecules. The reactive particles formed from water attack other molecules essential to proper cell function. Ionizing radiation damages living tissue by this route.

Techniques for measuring radiation range from the simple to the sophisticated. Individuals who work with radioactive materials wear **film badges** on their pockets or at their waist or as rings (Figure 4.9). The film in these badges reacts to radiation from radioactive isotopes or

FIGURE 4.9
Film badges are worn by people working around radioactive materials. [Courtesy of Tech/Ops Landauer, Inc., Glenwood, IL.]

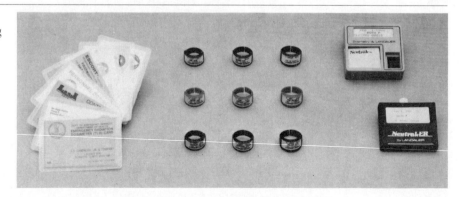

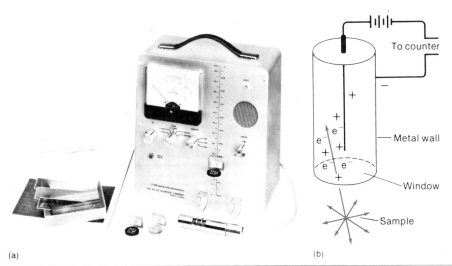

FIGURE 4.10
The Geiger counter, schematic diagram at right. [Courtesy of Sargent-Welch Scientific Company, Skokie, IL.]

X-ray sources just as photographic film reacts to light. A certain dose of radiation will cause the film in the badge to become exposed, alerting the wearer to the potential danger. Sophisticated electronic devices, like the **Geiger counter** (Figure 4.10), measure the ionizing effects of radiation and translate these into an observable signal (a meter reading or a clicking sound or a light flash. Other detectors provide a permanent visual record of the intensity of the radiation. Such detectors play an important role in medical diagnosis (Section 4.10).

4.6 Half-life

Radioactivity results when nuclei decay. We cannot predict when a particular nucleus will decay, but we can accurately predict the **rate** of decay of large numbers of radioactive atoms. (Life insurance companies cannot tell exactly when you will die, but their business depends on being able to predict how many of their clients will die in a particular period of time.)

A radioactive isotope can be characterized by a quantity called its **half-life**. The half-life of a radioactive isotope is that period of time in which one-half of the radioactive atoms present undergo decay. Suppose, for example, that you had 4 billion atoms of the radioactive hydrogen isotope, ^3_1H. The half-life of this isotope is 12.3 years. This means that in 12.3 years, one-half, or 2 billion, of the original 4 billion atoms would have undergone radioactive decay (Figure 4.11). In another 12.3 years, half of the remaining 2 billion atoms would have decayed. That is, after two half-lives, 1 billion atoms or one-fourth of the original atoms would remain unchanged. Two half-lives do not make a whole! The concept of half-life is shown graphically in Figure 4.12.

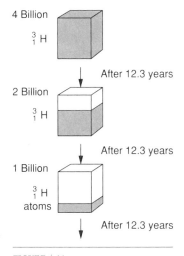

FIGURE 4.11
Life span of ^3_1H.

Half-life ($t_{1/2}$): the time in which one-half of the radioactive atoms present undergo decay.

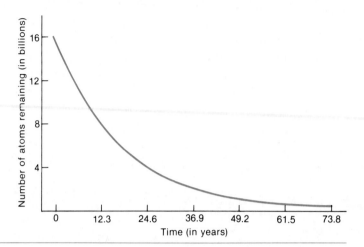

FIGURE 4.12
The radioactive decay of ^3_1H.

It is impossible to say exactly when *all* the ^3_1H will have decayed. For practical purposes one may assume that nearly all the radioactivity is gone after about 10 half-lives. For the tritium sample considered here, 10 half-lives would be 123 years, at which time only a fraction of a percentage of the original atoms would still be present.

The half-life of an isotope may be very long, as with $^{238}_{92}U$ (which has a half-life of 4 500 000 000 years), or very short, as with $^{21}_{12}Mg$ (with a half-life of 0.121 s).

The half-life of a carbon isotope can be used to date old artifacts. Such dating is based on the fact that there is a constant amount of carbon-14 in the ecosphere. Even though some is always lost as it decays, more is regenerated through other reactions so that an equilibrium amount of carbon-14 is maintained. As long as the plant or animal is alive and therefore a part of the ecosphere, its carbon-14 content will be the same as outside its system. Upon death, however, the exchange with the ecosphere ceases, and the carbon-14 slowly diminishes, making it possible to measure the organism's age.

EXAMPLE 4.4

A bone sample of the extinct American saber-toothed tiger was found to have 4 counts per minute (cpm) compared with 16 cpm for fresh bone. Estimate the age of the remains. The carbon-14 half-life is 5370 years.

One half-life ago, the sample would have had 8 cpm, and another half-life back it would have had 16 cpm. The age is then two half-lives: $2 \times 5370 = 10\ 740$ years.

By the same technique, the Shroud of Turin (a religious artifact believed by many to have been the burial cloth of Jesus Christ) was recently determined to be no older than the Middle Ages. Laboratories in England, Switzerland, and Arizona were each given three unmarked pieces of cloth without being told which, if any, was a sample of the shroud. Their independent conclusions were that it dates from about A.D. 1200.

Studies can also be forward in time.

EXAMPLE 4.5

Phosphorus-30 has a half-life of 3 min. How much of a 16.0-mg sample would be left after 15 min?

There are five half-lives in 15 minutes. Therefore the 16-mg sample would decrease, by half, five consecutive times: $16/2 = 8$; $8/2 = 4$; $4/2 = 2$; $2/2 = 1$; $1/2 = 0.5$ mg is left at the end of 15 min.

Much of the concern over nuclear power centers on the long half-lives of some isotopes that can be released in a nuclear accident or that are simply left as by-products of the normal operation of a nuclear reactor. In the first instance, the fear is that large areas could be rendered uninhabitable for thousands of years if an explosion released long-lived radioisotopes into the atmosphere. Even with no accidents, normal operation of a reactor produces nuclear wastes that must be safely stored for thousands of years.

4.7 Artificial Transmutation and Induced Radioactivity

The forms of radioactivity encountered thus far occur in nature. The helium we use to fill balloons has accumulated over billions of years from the alpha, $^4_2$He, decay of radioactive elements in the earth's crust (Figure 4.13). It is possible to bring about nuclear reactions not encountered in nature. Such reactions are referred to as **artificial transmutations** because in the process one element is changed into another. These reactions are brought about by bombardment of stable nuclei with alpha particles, neutrons, or other subatomic particles. These particles, given sufficient energy, penetrate the target nucleus and trigger a nuclear reaction.

Ernest Rutherford studied the bombardment of the nuclei of a variety of light elements with alpha particles. One such experiment, in which

Artificial transmutation:
Artificial—not occurring in nature
Transmutation—one element changes into another

he bombarded nitrogen nuclei, resulted in the production of protons.

$$\overset{18}{\underset{9}{{}^{14}_{7}N + {}^{4}_{2}He}} \longrightarrow \overset{18}{\underset{9}{{}^{17}_{8}O + {}^{1}_{1}H}}$$

(Recall that the hydrogen nucleus is a proton; hence the alternative symbol for the proton is ${}^{1}_{1}H$.) Notice that the sum of the mass numbers on the left of the equation equals the sum of the mass numbers on the right of the equation. The atomic numbers are also balanced.

Irène Curie (daughter of Marie and Pierre) and her husband, Frédéric Joliot (Figure 4.14), studied the bombardment of aluminum nuclei with alpha particles. The reaction yielded neutrons and an isotope of phosphorus.

$${}^{27}_{13}Al + {}^{4}_{2}He \longrightarrow {}^{30}_{15}P + {}^{1}_{0}n$$

Much to the surprise of the Joliot-Curies,* the target continued to emit particles after the bombardment was halted. The phosphorus isotope was radioactive. It emitted particles equal in mass but opposite in charge to the electron.

$${}^{30}_{15}P \longrightarrow {}^{0}_{+1}e + {}^{30}_{14}Si$$

The ${}^{0}_{+1}e$ particle is called a **positron**. Once again the question arises: if the nucleus contains only protons and neutrons, where does this particle come from?

Previously we accounted for a beta particle (an electron) popping out of a nucleus by saying that a neutron changed into a proton and an electron. Perhaps a similar happening can account for the appearance

*Frédéric changed his name to Joliot-Curie when he married Irène.

FIGURE 4.13
The helium that fills a blimp was originally formed over millions of years through the alpha decay of radioisotopes in the Earth's crust. [Courtesy of Fuji Photo Film U.S.A.]

FIGURE 4.14
Frédéric and Irène Joliot-Curie discovered artificially induced radioactivity in 1934. [Reprinted with permission from Mary E. Weeks, *Discovery of the Elements*, Chemical Education Publishing Company, Easton, PA, 1968. Copyright © 1968 by the Chemical Education Publishing Company.]

4.7 Artificial Transmutation and Induced Radioactivity

of a positron. Imagine a proton in the nucleus changing into a neutron and a positron. (Remember that a proton is the same as a hydrogen nucleus.)

$$^{1}_{1}H \longrightarrow \,^{0}_{+1}e + \,^{1}_{0}n$$

The equation balances nicely. When the positron is emitted, the nucleus suddenly has one less proton and one more neutron than before. Therefore, the weight of the product nucleus is the same, but its atomic number is one less than that of the original nucleus. Figure 4.5b presents another example of nuclear decay involving positron emission.

EXAMPLE 4.6

When potassium-39 is bombarded with neutrons, chlorine-36 is produced. What other particle is emitted?

$$^{39}_{19}K + \,^{1}_{0}n \longrightarrow \,^{36}_{17}Cl + ?$$

To balance the equation, a particle with a mass number of 4 and an atomic number of 2 is required—that is, an alpha particle.

$$^{39}_{19}K + \,^{1}_{0}n \longrightarrow \,^{36}_{17}Cl + \,^{4}_{2}He$$

EXAMPLE 4.7

Carbon-10, a radioactive isotope, emits a positron when it decays. Write an equation for this process.

$$^{10}_{6}C \longrightarrow \,^{0}_{+1}e + ?$$

To balance the equation, a particle with a mass number of 10 and an atomic number of 5 (boron) is required.

$$^{10}_{6}C \longrightarrow \,^{0}_{+1}e + \,^{10}_{5}B$$

Nuclear medicine depends on the availability of a broad range of radioisotopes, and many of these are artificially produced. Later in this chapter we will look into some aspects of nuclear medicine.

4.8 Fission and Fusion

In nuclear fission the nucleus breaks apart to form two nuclei.

In the nuclear reaction called **nuclear fission**, a large unstable nucleus is bombarded with relatively slow-moving neutrons. In the resulting nuclear reaction, the large nucleus does not simply emit a small particle; it breaks apart, leaving two medium-sized nuclei and releasing more neutrons. A typical fission reaction is

$$^{235}_{92}U + ^{1}_{0}n \longrightarrow ^{90}_{38}Sr + ^{143}_{54}Xe + 3 ^{1}_{0}n$$

Because it was bombardment with neutrons that triggered the reaction in the first place, the product neutrons can go on to cause more of the large nuclei to fission (break apart). Neutrons produced from this second wave of reactions will trigger more reactions, and so on. Thus, nuclear fission is a chain reaction (Figure 4.15). Each reaction in the chain releases energy. If the chain of reactions is carried out in a controlled manner, then the energy released can be used to generate electric power, such as is done in a nuclear reactor (Figure 4.16). The reaction can also be carried out in such a way that all the energy is released in one gigantic explosion. What one then has is a bomb—a nuclear bomb (Figure 4.17). It is not possible for a nuclear reactor to give a nuclear-bomblike explosion because of the special conditions required for an effective bomb. The products of fission, however, whether from a bomb or a reactor, are radioactive. When the bomb explodes, these products are thrown into the atmosphere and eventually reach the ground as dangerous "fallout." In a reactor, these products must be periodically removed and transferred to safe storage facilities.

FIGURE 4.15
The splitting of a uranium atom. The neutrons produced in this fission can split other uranium atoms, thus sustaining a chain reaction. The splitting of one uranium-235 atom yields 8.9×10^{-18} kWh of energy. Fission of a mole of uranium-235 (6.02×10^{23} atoms) produces 5 300 000 kWh of energy.

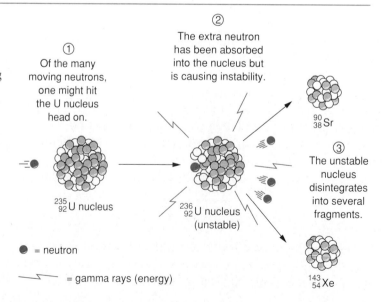

4.8 Fission and Fusion

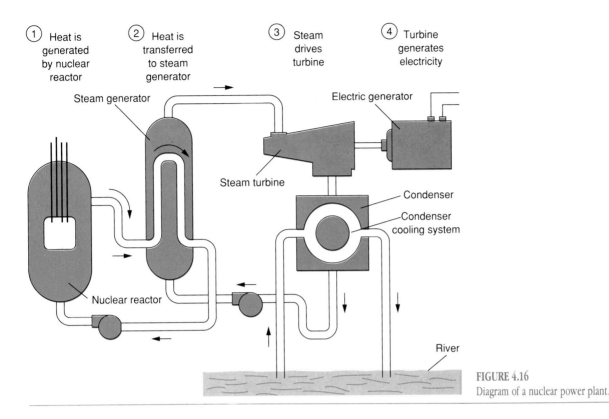

FIGURE 4.16
Diagram of a nuclear power plant.

A second important nuclear reaction is called **fusion**. In this case, two smaller nuclei are combined or fused into a larger nucleus. Again, this process is accompanied by the release of vast amounts of energy. A typical fusion reaction is

$$^2_1H + {}^3_1H \longrightarrow {}^4_2He + {}^1_0n$$

Our sun is powered by fusion reactions occurring in its core. Fusion reactions also generate the awesome destructive energy of a thermonuclear or hydrogen bomb. Scientists are currently attempting to bring fusion reactions under sufficient control to permit them to be used to generate power for peaceful uses. The problem is that to get two positively charged nuclei together, one has to overcome their repulsion for one another (like charges repel). One does this by slamming them into one another at high speed—and high speed means high temperature, in this case millions of degrees. The need to produce and contain these high temperatures is slowing the development of controlled fusion. (Note that this problem does not exist in fission reactions, which are triggered by neutrons, lasers, etc.; these neutral particles are not repulsed by the fissionable nucleus.)

In nuclear fusion, several nuclei combine to form one nucleus.

FIGURE 4.17
The now-familiar mushroom cloud that follows a nuclear explosion. [Courtesy of The Smithsonian Institution, Washington, DC.]

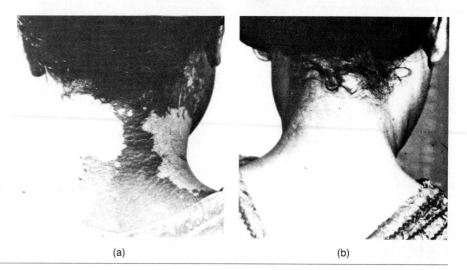

FIGURE 4.18
Burns from beta rays cause changes in skin pigmentation. (a) The skin one month after exposure. (b) The skin after full recovery, one year after exposure. [Reprinted from Normal A. Frigeria, *Your Body and Radiation*. Oak Ridge, TN: U.S. Department of Energy, 1967.]

4.9 Radiation and Living Things

High doses of radiation (on the order of 2000 rads) cause gross destruction of tissue (Figure 4.18). A person exposed to about 2000 rads passes into shock and dies in a few hours. With an exposure of 500 rads, the person usually survives long enough to exhibit the several phases of acute radiation syndrome.

1. A short latent period (a few hours) when no effects are observed.
2. A period of nausea and vomiting (which usually ends in 24 hr) and a drop in white blood cell count.
3. A period of few symptoms. A low-grade fever may persist.
4. The last period. Loss of hair begins rather abruptly anywhere from about the seventeenth through the twenty-first day. Increased discomfort sets in, with loss of appetite and diarrhea. The body temperature rises, and the patient complains of pain in the throat and gums. Emaciation sets in, the general condition deteriorates, and the person dies.

With smaller doses, recovery can take place, but such recovery may not be complete. Malignant growths may appear, even years later, and genetic effects may appear in succeeding generations.

Some cells are more susceptible to radiation than others. Those that are being constantly and rapidly replaced are affected most. These include the intestinal mucosa, germ cells, embryonic cells, blood cells, and the organs responsible for producing blood cells, such as the bone marrow. Damage to reproductive cells will show up as abnormalities in the descendants of affected persons.

The **rem** (*r*oentgen *e*quivalent in *m*an) is a measure of the relative biological damage produced by a particular dose of radiation. For our purposes, the roentgen and the rad are about equivalent since 1 R

TABLE 4.2 $LD_{50}/30$ Days Values for Various Species

Organism	LD_{50} (in rems)
Dog	310
Pig	375
Human	400–450 (estimated)
Monkey	600
Rat	790
Yeast	10 000
Bacterium	100 000
Virus	1 000 000

$LD_{50}/30$ *days* is the dosage required to kill (the *l*ethal *d*ose) 50% of the individuals in a large group within 30 days.

Adapted from K. Williams, C. L. Smith, and H. D. Chalke, *Radiation and Health*. Boston: Little, Brown, 1962.

> **TABLE 4.3 Protection from Nuclear Radiation**
>
> 1. Distance: The more distant the source, the greater the safety.
> 2. Sample size: The smaller the radiating sample, the greater the safety.
> 3. Radiation: The less penetrating the radiation, the greater the safety. Thus, for external sources safety increases in the order $\gamma < \beta < \alpha$.
> 4. Half-life: The longer the half-life, the greater the safety.
> 5. Time: Generally, the shorter the time of exposure, the greater the safety.
> 6. Frequency: The fewer the exposures, the greater the safety.

generates about 1 rad of energy when absorbed in muscle tissue. It is estimated that a whole body exposure of about 500 rads would kill most of us. Table 4.2 lists lethal doses of radiation in rem. For comparison, the International Commission on Radiological Protection recommends that adults whose occupations expose them to ionizing radiation limit their exposure to 5 rem in any one year.

People working with radioactive materials can do several things to protect themselves (Table 4.3). The simplest is to move away from the source, for intensity of radiation decreases with distance from the source. Workers can also be protected by shielding. A sheet of paper will stop most alpha particles. A block of wood or a thin sheet of aluminium will stop beta particles. But it takes several meters of concrete or several centimeters of lead to stop gamma rays.

4.10 Nuclear Medicine

Nuclear medicine involves two distinct uses of radioisotopes: therapeutic and diagnostic. In radiation therapy, an attempt is made to treat or cure disease with radiation. The diagnostic use of **radioisotopes** is aimed at obtaining information about the state of an individual's health. We will consider first the therapeutic applications of nuclear chemistry.

Cancer is not one disease but many. Some forms are particularly susceptible to radiation therapy. Radiation is carefully aimed at the cancerous tissue, and exposure of normal cells is minimized (Figure 4.19). If the cancer cells are killed by the destructive effects of the radiation, the malignancy is halted. But persons undergoing radiation therapy often get quite sick from the treatment. Nausea and vomiting are the usual symptoms of radiation sickness. (Remember that the intestinal mucosa is particularly susceptible to radiation.) Thus, the aim of radiation therapy is to destroy the cancerous cells before too much damage is done to healthy tissue.

The therapeutic use of a radioisotope is intended to treat or cure a disease. Radioisotopes are also used for diagnostic purposes, to help

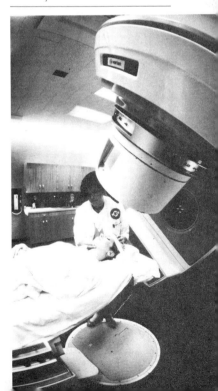

FIGURE 4.19
A cobalt-60 unit for radiation therapy. [Courtesy of the National Cancer Institute.]

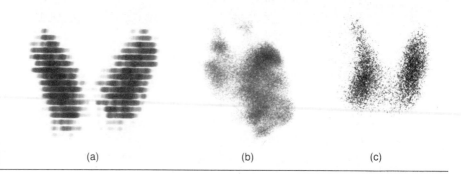

FIGURE 4.20
A linear photoscanner produced these pictures of (a) a normal thyroid, (b) a multinodal goiter, and (c) a thyroid adenoma. [Photos by Joseph J. Mentrikoski, Department of Medical Photography, Geisinger Medical Center.]

provide information about the type or extent of illness. Radioactive iodine-131 ($^{131}_{53}I$) is used to determine the size, shape, and activity of the thyroid gland as well as to treat cancer located in this gland and to control a hyperactive thyroid. First, one drinks a solution of potassium iodide, KI, incorporating iodine-131. The body concentrates iodide in the thyroid. Large doses are used for treatment of thyroid cancer; the radiation from the isotope concentrates in the thyroid cancer cells even if the cancer has spread to other parts of the body. For diagnostic purposes, however, only a small amount is needed. A radiation detector records the uptake of the isotope by body tissue and translates this information into a permanent visual record. The "picture" that results is referred to as a **photoscan** (Figure 4.20).

One can imagine an ideal isotope for diagnostic scanning. The isotope should be a gamma emitter, because gamma rays have high penetrating power. For diagnosis, radiation must escape from the body so that the scanner can detect it. It would be best if neither alpha nor beta particles were emitted. These are not needed for detection and would simply do unnecessary damage to the body. The gamma rays should have just the right amount of energy: they should not be so weak that we would have to wait around forever to get enough of a response from our detector, nor so strong that they would impart unnecessary energy to the body in making their way out. The half-life also should be just right: not so short that the activity is gone before we can measure it, nor so long that significant activity remains long after the scan has been obtained.

Happily, an isotope is available much like this ideal one. It is the synthetic element technetium-99m ($^{99m}_{43}Tc$). The m stands for **metastable**, which means that this isotope will give up some energy to become a more stable version of the same isotope (same atomic number, same mass number). The energy it gives up is the gamma ray we need for detecting the isotope while scanning.

$$^{99m}_{43}Tc \longrightarrow {}^{99}_{43}Tc + \gamma$$

4.11 Medical Imaging

TABLE 4.4 Some Radioisotopes and Their Application in Medicine

Isotope	Name	Use(s)
^{74}As	Arsenic-74	Location of brain tumors
^{51}Cr	Chromium-51	Determination of volume of red blood cells and total blood volume
^{58}Co	Cobalt-58	Determination of uptake of vitamin B_{12}
^{60}Co	Cobalt-60	Radiation treatment of cancer
^{131}I	Iodine-131	Detection of thyroid malfunction; measurement of liver activity and fat metabolism; treatment of thyroid cancer
^{59}Fe	Iron-59	Measurement of rate of formation and lifetime of red blood cells
^{32}P	Phosphorus-32	Detection of skin cancer or cancer of tissue exposed by surgery
^{226}Ra	Radium-226	Radiation therapy for cancer
^{24}Na	Sodium-24	Detection of constrictions and obstructions in the circulatory system
^{99m}Tc	Technetium-99m	Imaging of brain, thyroid, liver, kidney, bone, lung, and cardiovascular system
^{3}H	Tritium	Determination of total body water

The energy of the gamma ray is just about right, and so is the half-life, about 6 hr. Much effort has gone into making technetium-99m available in a variety of preparations. In some forms it is useful for kidney scanning; in other forms it will be concentrated by the liver and spleen. It can be used for brain scans or for lung scans. Thus, this isotope can replace a number of less-than-ideal radioisotopes.

Table 4.4 lists a selection of radioisotopes in common use in medicine.

4.11 MEDICAL IMAGING

Medical imaging provides a means of looking at internal organs without resorting to surgery. The history of medical imaging dates back to the discovery of X-rays at the turn of the century. Penetrating X-rays were used almost immediately to visualize skeletal structure. Radiation from an external X-ray source passes through the body (except where it is absorbed by more dense structures, such as bone) and exposes film, thus providing a picture that distinguishes the more dense structures from less dense tissue. Softer tissue can be visualized by introducing material that absorbs X-rays into the area to be studied. For example, compounds of barium are used to visualize portions of the digestive tract (Figure 4.21).

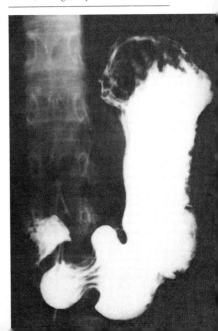

FIGURE 4.21
Barium sulfate ($BaSo_4$) is insoluble in water and opaque to X-rays. When swallowed, this salt can be used to outline the stomach for X-ray photographs. [Courtesy of B. Levin, Michael Reese Hospital and Medical Center, Chicago, IL.]

Radioisotopes can also be used to visualize internal organs. In this technique, the source of the radiation is inside the body, and the radiation (usually gamma rays) is detected as it emerges from the body.

Both X-ray technology and nuclear imaging have been coupled with computer technology to provide versatile and powerful imaging techniques. In computed tomography (referred to as CT or, sometimes, CAT scanning), many X-ray readings are obtained, processed by a computer, and then displayed. The resulting pictures present cross-sectional slices of a portion of the body. A series of these pictures gives a three-dimensional view of organs such as the brain.

Positron emission tomography (PET) can be used to measure dynamic processes occurring in the body, such as blood flow or the rate at which oxygen or glucose is being metabolized. PET scans are used to pinpoint the area of the brain damage that triggers severe epileptic seizures. Compounds incorporating positron-emitting isotopes, such as carbon-11, are inhaled or injected prior to the scan. Before the emitted positron can travel very far in the body, it encounters an electron (in any ordinary matter there are numerous electrons), and two gamma rays are produced.

$$^{11}_{6}C \longrightarrow ^{11}_{5}B + ^{0}_{+1}e$$

$$^{0}_{+1}e + ^{0}_{-1}e \longrightarrow 2\gamma$$

These exit from the body in exactly opposite directions. Detectors positioned on opposite sides of the patient record the gamma rays. If the recorders are set so that two simultaneous gamma rays must be "seen," gamma rays resulting from natural background radiation are ignored. A computer is then used to calculate the point within the body at which the annihilation of the positron and electron occurs, and an image of that area is produced.

Both X-rays and nuclear radiation are ionizing radiations, which means there is always some tissue damage involved. Modern techniques keep this damage to a minimum. Other imaging techniques use nonionizing radiation.

Perhaps you are familiar with ultrasonography. In this technique, high-frequency sound waves are bounced off tissue, and the echo of the sound wave is recorded. Once again, a computer is used to process the data and produce an image of the tissue. (The technique is related to sonar detection of submarines.) Because no ionizing radiation is involved, ultrasonography is used extensively in obstetrics for following fetal development (Figure 4.22). A 1984 conference on the use of ultrasonography (sponsored by the National Institutes of Health) concluded, however, that even this technique should not be used casually.

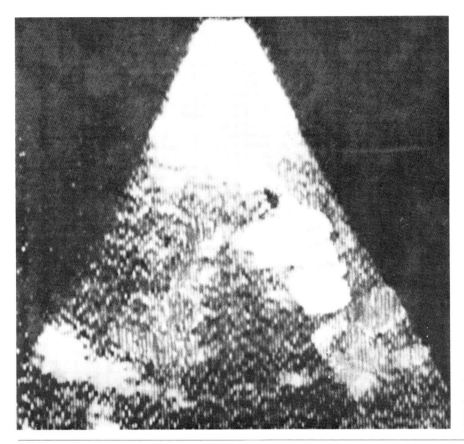

FIGURE 4.22
An ultrasonogram showing the face of a five-month-old fetus. [Courtesy of National Institutes of Health.]

The conference recommended that an ultrasonogram be obtained only if there is a good medical reason for doing so (for example, to evaluate fetal growth in mothers who have diabetes or hypertension).

Another nonionizing technique is magnetic resonance imaging (MRI). This technique depends on the following property of nuclei: Some nuclei behave as if they were little magnets. If they are placed in a strong magnetic field (for example, between the poles of a much more powerful magnet), the nuclei line up in a certain manner. When supplied with the right amount of energy, the nuclei absorb the energy and flip over in the magnetic field. The energy required to flip the nuclei is provided by nonionizing radiation. The absorption of the energy by the nuclei can be detected by appropriate equipment, and an image of the tissue in which the nuclei reside is produced. (Again, a computer is used to process the image.) The technique has demonstrated its potential for providing not only images of organs but also information about the metabolic activity in particular tissues.

The cost of most of the newer, computer-based technologies is very high, ranging into the millions of dollars for a single installation (Figure 4.23). The great advantage of these techniques is that they pro-

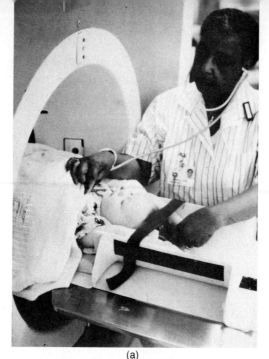
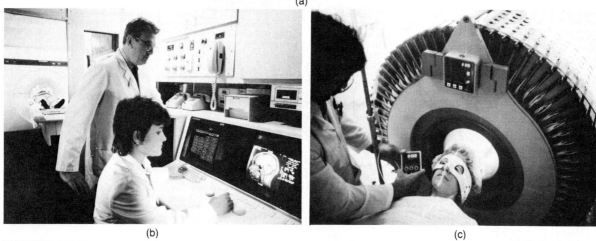

FIGURE 4.23
The mighty machines of modern medicine. (a) CT scanner. (b) MRI scanner. (c) PET scanner. [(a) and (c) courtesy of The University of Chicago Medical Center; (b) William Strode/Woodfin Camp & Associates.]

vide information that could otherwise be obtained only by subjecting the individual to the risks of surgery. In some instances, the information provided by the newer techniques is unavailable by any other route.

4.12 OTHER APPLICATIONS

In the last several decades nuclear research has emphasized peaceful rather than military applications. Aside from the uses already mentioned, many other applications have been found.

Elements that do not appear in nature can be prepared through transmutations. Examples of synthetic elements include elements 43, 61, and 85 and the transuranium elements. Chemical reaction mecha-

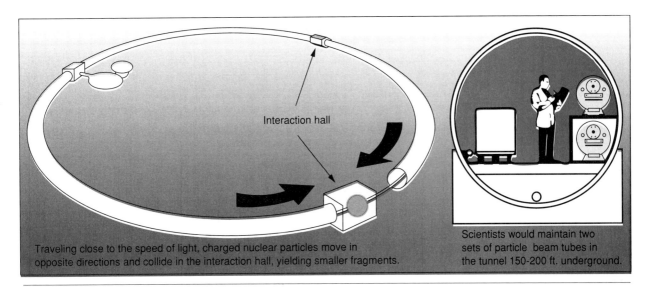

FIGURE 4.24
The supercollider permits new studies of the nucleus.

nisms (the manner in which elements interact to make a new product) can often be determined only with radioactive tracers. In photosynthesis, for instance, it has been found that carbon dioxide does indeed play a major role. The calcium uptake in metabolism has been found to be 90% in the young but only 40% in older animals and humans. Polycythemia vera (formation of too many red blood cells) has been studied with the aid of iron-59. It was found that ten times as much iron as the body can use is assimilated by patients with this disease. Through similar studies it has also been learned that the uptake of trace elements by trees is quite pronounced in the winter. Thus, zinc moves up the trunk of a tree at the rate of about 2 ft/day. The effectiveness of industrial lubricants can be measured by monitoring the concentration of metal residues in the oils. By tagging the metal with radioactive isotopes, concentrations as low as 10^{-19} g/L can be detected.

The accelerator (Figure 4.24) is an application of nuclear studies. Built in a long oval tunnel, it accelerates atoms in opposite directions, causing them to collide with speed-of-light intensity and break apart. This is a tool for studying the nature of matter.

4.13 The Nuclear Age Revisited

We live in an age in which fantastic forces have been unleashed. The threat of nuclear war has been a constant specter in our time. Nuclear bombs have been used to destroy cities—and men, women, and children. Science and scientists have been greatly involved in it all. Would the world be a better place had we not discovered the secrets of the atomic nucleus? Consider this: More lives have been saved through nuclear medicine than have been destroyed by nuclear bombs. Also, nuclear power, with all its attendant problems, still may be one of our best hopes for a plentiful energy supply until well into the twenty-first century. Imagine yourself to be Dalton. Could this gentle Quaker, whose formal education ended when he was 11 years old, have anticipated the chain of scientific developments that would follow from his original speculations on the nature of matter?

CHAPTER SUMMARY

Nuclear reactions involve the nucleus rather than the valence electrons in an atom. Such reactions are described using symbols ($^A_Z X$) that include the atomic number (Z) and mass number (A) of the isotope involved. The mass number of an isotope is the sum of the number of protons and the number of neutrons in an atom of the isotope. Atomic mass is the mass of any atom relative to the mass of an atom of carbon-12, which is assigned a value of 12 amu. The atomic weight of an element is the average mass of the atoms in a representative sample of the element.

Radioactivity is a phenomenon in which unstable nuclei revert to more stable forms by emitting radiation of three kinds: alpha, beta, and gamma rays. Alpha rays are actually subatomic particles equivalent to the helium nucleus ($^4_2 He$). Beta rays are streams of electrons ($^{\ 0}_{-1}e$), and gamma rays are a form of radiant energy. X-rays, which were discovered slightly before radioactivity, are also a form of radiant energy but are an electronic, rather than a nuclear, phenomenon.

By bombarding certain nuclei with subatomic particles, scientists found that they could change one element into another in a process called artificial transmutation. By these reactions, scientists could produce new radioactive products, some of which emitted previously unobserved particles, such as the positron ($^{\ 0}_{+1}e$).

Radioactive elements are characterized by their half-lives, the time required for half of the nuclei in a sample to decay to the more stable form. The radiation emitted by radioactive nuclei is characterized by its penetrating power. Smaller particles (such as beta particles) can penetrate matter farther than larger particles (such as alpha particles). Pure energy forms (such as gamma rays) are most penetrating. Low penetrating power represents a greater risk to health if the source is internal, whereas high penetrating power is a greater danger in an external source. Background radiation is the radiation (both internal and external) one is normally exposed to from natural sources.

Units of radiation measurement include the curie (which measures sample activity), the roentgen (which measures the disruptive energy of X-rays or gamma rays), the rad (which measures the energy absorbed by matter), and the rem (which measures relative biological damage). The various forms of nuclear radiation and X-rays are called ionizing radiation because the energy they transfer to matter results in the formation of damaging ions. A lethal dose of radiation for humans is a whole-body exposure of about 500 rads. The extent to which one has radiation syndrome and the chance for recovery depend on the level of exposure to the radiation.

In medicine, nuclear radiation is used for diagnostic and therapeutic purposes. In therapy, radiation is used to destroy tissue, for example, malignant tumors, and the radiation source may be either external or internal. A diagnostic use of increasing importance is organ imaging. When radioisotopes are used for this purpose, they are taken internally. A widely used isotope for this purpose is technetium-99m. X-rays from an external source are also used in medical imaging. PET scanning (using radioisotopes) and CT scanning (using X-rays) are newer techniques that use computers to process and present data. Imaging techniques that do not involve ionizing radiation include <u>ultrasonography</u> and <u>MRI</u>.

Fission and fusion are two forms of nuclear reactions that result in the release of enormous quantities of energy. In <u>fission</u>, a large unstable nucleus (such as $^{235}_{92}U$) is bombarded with neutrons and splits into two medium-size nuclei and more neutrons. The product neutrons strike other fissionable nuclei to cause a chain reaction that releases energy for use in a bomb or for the generation of electric power. The radioactive isotopes thereby produced, particularly as applied in nuclear bombs, are referred to as <u>fallout</u>. In a <u>fusion</u> reaction, two small nuclei are fused together at extremely high temperatures. The destructive force of this reaction is used in hydrogen bombs, but the means for controlling the release of fusion energy for peaceful purposes has not yet been developed.

CHAPTER QUIZ

1. The number of neutrons in $^{7}_{3}Li$ is
   a. 3   b. 4   c. 7   d. 10
2. Carbon-14 disintegrates with beta decay. The key product is
   a. nitrogen-14   b. boron-10
   c. carbon-13   d. boron-14
3. The most penetrating radiation is
   a. alpha   b. beta
   c. gamma   d. proton
4. A positron is
   a. $^{1}_{1}p$ b. $^{0}_{+1}e$ c. H   d. He^{2+}
5. If 100 mg of an isotope decays to about 12 mg in 12 hr, its half-life is ____ hr.
   a. 3   b. 4   c. 8   d. 12
6. Positrons and electrons combine to form
   a. α b. β c. γ d. n
7. The designation mCi stands for
   a. the amount of an element
   b. the effect of radiation
   c. ionizing radiation
   d. the rate of disintegration
8. "Transmutation" refers to
   a. nuclear radiation
   b. natural radioactivity
   c. new element formation
   d. gamma decay

9. An isotope with a long half-life
   a. has very intense radiation
   b. has alpha, beta, and gamma rays
   c. is safer than one with a short half-life
   d. is larger than one with a short half-life
10. The isotope $^{A}_{Z}X$ was produced through positron emission of $^{30}_{15}P$. X, Z, and A are, respectively,
    a. P, 15, 31
    b. Si, 14, 30
    c. S, 16, 29
    d. Al, 13, 28

PROBLEMS

1. Define or identify each of the following.
   a. isotope
   b. alpha particle
   c. beta particle
   d. gamma ray
   e. half-life
   f. positron
   g. curie (Ci)
   h. roentgen (R)
   i. rad
   j. rem
   k. film badge
   l. Geiger counter
   m. background radiation
   n. ionizing radiation
   o. fission
   p. fusion
   q. radioisotope
   r. artificial transmutation

4.1 Nuclear Arithmetic

2. Draw the nuclear symbols for protium, deuterium, and tritium (which are hydrogen-1, hydrogen-2, and hydrogen-3, respectively).
3. Draw the nuclear symbol for an isotope with a mass number of 8 and an atomic number of 5.
4. Draw the nuclear symbol for an isotope with $Z = 35$ and $A = 83$.
5. Draw the nuclear symbol for an isotope with 53 protons and 72 neutrons.

6. Draw the nuclear symbols for the following isotopes. You may refer to the periodic table.
   a. gallium-69
   b. molybdenum-98
   c. molybdenum-99
   d. technetium-98

7. Indicate the number of protons and the number of neutrons in atoms of the following isotopes.
   a. $^{62}_{30}Zn$
   b. $^{241}_{94}Pu$
   c. $^{99m}_{43}Tc$
   d. $^{81m}_{36}Kr$

8. Which of the following sets represent isotopes?
   a. $^{70}_{34}X$, $^{70}_{33}X$
   b. $^{57}_{28}X$, $^{66}_{28}X$
   c. $^{186}_{74}X$, $^{186}_{74}X$
   d. $^{8}_{2}X$, $^{6}_{4}X$
   e. $^{22}_{11}X$, $^{44}_{22}X$

9. The two principal isotopes of lithium are lithium-6 and lithium-7. The atomic weight of lithium is 6.9 amu. Which is the predominant isotope of lithium?

10. Out of every five atoms of boron, one has a mass of 10 amu and four have a mass of 11 amu. What is the atomic weight of boron? Use the periodic table only to check your answer.

11. Place a marble in an open field. Now step off 70 big steps away from it, and you will see the approximate radius of an atom with the marble as the nucleus.

12. Fill in the blanks for the six elements listed.

	Atomic No.	No. of Protons	No. of Electrons	No. of Neutrons	Mass No.
Pb				126	208
Sr		38		50	
N	7				14
Cr			24		52
Ag		47		60	
As			33	42	

4.3 Types of Radioactivity

13. How are X-rays and gamma rays similar? How are they different?

14. Draw the nuclear symbols for the following subatomic particles.
    a. alpha particle
    b. beta particle
    c. neutron
    d. positron

15. When a nucleus emits a beta particle, what changes occur in the mass number and atomic number of the nucleus?

16. Lead-209 undergoes beta decay. Write a balanced equation for this reaction.

17. When a nucleus emits an alpha particle, what changes occur in the mass number and atomic number of the nucleus?

18. Thorium-225 undergoes alpha decay. Write a balanced equation for this reaction.

19. When a nucleus emits a gamma ray, what changes occur in the mass number and the atomic number of the nucleus?

20. Write an equation representing the emission of a gamma ray by gold-186.

21. Radioactive radon gas in homes could be a lung cancer risk. One set of reactions involves two successive alpha decays from radon-222. What are the products formed?

22. Selenium-82 undergoes a rare two-beta emission reaction. What is the product?

4.6 Half-life

23. C. E. Bemis and colleagues at Oak Ridge National Laboratory confirmed the synthesis of element 104, the half-life of which was only 4.5 s. Only 3000 atoms of the element were created in the tests. How many atoms were left after 4.5 s? After 9.0 s?

24. Krypton-81m is used for lung ventilation studies. Its half-life is 13 s. How long will it take the activity of this isotope to reach one-fourth of its original value?

4.7 Artificial Transmutation and Induced Radioactivity

25. When a nucleus emits a positron, what changes occur in the mass number and the atomic number of the nucleus?

26. Write a balanced equation for the emission of a positron by sulfur-31.

27. When a nucleus emits a neutron, what changes occur in the mass number and the atomic number of the nucleus?

28. Write a balanced equation for the emission of a neutron by bromine-87.

29. When a nucleus emits a proton, what changes occur in the mass number and the atomic number of the nucleus?

30. Write a balanced equation for the emission of a proton by magnesium-21.

31. Complete the following equations.
    a. $^{179}_{79}Au \rightarrow ^{175}_{77}Ir + ?$
    b. $^{23}_{10}Ne \rightarrow ^{23}_{11}Na + ?$

32. Complete the following equations.
    a. $^{10}_{5}B + ^{1}_{0}n \rightarrow ^{4}_{2}He + ?$
    b. $^{12}_{6}C + ^{2}_{1}H \rightarrow ^{13}_{6}C + ?$
    c. $^{121}_{51}Sb + ? \rightarrow ^{121}_{52}Te + ^{1}_{0}n$
    d. $^{154}_{62}Sm + ^{1}_{0}n \rightarrow 2\,^{1}_{0}n + ?$

33. When magnesium-24 is bombarded with a neutron, a proton is ejected. What new element is formed? (*Hint:* Write a balanced nuclear equation.)

34. A radioactive isotope decays to give an alpha particle and bismuth-211. What was the original element?

4.8 Fission and Fusion

35. Compare nuclear fission and nuclear fusion.

36. Which subatomic particles are responsible for carrying on the chain of reactions that are characteristic of nuclear fission?

37. Why are such high temperatures required for nuclear fusion reactions?

38. What dangers are there in using nuclear fission to generate power?

39. Abandoned salt mines are often cited as good places to store nuclear waste. What are the pros and cons of such disposal?

4.9 Radiation and Living Things

40. List two ways in which workers can protect themselves from the radioactive materials with which they work.

41. A pair of gloves would be sufficient to shield the hands from which type of radiation: the heavy alpha particles or the massless gamma rays?

42. Heavy lead shielding is necessary as protection from which type of radiation: alpha, beta, or gamma?

43. Plutonium is especially hazardous when inhaled or ingested because it emits alpha particles. Why would alpha particles cause more damage to tissue than beta particles?

4.10 Nuclear Medicine

44. Explain how radioisotopes can be used for therapeutic purposes.

45. Which radioisotope has been used extensively for treatment of overactive or cancerous thyroid glands?

46. Describe the use of a radioisotope as a diagnostic tool in medicine.

47. What are some of the characteristics that make technetium-99m such a useful radioisotope for diagnostic purposes?

48. The activity of a radiation source is 500 Ci. The activity of another source is 10 mCi. To be used properly, one source is taken into the body and the other remains outside the body during treatment. Which is likely to be the internal source and which the external?

4.11 Medical Imaging

49. Patient A takes internally an iodine-131 sample with an activity of 150 mCi. Patient B takes internally a dose of 15 μCi. In which patient is the iodine-131 being used to treat a malignancy, and in which patient is the isotope used for imaging the thyroid gland?

50. About 2 mCi of thallium-201 is given by intravenous administration for imaging the heart. It is estimated that the total body radiation dose in humans is about 0.07 rad per mCi of thallium-201. How does the radiation dose from this procedure compare to the lethal dose for humans?

51. What form of radiation is detected in CT scans?

52. What form is detected in PET scans?

53. What is the advantage of using nonionizing radiation for medical imaging?

54. Name two imaging techniques that do not use ionizing radiation.

4.13 The Nuclear Age Revisited

55. Discuss the impact of nuclear science on the following.
    a. war and peace
    b. medicine
    c. our energy needs

56. Cesium is one of the components of nuclear fallout. Why is it a particularly dangerous threat to the environment?

57. Nuclear wastes typically need to be stored for 20 half-lives to be safe. This translates into hundreds of years of storage time. (For instance, cesium-137 and strontium-90 have half-lives of about 30 years.) Would the shooting of such wastes into outer space be a responsible solution?

58. The following coded telephone conversation took place at 3:25 p.m. on December 2, 1942, between Dr. Compton from the University of Chicago and Dr. Conant at Harvard University.

    Dr. Compton: "The Italian navigator has landed in the new world."
    Dr. Conant: "How were the natives?"
    Dr. Compton: "Very friendly."

    Can you unravel the code?

5

GASES

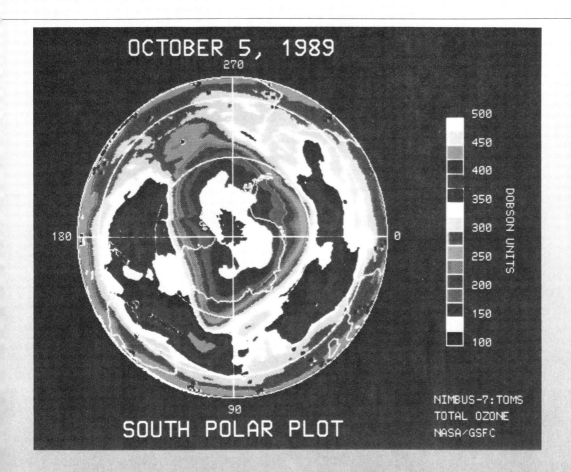

DID YOU KNOW THAT...

- the atmosphere of our planet Earth is unique?
- iron can be made gaseous, and air can be made solid?
- gas molecules occupy essentially no space compared to the volume they are moving in?
- your breathing process makes use of gas laws?
- gases are extraordinarily sensitive to changes in pressure, temperature, and volume?
- dew settles during summer nights because the humidity is high and the temperature is low?
- diving "bends" occur because of bubbles in the blood?
- weather balloons are limp and skinny when launched but expand fully once they are in the upper atmosphere?
- the popping of ears when changing altitude is the adjusting of the internal pressure to outside pressure?
- many gases rise, but some sink in the atmosphere?

Our astronauts have seen firsthand the barren, airless moon. Our spaceships have photographed the desolation of Mercury from a few kilometers up and measured the inhospitably high temperatures of Venus. They have given us close-up portraits of the crushing, turbulent atmospheres of Jupiter, Saturn, Uranus, and Neptune. Our experiments on the harsh Martian surface failed to confirm the presence of life there. As we enter the final years of the twentieth century, it is becoming increasingly clear that the Earth, a small island of green and blue in the vastness of space, is uniquely equipped to serve the needs of the life that inhabits it (Figure 5.1).

FIGURE 5.1
(a) Looking like the moon, the scarred planet Mercury is closest to the sun and has no atmosphere to protect its surface from meteors or intense solar radiation. (b) The crushing atmosphere of Venus traps the sun's energy and produces temperatures at the planet's surface high enough to melt lead. (c) The four dark spots are the peaks of gigantic Martian volcanoes reaching above hazy clouds in an atmosphere too thin to support human life. (d) As befits the largest of the planets, Jupiter possesses a turbulent atmosphere with cyclonic storms that could swallow several planets the size of the Earth. (e) Saturn is one of the beauties of our solar system, but its thick, poisonous atmosphere would be deadly to humans. (f) Blessed with abundant water and an atmosphere rich in oxygen, Earth offers its inhabitants a safe refuge for as long as they respect its limits. [NASA photographs.]

5.1 Air: A Mixture of Gases   141

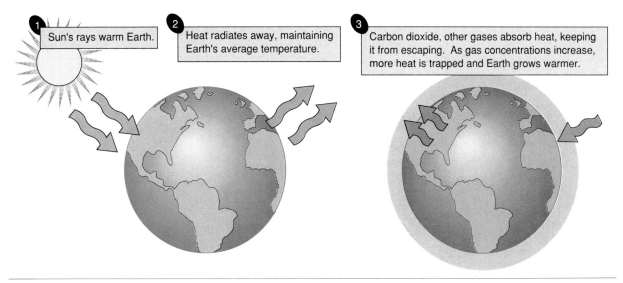

FIGURE 5.2
The greenhouse effect.

The life-support system of Spaceship Earth consists in part of a thin blanket of gases called the atmosphere. It is difficult to measure just how deep the atmosphere is. It does not end abruptly but gradually fades as the distance from the surface of the earth increases. We do know, though, that 99% of the atmosphere lies within 30 km of the surface of the Earth. That thin layer of air—it is relatively thinner than an apple peel is to an apple—is all that stands between us and the emptiness of space. That thin layer of gases—renewable but not inexhaustible—is essential to our life on this planet. The atmosphere contains, among other things, ozone (O_3), which protects living things from harmful ultraviolet rays. The atmosphere also controls the Earth's temperature. Changes in its composition can have negative results such as what is known as the *greenhouse effect* (Figure 5.2).

5.1 Air: A Mixture of Gases

Air is a mixture of gases. Dry air is (by volume) about 78% nitrogen (N_2), 21% oxygen (O_2), and 1% argon (Ar) (Table 5.1). The amount of water vapor varies up to about 4%. There are a number of minor constituents, the most important of which is carbon dioxide (CO_2).

Air is so familiar, and yet so nebulous, that it is difficult to think of it as matter. But it is matter—matter in the gaseous state. Like other forms of matter, gases obey certain physical laws. Because, as we noted in Chapter 3, the forces of attraction between particles of a gas are weak, almost nonexistent, the laws that govern gases are straightforward. Therefore, although changes in pressure, temperature, and volume affect all matter, at least to some extent, gases respond to such changes most dramatically. For gases we do not have to take into

TABLE 5.1 Composition of the Earth's Dry Atmosphere

Substance	Formula	Number of Molecules[a]
Nitrogen	N_2	7800
Oxygen	O_2	2100
Argon	Ar	93
Carbon dioxide	CO_2	3
All others	—	4
Total		10 000

[a] In molecules of each substance per 10 000 molecules.

account the strength of various interactions between particles. That makes it easy to calculate the changes that occur when some property of the gas is varied. By studying the laws that govern the behavior of gases, we gain considerable insight into certain vital processes—breathing, for example.

5.2 THE KINETIC MOLECULAR THEORY

We have mentioned the **kinetic molecular theory** previously (Section 3.10). Now we are going to go into more detail in order to give you some idea of how this model enables scientists to visualize gases and to understand the gas laws.

The theory treats gases as collections of individual particles in rapid motion (hence the term *kinetic*). The particles of nitrogen gas, for example, are molecules (N_2); those of argon gas (Ar) are atoms. The distances between the particles are large compared to the dimensions of the particles themselves. This fact explains why gases, by comparison with solids and liquids, can be so readily compressed. According to the theory, the individual particles of a solid or liquid are in contact with one another. They cannot be pushed much closer together (they cannot be compressed very much) because they are already touching.

The particles of a gas are in such rapid motion and are so light that gravity has little effect on them. They move up and down and sideways with ease and will not fall to the bottom of a container (as a liquid will). Any container of gas is completely filled. By *filled*, we do not mean that the gas is packed tightly, but rather that it is distributed throughout the container's entire volume. A particle moves along a straight path unless it strikes something (another particle or the walls of the container). Then it may bounce off at an angle and travel from the point of collision along a straight path until it hits something else. These collisions are **perfectly elastic**. That means that there is no tendency for the collection of particles to slow down and eventually stop. Two particles that are about to collide have a certain combined kinetic energy. After the collision, the sum of their kinetic energies is the same. One of the particles may have been slowed down by the collision, but the other will have been speeded up just enough to compensate (Figure 5.3).

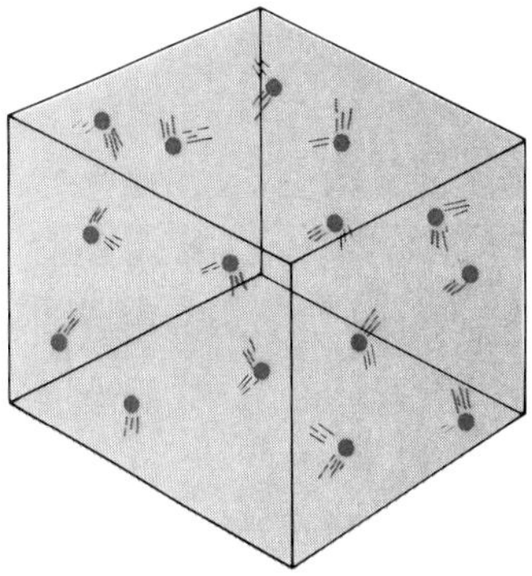

The kinetic molecular theory explains what it is that we are measuring when we measure temperature. According to the theory, temperature is a reflection of the kinetic energy of the gas particles. The higher the kinetic energy, that is, the faster the particles are moving, the higher the temperature of the sample. On the average, the particles of a cold sample are moving more slowly than the particles of a hot sample. In any single sample, some particles are moving faster than others. Temperature reflects the *average* kinetic energy of the particles.

As a particle strikes a wall of its container, it gives the wall a little push. (If you were hit with a baseball or a brick, you would feel a push.) The sum of all these tiny pushes over a given area of the wall is what we call **pressure**. Thus, if your car's tire pressure is 28 lb/in.^{2} (psi), then every square inch of your tire is experiencing a force of 28 lb above atmospheric pressure because billions of gas molecules are striking the tire's walls.

FIGURE 5.3
According to the kinetic molecular theory, particles of a gas are in constant motion, occasionally bouncing off one another and off the walls of their container.

5.3 ATMOSPHERIC PRESSURE

Molecules of air are constantly bouncing off each of us. Molecules are so tiny, though, that we do not feel the impact of individual ones. In fact, at ordinary altitudes we do not feel the molecules pushing on our skin because there are molecules on the inside pushing out just as hard. When we increase our altitude rapidly by driving up a mountain or riding an elevator to the top of a tall building, however, our ears pop because there are fewer molecules (because of the thinner air) on the outside pushing in than there are on the inside pushing out. Once we are at the top of the mountain or building, the pressures soon equalize, and the popping stops.

The pressure of the atmosphere is measured by a device called a **barometer**. The simplest type of barometer is a long glass tube, closed at one end, filled with mercury, and inverted in a shallow dish containing mercury (Figure 5.4). Suppose that the tube is 1 m long. Some of the mercury in the tube will drain into the dish, but not all of it. The mercury will drain out only until the pressure exerted by the mercury remaining in the tube exactly balances the pressure exerted by the atmosphere on the surface of the mercury in the dish. The mercury in the tube is trying to push its way out under the influence of

A barometer measures the pressure of the atmosphere.

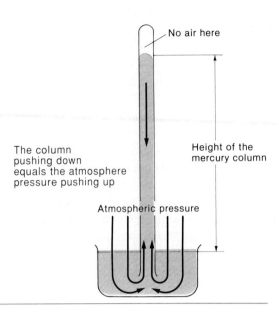

FIGURE 5.4
A mercury barometer.

1 atm = 760 mm Hg = 76.0 cm Hg = 760 torr

1 cm H₂O = 0.74 mm Hg

1.00 atm = 29.9 in. Hg = 14.7 psi

1 Pa = 1 N/m² = 1 × 10⁻⁵ atm

gravity, and the air pressure is pushing it back in. At some point the competing pressures reach a stalemate. Mercury is a very heavy (dense) liquid. On the average, at sea level, a column of mercury 760 mm high balances the push of a column of air many kilometers high. The average atmospheric pressure at sea level is thus said to be 760 mm Hg (**millimeters of mercury**). This pressure is called 1 **atmosphere** (atm) or standard pressure. The unit millimeter of mercury is also called a **torr** (after the Italian physicist Evangelista Torricelli, the inventor of the mercury barometer).

If water were the fluid used in the barometer, the height of the water column required to balance 1 atm of pressure would be almost 10 m (as high as a three-storey building). Respiratory therapists occasionally use a unit of pressure that does assume that water is the liquid in the barometer. The unit is **centimeters of water**, and it is particularly useful for the measurement of relatively small differences in pressure. Two other widely used units of pressure are **inches of mercury** (used in weather reports) and **pounds per square inch** (psi, used by engineers).

The SI unit for pressure is the **pascal** (Pa), a unit that is interpreted in terms of newtons (N). (A newton is the force that, when applied to a mass of 1 kg, will give that mass an acceleration of 1 m/sec².)

5.4 Boyle's Law: Pressure–Volume Relationship

The relationship between the volume and the pressure of a gas was first determined by Robert Boyle in 1662 (Figure 5.5). This relationship,

5.4 Boyle's Law: Pressure–Volume Relationship

called **Boyle's law**, states that, *for a given mass of gas at constant temperature, the volume varies inversely with the pressure.* $V \propto 1/P$, which says that when the pressure goes up, the volume goes down. When the pressure goes down, the volume goes up (Figure 5.6). Think of gases as pictured in the kinetic molecular theory. A sample of gas in a container exerts a certain pressure because the particles are bouncing against the walls at a certain rate with a certain force. If the volume of the container is expanded, the particles will have to travel longer distances before they strike the walls. Also, the surface area of the walls increases as the volume increases, so each unit of area is struck by fewer particles. Thus, the rate of these strikes (these "pushes") decreases, and the pressure measurement reflects this.

Mathematically, Boyle's law is written

$$VP = k$$

This is an elegant and precise, if somewhat abstract, way of summarizing a lot of experimental data. The equation $VP = k$ says what Figure 5.6 says. If the product of V (volume) times P (pressure) is to have a constant value (k), then when V increases P must decrease, and vice versa. A typical Boyle's law experiment is diagrammed in Figure 5.7a. Gas is enclosed in a cylinder fitted with a movable piston. In the example, the gas occupies a volume of 1 L under 1 atm of pressure. When the pressure is increased to 2 atm, the volume is reduced to 0.5 L. When the pressure is 4 atm, the volume becomes 0.25 L. An extended mathematical expression of this is

$$\frac{V_1}{V_2} = \frac{P_2}{P_1}$$

which can be shown graphically as in Figure 5.7b.

Boyle's law has a number of practical, as well as theoretical, applications. These are perhaps best illustrated by a number of examples. Note that in applications of Boyle's law any units for pressure or volume can be used, as long as the use is consistent.

Volume is inversely proportional to pressure,

$$V \propto \frac{1}{P}$$

$\propto$ means "proportional to."

FIGURE 5.5
Robert Boyle. [Courtesy of The Smithsonian Institution, Washington, DC.]

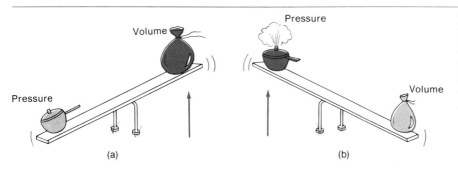

FIGURE 5.6
The pressure–volume relationship is like a seesaw. (a) When the pressure goes down, the volume goes up. (b) When the pressure goes up, the volume goes down. [From an idea by Cindy Hill.]

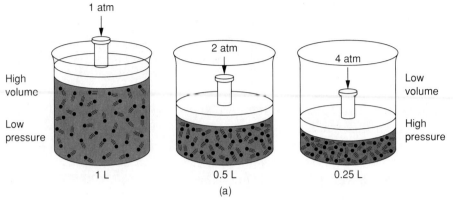

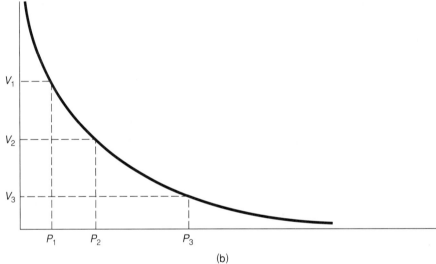

FIGURE 5.7
(a) A diagram illustrating the effect of different pressures on the volume of a gas. (b) A graph of the relationship.

EXAMPLE 5.1

A cylinder of oxygen has a volume of 2.00 L. The pressure of the gas is 1470 psi at 20 °C. What volume will the oxygen occupy at standard atmospheric pressure (14.7 psi) assuming no temperature change?

It is most helpful to separate the initial from the final condition

Initial	Final	Change
$P_1 = 1470$ psi	$P_2 = 14.7$ psi	↓
$V_1 = 2.00$ L	$V_2 = ?$	↑

While we could just use the formula $V_1/V_2 = P_2/P_1$, it is much safer (and more satisfying) to think about the process.

5.4 Boyle's Law: Pressure–Volume Relationship

The best way to work problems of this type is to evaluate them as follows. You are asked for the volume that results when the pressure is changed from 1470 psi to 14.7 psi. The pressure has been decreased; therefore, the volume must increase. But by how much? To find out, multiply by a fraction made up of the two pressures. To make the fraction greater than 1 (so that the volume will increase), arrange the pressures with the larger one on top, like this:

$$\frac{1470 \text{ psi}}{14.7 \text{ psi}}$$

Then multiply the original volume by this fraction.

$$2.00 \text{ L} \times \frac{1470 \text{ psi}}{14.7 \text{ psi}} = 200 \text{ L}$$

Gases are stored for use (in hospitals, for example) under high pressure, even though they will be used at atmospheric pressure. This arrangement allows much gas to be stored in a small area.

EXAMPLE 5.2

A space capsule is equipped with a tank of air that has a volume of 0.10 m³. The air is under a pressure of 110 atm. After a space walk, during which the cabin pressure drops to zero, the cabin is closed and filled with the air from the tank. What will be the final pressure if the volume of the capsule is 11 m³?

Initial	Final	Change
$V_1 = 0.10 \text{ m}^3$	$V_2 = 11 \text{ m}^3$	↑
$P_1 = 110 \text{ atm}$	$P_2 = ?$	↓

Since the volume in which the air is confined increases, the pressure must decrease; therefore, the multiplier must be a fraction less than 1.

$$110 \text{ atm} \times \frac{0.10 \text{ m}^3}{11 \text{ m}^3} = 1.0 \text{ atm}$$

EXAMPLE 5.3

A weather balloon is partially filled with helium gas. On the ground, where the atmospheric pressure is 740 mm Hg, the volume of the balloon is 10 m³. What will the volume be when the balloon reaches an altitude where the pressure is 370 mm Hg? Assume that the temperature is constant.

Initial	Final	Change
$P_1 = 740$ mm Hg	$P_2 = 370$ mm Hg	↓
$V_1 = 10$ m³	$V_2 = ?$	↑

Since the pressure decreases, the volume increases.

$$10 \text{ m}^3 \times \frac{740 \text{ mm Hg}}{370 \text{ mm Hg}} = 20 \text{ m}^3$$

The pressure–volume relationship can be used to explain the mechanics of breathing. When we breathe in (inspire), the diaphragm is lowered and the chest wall is expanded (Figure 5.8). This increases the volume of the chest cavity. According to Boyle's law, the pressure inside the cavity must decrease. Outside air enters the lungs because it is at a higher pressure than that in the chest cavity. When we breathe out (expire), the diaphragm rises, and the chest wall contracts. This decreases the volume of the chest cavity. The pressure is increased, and some of the air is forced out.

During normal inspiration the pressure inside the lungs drops to

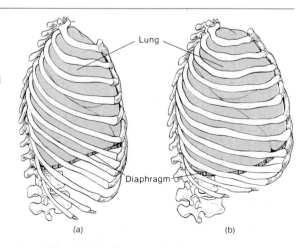

FIGURE 5.8
The mechanics of breathing. (a) Expiration. The diaphragm is relaxed, and the rib cage is down. (b) Inspiration. The diaphragm is pulled down, and the rib cage is lifted up and out, increasing the volume of the chest cavity.

5.4 Boyle's Law: Pressure–Volume Relationship

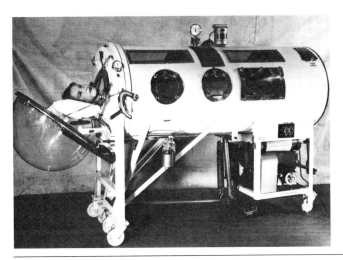 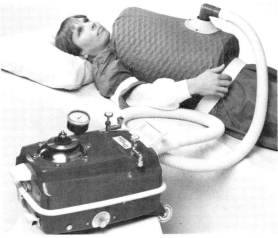

FIGURE 5.9
An iron lung uses changes in pressure to force air into and out of the lungs. (a) The older version enclosed the entire body except for the head. (b) A modern iron lung encloses only the chest. [Courtesy of J. H. Emerson Company, Cambridge, MA.]

about 3 mm Hg *below* atmospheric pressure. During expiration the internal pressure is about 3 mm Hg *above* atmospheric pressure. About half a liter of air is moved in and out of the lungs in this process, and this normal breathing volume is referred to as the **tidal volume**. The **vital capacity** is the maximum amount of air that can be forced from the lungs and ranges in volume from 3 to 7 L depending on the individual. A pressure inside the lungs 100 mm Hg *greater* than the external pressure is not unusual during such a maximum expiration.

The lungs are never emptied completely, however. The space around the lungs is maintained at a slightly lower pressure than are the lungs themselves, causing the lungs to be kept partially inflated by the higher pressure within them. If a lung, the diaphragm, or the chest wall is punctured, allowing the two pressures to equalize, the lung collapses. Sometimes a lung will be collapsed intentionally to give it time to heal. Closing the opening reinflates the lung.

People were breathing long before Boyle formulated his law, but it is satisfying to understand how the process works. We get more than just satisfaction out of science, however. An understanding of the pressure–volume relationship has enabled us to keep people alive. When paralysis prevents people from being able to breathe, they can be kept alive by artificial respirators. The iron lung, which kept many polio victims alive during the 1950s, is a sealed chamber connected to a compressor and bellows (Figure 5.9a). The pressure in the chamber is varied rhythmically. When the bellows is moved out, the pressure in the chamber is reduced. The pressure around the nose and mouth (outside the tank) is greater than the pressure on the chest (inside the tank), so air flows in and fills the lungs. When the bellows is moved in, pressure in the tank increases and air is expelled from the lungs.

The iron lung, designed to enclose the patient completely (except for the head), was cumbersome and uncomfortable. It has been replaced by respirators that enclose the chest only (Figure 5.9b). In fact, the whole area of respiratory therapy has become far more complicated in recent years. Specialists in this area are rapidly becoming an indispensable part of the medical team. In many respects, the increasing sophistication of respiratory therapy resembles the earlier development of anesthesiology.

5.5 Charles's Law: Volume-Temperature Relationship

In 1787, the French physicist J. A. C. Charles studied the relationship between the volume and temperature of gases. When a gas is cooled at constant pressure, the volume of the gas decreases. When a gas is heated, the volume increases. Aha!—we say, as perhaps did Charles— what we have here is another tidy little law. *Temperature and volume vary directly*; that is, they rise or fall together. But this law required a bit more thought. If 1.0 L of gas is heated from 100 °C to 200 °C at constant pressure, the volume does not double but only increases to about 1.3 L. The law appears not to be as tidy as we had hoped.

Remember how temperature scales were defined? While zero pressure or zero volume really means zero—that is, there is zero (or no) pressure or volume to be measured, no matter what units you are using—zero degrees means only the freezing point of water (0 °C) or the temperature of a very cold mixture of salt and ice (0 °F). The zero points on these temperature scales were arbitrarily set.

What Charles did was to note a trend in the variation of volume with temperature. If you plot the change in temperature against the change in volume on a graph (Figure 5.10), you can extend the line, in theory,

> Temperature is directly proportional to volume.
>
> $T \propto V$

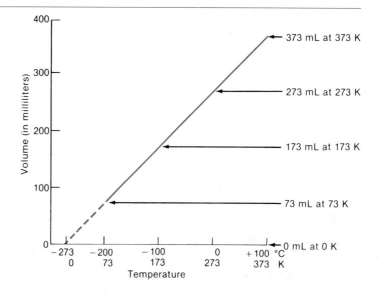

FIGURE 5.10
The volume–temperature relationship for gases (with pressure constant).

5.5 Charles's Law: Volume–Temperature Relationship

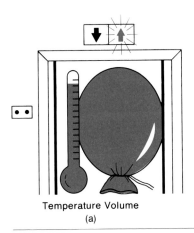
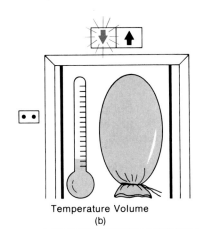

FIGURE 5.11
Temperature and volume are like passengers on an elevator. (a) When one goes up, the other must go up as well. (b) When one goes down, the other goes down too. [From an idea by Cindy Hill.]

to the point at which the volume of the gas hits zero. Before a gas ever reaches this point it liquefies, so this is an exercise for one's imagination. From the graph you can determine the temperature at which the volume of the imaginary gas would reach zero. If a different gas were used, the curve might have a different slope, but it too would point to the same temperature. That temperature turns out to be -273 °C. This value was made the zero point on a new temperature scale, the absolute temperature scale whose unit is the kelvin (K).

We can compare changes in the volume of a gas to changes in temperature on the absolute scale. When we do this, we find that a doubling of the temperature (from 100 K to 200 K, for example) resulted in a doubling in the volume (from 1 L to 2 L, for example). Here is the simple relationship we are looking for.

Charles's law states that the volume of a gas is directly proportional to its temperature (on the absolute scale, as indicated by the capital T) if pressure is held constant. Mathematically this relationship is expressed as

$$\frac{V}{T} = k \quad \text{or} \quad \frac{V_1}{V_2} = \frac{T_1}{T_2}$$

To keep V/T equal to a constant value, when the volume increases, the temperature must also increase. When volume decreases, temperature must decrease (Figure 5.11). Again let us emphasize that the values for temperature must be in kelvins, not degrees Celsius (or Fahrenheit). In the case of temperature (unlike that of volume or pressure), the units of measurement for gas law calculations are restricted.

The kinetic molecular model accounts for the volume–temperature relationship. If we heat a gas, we supply it with energy, and the particles of the gas begin moving faster. These speedier particles strike the walls of their container harder and more often. If the pressure is to remain constant, the volume of the container must increase. As the volume increases, the distance between the walls increases. The particles must spend more time traveling from one wall to another. The increase in volume also results in an increase in the area of the walls. Therefore, each unit of area will be hit less often. The pressure exerted by slower (low-temperature) particles confined within the smaller volume is the same as the pressure of the faster-moving (high-temperature) particles contained within the larger volume.

EXAMPLE 5.4

A balloon indoors, where the temperature is 27 °C, has a volume of 2.0 L. What will its volume be outside where the temperature is −23 °C? Assume no change in pressure.

First, convert all temperatures to the absolute scale. The initial temperature is

$$27 + 273 = 300 \text{ K}$$

The final temperature is

$$-23 + 273 = 250 \text{ K}$$

Initial	Final	Change
$t_1 = 27\,°C$	$t_2 = -23\,°C$	↓
$T_1 = 300\text{ K}$	$T_2 = 250\text{ K}$	↓
$V_1 = 2.0\text{ L}$	$V_2 = ?$	↓

As the temperature decreases, the volume must also decrease. Multiply the original volume by a fraction made up of the two temperatures. The fraction must be less than 1 because the volume must decrease.

$$2.0 \text{ L} \times \frac{250 \text{ K}}{300 \text{ K}} = 1.7 \text{ L}$$

EXAMPLE 5.5

What would be the final volume of the balloon in Example 5.4 if it were measured where the temperature was 47 °C? Assume no change in pressure.

The initial temperature is

$$27 + 273 = 300 \text{ K}$$

The final temperature is

$$47 + 273 = 320 \text{ K}$$

Initial	Final	Change
$t_1 = 27\ °C$	$t_2 = 47\ °C$	↑
$T_1 = 300\ K$	$T_2 = 320\ K$	↑
$V_1 = 2.0\ L$	$V_2 = ?$	↑

In this case, since the temperature increases, the volume must also increase.

$$2.0 \text{ L} \times \frac{320 \text{ K}}{300 \text{ K}} = 2.1 \text{ L}$$

5.6 Gay-Lussac's Law: Pressure–Temperature Relationship

At about the same time that Charles was doing the experiments on which his law is based, Joseph Gay-Lussac, a French chemist, was also working with gases. His experiments showed that, *at constant volume, pressure is directly proportional to absolute temperature.* Mathematically, the law of Gay-Lussac is expressed as

Pressure is directly proportional to temperature.

$$P \propto T$$

$$\frac{P}{T} = k$$

or in extended form,

$$\frac{P_1}{T_1} = \frac{P_2}{T_2}$$

The relationship is shown graphically in Figure 5.12.

At constant volume, as the temperature increases, so does the pressure. At the higher temperature, particles move faster and hit the walls of their container harder and more often. Thus, the pressure increases.

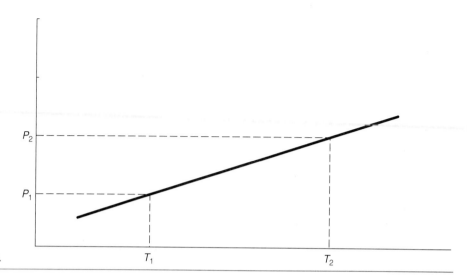

FIGURE 5.12
The pressure–temperature relationship.

EXAMPLE 5.6

Automobile tires are filled to a pressure of 30 psi at 20 °C. The tires become hot from high-speed driving and reach a temperature of 50 °C. What will the pressure be at that temperature? Assume that the volume remains constant.

The initial temperature is

$$20 + 273 = 293 \text{ K}$$

The final temperature is

$$50 + 273 = 323 \text{ K}$$

Initial	Final	Change
$t_1 = 20$ °C	$t_2 = 50$ °C	↑
$T_1 = 293$ K	$T_2 = 323$ K	↑
$P_1 = 30$ psi	$P_2 = ?$	↑

The temperature increases, so the pressure must also increase.

$$30 \text{ psi} \times \frac{323 \text{ K}}{293 \text{ K}} = 33 \text{ psi}$$

5.6 Gay-Lussac's Law: Pressure–Temperature Relationship

EXAMPLE 5.7

At 127 °C the pressure in an autoclave (Figure 5.13) is 2.0 atm. What will the pressure be at 27 °C?

The initial temperature is

$$127 + 273 = 400 \text{ K}$$

The final temperature is

$$27 + 273 = 300 \text{ K}$$

Initial	Final	Change
$t_1 = 127 \,°\text{C}$	$t_2 = 27 \,°\text{C}$	↓
$T_1 = 400 \text{ K}$	$T_2 = 300 \text{ K}$	↓
$P_1 = 2.0 \text{ atm}$	$P_2 = ?$	↓

The temperature decreases; therefore, the pressure must also decrease.

$$2.0 \text{ atm} \times \frac{300 \text{ K}}{400 \text{ K}} = 1.5 \text{ atm}$$

FIGURE 5.13
A hospital autoclave uses the principle of Gay-Lussac's law to achieve the high temperatures needed to sterilize surgical instruments. The trend today is toward disposable equipment that does not require resterilization. [Photo by Raymond C. Carballada, Department of Medical Photography, Geisinger Medical Center.]

If the amount of gas were included as a fourth parameter, we would discover additional laws such as "the volume is proportional to the amount of gas." Except for Boyle's law, all relationships are *direct* ones; in other words, as one parameter increases, so does the other.

5.7 Putting It All Together: The Combined Gas Law

We have seen that the volume of a gas varies with both temperature and pressure. If we want to compare two samples, we have to do so at identical temperatures and pressures for the comparison to be meaningful. We can use a combination of Boyle's law and Charles's law to calculate the volume of a gas at any given temperature and pressure provided we know its volume at any other temperature and pressure.

EXAMPLE 5.8

A balloon is partially filled with helium on the ground at 27 °C and 740 mm Hg pressure. Its volume is 10 m³. What would the volume be at a higher altitude where the pressure is 370 mm Hg and the temperature is −23 °C?

The temperature decreases from 27 °C to −23 °C, or from

$$27 + 273 = 300 \text{ K}$$

to

$$-23 + 273 = 250 \text{ K}$$

Initial	Final	Change
$t_1 = 27$ °C	$t_2 = -23$ °C	↓
$T_1 = 300$ K	$T_2 = 250$ K	↓
$P_1 = 740$ mm Hg	$P_2 = 370$ mm Hg	↓
$V_1 = 10$ m	$V_2 = ?$	

To work this problem, you need the original volume and the two temperatures and two pressures. The pressure decreases from 740 to 370 mm Hg. This will cause an increase in the volume of the gas.

$$10 \text{ m}^3 \times \frac{740 \text{ mm Hg}}{370 \text{ mm Hg}} = 20 \text{ m}^3$$

5.7 Putting It All Together: The Combined Gas Law

A temperature decrease will cause a decrease in the volume.

$$20 \text{ m}^3 \times \frac{250 \text{ K}}{300 \text{ K}} = 17 \text{ m}^3$$

You can do these two calculations in any order. The final answer is the same. It is usual to combine both changes in one equation (particularly in this day of the hand-held calculator).

$$10 \text{ m}^3 \times \frac{740 \text{ mm Hg}}{370 \text{ mm Hg}} \times \frac{250 \text{ K}}{300 \text{ K}} = 17 \text{ m}^3$$

In the above example, we have combined Boyle's law, $V_1/V_2 = P_2/P_1$, with Charles's law $V_1/V_2 = T_1/T_2$, to obtain the **combined gas law**:

$$\frac{V_1}{V_2} = \frac{P_2}{P_1} \times \frac{T_1}{T_2}$$

Because gases are so sensitive to changes in temperature and pressure, chemists have found it convenient to define *s*tandard conditions of *t*emperature and *p*ressure (referred to as **STP**) as 0 °C (273 K) and 1 atm (760 mm Hg).

Standard condition of temperature and pressure (STP)

$T = 273 \, K$

$P = 1 \, atm$

EXAMPLE 5.9

What is the volume at STP of a sample of carbon dioxide whose volume at 25 °C and 4.0 atm is 10 L?

Initial	Final	Change
$t_1 = 25\,°C$		
$T_1 = 298 \text{ K}$	$T_2 = 273 \text{ K}$	↓
$P_1 = 4.0 \text{ atm}$	$P_2 = 1 \text{ atm}$	↓
$V_1 = 10 \text{ L}$	$V_2 = ?$	

The pressure decreases from 4.0 atm to the standard pressure of 1.0 atm; this should cause a volume increase.

$$10 \text{ L} \times \frac{4.0 \text{ atm}}{1.0 \text{ atm}} = 40 \text{ L}$$

The temperature decrease from 298 K (25 °C) to 273 K (0 °C) should be accompanied by a volume decrease.

$$40 \text{ L} \times \frac{273 \text{ K}}{298 \text{ K}} = 37 \text{ L}$$

or

$$10 \text{ L} \times \frac{4.0 \text{ atm}}{1.0 \text{ atm}} \times \frac{273 \text{ K}}{298 \text{ K}} = 37 \text{ L}$$

5.8 Dalton's Law of Partial Pressures

Although John Dalton is most renowned for his atomic theory (Section 2.1), Dalton had wide-ranging interests, including meteorology. In trying to understand the weather, he did a number of experiments on water vapor in the air. In one experiment, he found that if he added water at a certain pressure to dry air, the pressure exerted by the air would increase by an amount equal to the pressure of the water vapor. On the basis of this and other experiments, Dalton concluded that each of the gases in a mixture behaves independently of the other gases. Each gas exerts its own pressure. *The total pressure of the mixture is equal to the sum of the* **partial pressures** *exerted by the separate gases* (Figure 5.14).

Mathematically, **Dalton's law** is stated just the way you would expect it to be.

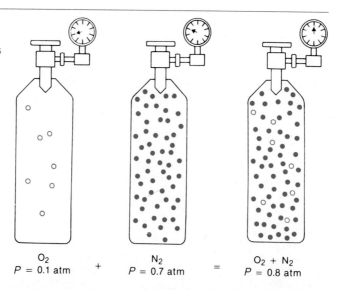

FIGURE 5.14
Dalton's law of partial pressures states that the pressure of a mixture of gases is equal to the sum of the pressures exerted by each of the gases.

O_2
$P = 0.1$ atm
+
N_2
$P = 0.7$ atm
=
$O_2 + N_2$
$P = 0.8$ atm

5.8 Dalton's Law of Partial Pressures

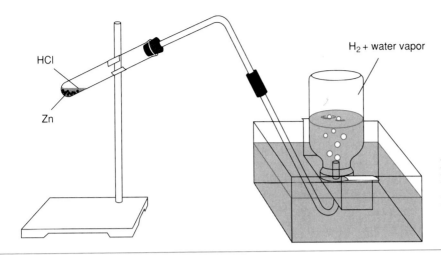

FIGURE 5.15
Diagram of an apparatus for the laboratory synthesis of hydrogen. Because it is collected over water, the hydrogen gas is saturated with water vapor.

$$P_{total} = P_1 + P_2 + P_3 + \cdots$$

The terms on the right side refer to the partial pressures of gases 1, 2, 3, and so on.

Gases such as oxygen, nitrogen, and hydrogen are nonpolar. They are only slightly soluble in water and are usually collected over water by the technique of displacement. Such gases always contain water vapor, and the total pressure in the collection vessel is that of the gas plus that of water vapor (Figure 5.15).

The vapor pressure of water depends upon the temperature of the water. The hotter the water, the higher the vapor pressure. If a gas is collected over water, we can make use of vapor pressure tables (such as Table 5.2) to calculate the pressure due to the gas alone. We need only subtract the vapor pressure of the water, as determined from the table, from the value for the total pressure within the collection vessel.

TABLE 5.2 Water Vapor Pressure at Various Temperatures

Temperature (°C)	Water Vapor Pressure (mm Hg)
0	5
10	9
20	18
30	32
40	55
50	93
60	149
70	234
80	355
90	526
100	760

EXAMPLE 5.10

Oxygen is collected over water at 20 °C. The pressure inside the jar is 740 mm Hg. What is the pressure due to oxygen alone?

$$P_{total} = P_{O_2} + P_{H_2O}$$

From Table 5.2 we find that the vapor pressure of water at 20 °C is 18 mm Hg. Since the total pressure is equal to 740 mm Hg, we have

$$740 \text{ mm Hg} = P_{O_2} + 18 \text{ mm Hg}$$

$$P_{O_2} = 740 - 18 = 722 \text{ mm Hg}$$

Humidity is a measure of the amount of water vapor in the air. **Relative humidity** compares the actual amount of water vapor in the air to the maximum amount the air could hold at the same temperature. If the temperature is 20 °C and the vapor pressure of water in the atmosphere is 12 mm Hg, then the relative humidity is

$$\frac{12 \text{ mm Hg}}{18 \text{ mm Hg}} \times 100 = 67\%$$

The 18 mm Hg in the denominator was obtained from Table 5.2. When the relative humidity is 100%, the air is saturated with water vapor. (*Note*: 100% relative humidity does not mean that the air is 100% water vapor, just that it is holding as much water as it can. At 20 °C and 100% relative humidity, only about 2 or 3 molecules in every 100 molecules of air are water.)

Since cool air can hold less water vapor than warm air, the atmosphere may become saturated as the temperature falls during the evening and early morning hours. The result is dew, water that has **condensed** from the air as the air became saturated.

The *heat index* relates temperature to relative humidity. Higher humidity inhibits evaporation of sweat, which normally cools the body, thus making it now feel hotter. This apparent temperature can be found from Figure 5.16.

Respiratory therapists must concern themselves with the humidity of the gases they administer to patients. Normally, as one breathes, the inspired air is saturated with moisture as it passes through the nose and respiratory passages. Oxygen as it comes from the tank is quite dry. If oxygen is being administered over a long period of time, it must be

FIGURE 5.16
The heat index.

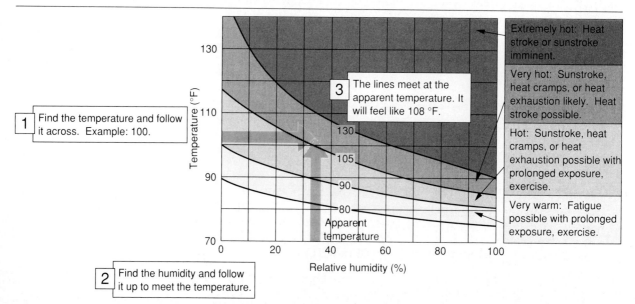

humidified to prevent it from irritating the mucous linings of the nasal passages and the lungs. If the oxygen or mixture of gases is conducted through the nose, the therapist may merely assist the normal body processes by imparting about 30% humidity to the inspired gases. If the breathing mixture is conducted directly to the trachea (bypassing the nose), the therapist saturates the gas mixture with water vapor.

The vapor pressure of liquids depends on the forces holding the liquid molecules to each other. Thus, ether molecules have much weaker molecule-to-molecule forces than water molecules. As a result, at a given temperature, the vapor pressure of ether is much higher than that of water. Ether reaches a vapor pressure of 760 mm Hg at 34.6 °C, which is its boiling point. Ethylene glycol (antifreeze) has a much stronger molecule–molecule interaction and therefore has a lower vapor pressure and a higher boiling point than water.

5.9 Henry's Law: Pressure–Solubility Relationship

In the 1760s, Joseph Priestley produced soda water by dissolving carbon dioxide gas in water. No doubt you have noticed the hissing sound and the formation of bubbles when you opened a bottle of soda (or beer or champagne, for that matter). Carbon dioxide is dissolved in the liquid, and the bottle is capped under pressure. William Henry, a close friend of John Dalton, spent a great deal of time studying the solubility of gases in liquids. In 1801, he summarized his findings in the law named for him. **Henry's law** states that *the solubility of a gas in a liquid at a given temperature is directly proportional to the pressure of the gas at the surface of the liquid.* To get back to the bottle of soda: when the bottle is capped under pressure, a certain amount of carbon dioxide dissolves in the liquid. When you open the bottle, the pressure is *reduced* (the hissing sound you hear is gas being released), and the solubility of the carbon dioxide is *reduced* (the bubbles of gas you notice are carbon dioxide escaping from solution).

The carbonated beverage industry is not the only group interested in Henry's law. Deep-sea divers can develop a condition called **the bends** (caisson disease) from nitrogen dissolved in the blood. Very little nitrogen dissolves in our blood at normal pressures. When a diver breathes air under greater pressure, appreciable amounts dissolve in the blood. The dissolved nitrogen has a narcotic effect, but it can be even more deadly. If the diver comes up too rapidly, the nitrogen comes out of solution as the pressure diminishes. Tiny bubbles of nitrogen form. These cause severe pain in the arms, legs, and joints, perhaps by disrupting nerve pathways. To prevent the bends, divers sometimes use a mixture of helium and oxygen rather than air. Very little of the helium dissolves, even under increased pressure, and the problem of bubble formation as the diver ascends is minimized.

Henry's law: solubility of a gas is directly proportional to the pressure of the gas.

The pressure–solubility relationship is also used in therapy. In cases of carbon monoxide poisoning (CO displaces oxygen from hemoglobin), the victim is placed in a hyperbaric (high-pressure) chamber. This chamber is a device that supplies oxygen at pressures of 3 or 4 atm. More oxygen is forced into the tissues at these pressures to compensate for the lack of oxygen that accompanies carbon monoxide poisoning. At atmospheric pressure, about 0.3 cm^3 of O_2 dissolves in 100 cm^3 of blood. In a hyperbaric chamber filled with pure oxygen at 4 atm, about 7 cm^3 of O_2 dissolves in 100 cm^3 of blood.

Hyperbaric chambers are also used to treat infections by anaerobic bacteria (bacteria that live without air). Gangrene is one such disease. The organisms that cause gangrene cannot survive in an oxygenated atmosphere. If sufficient oxygen can be forced into the diseased tissues, the infection can be arrested.

5.10 Partial Pressure and Respiration

When we breathe in, the inspired air becomes moistened and warmed to our body temperature of 37 °C. The air is drawn into our lungs, where it enters a highly branched system of tubes that end in tiny air sacs called alveoli. These thin-walled pouches are surrounded by blood vessels that are part of a circulatory system serving every cell in the body (Figure 5.17).

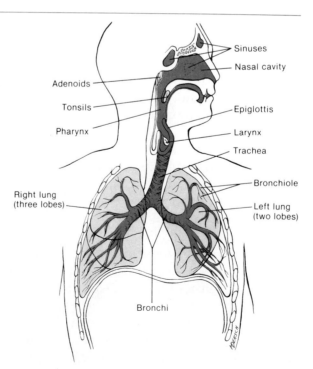

FIGURE 5.17
The respiratory system, showing the route of air through the nose, pharynx (throat), larynx (voice box), and trachea (windpipe) into the bronchi and bronchioles (bronchial tubes) and ending in the alveoli (air sacs).

5.10 Partial Pressure and Respiration

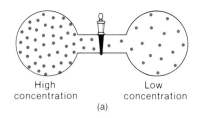

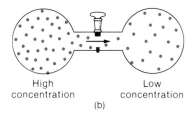

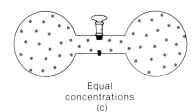

FIGURE 5.18
A gas (air in the lungs, for example) flows from an area of high concentration to an area of low concentration. (a) With the stopcock closed no flow is possible. (b) With the stopcock open there is a net flow of gases from the area of higher concentration on the left to the area of lower concentration on the right. (c) With the stopcock open there is no net flow of gas when the concentrations are equal.

Inspired air is rich in oxygen (P_{O_2} = 150 mm Hg) and poor in carbon dioxide (P_{CO_2} = 0.2 mm Hg). The fluid in our cells is poor in oxygen (P_{O_2} = 6 mm Hg) and rich in carbon dioxide (P_{CO_2} = 50 mm Hg). Our cells use up oxygen in metabolic reactions designed to produce energy. Carbon dioxide accumulates in the cells as a waste product of these reactions. To maintain life, we must transfer the oxygen in the inspired air to our cells. At the same time, the carbon dioxide waste in our cells must be transferred to our lungs and then exhaled to the atmosphere. The transfer of both gases occurs through the process of **diffusion**. In diffusion, gases flow from regions of higher concentration to regions of lower concentration (Figure 5.18). In our bodies, oxygen makes its way to the cells through a pressure gradient, that is, in a series of steps in which oxygen diffuses from areas where its concentration is higher into areas where its concentration is lower. By the same method, carbon dioxide moves in the opposite direction. It makes its way from the cells, where its partial pressure is high, to the atmosphere, where its partial pressure is low. Figure 5.19 shows the steps in the gradient for both gases. Thus, given the mechanical action of the chest and diaphragm (Section 5.4) to get air into and out of the lungs, respiration is all downhill, so to speak, as far as the gases are concerned.

Diffusion: the flow of gas from a region of higher concentration to a region of lower concentration.

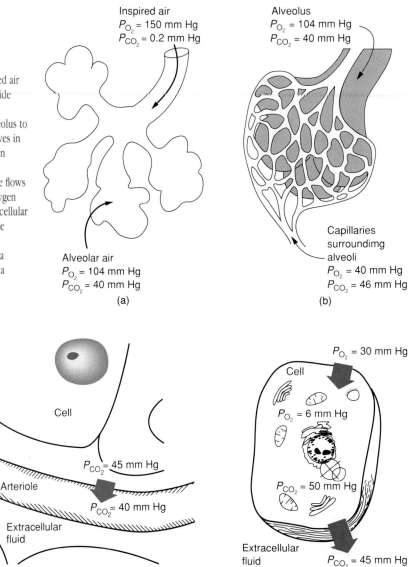

FIGURE 5.19
(a) Oxygen flows from the inspired air into the alveolar air. Carbon dioxide flows in the opposite direction.
(b) Oxygen diffuses from the alveolus to the capillary. Carbon dioxide moves in the opposite direction. (c) Oxygen diffuses from an arteriole into the extracellular fluid. Carbon dioxide flows in the opposite direction. (d) Oxygen diffuses into a cell from the extracellular fluid. Carbon dioxide moves in the opposite direction.

In each case the flow is from a region of high partial pressure to a region of low partial pressure.

Normally, the carbon dioxide level in the blood (not the oxygen level) acts as a trigger for the breathing process. To oversimplify, when carbon dioxide levels build up, we take a breath; if they get too low, we do not. Thus, one of the concerns of a therapist is the partial pressure of carbon dioxide in the blood. Under certain unusual conditions, it is possible for the level of carbon dioxide to fall so low that it fails to trigger the breathing mechanism. A person with access to a plentiful supply of oxygen suffocates because he or she simply stops breathing.

TABLE 5.3 Ten Compressed Gases Used in Medicine

Gas	Chemical Formula	Use
Air	N_2 and O_2	Life support
Carbon dioxide	CO_2	Laboratory tests; lung function tests
Carbon dioxide–oxygen mixtures	CO_2 and O_2	Diagnosis, respiratory therapy
Cyclopropane	C_3H_6	Anesthetic
Helium	He	Laboratory analyses
Helium–oxygen mixtures	He and O_2	Respiratory therapy, diagnostic tests
Nitrogen	N_2	Diagnostic testing; respiratory therapy
Nitrous oxide	N_2O	Anesthetic
Oxygen	O_2	Life support, medical emergencies, adjunct to anesthetics
Oxygen–nitrogen mixtures	O_2 and N_2	Treatment of obstructive lung diseases

A number of commercially supplied gases are used in medicine, the majority in respiratory therapy. Table 5.3 lists these gases and some of their uses.

CHAPTER SUMMARY

The atmosphere is a thin layer of gases at the surface of the Earth consisting of nitrogen (78%), oxygen (21%), argon (1%), carbon dioxide (<1%), and varying amounts of water vapor.

The kinetic molecular theory pictures gases as collections of rapidly and independently moving particles. Collisions between particles or between a particle and the wall of a container are perfectly elastic, meaning there is no net loss of kinetic energy in the collision. <u>Temperature</u> is a measure of the average kinetic energy of the particles, and <u>pressure</u> is a measure of the impacts of the particles against a surface. A <u>barometer</u> is an instrument used to measure pressure, particularly the pressure exerted by the gases in the atmosphere.

Units of pressure include the atmosphere (atm), millimeters of mercury (mm Hg), inches of mercury (in. Hg), centimeters of water (cm H_2O), torr, pounds per square inch (psi), and pascal (Pa).

$$1.00 \text{ atm} = 760 \text{ mm Hg}$$
$$= 760 \text{ torr}$$
$$= 29.9 \text{ in. Hg}$$
$$= 14.7 \text{ psi}$$
$$= 1.01 \times 10^5 \text{ Pa}$$

$$1.0 \text{ cm } H_2O = 0.74 \text{ mm Hg}$$

The behavior of gases is quantitatively described in a series of scientific laws. <u>Boyle's law</u> indicates that, at constant temperature, the volume and the pressure of a gas vary inversely ($VP = k$). <u>Charles's law</u> states that, at constant pressure, the volume and temperature of a gas vary directly ($V/T = k$). According to Gay-Lussac's law, with the volume held constant, the pressure and temperature of a gas vary directly ($P/T = k$). Application of the combined gas law allows us to detail the changes that occur in gases when two of the three variables (volume, temperature, and pressure) are changed. In applying the gas laws, temperature must always be in kelvins.

Boyle's law explains the mechanics of breathing, since inspiration of air results from an increase in the volume of the lungs (and thus a decrease in the pressure within the lungs). Air flows from the region of higher pressure to a region of lower pressure. The reverse process (decrease in lung volume with consequent increase in pressure) occurs in expiration.

Dalton's law specifically describes the behavior of mixtures of gases. According to this law, the total pressure of a mixture of gases is equal to the sum of the pressures of the individual gases considered separately ($P_{total} = P_1 + P_2 + P_3 + \cdots$). The pressure of an individual gas in a mixture is called its partial pressure.

Relative humidity compares the partial pressure of water vapor in an air sample to the partial pressure of water vapor in a saturated air sample at the same temperature. Air as it normally reaches the lungs is saturated with water vapor. Therefore, the humidity of gases used in respiratory therapy is of concern in medicine.

Henry's law describes the behavior of gases in solution. The solubility of a gas in a liquid is directly proportional to the pressure of the gas above the liquid. The bends is a condition appearing in deep-sea divers when nitrogen, which dissolves in their blood under high pressure, is released in the ascent to lower pressures nearer the surface. The high pressures used in hyperbaric chambers can be used to reverse the process. Hyperbaric chambers are also used to increase the level of dissolved oxygen in body tissues for the treatment of carbon monoxide poisoning or gangrene.

The physiological process of respiration can be explained by the diffusion of gases from high-pressure areas to low-pressure areas. The partial pressure of oxygen is highest in the alveoli of the lungs and lowest in the cells. Therefore, oxygen diffuses from the alveoli into the blood and then into the cells. The partial pressure of carbon dioxide (a waste product of metabolism) is highest in the cells and lowest in the air in the lungs. Carbon dioxide diffuses from cells to blood to lungs, where it is exhausted to the atmosphere. An increase in the carbon dioxide level in the blood normally triggers the breathing reflex.

CHAPTER QUIZ

1. In addition to oxygen, the primary constituent of air is
   a. Ar   b. N_2 c. CO_2 d. CO

2. The kinetic molecular theory treats gases as particles that
   a. are in motion
   b. are connected to each other
   c. have polar attraction
   d. are incompressible

3. The STP pressure is
   a. 10 m water   b. 76 mm Hg
   c. 760 cm Hg   d. 1 atm

4. The pressure of a liter of gas is 1 atm at room temperature. What will it be if the volume is tripled at the same temperature?
   a. 3 atm   b. 1 atm   c. $\frac{1}{3}$ atm   d. 2.0 atm

5. If the temperature of 173 mL of a gas is 173 K, what would be the extrapolated temperature reading at a hypothetical 0 mL?
   a. 273 K   b. 173×2 K
   c. 0 °C   d. -273 °C

6. Find the pressure of a gas, initially at 2.0 atm and 37 °C, after the temperature is changed to 27 °C.
   a. 1.5 atm   b. 1.9 atm
   c. 2.1 atm   d. 2.7 atm

7. What will be the volume of a gas that occupies 2.0 L at 1.7 atm and 15 °C if conditions are changed to STP?
   a. 1.1 L   b. 3.2 L   c. 3.6 L   d. 5.1 L

8. What is the partial pressure of oxygen in air at 760 mm Hg? (Use the data in Table 5.1.)
   a. 600 mm   b. 550 mm
   c. 210 mm   d. 160 mm

9. It was found that on a hot 40 °C summer day the relative humidity was 92%. What was the water vapor pressure?
   a. 40 mm
   b. 51 mm
   c. 55 mm
   d. 92 mm
10. Emergency treatment of patients in shock sometimes includes a paper bag over the nose and mouth. The purpose is to
    a. avoid loss of oxygen
    b. increase the partial pressure of CO_2
    c. increases the partial pressure of O_2
    d. encourage breathing

PROBLEMS

1. Define or explain the following terms.
   a. kinetic molecular theory
   b. mm Hg
   c. cm H_2O
   d. torr
   e. atmosphere
   f. pascal
   g. Boyle's law
   h. Charles's law
   i. Gay-Lussac's law
   j. Dalton's law
   k. Henry's law
   l. STP
   m. partial pressure
   n. diffusion
   o. relative humidity
   p. tidal volume
   q. vital capacity

5.1 Air

2. List the four major gases found in dry air. Which of these are important in respiration?

5.2 Kinetic Molecular Theory

3. According to the kinetic molecular theory, what change in temperature is occurring if the particles of a gas begin to move more slowly, on the average?

4. According to the kinetic molecular theory, what change in pressure occurs when particles of a gas strike the walls of the container less often?
5. Container A has twice the volume but holds twice as many gas particles as container B. Using the kinetic molecular theory as the basis for your judgment, compare the pressures in the two containers.
6. Density is defined as mass per unit volume. For each of the following, indicate which sample you would expect to exhibit the higher density. (*Hint*: How many particles are there in equivalent volumes of the sample?)
   a. Containers A and B have the same volume and are at the same temperature, but the gas in container A is at a higher pressure.
   b. Containers A and B are at the same pressure and temperature, but the volume of container A is greater than that of container B.
   c. Containers A and B are at the same pressure and volume, but the temperature of the gas in container A is higher. (*Hint*: A *hot* air balloon floats in the air.)

5.3 Atmospheric Pressure

7. Carry out the following conversions.
   a. 2.0 atm to mm Hg
   b. 0.50 atm to torr
   c. 0.0030 atm to Pa
8. Carry out the following conversions.
   a. 76 torr to atm
   b. 320 torr to mm Hg
   c. 0.076 torr to Pa
9. Carry out the following conversions.
   a. 10 atm to psi
   b. 10 mm Hg to cm H_2O
   c. 3.0 in. Hg to atm
10. Why is atmospheric pressure greater at sea level than on the top of a high mountain?
11. Explain how a mercury barometer works.

5.4 Boyle's Law

12. A tank contains 500 mL of helium at 1500 torr. What volume will the helium occupy at 750 torr, assuming no temperature change?

13. A hyperbaric chamber with a volume of 10 m^3 operates at an internal pressure of 4.0 atm. What volume would the air inside the chamber occupy at normal atmospheric pressure?
14. Oxygen used in respiratory therapy is stored under a pressure of 2200 psi in gas cylinders with a volume of 60 L.
    a. What volume would the gas occupy at normal atmospheric pressure?
    b. If oxygen flow to the patient is adjusted to 8.0 L per min, how long will the tank of gas last?
15. The pressure within a balloon with a volume of 2.5 L is 1.0 atm. If the volume of the balloon increases to 7.5 L, what will be the final pressure within the balloon?
16. During inhalation, does the chest cavity expand or contract? Is the pressure inside the lungs decreased or increased? What happens during exhalation?
17. The cough reflex is designed to keep air passages clear. Typically, when a person coughs, he or she first inhales about 2.0 L of air. The epiglottis and the vocal cords then shut, trapping the air in the lungs. The air in the lungs is then compressed to a volume of about 1.7 L by the action of the diaphragm and chest muscles. The sudden opening of the epiglottis and vocal cords releases this air explosively. Just prior to the release, what is the pressure of the gas inside the lungs?

5.5 Charles's Law

18. Carry out the following conversions.
    a. 37 °C to K   b. 420 K to °C
    c. −82 °C to K
19. A gas at a temperature of 100 K occupies a volume of 100 mL. What will the volume be at a temperature of 10 K assuming no change in pressure?
20. A balloon is filled with helium. Its volume is 5.0 L at 27 °C. What will be its volume at −73 °C assuming no pressure change?

5.6 Gay-Lussac's Law

21. An automobile tire is inflated to 30 psi at 27 °C. What will be the pressure at 127 °C assuming no volume change?
22. A gas at a temperature of 300 K and a pressure of 1.0 atm is cooled to 250 K. Assuming no change in volume, calculate the change in pressure.
23. A sealed can with an internal pressure of 720 mm Hg at 25 °C is thrown into an incinerator operating at 750 °C. What will the pressure be inside the heated can, assuming the container remains intact during incineration?

5.7 Combined Gas Law

24. If a gas occupies 4.0 L at a temperature of 25 °C and a pressure of 2.0 atm, what volume will it occupy at a temperature of 200 °C and a pressure of 1.0 atm?
25. If a gas occupies 2.5 m^3 at a temperature of −15 °C and a pressure of 190 mm Hg, what volume will it occupy at a temperature of 25 °C and 1140 mm Hg pressure?
26. What volume will 500 mL of a gas, measured at 27 °C and 720 mm Hg, occupy at STP?
27. If a gas has a volume of 55 cc at STP, what volume will it occupy at 100 °C and 76 torr?
28. If a gas has a volume of 732 mL at 760 torr and 25 °C, what volume will it occupy at 1.00 atm and 298 K?
29. What effect will the following changes have on the volume of a gas?
    a. an increase in pressure
    b. a decrease in temperature
    c. a decrease in pressure coupled with an increase in temperature
30. What effect will the following changes have on the pressure of a gas?
    a. an increase in temperature
    b. a decrease in volume
    c. an increase in temperature coupled with a decrease in volume

5.8 Dalton's Law

31. Referring to Table 5.2, prepare a plot of temperature versus water vapor pressure.
32. A container holds oxygen at a partial pressure of 0.25 atm, nitrogen at a partial pressure of

0.50 atm, and helium at a partial pressure of 0.20 atm. What is the pressure inside the container?

33. A container is filled with equal numbers of nitrogen, oxygen, and carbon dioxide molecules. The total pressure in the container is 750 mm Hg. What is the partial pressure of nitrogen in the container?

34. Oxygen is collected over water. If the temperature is 30 °C and the collected sample has a pressure of 740 mm Hg, what is the partial pressure of the oxygen in the container? (You may refer to Table 5.2.)

35. The pressure of the atmosphere on the surface of Venus is about 100 atm. Carbon dioxide makes up about 97% of the atmospheric gases. What is the partial pressure of carbon dioxide in the atmosphere of Venus?

36. Atmospheric pressure on the surface of Mars is about 6.0 torr. The partial pressure of carbon dioxide is 5.7 torr. What percent of the Martian atmosphere is carbon dioxide?

37. When air is inspired it becomes fully saturated with water vapor as it passes through the trachea on its way to the lungs. What is the approximate partial pressure of water vapor in the air in the alveoli? (*Hint*: At what temperature will the air be?)

5.9 Henry's Law

38. If the P_{H_2O} in air is 12 mm Hg on a day when the temperature is 20 °C, what is the relative humidity?

5.10 Partial Pressures and Respiration

39. Two flasks are connected. Flask A contains only oxygen at a pressure of 460 mm Hg. Flask B has oxygen at a partial pressure of 320 mm Hg and nitrogen at a partial pressure of 240 mm Hg.
    a. Which direction will the net flow of oxygen take?
    b. Which direction will the net flow of nitrogen take?

40. In which net direction, cells to lungs or lungs to cells, does oxygen move? What about carbon dioxide?

41. Why does oxygen flow from the alveoli to the pulmonary capillaries? Why does carbon dioxide flow in the reverse direction?

6
CHEMICAL REACTIONS

DID YOU KNOW THAT...

- chemical reactions involve electron rearrangements?
- all life processes are based on chemical reactions?
- in a reaction, reagents interact in whole-number units, not masses?
- the mole concept is a tool to help us count units by weighing masses?
- no matter what size, type, or shape gas molecules are, they will always fill a given container to the same extent?
- our bodies burn chemicals to obtain energy?
- a catalyst affects how fast a reaction goes but not how far?
- enzymes are catalysts?
- a chemical equilibrium can be way "off center"?
- oxidation can take place without oxygen?
- life and energy ultimately originate from the sun?
- European cities do not chlorinate their water?

A reaction is a change. In a nuclear reaction, changes occur in the nuclei of atoms. Chemical reactions involve rearrangements in the outer electrons of atoms. We first introduced the concept of a chemical reaction in a discussion of bond formation (Chapter 3).

In this chapter, we shall give chemical reactions our full attention. For the most part we will deal with simple nonliving systems. The principles developed for such systems are equally valid when applied to the complex chemical processes in living cells. Our study of the basic principles governing chemical reactions will include a consideration not only of the changes that occur in chemical composition but also of the changes in energy that accompany chemical reactions.

Life can be regarded as a balance between two complementary chemical processes—*oxidation* and *reduction*. Oxidation–reduction reactions occurring in green plants provide us with energy-rich compounds. Oxidation–reduction reactions also enable us to use the energy stored in these substances for our survival. We will conclude this chapter with a consideration of this important class of chemical reactions.

6.1 THE CHEMICAL EQUATION

Carbon reacts with oxygen to form carbon dioxide. In chemical shorthand this reaction is written

$$C + O_2 \longrightarrow CO_2$$

Reactants → products

The plus sign (+) indicates addition of carbon to oxygen (or vice versa) or a mixing of the two in some manner. The arrow (→) is often read "yields." Substances to the left of the arrow are **reactants** or **starting materials**. Those to the right are the **products of the reaction**. The conventions here are like those we used in writing nuclear equations (Chapter 4).

Chemical equations have meaning on the atomic and molecular level. The equation

$$C + O_2 \longrightarrow CO_2$$

means that one atom of carbon (C) reacts with one molecule of oxygen (O_2) to produce one molecule of carbon dioxide (CO_2). In other words, the reaction time goes from left (start) to right (end).

Not all chemical reactions are so simply represented. Hydrogen reacts with oxygen to form water. We can write this reaction as

$$H_2 + O_2 \longrightarrow H_2O \quad \text{(not balanced)}$$

6.1 The Chemical Equation

This representation, however, is not consistent with the law of conservation of matter. The **law of conservation of matter** states that matter is neither created nor destroyed in chemical reactions. Two oxygen atoms are shown among the reactants (as O_2) and only one among the products (in H_2O). For the equation to correctly represent what happens chemically, it must be *balanced*. To balance the oxygen atoms we need to place the coefficient 2 in front of the formula for water.

Matter is neither created nor destroyed.

$$H_2 + O_2 \longrightarrow 2\,H_2O \quad \text{(not balanced)}$$

This coefficient means that there are *two* molecules of water involved. A coefficient of 1 is understood where no other number appears. A coefficient preceding a formula is an instruction to multiply everything in the formula by that number. In the second equation, the coefficient 2 not only increases the number of oxygen atoms to two, but also increases the number of hydrogen atoms to four. But the equation is still not balanced. In taking care of the oxygen, we have unbalanced the hydrogen. To balance hydrogen, we place the coefficient 2 in front of the H_2.

$$2\,H_2 + O_2 \longrightarrow 2\,H_2O \quad \text{(balanced)}$$

Equations are balanced by adjusting coefficients only.

Now there are enough hydrogen atoms on the left. In fact, there are four hydrogen atoms and two oxygen atoms on each side of the equation. Atoms are conserved: the equation is balanced (Figure 6.1).

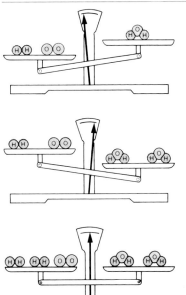

FIGURE 6.1
To balance the equation for the reaction of hydrogen with oxygen that forms water the same number of each kind of atom must appear on each side (atoms are conserved). When the equation is balanced, there are four hydrogen atoms and two oxygen atoms on each side.

We have used a chemical bookkeeping system to help us balance the equation.

Note that we could *not* balance the equation by changing the subscript for oxygen in water.

$$H_2 + O_2 \longrightarrow H_2O_2 \quad \text{(not correct)}$$

The equation would be balanced, but it would not mean "hydrogen reacts with oxygen to form *water*." The formula H_2O_2 represents *hydrogen peroxide*, which is a different compound. We cannot change the compounds in a chemical equation merely for the convenience of balancing chemical equations. Once products and reagents are defined, we do not change subscripts.

EXAMPLE 6.1

Balance the following equation.

$$N_2 + H_2 \longrightarrow NH_3 \quad \text{(not balanced)}$$

For this sort of problem we will again find the concept of the least common multiple useful. We will balance the hydrogen first. There are two hydrogen atoms on the left and three on the right. The least common multiple of 3 and 2 is 6. Six will be the smallest number of hydrogen atoms that can be evenly converted from reactants to products. Thus, we need three molecules of H_2 and two of NH_3.

$$N_2 + 3\,H_2 \longrightarrow 2\,NH_3 \quad \text{(balanced)}$$

We have balanced the hydrogen atoms and, in the process, the nitrogen atoms as well. There are two nitrogen atoms on the left and two on the right. The entire equation is balanced! (See Figure 6.2.)

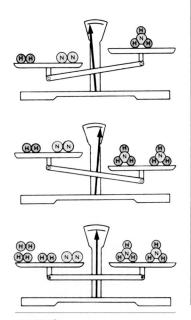

FIGURE 6.2
To balance the equation for the reaction of nitrogen with hydrogen that forms ammonia, the same number of each kind of atom must appear on each side of the equation. When the equation is balanced, there are two nitrogen atoms and six hydrogen atoms on each side.

EXAMPLE 6.2

Balance the following equation.

$$Fe + O_2 \longrightarrow Fe_2O_3 \quad \text{(not balanced)}$$

Begin by balancing the oxygen atoms. The least common multiple is 6. We need three molecules of O_2 and two of Fe_2O_3.

6.1 The Chemical Equation

$$Fe + 3\,O_2 \longrightarrow 2\,Fe_2O_3 \quad \text{(not balanced)}$$

We now have four atoms of iron on the right side. We can get four on the left by placing the coefficient 4 in front of Fe.

$$4\,Fe + 3\,O_2 \longrightarrow 2\,Fe_2O_3 \quad \text{(balanced)}$$

The equation is now balanced.

EXAMPLE 6.3

Balance the following equation.

$$CH_4 + O_2 \longrightarrow CO_2 + H_2O \quad \text{(not balanced)}$$

In this equation, oxygen appears in two different products, so we will leave the oxygen for last and balance the other two elements first. Carbon is already balanced, with one atom on each side of the equation. The least common multiple of 2 and 4 is 4, so to balance hydrogen we place the coefficient 2 in front of H_2O and we have four hydrogen atoms on each side.

$$CH_4 + O_2 \longrightarrow CO_2 + 2\,H_2O \quad \text{(not balanced)}$$

Now for the oxygen. There are four oxygen atoms on the right. If we place a 2 in front of O_2, the oxygen atoms balance.

$$CH_4 + 2\,O_2 \longrightarrow CO_2 + 2\,H_2O \quad \text{(balanced)}$$

The equation is balanced.

EXAMPLE 6.4

Balance the following equation

$$H_2SO_4 + NaCN \longrightarrow HCN + Na_2SO_4 \quad \text{(not balanced)}$$

Here we have an equation that involves compounds incorporating polyatomic ions. The SO_4 group should be treated as a unit and balanced as a whole. The same is true of the CN group. As the equation is presently written, the SO_4 groups and the CN groups are balanced, but the hydrogen atoms and the sodium atoms are not. The least common multiple for sodium is 2, so to get two sodium atoms on the left we place a 2 before NaCN. The least common multiple for hydrogen is 2, so to get two hydrogen atoms on the right we place a 2 before HCN.

$$H_2SO_4 + 2\ NaCN \longrightarrow 2\ HCN + Na_2SO_4 \quad \text{(balanced)}$$

It turns out that the same coefficients balance the CN groups and the SO_4 groups, and the equation as a whole is balanced.

We have made the task of balancing equations deceptively easy by considering simple reactions. It is more important for you to understand the principle, however, than for you to be able to balance complicated equations. You should know what is meant by a balanced equation and be able to handle simple systems.

6.2 Chemical Arithmetic and the Mole

An equation is meaningful on the atomic and molecular level. The equation

$$C + O_2 \longrightarrow CO_2$$

says that one carbon atom reacts with one oxygen molecule to produce one carbon dioxide molecule. One carbon atom has a mass of 12 amu. One oxygen molecule has a mass of 32 amu (2×16 amu per oxygen atoms). Thus the equation also indicates that 12 amu of carbon reacts with 32 amu of oxygen. In fact, 12 mass units of carbon reacts with 32 mass units of oxygen no matter what the mass or weight units are (amu, grams, pounds, anything). If 100 atoms of carbon react with 100 molecules of oxygen, or if one million react with one million, or if 10^{20} react with 10^{20}, or if 7328 react with 7328, the ratio of the mass of the carbon to the mass of the oxygen will be 12 to 32 because the ratio of the masses of the individual particles of reactants is 12 to 32 (Figure 6.3).

Carbon and oxygen can react to form a different product, carbon monoxide.

$$2\ C + O_2 \longrightarrow 2\ CO$$

In this reaction carbon reacts with oxygen in a mass ratio of 24 to 32 because two carbon atoms, each with a mass of 12 amu, react with one oxygen molecule.

6.2 Chemical Arithmetic and the Mole

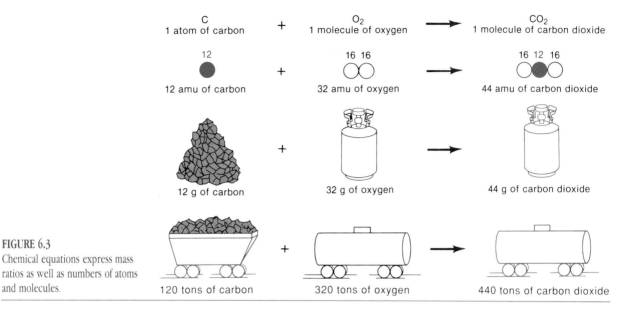

FIGURE 6.3
Chemical equations express mass ratios as well as numbers of atoms and molecules.

It should be clear by now that the atomic or **formula weights** of the species involved in reactions are important, and you should be able to calculate formula weights.

Formula weight is the sum total of all atomic weights comprising the formula.

EXAMPLE 6.5

What is the formula weight of ozone, O_3?
  Formula weights are calculated by adding together the atomic weights of the constituents atoms. The ozone molecule contains three oxygen atoms. The atomic weight of oxygen is 16 amu, and therefore the formula weight of ozone is

$$3 \times 16 = 48 \text{ amu}$$

EXAMPLE 6.6

What is the formula weight of ammonia, NH_3?
  The atomic weight of nitrogen is 14 amu and that of hydrogen is 1 amu. There are three hydrogen atoms in the molecule; therefore, the formula weight is

$$14 + (3 \times 1) = 17 \text{ amu}$$

EXAMPLE 6.7

What is the formula weight of glucose, $C_6H_{12}O_6$?

The atomic weight of carbon is 12 amu; that of hydrogen is 1 amu; that of oxygen, 16 amu. The formula weight of glucose is

$$(6 \times 12) + (12 \times 1) + (6 \times 16) =$$
$$72 + 12 + 96 = 180 \text{ amu}$$

EXAMPLE 6.8

What is the formula weight of ammonium sulfate, $(NH_4)_2SO_4$?

Now it is imperative that you understand the meaning of parentheses in a formula. $(NH_4)_2$ means that everything within the parentheses is doubled. Ammonium sulfate contains

$$\begin{aligned} 2\,N &= 2 \times 14 = 28 \\ 8\,H &= 8 \times 1 = 8 \\ 1\,S &= 1 \times 32 = 32 \\ 4\,O &= 4 \times 16 = 64 \\ \hline (NH_4)_2SO_4 &= 132 \text{ amu} \end{aligned}$$

The mass of 1 mole is the atomic, molecular, or formula weight expressed in grams.

Laboratory balances are not sensitive enough to measure the mass of units as small as individual atoms or molecules. Therefore, scientists have defined a unit of matter, the *mole*, that is larger and more convenient to manipulate in a laboratory. Officially the **mole** (1 mol) is defined as *the amount of a substance containing as many elementary units as there are atoms in exactly 12 g of the carbon-12 isotope.* That probably does not sound like a convenient unit to you, but you may change your mind when you realize that, in practice, it means that the mass of a mole of any chemical is simply the atomic or formula weight expressed in grams. One mole of carbon has a mass of 12 g; 1 mol of oxygen molecules has a mass of 32 g; 1 mol of the compound ammonium sulfate (Example 6.8) has a mass of 132 g.

$$\text{number of moles} = \frac{\text{mass of substance}}{\text{molar mass}}$$

How many carbon atoms are in 12.0 g of carbon? The number of atoms in 12.0 g of carbon is unimaginably large. In exponential form it is

6.2 Chemical Arithmetic and the Mole

written 6.02×10^{23} and is called **Avogadro's number**. Scientists love to find ways of making very large numbers meaningful. Here is one: if you had a mole of dollars, that is, 6.02×10^{23} dollars, you could spend a billion dollars each second of your entire life and have used only about 0.001% of your money. If you had 6.02×10^{23} peas (1 mol of peas), they would cover the surface of the Earth to a depth of 15 m. All the people now on Earth equal less than 0.000 000 000 000 01 mol of people. A stack containing 6.02×10^{23} copies of this textbook would reach not just from here to the moon or to the sun or even to the nearest star, but from our galaxy to our neighbor, the Andromeda galaxy. A mole, that is, 6.02×10^{23}, of anything is an enormous number of anything. Yet atoms and molecules are so small that 1 mol of carbon atoms (12 g) could be mailed with a single stamp and a mole of water molecules is no more than a few sips.

Avogadro's number = 6.02×10^{23}

The mole used in chemistry is something like the dozen we use every day. You can have a dozen eggs, a dozen strips of bacon, or a dozen breakfasts. In each case, you have 12 of whatever you have specified. A mole simply means 6.02×10^{23} of whatever you are talking about. Consider the following analogies.

1 dozen wheat kernels	contain 12 kernels	weigh 3 g
1 dozen eggs	contain 12 eggs	weigh 140 g
1 dozen oranges	contain 12 oranges	weigh 500 g
1 mol sodium	contains 6.02×10^{23} Na atoms	weighs 23 g
1 mol chlorine	contains 6.02×10^{23} Cl atoms	weighs 35.5 g
1 mol sodium chloride	contains 6.02×10^{23} molecules	weighs 58.5 g

Generally, then, 1 mol = 6.02×10^{23} units and mass of 1 mol = 1 gram formula weight

Study Figures 6.4 and 6.5 to further understand the concept of the mole and the relationship between units and masses.

Back to our favorite equation:

$$C + O_2 \longrightarrow CO_2$$

It now says that

1 atom of carbon reacts with 1 molecule of oxygen to yield 1 molecule of carbon dioxide

or that

12 amu of carbon reacts with 32 amu of oxygen to yield 44 amu of carbon dioxide

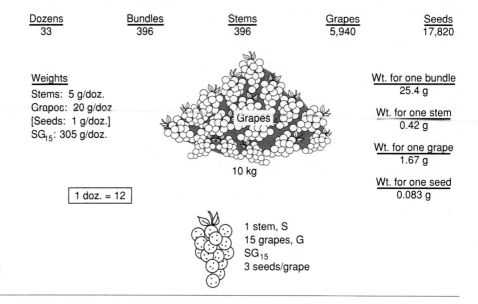

FIGURE 6.4
Numerical relationships in 19 kg of grapes in which each bundle contains 15 grapes with three seeds per grape.

or that

12 g of carbon reacts with 32 g of oxygen to yield 44 g of carbon dioxide

or that

1 mol of carbon atoms reacts with 1 mol of oxygen molecules to yield 1 mol of carbon dioxide molecules

(See Figure 6.6.)

We have previously referred to a second reaction of carbon with oxygen that produced carbon monoxide.

$$2\,C + O_2 \longrightarrow 2\,CO$$

FIGURE 6.5
Numerical relationships in 10 kg of $AlCl_3$.

Moles	Molecules	Al	Cl	Protons
75	4.5×10^{25}	4.5×10^{25}	13.5×10^{25}	288×10^{25}

Weights
$_{13}Al$: 27 g/mol
$_{17}Cl$: 35.5 g/mol
[Protons: 1 g/mol]
$AlCl_3$: 133 g/mol

$AlCl_3$

10 kg

1 mole = 6×10^{23}

Cl
\
Al — Cl
/
Cl

Wt. of one $AlCl_3$ molecule
22×10^{-23} g

Wt. of one Al atom
4.5×10^{-23} g

Wt. of one Cl atom
5.9×10^{-23} g

Wt. of one proton
0.17×10^{-23} g

6.2 Chemical Arithmetic and the Mole

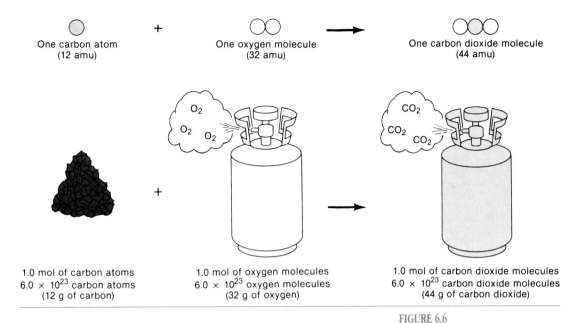

FIGURE 6.6
We cannot weigh single atoms or molecules, but we can weigh equal numbers of these fundamental particles.

This equation now tells us that

> 2 atoms of carbon react with 1 molecule of oxygen to yield 2 molecules of carbon monoxide

or that

> 24 amu of carbon reacts with 32 amu of oxygen to yield 56 amu of carbon monoxide (2 molecules of CO weighing 28 amu each)

or that

> 24 g of carbon reacts with 32 g of oxygen to yield 56 g of carbon monoxide

or that

> 2 mol of carbon atoms reacts with 1 mol of oxygen molecules to yield 2 mol of carbon monoxide molecules

Notice that the coefficients in the equation give you directly the combining ratio of atoms and molecules or the combining ratio of moles, but they do not give directly the combining ratio in grams. The formula weights must always be taken into account when dealing with grams.

Another point—although we cannot work with fractions of a molecule or atom, we can work with fractions of a mole. Even though there is no such things as half an atom of carbon, half a mole (like half a pound) is perfectly reasonable. Half a mole of carbon atoms has a mass of 6 g and contains 3.01×10^{23} atoms.

EXAMPLE 6.9

What is the mass of 0.50 mol of carbon dioxide, CO_2?

$$1 \text{ mol} = 1 \text{ gram formula weight}$$
$$1 \text{ mol } CO_2 = 44 \text{ g } CO_2$$

This equivalence can be used as a conversion factor.

$$0.50 \text{ mol } CO_2 \times \frac{44 \text{ g } CO_2}{1 \text{ mol } CO_2} = 22 \text{ g } CO_2$$

EXAMPLE 6.10

What is the mass of 3.0 mol of carbon monoxide?
The formula weight of carbon monoxide, CO, is 28. Therefore,

$$1 \text{ mol CO} = 28 \text{ g CO}$$

and

$$3.0 \text{ mol CO} \times \frac{28 \text{ g CO}}{1 \text{ mol CO}} = 84 \text{ g CO}$$

EXAMPLE 6.11

How many moles of carbon monoxide are in 7.0 g of carbon monoxide?
You need only invert the conversion factor used in Example 6.12 to convert from grams to moles.

$$7.0 \text{ g CO} \times \frac{1 \text{ mol CO}}{28 \text{ g CO}} = 0.25 \text{ mol CO}$$

EXAMPLE 6.12

How many moles of ammonium sulfate are in 1.32 g of the compound?

From Example 6.8 we know that the formula weight of ammonium sulfate, $(NH_4)_2SO_4$, is 132. Thus

$$1 \text{ mol } (NH_4)_2SO_4 = 132 \text{ g } (NH_4)_2SO_4$$

$$1.32 \text{ g} \times \frac{1 \text{ mol}}{132 \text{ g}} = 0.0100 \text{ mol } (NH_4)_2SO_4$$

Once again let us consider the equation

$$C + O_2 \longrightarrow CO_2$$

Carbon and oxygen combine in a one-to-one (1:1) ratio. That means 1 mol to 1 mol or 0.5 mol to 0.5 mol or 10 mol to 10 mol. A 1:1 ratio just means that the molar amounts must be equal.

For

$$2\,C + O_2 \longrightarrow 2\,CO$$

the mole ratio of reactants is 2:1, that is, 2 mol of carbon to 1 mol of oxygen, or 0.50 mol of carbon to 0.25 mol of oxygen, or 10 mol to 5 mol.

EXAMPLE 6.13

According to the equation

$$2\,C + O_2 \longrightarrow 2\,CO$$

how many moles of carbon will react with 0.2 mol of oxygen?

The coefficients in the equation can be used to provide a convenient conversion factor.

$$2 \text{ mol C} \quad \overset{\text{reacts with}}{\rightleftharpoons} \quad 1 \text{ mol } O_2$$

$$0.2 \text{ mol } O_2 \times \frac{2 \text{ mol C}}{1 \text{ mol } O_2} = 0.4 \text{ mol C}$$

EXAMPLE 6.14

According to the equation

$$C_3H_8 + 5\,O_2 \longrightarrow 3\,CO_2 + 4\,H_2O$$

how many moles of CO_2 will be produced in the reaction of 1.0 mol of O_2? How many grams?

$$5\text{ mol }O_2 \overset{\text{yields}}{\rightleftharpoons} 3\text{ mol }CO_2$$

$$1.0\text{ mol }O_2 \times \frac{3\text{ mol }CO_2}{5\text{ mol }O_2} = 0.60\text{ mol }CO_2$$

To find the number of grams, just use the gram formula weight (1 mol CO_2 = 44 g CO_2).

$$0.60\text{ mol }CO_2 \times \frac{44\text{ g }CO_2}{1\text{ mol }CO_2} = 26\text{ g }CO_2$$

EXAMPLE 6.15

According to the equation in Example 6.14, how many grams of propane, C_3H_8, will react with 3.2 g of O_2?

Always break up problems like this into workable parts (Figure 6.7). To use the equation you must deal in moles, but you are both given and asked for amounts in grams. First, the amount given in grams must be converted to moles, using the gram formula weight. According to the problem, 3.2 g of O_2 reacts. The formula weight of O_2 is 32. Therefore,

$$3.2\text{ g }O_2 \times \frac{1\text{ mol }O_2}{32\text{ g }O_2} = 0.10\text{ mol }O_2$$

Now you use the coefficients from the equation to construct a conversion factor relating moles of O_2 to moles of propane. According to the equation,

$$1\text{ mol }C_3H_8 \overset{\text{reacts with}}{\rightleftharpoons} 5\text{ mol }O_2$$

$$0.10\text{ mol }O_2 \times \frac{1\text{ mol }C_3H_8}{5\text{ mol }O_2} = 0.020\text{ mol }C_3H_8$$

6.2 Chemical Arithmetic and the Mole

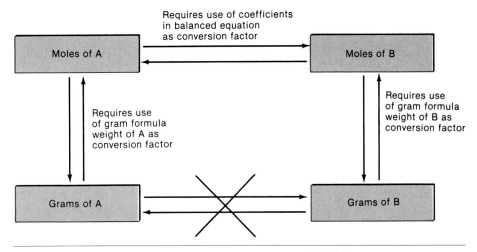

FIGURE 6.7
Outline of the procedure for using chemical equations to relate amounts of reactants and products. Direct gram-to-gram conversion is not possible.

Finally, you can convert the moles of C_3H_8 to grams, using the gram formula weight of C_3H_8:

$$1 \text{ mol } C_3H_8 = 44 \text{ g } C_3H_8$$

Therefore,

$$0.020 \text{ mol } C_3H_8 \times \frac{44 \text{ g } C_3H_8}{1 \text{ mol } C_3H_8} = 0.88 \text{ g } C_3H_8$$

EXAMPLE 6.16

How many grams of water are produced in the reaction of hydrogen sulfide with 6.4 g of oxygen according to the following equation?

$$2 \text{ H}_2\text{S} + 3 \text{ O}_2 \longrightarrow 2 \text{ SO}_2 + 2 \text{ H}_2\text{O}$$

You are given the mass of oxygen that reacts. Convert this to moles using the gram formula weight of O_2 (32 g/mol) as a conversion factor.

$$6.4 \text{ g } O_2 \times \frac{1 \text{ mol } O_2}{32 \text{ g } O_2} = 0.20 \text{ mol } O_2$$

Next convert moles of O_2 to moles of H_2O using the appropriate coefficients from the equation.

$$0.20 \text{ mol } O_2 \times \frac{2 \text{ mol } H_2O}{3 \text{ mol } O_2} = 0.13 \text{ mol } H_2O$$

Now convert the moles of water to grams of water using the formula weight of water (18 g/mol) as the conversion factor.

$$0.13 \text{ mol H}_2\text{O} \times \frac{18 \text{ g H}_2\text{O}}{1 \text{ mol H}_2\text{O}} = 2.3 \text{ g H}_2\text{O}$$

EXAMPLE 6.17

A compound is found to contain three elements: Na, Cl, and O. A mass analysis shows 18.82% Na, 29.0% Cl, and 52.2% O. What is the formula of the compound?

The question really is about the subscripts x, y, and z in the formula $\text{Na}_x\text{Cl}_y\text{O}_z$. These are unit ratios, not masses. So the percent masses given in the problem must be changed to moles.

For Na,

$$\frac{18.8 \text{ g}}{23 \text{ g/mol}} = 0.817 \text{ mol Na}$$

For Cl,

$$\frac{29.0 \text{ g}}{35.5 \text{ g/mol}} = 0.817 \text{ mol Cl}$$

For O,

$$\frac{52.2 \text{ g}}{16.0 \text{ g/mol}} = 3.26 \text{ mol O}$$

The formula could be written $\text{Na}_{0.817}\text{Cl}_{0.817}\text{O}_{3.26}$ because those are the proper mole ratios. However, these ratios can be simplified by dividing through by the smallest number, 0.817. This yields NaClO_4.

6.3 Gases Revisited

Avogadro's Law

When liquids and solids fill a container, they do so by placing the molecules right next to each other, filling the container from the bottom up. In all directions the molecules touch each other. Since molecules are of different sizes, it takes different numbers of molecules of different materials to fill a given container (Figure 6.8). Thus, the relatively large sugar molecule, $C_6H_{12}O_6$, will fill a given container with fewer molecules than is the case with the smaller H_2O molecule.

An amazing thing happens when gas molecules fill a container. They do not stack next to each other but spread as far apart from each other as possible, filling the container in every direction. In fact, they are spread so far apart that the space the molecules themselves occupy is insignificant compared to the vast distances between molecules. Whether the gas molecules are small, as for H_2, or large, as for Br_2, or whether the gas molecules are concentrated or not, the distances between them are the important factor. This observation is summarized in Avogadro's law; given constant temperature and pressure, *gas volumes are proportional to the number (moles) of gas molecules*, $V \propto n$. That is, a given container at fixed temperature and pressure will always hold the same number of gas molecules, *no matter what the gas is*.

FIGURE 6.8
It takes relatively few large molecules (a) and more small molecules (b) to fill a container.

The Ideal Gas Law

If we combine the various laws containing the four parameters V, n, P, and T, we obtain the ideal gas law.

$$V \propto 1/P$$
$$V \propto n$$
$$V \propto T$$

yields

$$V \propto \frac{1}{P}nT \quad \text{or} \quad PV = nRT$$

This relationship holds for "ideal" gases—those without any attraction between gas particles. If P is kept in atmospheres, V in liters, n in moles, and T in kelvins, then $R = 0.082$ L·atm/mol·K).

EXAMPLE 6.18

How many moles of gas are present in a 5.0-L flask containing oxygen at 750 mm Hg pressure and standard temperature? (Note that 750 mm Hg is equal to 750/760 = 0.987 atm.)

Solving $PV = nRT$ for moles gives

$$n = \frac{PV}{RT} = \frac{0.987 \text{ atm} \times 5.0 \text{ L}}{[0.082 \text{ L·atm}/(\text{mol·K})] \times 273 \text{ K}}$$

$$= 0.22 \text{ mol}$$

6.4 Energy Changes in Chemical Reactions

There are often many reasons why reactions do proceed. If a reaction product is constantly removed (say, through precipitation or gas evolution) as soon as it is formed, then the reaction proceeds to produce more of what is lost. Also, if the number of product molecules formed is greater than the number of reagents, then the reaction tends to proceed (it is easier to break things into smaller fragments than to build something from many pieces). Electron configuration also comes into play. Some electron configurations are more stable than others. Sodium ions and chloride ions, arranged in a crystal, are less reactive than sodium atoms and chlorine molecules. If sodium metal and chlorine gas are mixed, a vigorous reaction occurs.

$$2\,Na + Cl_2 \longrightarrow 2\,NaCl$$

A great deal of energy in the form of heat and light is produced during the reaction. Sometimes this energy is listed as one of the products in the chemical equation.

$$2\,Na + Cl_2 \longrightarrow 2\,NaCl + energy$$

Exothermic reactions release heat.

Chemical reactions that result in the net release of heat are said to be **exothermic**.

Other reactions, such as the decomposition of water, have a net energy requirement.

$$2\,H_2O + energy \longrightarrow 2\,H_2 + O_2$$

Endothermic reactions require heat.

If the energy is supplied as heat, such reactions are said to be **endothermic**.

In exothemic reactions, reactants give up potential energy to form more stable products. In endothermic reactions, reactants absorb energy and convert it to potential energy stored in the bonds of the less stable products. The general tendency in nature is to achieve greater stability, which is another way of saying that exothermic reactions are usually favored. If energy is available, however, endothermic reactions do occur (and one of these is responsible for the maintenance of all life on Earth, as we shall soon see).

To say that exothermic reactions are generally favored does not necessarily mean that the reactions occur as soon as the reactants are brought into contact with one another. Before products can begin to form, bonds in reactant molecules must be broken. Energy must be supplied to the reactant molecules in both exothermic and endothermic reactions to begin this pulling-apart process. The minimum amount of energy necessary to initiate a reaction is called the **energy of activation**.

> Energy of activation—the energy needed to cause the reaction to occur.

Coal, a form of carbon, reacts with oxygen in a strongly exothermic reaction, but to initiate the reaction the coal must first be heated to several hundred degrees. Eventually, the energy released by the burning coal exceeds the amount of energy needed to start the reaction. The excess energy is what makes the reaction exothermic. In an endothermic reaction, more energy goes in than comes out.

Figure 6.9 illustrates these energy changes. In Figure 6.9a, an exothermic reaction is depicted. The reactants are in the valley at the left, and the products are in the lower energy valley at the right. The distance from the reactant valley to the top of the energy hill represents the activation energy, and the difference in the height of the two valleys represents the heat released in this exothermic reaction. Figure 6.9b diagrams an endothermic reaction. Now the products are in a higher energy state than the reactants. The difference in the heights of the reactant and product valleys represents the net energy that had to be added to the reactants to raise them to the product level. Note that, as in the diagram for the exothermic reaction, the difference in height between the reactant valley and the top of the energy hill is the activation energy. Note also that we really only needed to draw one diagram. Look at the first diagram and ignore the labels for the moment. If you begin at the left and move to the right, you start from a high energy state and end up at a lower energy state (an exothermic reaction). If you read the diagram in reverse, from right to left, you start at a low energy state and end up at a higher energy state (an endothermic reaction).

Many reactions are actually reversible. Living cells take in oxygen and "burn" a sugar called glucose. This exothermic reaction provides the energy the cell needs to survive.

$$C_6H_{12}O_6 + 6\,O_2 \longrightarrow 6\,CO_2 + 6\,H_2O + \text{energy}$$
Glucose

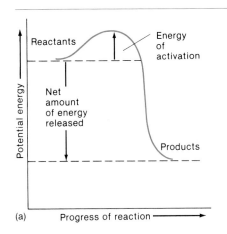
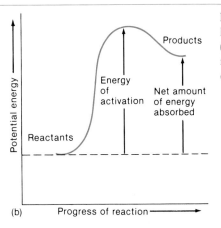

FIGURE 6.9
Energy changes in chemical reactions. (a) Energy diagram for an exothermic reaction. (b) Energy diagram for an endothermic reaction.

In photosynthesis, energy from sunlight is used to convert CO_2 and H_2O to $C_6H_{12}O_6$ and O_2.

Green plants carry out the reverse reaction (an endothermic reaction called **photosynthesis**) by using energy from sunlight to convert carbon dioxide and water to glucose and oxygen.

$$6\ CO_2 + 6\ H_2O + \text{energy} \longrightarrow C_6H_{12}O_6 + 6\ O_2$$

Life on our planet is based on this reversible reaction, which we will encounter again later in this chapter.

6.5 Reaction Rates

Before atoms, molecules, or ions can react, they must first get together—they must collide with proper intensity and proper alignment. For both endothermic and exothermic reactions, the collisions must provide a certain minimum of energy, which we have called the *energy of activation*. The **rate** of a chemical reaction—how fast it proceeds can be influenced by factors that change the number of effective collisions. We will consider three such factors: temperature, concentration, and catalysis.

The Effect of Temperature on Reaction Rates

As the temperature of a system increases, the rate of reaction increases.

The rate of a reaction increases as temperature increases. At higher temperatures, the rapidly moving molecules collide more frequently and strike one another harder. Such collisions are more likely to supply the activation energy needed to get the reaction going.

In our daily lives we make use of our knowledge of the effect of temperature on chemical reactions. For example, we freeze foods to retard the chemical reactions that lead to spoilage. On the other hand, if we want to cook our food more rapidly, we turn up the heat. The chemical reactions that occur in our bodies do so at a constant temperature of 37 °C (98.6 °F). A few degrees' rise in temperature (fever) leads to an increase in respiration, pulse rate, and other physiological reactions. A drop in body temperature of a few degrees slows these same processes considerably, as is exemplified by the slowed metabolism of hibernating animals. Use is made of this phenomenon in some surgical procedures. In some cases of heart surgery, the body temperature of the patient is lowered to about 15 °C (60 °F). Ordinarily, the brain is permanently damaged when its oxygen supply is interrupted for more than 5 min. But at the lower temperature, metabolic processes slow considerably, and the brain may survive much longer periods of oxygen deprivation. The surgeon can stop the heartbeat, perform an hours-long surgical procedure on the heart, and then restart the heart and bring the patient's temperature back to normal

6.5 Reaction Rates

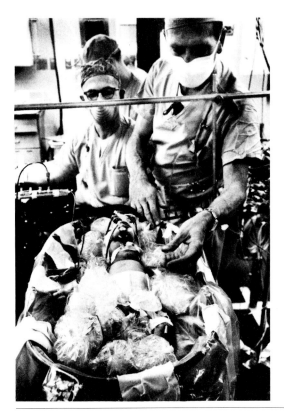

FIGURE 6.10
Heart surgery is often performed with the patient's temperature lowered to slow metabolic processes and thus minimize the possibility of brain damage. With infants the body is cooled using ice packs. [Photo © AP/Wide World Photos.]

(Figure 6.10). An often-used rule of thumb says that the rate of a reaction is doubled for every 10 °C temperature increase.

The Effect of Concentration on Reaction Rates

Another factor affecting the rate of a chemical reaction is the concentration of reactants. The more reactant molecules there are in a volume of space, the more collisions will occur. The more collisions there are, the more reactions occur.

An increase in concentration (more reactant molecules) causes more collisions to occur, thus causing more reactions.

For reactions in solution, the concentration of reactants can be increased if greater amounts are dissolved. With gases, concentration can also be increased by increasing the pressure.

Materials burn (i.e., react with oxygen) at a certain rate in air, which is only 21% oxygen. The same materials burn much more rapidly in pure oxygen. The prohibition against smoking in hospital rooms where patients are undergoing oxygen therapy acknowledges this effect of concentration on reaction rates (Figure 6.11).

Catalysts and Reaction Rates

A third way to influence the rate of a reaction is through the use of a catalyst. A **catalyst** is a substance that changes the rate of a chemical reaction without itself being permanently changed. Catalysts act by

A catalyst changes the rate of reaction and is recovered unchanged.

FIGURE 6.11
The "No smoking" sign is not simply a concession to the sensitivity of the patient undergoing oxygen therapy. It is a warning about the increased danger from fires that an oxygen-rich atmosphere presents.

lowering the activation energy of the reaction (Figure 6.12). If the activation energy is lower, then more collisions will be effective. Collisions of slower-moving molecules may not supply the activation energy for the uncatalyzed reaction, but such collisions are effective in the catalyzed reaction. In a catalyzed reaction the energy of activation is lowered because the catalyst changes the mechanism of the reaction. Although overall the same reactant bonds are broken and the same product bonds are formed, the precise way in which these changes are accomplished is different when a catalyst is present. (You can sew a dress by hand or with a sewing machine. In both cases, the thread and material you start with end up as a garment, but with the sewing machine the job is accomplished more quickly.) A catalyst actually does get involved in the reaction, but it eventually emerges in its

FIGURE 6.12
Effect of catalyst on activation energy.

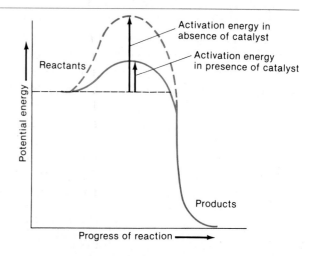

original form. (The sewing machine is involved in the construction of the garment, but while thread and material become a dress, the sewing machine is just as it was in the beginning—ready to sew another dress.)

Catalysts are extremely important in living organisms, where raising the temperature by 100 °C is not a feasible way of increasing the rate of critical reactions. If we were to raise our body temperatures by 100 °C, we would boil our blood, among other things. Biological catalysts, called **enzymes**, mediate nearly all the chemical reactions that take place in living systems. So important are enzymes to life that we shall discuss them more fully in Chapter 15.

Biological catalysts are called enzymes.

6.6 Equilibrium in Chemical Reactions: Le Châtelier's Principle

At the start of a reaction, reactants produce products at a certain rate. As the reaction progresses and reactant molecules are converted to products, fewer reactant molecules remain to react, and the rate of the reaction decreases. On the other hand, if the reaction is reversible, the rate of the reverse reaction increases as the concentration of product increases. Eventually the rates of the forward and reverse reactions become equal, and a condition called **equilibrium** is established (Figures 6.13 and 6.14).

At equilibrium there *appears* to be no further reaction, but appearances are deceiving. Reactants are still changing to products, and products are still changing back to reactants. For every reactant molecule lost through the forward reaction, however, one is gained through the reverse reaction. Once equilibrium is established, we can measure no net change in the concentration of reactants or products.

The concentration of reactants does not necessarily equal the concentration of products at equilibrium. These relative concentrations depend on the particular system considered. What is necessarily true at equilibrium is that each time some reactant changes to a product, an equal amount of product changes back to a reactant. Because molecules are still reacting, even though their concentrations do not change, we say that equilibrium is a *dynamic* situation. Therefore, an equilibrium may be disturbed.

In the nineteenth century, scientists subjected equilibrium systems to changes in concentration and temperature, among other things. A French chemist, Henri Louis Le Châtelier, summarized the effects of these variations. His rule, still called **Le Châtelier's principle**, states that if a stress is applied to a system in equilibrium, the system adjusts in such a way as to minimize the stress.

Le Châtelier's principle: as a stress is applied to a system in equilibrium, the system adjusts to minimize the stress.

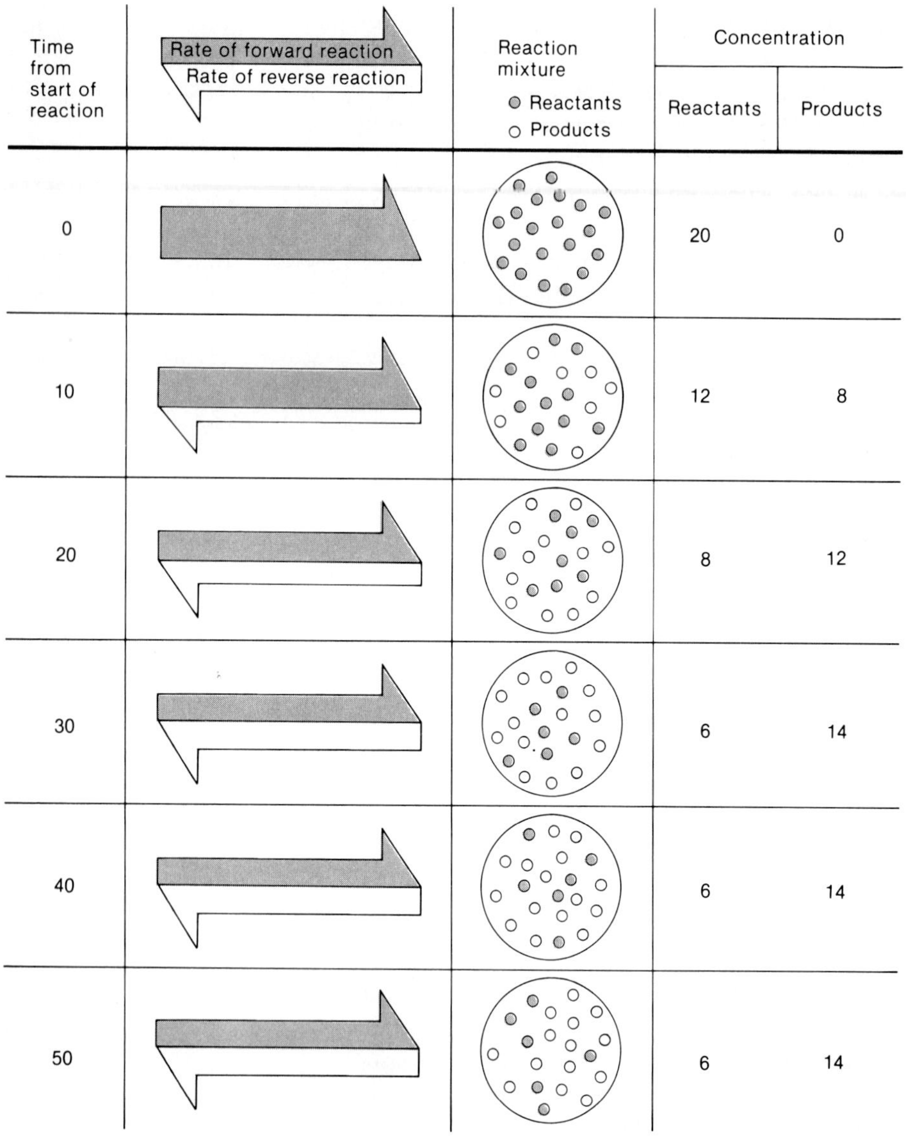

Time from start of reaction	Rate of forward reaction / Rate of reverse reaction	Reaction mixture ● Reactants ○ Products	Concentration	
			Reactants	Products
0			20	0
10			12	8
20			8	12
30			6	14
40			6	14
50			6	14

FIGURE 6.13
Progress of reaction toward achieving equilibrium.

FIGURE 6.14
Reaction rate is time dependent—equilibrium is not.

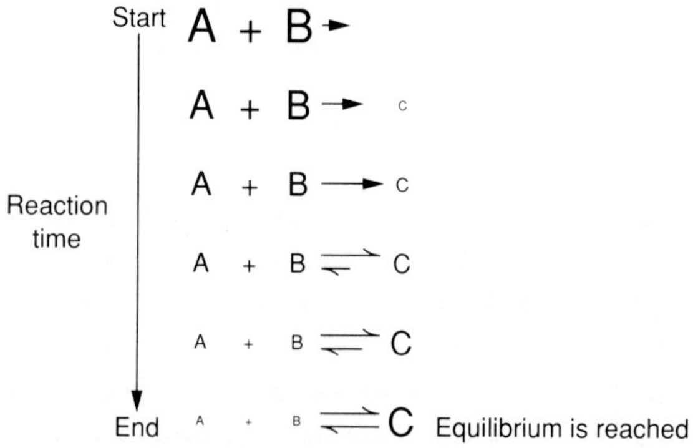

Start: A + B →
A + B → c
A + B ⟶ C
A + B ⇌ C
A + B ⇌ C
End: A + B ⇌ C Equilibrium is reached

Reaction time

6.6 Equilibrium in Chemical Reactions: Le Châtelier's Principle

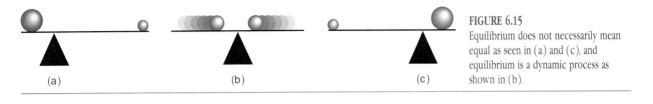

FIGURE 6.15
Equilibrium does not necessarily mean equal as seen in (a) and (c), and equilibrium is a dynamic process as shown in (b).

Consider this equation and assume that the reaction is at equilibrium.

$$N_2 + 3 H_2 \rightleftharpoons 2 NH_3 + energy$$

(The double arrow indicates that this is a reversible reaction.) If the system is heated, the rate of the reverse reaction will increase, using up heat (energy). If more nitrogen (N_2) is added to the system, the reaction will go to the right, using up nitrogen. If ammonia (NH_3) is removed from the system, the equilibrium will shift to the right, forming more ammonia. If hydrogen (H_2) is removed from the system, the equilibrium shifts to the left and more hydrogen is formed.

We should note that a catalyst will not shift an equilibrium system. Catalysts change the rate of both the forward and the reverse reactions by lowering the activation energy for both reactions (refer to Figure 6.12). The position of the equilibrium, that is, the equilibrium concentrations of reactants and products, is not changed.

EXAMPLE 6.19

What effect, if any, will each of the following changes have on the equilibrium of the reaction below?

$$2 CO + O_2 \rightleftharpoons 2 CO_2 + heat$$

a. addition of CO
b. removal of O_2
c. cooling the reaction system
d. addition of a catalyst

The effects of the changes are as follows.

a. The equilibrium shifts to the right to use up the added CO.
b. The equilibrium shifts to the left to replace the O_2 that is removed.
c. The equilibrium shifts to the right to replace the lost heat.
d. A catalyst has no effect on the position of equilibrium.

6.7 Oxidation-Reduction Reactions

Chemical reactions are classified in many ways (for instance, as precipitation, decomposition, oxidation, reduction, endothermic, or exothermic). For living systems, however, there are two particularly important types of chemical processes: acid–base reactions and oxidation–reduction reactions. We will take up the topic of acid–base chemistry in Chapter 8. We have already looked at a number of oxidation–reduction reactions, so we will conclude this chapter with a brief review of this chemistry.

A number of the reactions we have used as examples in this chapter involved reaction with oxygen.

$$C + O_2 \longrightarrow CO_2$$
$$2\,C + O_2 \longrightarrow 2\,CO$$
$$C_3H_8 + 5\,O_2 \longrightarrow 3\,CO_2 + 4\,H_2O$$
$$C_6H_{12}O_6 + 6\,O_2 \longrightarrow 6\,CO_2 + 6\,H_2O$$

Carbon and its compounds are not the only substances that react with oxygen (although the compounds of carbon are especially important in living systems). We are all familiar with the rusting of iron, and that, too, is a reaction with oxygen.

$$4\,Fe + 3\,O_2 \longrightarrow 2\,Fe_2O_3$$
Rust

The combination of elements or compounds with oxygen is one definition of oxidation.

The combination of elements and compounds with oxygen is called **oxidation**. The substances that combine with oxygen are said to be **oxidized**.

Hydrogen reacts with a variety of metal oxides to remove oxygen and give the free metallic element. For example, when hydrogen gas is passed over heated copper(II) oxide, metallic copper and water are formed.

$$CuO + H_2 \longrightarrow Cu + H_2O$$

The process in which a metal oxide is changed to the metallic element is one definition of reduction.

This process in which a metal oxide is changed, or *reduced*, to the metallic element is called **reduction**.

As scientists expanded their studies of chemical reactions, the definitions of oxidation and reduction grew more complicated. Look again at the reduction of copper(II) oxide by hydrogen. The reaction involves the removal of oxygen from copper(II) oxide, so we can define reduction as the removal of oxygen. Or consider what is happening to

6.7 Oxidation–Reduction Reactions

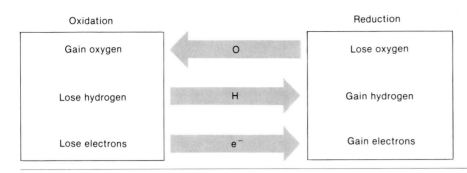

FIGURE 6.16
Three different definitions of oxidation and reduction.

the electronic structure of copper. Think of copper(II) oxide as an ionic compound composed of copper(II) cations (Cu^{2+}) and oxygen anions (O^{2-}). Now the reduction of copper can be viewed as a gaining of electrons: the positive copper ion changes to the neutral copper atom by picking up two electrons (from hydrogen). All these definitions are used for reduction (Figure 6.16).

Sometimes hydrogen reacts by combining with a compound, as shown for the compound ethene.

$$C_2H_4 + H_2 \xrightarrow{\text{catalyst}} C_2H_6$$

This reaction of hydrogen is also regarded as a reduction reaction because the oxidation number of C in C_2H_4 is changed from -2 to -3 in C_2H_6 (see page 200).

The definition of oxidation has been similarly expanded. Something is oxidized if it gains oxygen or loses hydrogen or loses electrons. Note that neither oxygen nor hydrogen has to be involved in a reaction for oxidation or reduction to take place.

$$Cu + Cl_2 \longrightarrow CuCl_2$$

Here neutral, elemental copper loses electrons (is oxidized) to become copper(II) ion.

$$Cu \longrightarrow Cu^{2+} + 2\ e^-$$

The neutral chlorine molecule picks up the electrons (is reduced) and becomes chloride ions.

$$Cl_2 + 2\ e^- \longrightarrow 2\ Cl^-$$

Oxidation and reduction go hand in hand. You cannot have one without the other. When one substance is oxidized, another must be reduced. In the copper–chlorine reaction, copper is oxidized and chlorine is reduced. If the chlorine were not there to pick up electrons

Oxidizing agent: a substance that causes another to be oxidized and is itself reduced.

Reducing agent: a substance that causes another to be reduced and is itself oxidized.

Oxidation: loss of electrons
Reduction: gain of electrons

from copper, then the copper would not be oxidized. Therefore, chlorine is called the **oxidizing agent**, that is, the agent that brings about the oxidation of another substance. Copper is the **reducing agent**; it supplies the electrons that cause chlorine to be reduced. Remember—*the reducing agent is the substance being oxidized; the oxidizing agent is the substance being reduced.* Because reduction and oxidation are always paired, the term **redox** is often used to describe these reactions.

Oxidation;
copper is being oxidized;
Cu is the reducing agent.

$$\text{Cu} + \text{Cl}_2 \longrightarrow \text{Cu}^{2+} + 2\,\text{Cl}^-$$

Reduction;
chlorine is being reduced;
Cl_2 is the oxidizing agent.

FIGURE 6.17
Corrosion of the copper statue is the result of oxidation enhanced by an acid environment. [Ezio Peterson/UPI/Bettmann Newsphotos.]

6.7 Oxidation–Reduction Reactions

EXAMPLE 6.20

Circle the oxidizing agent and underline the reducing agent in the following redox reactions.

a. $2\,C + O_2 \longrightarrow 2\,CO$
b. $N_2 + 3\,H_2 \longrightarrow 2\,NH_3$
c. $SnO + H_2 \longrightarrow Sn + H_2O$
d. $Mg + Cl_2 \longrightarrow Mg^{2+} + 2\,Cl^-$

The answers can be determined using the definitions previously given.

a. $\underline{C}$ + (O_2)

C gains oxygen and is oxidized, so it must be the reducing agent. Therefore, O_2 is the oxidizing agent.

b. (N_2) + $\underline{H_2}$

N_2 gains hydrogen and is reduced, so it is the oxidizing agent. Therefore, H_2 is the reducing agent.

c. (SnO) + $\underline{H_2}$

SnO loses oxygen and is reduced, so it is the oxidizing agent. H_2 is the reducing agent.

d. $\underline{Mg}$ + (Cl_2)

Mg loses electrons and is oxidized, so it is the reducing agent. Cl_2 is the oxidizing agent.

TABLE 6.1 Summary of Rules for Determining Oxidation Numbers

Rule	Examples
1. Free elements = 0	He, N_2, O_3, S_8
2. In compounds	
a. Group IA = +1	NaOH, $KMnO_4$, Li_2O
b. Group IIA = +2	$CaCl_2$, MgO, $BaSO_4$
3. H = +1 (most commonly)	H_2O, NaOH, $HC_2H_3O_2$
4. In compounds	
a. Oxygen = −2	H_2O, H_2SO_4, $Al(OH)_3$
b. Except, oxygen = −1 in peroxides	H_2O_2
5. In compounds, other elements are assigned numbers so that the sum of the oxidation numbers = 0	$\overset{-1}{Na}Cl$, $\overset{+7}{KMn}O_4$, $Ba\overset{+6}{S}O_4$
6. In ions, other elements are assigned numbers so that the sum of the oxidation numbers = ionic charge	$\overset{-1}{Br^-}$, $\overset{+5}{P}O_4^{3-}$, $C_2\overset{0}{H}_3O_2^-$

Oxidation numbers are agreed-upon values assigned to elements for bookkeeping purposes.

As it became clear that the early, simple definitions of oxidation and reduction were no longer adequate to cover the variety of reactions that fall under this classification, chemists resorted to a bookkeeping tool to deal with redox reactions. **Oxidation numbers** are numerical values assigned to elements in compounds according to the following (abbreviated) rules. (See also Table 6.1.)

1. The oxidation number of free (uncombined) elements is 0.
2. In compounds, Group IA and Group IIA elements are assigned oxidation numbers of +1 and +2, respectively.
3. In most compounds, hydrogen has an oxidation number of +1.
4. In most compounds, oxygen has an oxidation number of −2. (The exception to this rule that we are likely to encounter is oxygen in hydrogen peroxide, H_2O_2, in which oxygen is assigned an oxidation number of −1.)
5. For compounds, the algebraic sum of all the oxidation numbers must be 0.
6. For ions, the algebraic sum of all the oxidation numbers must equal the charge on the ion.

The oxidation numbers of other elements in compounds or ions can be calculated using these rules, as demonstrated in the following examples.

EXAMPLE 6.21

What is the oxidation number of sulfur in SO_2?

Oxygen is −2 (rule 4). There are two oxygen atoms, for a total of −4. Since the sum must be 0 (rule 5), the sulfur must be +4.

EXAMPLE 6.22

What is the oxidation number of sulfur in H_2SO_4?

Hydrogen is +1 (rule 3), and oxygen is −2 (rule 4). There are two hydrogen atoms, for a total of +2, and four oxygen atoms, for a total of −8. Since the sum must be 0 (rule 5), we can write

$$2(+1) + 4(-2) + x = 0$$

where x is the unknown oxidation state of sulfur. Solving the equation we get

$$(+2) + (-8) + x = 0$$
$$-6 + x = 0$$
$$x = +6$$

You need not use algebra if you can see (without it) that sulfur must be +6.

EXAMPLE 6.23

What is the oxidation number of nitrogen in the nitrate ion (NO_3^-)?

Oxygen is −2 (rule 4). There are three oxygen atoms, for a total of −6. Since the sum of oxidation numbers must be −1 (rule 6), we can write

$$(-6) + x = -1$$
$$x = +5$$

Nitrogen must be +5.

EXAMPLE 6.24

What is the oxidation number of nitrogen in the ammonium ion (NH_4^+)?

Hydrogen is +1 (rule 3). There are four hydrogen atoms, for a total of +4. Since the sum of oxidation numbers must be +1 (rule 6), we can write

$$(+4) + x = +1$$
$$x = -3$$

Nitrogen is -3.

EXAMPLE 6.25

What is the oxidation number of carbon in C_2H_4O?

Hydrogen is +1. Four hydrogen atoms total +4. One oxygen atom is -2. Since the sum must be 0,

$$(+4) + (-2) + x = 0$$
$$x = -2$$

To balance the oxidation numbers, we need -2, so the carbon atoms total -2. But there are two carbon atoms to share the burden, so each carbon atom has an oxidation number of -1.

EXAMPLE 6.26

Refer again to the redox equations shown in Example 6.22. Use oxidation numbers to determine which element is oxidized and which is reduced in each reaction.

a. $$2\,C + O_2 \longrightarrow 2\,CO$$

The oxidation numbers for carbon are

$$\overset{0}{2\,C} + O_2 \longrightarrow \overset{+2}{2\,CO}$$

Oxidation: increase in oxidation number

Carbon increases in oxidation number, which means that it is oxidized. The oxidation numbers for oxygen are

$$\overset{0}{2\,C} + \overset{0}{O_2} \longrightarrow 2\,\overset{-2}{CO}$$

The oxidation number decreases; therefore, oxygen is reduced.

Reduction: decrease in oxidation number

b. $$\overset{0}{N_2} + 3\,\overset{0}{H_2} \longrightarrow 2\,\overset{-3}{N}\overset{+1}{H_3}$$

Nitrogen is reduced; hydrogen is oxidized.

c. $$\overset{+2}{Sn}O + \overset{0}{H_2} \longrightarrow \overset{0}{Sn} + \overset{+1}{H_2}O$$

Tin (Sn) is reduced; hydrogen is oxidized. The oxidation number for oxygen is -2 in SnO and in H_2O. Therefore, oxygen is neither oxidized nor reduced.

d. $$\overset{0}{Mg} + \overset{0}{Cl_2} \longrightarrow \overset{+2}{Mg^{2+}} + 2\,\overset{-1}{Cl^-}$$

Magnesium is oxidized; chlorine is reduced.

Not surprisingly, the use of oxidation numbers gives the same answers as the use of other definitions for oxidation and reduction.

Living organisms obtain energy for physical and mental activities by the slow, multistep oxidation of food. We shall consider this complex process in more detail in Chapter 16. For the moment we will use the oxidation of the carbohydrate glucose to represent the process.

$$C_6H_{12}O_6 + 6\,O_2 \longrightarrow 6\,CO_2 + 6\,H_2O + \text{energy}$$

In this reaction, carbon is oxidized (its oxidation number goes from 0 to $+4$) and oxygen is reduced (its oxidation number changes from 0 to -2).

FIGURE 6.18
The food we eat is oxidized to provide energy for our activities. That energy originates in the sun and is trapped by plants through photosynthetic reactions that reduce carbon dioxide to carbohydrates. [Merri Cyr]

In photosynthesis, the above recation is reversed.

$$6\ CO_2 + 6\ H_2O + \text{energy} \longrightarrow C_6H_{12}O_6 + 6\ O_2$$

As we noted earlier, the first reaction is exothermic, the second is endothermic. Green plants make life on Earth possible. The plants carry out the oxidation–reduction reaction in which the energy of the sun is stored in chemical bonds. Animals release the stored energy by reversing this redox reaction. Your body runs on the energy of sunlight, which has been transformed into a usable fuel by oxidation and reduction reactions (Figure 6.18).

6.8 Oxidation and Antiseptics

Antiseptics are used to limit or prevent the growth of microorganisms.

Many common **antiseptics** (compounds that are applied to living tissue to kill or prevent the growth of microorganisms) are mild oxidizing agents. Hydrogen peroxide (H_2O_2), in the form of a 3% aqueous solution, finds use in medicine as a topical antiseptic to treat minor cuts and abrasions on certain parts of the body. Potassium permanganate ($KMnO_4$), in concentrations ranging from 0.01% to 0.2%, can be used as a topical antiseptic and an astringent. Potassium chlorate ($KClO_3$) was once used as an antiseptic for skin and mucous membranes. Its use has been largely discontinued because it is rather irritating, somewhat toxic, and only marginally effective.

Sodium hypochlorite (NaOCl), available in aqueous solution as laundry bleach (Purex, Clorox, and the like), finds some use in the irrigation of wounds and for bladder infections. It is also used as a disinfectant and deodorizer. Solutions of iodine are frequently used as antiseptics. In preparation for surgery, skin may be painted with these brown antiseptic solutions.

The exact method of operation of these oxidizing agents as antiseptics is unknown. Their action is rather indiscriminate; they attack human cells as well as micoorganisms. For many purposes, these compounds have been replaced by less poisonous materials.

Other oxidizing agents are used as **disinfectants**—chemicals that destroy harmful bacteria. Calcium hypochlorite [$Ca(OCl)_2$], called bleaching powder, is used to disinfect clothing and bedding. Chlorine (Cl_2) is used to kill pathogenic (disease-causing) microorganisms in drinking water. Wastewater is usually treated with chlorine also, before it is returned to a stream or lake. Such treatment has been quite effective in preventing the spread of infectious diseases such as typhoid fever.

Use of chlorine in this manner has been criticized. Researchers have shown that the chlorine reacts with some compounds (presumably from industrial wastes) to form toxic chlorinated compounds.

Ozone has also been used to disinfect drinking water. Many European cities, including Paris and Moscow, use ozone to treat their drinking water. Ozone is more expensive than chlorine, but less of it is needed. An added advantage is that ozone kills viruses on which chlorine has little, if any, effect. Tests in the Soviet Union have shown ozone to be a hundred times as effective as chlorine for killing polio virus. The oxidized contaminants from ozone treatment are thought to be less toxic than the chlorinated ones. In addition, ozone imparts no "chemical taste" to water. We may well see a shift from chlorine to ozone in the treatment of our drinking and wastewater.

Natural bodies of water are purified through interaction with the oxygen in the air. It is calculated that a stream is regenerated through oxygen disinfection every 10 miles as it tumbles down the mountain.

CHAPTER SUMMARY

Chemical reactions are changes involving the outermost electrons of elements and compounds. A chemical equation is used to describe reactions in the following manner.

$$\text{Reactants (starting materials)} \xrightarrow{\text{yield}} \text{products}$$

Such equations must be balanced to be consistent with the law of conservation of matter, which states that matter is neither created nor destroyed in chemical reactions. In a balanced equation the number of atoms of a particular element distributed among the reactants equals the number among the products.

The coefficients in a balanced equation give directly the correct ratio of reacting units (molecules or atoms). If the formula weights of the reactants are known, the equation can also be used to calculate the combining weight ratios. Formula weights are determined by adding together the atomic weights of the atoms in the formula unit. A mole is a gram formula weight, that is, the formula weight expressed in grams. One mole of a substance contains Avogadro's number, 6.02×10^{23} units, of the substance. The coefficients in an equation also give directly the combining ratio of moles.

Chemical reactions may be exothermic (involving a net release of heat) or endothermic (involving a net absorption of heat). Exothermic reactions are generally favored in nature. Both exothermic and endothermic reactions require an initial input of energy (the activation energy) to initiate bond breaking. Energy diagrams are used to illustrate the changes in potential energy that occur during a reaction.

Temperature, concentration of reactants, and catalysts influence the rates of chemical reactions. Reaction rates increase with increases in temperature and concentration because both these changes increase the number of effective reaction collisions. Catalysts also increase the number of effective reaction collisions, but they do so by changing the mechanism of the reaction and lowering the activation energy. Although a catalyst is involved in a reaction, it undergoes no permanent change in the reaction. Enzymes are biological catalysts that mediate reactions in living systems.

A reversible reaction is one that can proceed in both a forward and a reverse direction. Reversible reactions may reach equilibrium, a condition in which the rates of the forward and reverse reactions are equal. Le Châtelier's principle states that if an equilibrium system is disturbed (by changes in concentration, temperature, and so on), the system will react in a way that will relieve the stress.

Oxidation–reduction (redox) reactions are a major class of reactions of great importance in living systems. Photosynthesis is a redox reaction, and the metabolic conversion of carbohydrates to carbon dioxide is the reverse redox reaction. Oxidation is defined as the addition of oxygen, the removal of hydrogen, the removal of electrons, or an increase in oxidation number. Reduction is defined as the reverse of these processes. Oxidation numbers are calculated according to the rules given in Table 6.1.

Oxidation and reduction are always coupled. A substance that is oxidized in a reaction is called a reducing agent, and a substance that is reduced is referred to as an oxidizing agent. A number of oxidizing agents are used as antiseptics and disinfectants.

CHAPTER QUIZ

1. When the equation $Na + H_2O \rightarrow H_2 + NaOH$ is balanced, the sum of all coefficients is
   a. 4   b. 6   c. 7   d. 8
2. How many moles are there in 11 g of CO_2?
   a. 0.25   b. 11   c. 44   d. 484
3. How many grams are in 3.0×10^{-2} mol of H_2O?
   a. 0.030   b. 0.54   c. 0.60   d. 1.0
4. According to the equation $2 C + O_2 \rightarrow 2 CO$, how many grams of oxygen gas are needed to produce 42 g of CO?
   a. 21   b. 24   c. 32   d. 84
5. What is the formula for a molecule that contains 60% O and 40% S by mass?
   a. SO   b. S_2O_3 c. SO_2 d. SO_3
6. Which does *not* affect the reaction rate?
   a. catalyst
   b. concentration
   c. temperature
   d. color
7. Which does *not* affect equilibrium shifts of a reaction?
   a. concentration of products
   b. concentration of reactants
   c. temperature
   d. catalyst
8. What is the oxidation number for Mn in MnO_4^-?
   a. +1   b. +7   c. −5   d. −1
9. Which is reduced in the reaction $4 Fe + O_2 \rightarrow 2 Fe_2O_3$?
   a. Fe   b. O_2
   c. Fe_2O_3 d. nothing
10. Which is the oxidizing agent in the reaction $H_2 + Cl_2 \rightarrow 2 HCl$?
    a. H_2 b. Cl_2 c. HCl   d. none

PROBLEMS

1. Define or illustrate each of the following.
   a. formula weight
   b. mole
   c. Avogadro's number
   d. exothermic
   e. endothermic
   f. energy of activation
   g. catalyst
   h. reversible reaction
   i. equilibrium
   j. Le Châtelier's principle

6.1 The Chemical Equation

2. Write the following sentences as chemical equations.
   a. Coal burns in oxygen to produce carbon dioxide.
   b. Calcium chloride reacts with sulfuric acid to yield calcium sulfate and hydrogen chloride.

3. Relate the law of conservation of matter to the need to work with balanced chemical equations.

4. How many hydrogen atoms are there per formula unit in each of the following?
   a. NH_4NO_3 b. $(NH_4)_2HPO_4$

5. How many atoms of each kind (Al, C, H, and O) are included in the notation $2\,Al(C_2H_3O_2)_3$?

6. Indicate whether the equations are balanced. You need not balance the equation. Just determine whether it is balanced as written.
   a. $Mg + H_2O \rightarrow MgO + H_2$
   b. $FeCl_2 + Cl_2 \rightarrow FeCl_3$
   c. $F_2 + H_2O \rightarrow 2\,HF + O_2$
   d. $Ca + 2\,H_2O \rightarrow Ca(OH)_2 + H_2$
   e. $2\,LiOH + CO_2 \rightarrow Li_2CO_3 + H_2O$

7. Indicate whether the equations are balanced as written.
   a. $2\,KNO_3 + 10\,K \rightarrow 6\,K_2O + N_2$
   b. $2\,NH_3 + O_2 \rightarrow N_2 + 3\,H_2O$
   c. $4\,LiH + AlCl_3 \rightarrow 2\,LiAlH_4 + 2\,LiCl$
   d. $SF_4 + 3\,H_2O \rightarrow H_2SO_3 + 4\,HF$
   e. $4\,BF_3 + 3\,H_2O \rightarrow H_3BO_3 + 3\,HBF_4$

8. Indicate whether the equations are balanced as written.
   a. $2\,Sn + 2\,H_2SO_4 \rightarrow 2\,SnSO_4 + SO_2 + 2\,H_2O$
   b. $3\,Cl_2 + 6\,NaOH \rightarrow 5\,NaCl + NaClO_3 + 3\,H_2O$

9. Balance the following chemical equations.
   a. $Al + O_2 \rightarrow Al_2O_3$
   b. $C + O_2 \rightarrow CO$
   c. $N_2 + O_2 \rightarrow NO$
   d. $SO_2 + O_2 \rightarrow SO_3$
   e. $NO + O_2 \rightarrow NO_2$

10. Balance the following chemical equations.
    a. $Zn + HCl \rightarrow ZnCl_2 + H_2$
    b. $H_2S + O_2 \rightarrow H_2O + S$
    c. $Al_2(SO_4)_3 + NaOH \rightarrow Al(OH)_3 + Na_2SO_4$
    d. $Zn(OH)_2 + HNO_3 \rightarrow Zn(NO_3)_2 + H_2O$
    e. $NH_4OH + H_3PO_4 \rightarrow (NH_4)_3PO_4 + H_2O$

11. Balance the following chemical equations.
    a. $Cu + H_2SO_4 \rightarrow SO_2 + CuSO_4 + H_2O$
    b. $NH_4Cl + CaO \rightarrow CaCl_2 + H_2O + NH_3$

6.2 Chemical Arithmetic and the Mole

12. Calculate the formula weight (to the nearest whole number) for each of these compounds.
    a. CH_4 b. AlF_3 c. UF_6

13. Calculate the formula weight (to the nearest whole number) for the following.
    a. NH_4NO_3 b. $BaSO_4$
    c. H_3PO_4 d. $KClO_3$

14. Calculate the formula weight (to the nearest whole number) for the following.
    a. $Ca(NO_3)_2$ b. $Mg(OH)_2$
    c. $(NH_4)_2SO_4$

15. Calculate the number of moles of compound for each of the following.
    a. 32 g of CH_4 b. 336 g of AlF_3
    c. 17.6 g of UF_6

16. Calculate the number of moles of compound for each of the following.
    a. 8.0 g of NH_4NO_3 b. 0.233 g of $BaSO_4$
    c. 980 g of H_3PO_4 d. 3.69 g of $KClO_3$

17. Using the examples of Problem 16, determine the number of moles of
    a. N in NH_4NO_3 b. O in $BaSO_4$
    c. H in H_3PO_4 d. K in $KClO_3$

18. Calculate the number of moles of compound for each of the following.
    a. 1.64 g of $Ca(NO_3)_2$
    b. 5.8 g of $Mg(OH)_2$
    c. 0.0132 g of $(NH_4)_2SO_4$

19. Calculate the number of grams in each of these samples.
    a. 0.0010 mol of CH_4
    b. 6.00 mol of AlF_3
    c. 40.0 mol of UF_6

20. Calculate the number of grams in each of these samples.
    a. 0.05 mol of NH_4NO_3
    b. 3.00 mol of $BaSO_4$
    c. 10 mol of H_3PO_4
    d. 0.0200 mol of $KClO_3$

21. Calculate the number of grams in each of these samples.
    a. 1.50 mol of $Ca(NO_3)_2$
    b. 5.8 mol of $Mg(OH)_2$
    c. 0.25 mol of $(NH_4)_2SO_4$

22. Consider the reaction
    $$S + O_2 \rightarrow SO_2$$
    a. How many moles of SO_2 would be formed by the burning of 4 mol of sulfur?
    b. How many moles of SO_2 would be formed in the reaction of 0.6 mol of O_2?
    c. How many moles of sulfur would be required to react with 0.0684 mol of O_2?

23. Answer the following questions based on the equation in Problem 22.
    a. How many moles of SO_2 would form in the reaction of 32 g of S?
    b. How many moles of S would react with 32 g of O_2?

24. Again using the equation in Problem 22, answer the following questions.
    a. How many grams of SO_2 would form in the reaction of 16 g of O_2?
    b. How many grams of SO_2 would be formed by the burning of 8.0 g of sulfur?
    c. How many grams of sulfur would be needed to produce 3.2 g of SO_2?

25. A 1.785-g sample of a hydrate of nickel chloride ($NiCl_2 \cdot xH_2O$) is heated to drive off the water. The anhydrous $NiCl_2$ remaining weighs 0.975 g. Determine the value for x.

26. Consider the reaction
    $$CH_4 + 2 O_2 \rightarrow CO_2 + 2 H_2O$$
    a. How many moles of water are produced in the reaction of 0.4 mol of CH_4?
    b. How many moles of O_2 are required to produce 25 mol of CO_2?
    c. How many moles of CH_4 are required to react with 10 mol of O_2?
    d. If 0.86 mol of O_2 reacts, how many moles of H_2O form?

27. Answer the following questions based on the reaction used in Problem 26.
    a. If 9 g of H_2O forms, how many moles of CH_4 react?
    b. If 2.0 mol of O_2 reacts, how many grams of CO_2 form?

28. Again consider the reaction
    $$CH_4 + 2 O_2 \rightarrow CO_2 + 2 H_2O$$
    a. How many grams of O_2 are required to react completely with 8.0 g of CH_4?
    b. How many grams of CO_2 are produced if 180 g of H_2O is formed?

29. Consider the equation
    $$C_3H_8 + 5 O_2 \rightarrow 3 CO_2 + 4 H_2O$$
    a. How many moles of CO_2 are produced in the reaction of 2 mol of C_3H_8?
    b. How many moles of H_2O are produced in the reaction of 20 mol of O_2?

30. Again, use the equation
    $$C_3H_8 + 5 O_2 \rightarrow 3 CO_2 + 4 H_2O$$
    a. How many moles of O_2 react with 22 g of C_3H_8?
    b. If 1.8 g H_2O is produced, how many moles of CO_2 are formed?

31. Answer the following questions based on the same reaction as in Problems 29 and 30.
    a. If 16 g of O_2 reacts, how many grams of C_3H_8 react?
    b. If 320 g of O_2 reacts, how many grams of CO_2 are formed?

6.3 Gases Revisited

32. A container is found to have 3.2 mol of H_2. If all the gas were evacuated and the container refilled with Cl_2 gas at the same temperature and pressure, how many moles of Cl_2 gas would there be?
33. How many moles of gas are there in a 2.0-L container at 740 mm Hg and standard temperature?
34. If the gas in Problem 33 weighs 3.42 g, what is its molecular weight? What might the gas be?

6.4 Energy Changes in Chemical Reactions

35. What work is accomplished by the activation energy required in both exothermic and endothermic reactions?
36. Refer to the following reaction diagram in answering the questions.

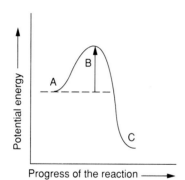

   a. Which letter in the diagram refers to the products?
   b. Which letter refers to the activation energy?
   c. Which letter refers to the reactants?
37. In the diagram for Problem 36, indicate
   a. where an activation complex is found
   b. the amount of heat evolved
38. Is the reaction diagrammed in Problem 36 endothermic or exothermic?
39. If the reaction diagrammed in Problem 36 were reversible, would the reverse reaction be endothermic or exothermic?

6.5 Reaction Rates

40. What effect does changing the temperature have on the rate of a reaction?
41. What is the effect of a change in concentration on the rate of a reaction?
42. If all other conditions were kept constant, what effect would decreasing the temperature of a reaction have on the rate of a reaction?
43. If all other conditions were kept constant, what effect would decreasing the concentration of reactants have on the rate of the reaction?
44. If all other conditions were kept constant, what effect would adding a catalyst have on the rate of the reaction?
45. Why does a catalyst increase the rate of a chemical reaction?

6.6 Equilibrium in Chemical Reactions: Le Châtelier's Principle

46. According to Le Châtelier's principle, what effect will increasing the temperature have on the following equilibria?
   a. $H_2 + Cl_2 \rightleftharpoons 2\,HCl + energy$
   b. $2\,CO_2 + energy \rightleftharpoons 2\,CO + O_2$
   c. $3\,O_2 + heat \rightleftharpoons 2\,O_3$
47. What effect would decreasing the concentration of O_2 have on the equilibrium in Problem 46b?
48. What would increasing the concentration of O_2 do to the equilibrium in Problem 46c?

6.7 Oxidation-Reduction Reactions

49. Define oxidation and reduction in terms of the following.
   a. oxygen atoms gained or lost
   b. hydrogen atoms gained or lost
   c. electrons gained or lost
   d. change in oxidation number
50. In which of the following is the reactant undergoing oxidation? (These are not complete chemical equations.)
   a. $Cl_2 \rightarrow 2\,Cl^-$
   b. $WO_3 \rightarrow W$
   c. $2\,H^+ \rightarrow H_2$
   d. $CO \rightarrow CO_2$
51. The following "equations" show only part of a chemical reaction. Indicate whether the reactant shown is being oxidized or reduced.
   a. $C_2H_4O \rightarrow C_2H_4O_2$
   b. $C_2H_4O \rightarrow C_2H_6O$ c. $Fe^{3+} \rightarrow Fe^{2+}$

52. Green grapes are exceptionally sour because of a high concentration of tartaric acid. As the grapes ripen, this compound is converted to glucose.

$$C_4H_6O_2 \longrightarrow C_6H_{12}O_6$$

Tartaric acid   Glucose

Is the tartaric acid being oxidized or reduced?

53. Circle the oxidizing agent and underline the reducing agent in these reactions.
    a. $4\ Al + 3\ O_2 \rightarrow 2\ Al_2O_3$
    b. $C_2H_2 + H_2 \rightarrow C_2H_4$
    c. $2\ SO_2 + O_2 \rightarrow 2\ SO_3$
    d. $2\ AgNO_3 + Cu \rightarrow Cu(NO_3)_2 + 2\ Ag$

54. Assign oxidation numbers to the underlined elements.
    a. $\underline{S}_8$
    b. $\underline{S}O_3$
    c. $H_2\underline{S}$
    d. K$\underline{I}$

55. Assign oxidation numbers to the underlined elements.
    a. $\underline{Al}_2O_3$
    b. $H\underline{N}O_2$
    c. $Na_2\underline{S}O_3$

56. Assign oxidation numbers to the underlined elements.
    a. $H\underline{C}O_3^-$
    b. $H\underline{P}O_2^-$

57. In the following reactions, which element is oxidized and which is reduced?
    a. $2\ HNO_3 + SO_2 \rightarrow H_2SO_4 + 2\ NO_2$
    b. $2\ CrO_3 + 6\ HI \rightarrow Cr_2O_3 + 3\ I_2 + 3\ H_2O$
    c. $5\ C_2H_6O + 4\ MnO_4^- + 12\ H^+ \rightarrow 5\ C_2H_4O_2 + 4\ Mn^{2+} + 11\ H_2O$

58. The dye indigo (used to color blue jeans) is formed from indoxyl by exposure of the latter to air.

$$2\ C_8H_7ON + O_2 \rightarrow C_{16}H_{10}N_2O_2 + 2\ H_2O$$

Indoxyl   Indigo

What substance is oxidized? What is the oxidizing agent?

59. Relate the chemistry of photosynthesis to the chemistry that provides energy for your heartbeat.

6.8 Oxidation and Antiseptics

60. Name some oxidizing agents used as antiseptics and disinfectants.

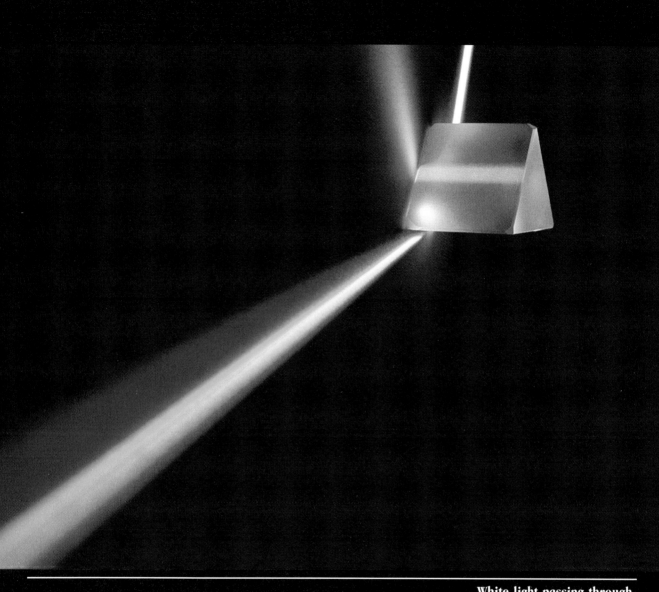

White light passing through a triangular prism is refracted into the spectrum. [David Parker/Science Photo Library/Photo Researchers, Inc.]

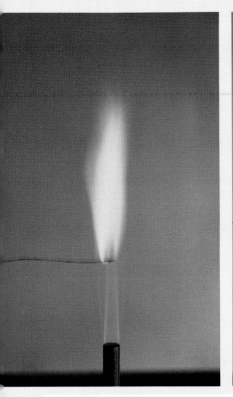 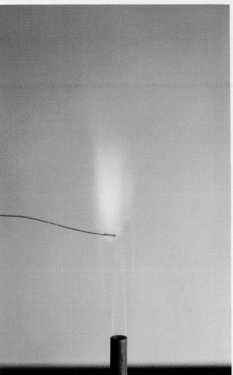 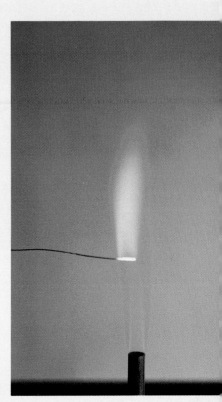

Metal ions produce characteristic flame colors: sodium flame (left), calcium flame (center), potassium flame (right).

Most inorganic compounds are solids, and many are crystalline. Galena, PbS, is a crystal with definite planes and angles.

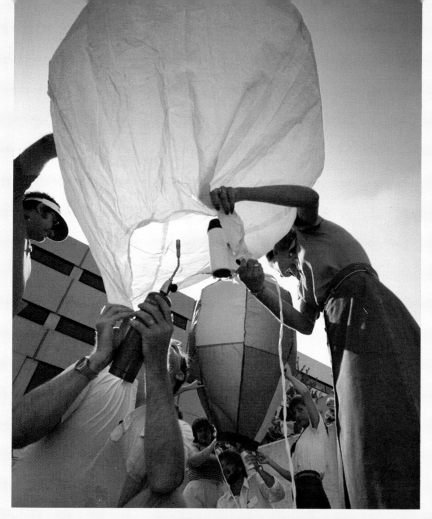

Gases expand as the temperature increases. Air trapped in a balloon is heated and expands, eventually lifting the balloon.

Top: A solution of cobalt chloride poured into a solution of potassium phosphate produces solid cobalt phosphate.

Above: A colorless solution of lead nitrate poured into a colorless solution of sodium iodide produces a yellow precipitate of lead iodide.

Below: Alkali metals are among the most reactive elements. Here a small ball of sodium metal reacts vigorously with water, producing hydrogen and sodium hydroxide. The intense heat also releases steam.

Bottom: A strip of magnesium metal dipped into a dilute solution of hydrochloric acid produces bubbles of hydrogen gas and water-soluble magnesium chloride.

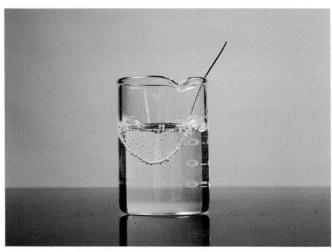

When treated with a drop of water, powdered aluminum and iodine react vigorously, giving off purple iodine vapors and producing aluminum iodide.

Below left: Steel wool burns when heated in a strong flame, producing iron oxide.

Below right: Ammonium dichromate, $(NH_4)_2Cr_2O_7$, undergoing the violent "volcano reaction."

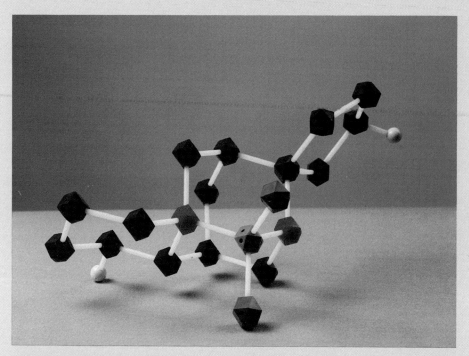

Many organic compounds have complex structures. The ball-and-stick model of sparteine (top) clearly shows the relationships of atoms to each other. The space-filling model of the same compound (left) shows the proper size relationships.

Many organic materials combust. Here ethanol is burning.

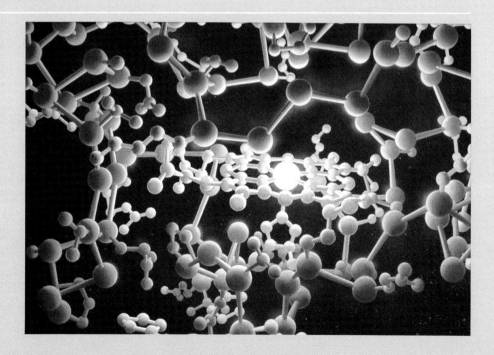

Computer-generated model of a cytochrome molecule, showing heme (in yellow) with its central iron atom (in white). [© Irving Geis/Science Source/Photo Researchers, Inc.]

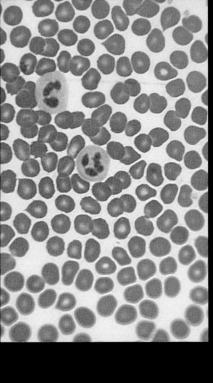
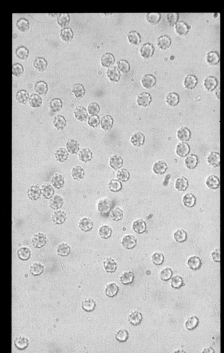

Blood cells in an isotonic solution (far left). Crenation, shrinking, of blood cells in a hypertonic solution (left). Hemolysis, swelling, of a blood cell in a hypotonic solution (below). [Eric Grave/Science Source/Photo Researchers, Inc.; Phil Harrington/Peter Arnold, Inc.; Custom Medical Stock Photo (respectively)]

7

Systems of Solvents, Solutes, and Solutions

Did You Know That...

- most chemical reactions do not take place between pure reagents but rather between *solutions* of the reagents?
- a single cubic mile of ocean water contains over $100 million in dissolved gold and silver?
- solutions can be solid, liquid, or gaseous?
- although water is a common substance, it has extremely unusual properties?
- if water were normal, it would be a gas at room temperature?
- the ocean is our largest body of solution, containing vast amounts of Zn, Mn, Ni, Cu, and Co ions?
- the ocean *is* the plausible place for humans to dispose of some of their wastes?

- some substances are infinitely soluble in each other?
- 80 mL of alcohol added to 20 mL of water gives a total volume of only 98 mL?
- icebergs are not salty like ocean water?
- carrots stay fresh in water and the ice on sidewalks melts with salt because of colligative properties?
- there are vast energy reserves in the osmotic pressure generated between seawater and incoming fresh river water?
- Benjamin Franklin recognized that lead in the environment is a health hazard?
- the Egyptians used osmosis to preserve the dead as mummies?

Our bodies are largely solutions of sodium ions, potassium ions, calcium ions, bicarbonate ions, chloride ions, glucose, amino acids, fatty acids, glycerol, fats, proteins, acetylcholine, and many other substances. We are not entirely in solution, however, or you would be a puddle instead of a person. Except for a few semisolid parts, such as skin, muscle, and bone, human beings are mostly water solutions. Floating around in those solutions are the chemicals of life.

Almost all living systems are made up of chemical "soups" in contact with membranes and small cellular parts called organelles. Life processes occur in solution and at the interface between solutions and in the semisolid organelles and membranes. These processes of life are largely chemical reactions, and for chemical reactions to occur several requirements must be met.

First, chemical reactions depend on the ability of reactants to collide with sufficient impact and with proper lineup for a bond to be formed. This immediately leads to a second requirement: it must be possible for the reactants to move with respect to each other. And one of the simplest ways to allow for motion of chemicals is to place them in a liquid environment. Many of the chemicals of life are solids and thus are essentially frozen in place (Figure 7.1). Fortunately, however, most of these chemicals are also water-soluble and thus become mobile. A third requirement is that of exposure to the environment. Because many of the collisions between chemicals will be unsuccessful owing to improper lineup or insufficient impact force, there must be provision for many, many collisions. Solutions provide for this. They divide the chemicals into the smallest possible components, thereby allowing each molecule or ion to participate in the reaction effort.

To understand life processes, therefore, one must understand the chemistry of solutions.

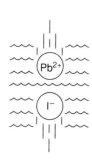

FIGURE 7.1
Chemical interactions between solids are virtually impossible (a); however, when placed in solution (b), reactions are facilitated (c) because of freedom of movement.

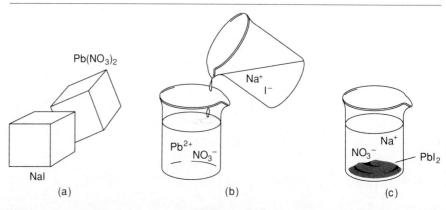

7.1 SOLUTIONS: A REVIEW OF SOME DEFINITIONS

In Chapter 3, **solutions** were defined as intimate, **homogeneous** (evenly distributed) mixtures of two or more components. (By way of distinction, **heterogeneous** systems are uneven mixtures of two or more components in which the components are visually separate phases.) The component that "dissolves" the other components— usually the one present in the greatest amount—is called the **solvent**. The components that are dissolved by the solvent are called **solutes**. Solutes, solvents, and solutions may all be liquid, gas, or solid. Vinegar is a solution of acetic acid (a liquid) in water. Champagne and soda pop have a gas (carbon dioxide) dissolved in water. Sugar and salt are examples of solids that dissolve in water. Blood plasma is a water solution of solids (e.g., salt), liquids (e.g., alcohol, if you have been drinking), and gases (e.g., oxygen and carbon dioxide). A solvent need not be water or even a liquid, as we said. Air is a solution of oxygen, argon, water vapor, and other gases in nitrogen gas. Steel is a solution of carbon (the solute) in iron (the solvent)—a solid in a solid.

In this chapter we shall deal primarily with **aqueous** solutions, those in which the solvent is water. Such solutions are the most important for living systems.

Solutions are homogeneous mixtures.

In an aqueous solution, water is the solvent.

7.2 WATER: SO COMMON, YET SO UNUSUAL

Water is the most abundant compound on our planet, covering three-fourths of the surface of the Earth. In fact, there are some 3.2×10^8 cubic miles of water. More than half again as much water exists in the Earth's crust in the form of hydrates. The importance of water has, of course, long been recognized. As early as the sixth century B.C., Thales of Miletus proposed a philosophy in which water was the basis of all things: "Everything comes from water, and everything goes back to it." However, water is also one of the most unusual substances on earth. Consider these characteristics.

▶ Water exists in nature in all three phases: solid, liquid, and gas.
▶ Like other substances, water expands upon heating and contracts upon cooling. However, as it cools near its freezing point, it reverses from contraction to sudden and rapid expansion. As a result, ice has an open structure (Figure 7.2) and floats on water.
▶ Water has unusually high boiling and freezing points. If these were "normal" relative to similar compounds, boiling would occur at $-80\ °C$ and freezing at $-100\ °C$.
▶ Water has unusually high heats of vaporization and fusion, meaning that much heat is needed to vaporize it or melt it.
▶ Water is a solvent par excellence—an almost universal solvent.

FIGURE 7.2
The structure of ice. When frozen, water molecules may form into hollow rings. The ice has low density and will float in liquid water. Water molecules are now held rigidly in place by the hydrogen bonds.

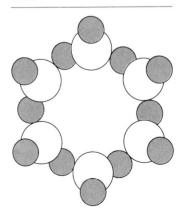

Life depends on the general availability of water but also on its many unusual properties. If it were not for the abnormalities in melting and boiling, water would be vapor in its normal state. If it were not for the unusual density reversal, lakes would freeze from the bottom up, killing marine life. If it were not for the high heat of vaporization, we would need to perspire vast amounts of water to release excess heat. The unusual properties of water can be traced back to its unusual hydrogen bonding, which is important in much of the chemistry of life (e.g., uniting the two strands of the DNA double helix).

These unusual properties can also be destructive. Frostbite results from the rupture of cells, which expand as their water content freezes much as water pipes burst when water in them freezes. Robert Frost spoke to this awesome power in his little poem "Fire and Ice."*

> Some say the world will end in fire,
> Some say in ice,
> From what I've tasted of desire
> I hold with those who favor fire.
> But if it had to perish twice,
> I think I know enough of hate
> To say that for destruction ice
> Is also great
> And would suffice.

7.3 Some Properties of Solutions

Solutions have a number of characteristic properties. The particles of a solution are the smallest possible: molecules, atoms, or ions. Once the solute and solvent are thoroughly mixed, the *solute does not settle out.* Molecular motion keeps the particles randomly and evenly distributed. Thus, if we were to sample the solution at the top, the side, the middle, or the bottom, we would find the composition to be always the same. Solutes cannot be removed from a true solution by passing the solution through a filter paper. The solute particles go through the pores of the paper as readily as the solvent particles (Figure 7.3).

Liquid solutions may be colored, but they are transparent. A beam of light will shine right through a true solution but will not be visible in the solution. When the path of light through a mixture is visible, then particles larger than those in solution are present (Section 7.13).

FIGURE 7.3
(a) This aqueous solution is colored by its solute, potassium permanganate. After filtering the color remains in the solution because the filter paper is unable to stop the solute particles. (b) A suspension of dirt in water can be separated by filtration. The suspended material is trapped by the filter paper, giving a colorless filtrate.

---

* From *The Poetry of Robert Frost* edited by Edward Connery Lathem. Copyright 1923, © 1969 by Holt, Rinehart and Winston. Copyright 1951 by Robert Frost. Reprinted by permission of Henry Holt and Company, Inc.

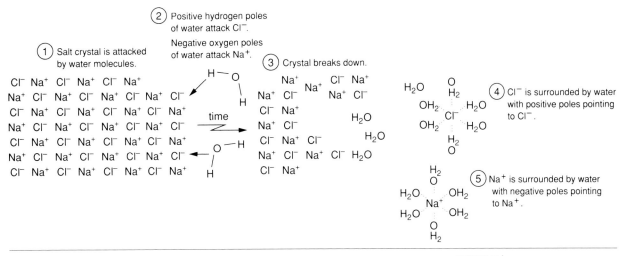

FIGURE 7.4
The process of dissolving salt in water.

7.4 QUALITATIVE ASPECTS OF SOLUBILITY

Table salt is **soluble** in water. Our first question might be: Why is this so? To review the reasons, reread Section 3.14 and carefully study Figure 7.4.

Next we might ask: How fast will a solute dissolve in a solvent? The **rate** of dissolving depends on the types of solute and solvent, on the temperature (higher temperatures increase the rate), on the particle size (the greater the surface area, the faster the dissolving), and on stirring. However, the rate of dissolution is totally independent of the amount that will eventually dissolve.

So the third question might be: How much can we dissolve: a pinch of salt in a cup of water? 10 teaspoonfuls or 100 teaspoonfuls in a cup of water? We know from experience that there is a limit to the amount of salt we can dissolve in a given volume of water. Usually, when we say that something is soluble we are just trying to indicate that an appreciable amount of the material dissolves.

A few substances can be mixed in any proportions to form solutions. Water and alcohol are familiar examples; we say that such substances are completely **miscible**.

At the other extreme of solubility are materials that are essentially **insoluble** in one another. Oil and water are **immiscible**. If we use a method of analysis that is sensitive enough, we would find that some sand does indeed dissolve in water when the two are mixed together. The amount is so minute, however, that for practical purposes we say that the sand is insoluble in water.

Most substances we will encounter in our study of chemistry are like salt, falling somewhere between the two extremes of complete miscibility and insolubility.

Completely miscible substances can be mixed in any proportions.

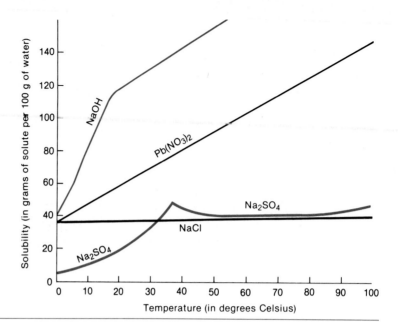

FIGURE 7.5
The effect of temperature on the solubility of several solids in water.

As we noted in Chapter 3, the **amount** of solute that dissolves in a given solvent (the *extent*, or *degree*, to which it is soluble) depends on the nature of the solute and solvent. Temperature and, in the case of gases, pressure are also factors. Solubility usually increases with increasing temperature (Figure 7.5). Thus a handbook might report the solubility of sugar as 179 g/100 g water at 0 °C.

Gaseous solutes are major exceptions to this rule, since the solubility of a gas *decreases* with increasing temperature (Figure 7.6). You have

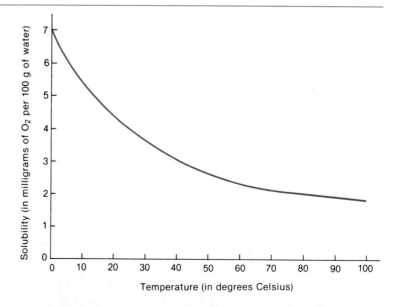

FIGURE 7.6
The solubility of oxygen at various temperatures at 1 atm of pressure.

7.4 Qualitative Aspects of Solubility

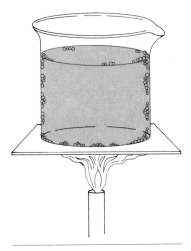

FIGURE 7.7
When a beaker of water is heated, bubbles of air form because air is less soluble in warm water.

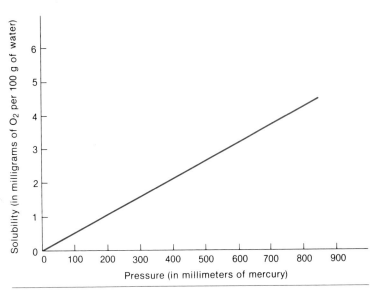

FIGURE 7.8
The effect of pressure on the solubility of oxygen in water at constant temperature.

probably observed this phenomenon when heating water. Long before the water boils, dissolved gases, primarily O_2 and N_2 (air), come out of solution and form bubbles that rise to the surface of the liquid and escape to the atmosphere (Figure 7.7). The solubility of gaseous solutes also varies with pressure, increasing with rising pressure and decreasing with falling pressure (Henry's law, Section 5.9, and Figures 7.8 and 7.9). The kinetic molecular theory offers a model that explains these effects, and that will be our focus in the next section.

FIGURE 7.9
Soft drinks are bottled under pressure so as to dissolve large amounts of carbon dioxide gas. [Edward L. Miller/Stock Boston.]

7.5 Dynamic Equilibrium and Saturated Solutions

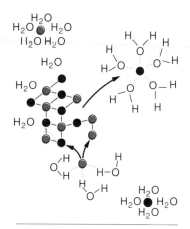

FIGURE 7.10
In a saturated solution the rate of escape from the crystal equals the rate of return to the crystal, establishing a dynamic equilibrium.

A saturated solution has all the solute dissolved in it that it can hold.

Assume that we have just mixed a large amount of crystalline sodium chloride with water. Initially, sodium ions and chloride ions leave the surface of the crystals and wander about at random through the solvent (Section 3.14). Some of the ions in their wanderings return to a crystal surface. These ions can even be trapped there, becoming once more a part of the crystal. As more and more salt dissolves, the number of "wanderers" increases, and thus the number of wanderers available for recapture by the crystal increases. Eventually the number of ions leaving the surface of the remaining undissolved crystals just equals the number returning during the same period. From this point on, the *net* amount of sodium chloride in solution does not change despite the fact that there is still a great deal of activity as ions come and go from the surface of the crystals. The net amount of undissolved crystals also remains constant, although individual crystals may change in shape and size as ions leave one part of the crystal to enter solution while others are deposited in another part of the crystal. Some crystals may even disappear as others grow larger, yet the net amount of undissolved salt does not change. The rate of dissolution just equals the rate of regrowth. A condition of *dynamic equilibrium* has been established (Figure 7.10).

A solution that contains all the solute that it can at equilibrium is said to be **saturated**. One that contains less than this amount is **unsaturated**.

An equilibrium is established at a given temperature. A change in temperature disrupts the balance between the rate of dissolution and the rate of recrystallization. As the temperature goes up, the motion of all the particles increases. More ions are knocked loose from the crystal and go into solution. Further, it is more difficult for the crystal to recapture the ions that return to its surface because they are moving at higher speeds. Thus, if a solution is at equilibrium and the temperature is then raised, more solute usually dissolves. If a solution at equilibrium is cooled, solute usually precipitates. In both cases a new equilibrium can be established at the new temperature. In contrast to solid solutes, gaseous ones can escape from the solution when they reach the surface of a liquid in an open container. At higher temperatures, the faster-moving gaseous solute particles escape more readily, and thus the solubility of a gas decreases with increasing temperature.

If a saturated solution in equilibrium with undissolved solid solute is heated to a sufficiently high temperature, the solution may become unsaturated. All the solid dissolves, and there is still room left for more in the solution. What happens when this solution is cooled? Eventually it will again reach the point of saturation, but in this case no excess

 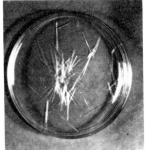 

FIGURE 7.11
Addition of a seed crystal induces rapid crystallization of excess solute from a supersaturated solution. [From Charles W. Keenan, Jesse H. Wood, and Donald C. Kleinfelter, *General College Chemistry*, 5th ed. New York: Harper & Row, 1976. Copyright © 1976 by Keenan, Wood, and Kleinfelter. Reprinted by permission of Harper & Row, Publishers, Inc.]

solid is present in the container. Will solute precipitate? Usually it does, but it might not. There is no equilibrium—no crystals to capture the wandering ions. The solution can be cooled to below the saturation temperature without the occurrence of precipitation. Such a solution, containing solute in excess of that which it could contain *if* it were at equilibrium, is said to be **supersaturated.** This system is not stable, because it is not at equilibrium. Solute may precipitate when the solution is stirred or if the inside of the container is scratched with a glass rod. Addition of a "seed" crystal of solute will nearly always result in the sudden precipitation of all the excess solute (Figure 7.11). Equilibrium is established rather rapidly when there is a crystal to which the ions can attach themselves.

Supersaturated solutions are not unknown in nature. Honey is one example: the solute is sugar. If honey is left to stand, the sugar crystallizes. We say, not very scientifically, that the honey has "turned to sugar." Supersaturated sugar solutions are fairly common in cooking. Jellies are another example. Sugar often crystallizes from jelly that has been standing for a long time.

7.6 Quantitative Aspects of Solubility

Percent Concentrations

Most reactions of interest to us, including those in our bodies, take place in solution. A good cook may well get by with concentrations expressed as "a pinch of salt in a pint of water," but scientific work

FIGURE 7.12
Correct procedure for preparing percent by volume solution.

1. Measure out solute
2. Pour solute into volumetric flask
3. Pour solvent into volumetric flask to the mark

generally requires more precise measurements of amounts. Nonetheless, even scientists find some imprecise terms, such as *weak* and *strong* or *dilute* and *concentrated*, useful on occasion. Thus, a solution may be described as **dilute** if it contains a little bit of solute in a lot of solvent. **Concentrated** may be used to designate a solution in which a lot of solute is dissolved in a relatively small amount of solvent.

The concentration of a solution can be expressed precisely in many ways. **Percent concentration** gives the amount of solute as a percentage of the solution. A 10% concentration means that there are 10 parts of solute in 100 parts of solution. Because the amount of solute and solution can be measured in volume units or in mass units, one can report **percent by volume** or **percent by mass** (commonly called *percent by weight*). If both solute and solvent are liquids, percent by volume is usually used.

FIGURE 7.13
A volumetric flask. When filled to the mark on the neck, the flask contains 1000 mL (1 L) of solution.

$$\% \text{ by volume} = \frac{\text{volume of solute}}{\text{volume of solution}} \times 100$$

Perhaps surprisingly, the mixing of 10 mL of solute and 90 mL of solvent does not necessarily produce 100 mL of solution. As the solute particles distribute themselves among the solvent particles, the interactions between particles change somewhat and so may the total volume. Usually the total volume is quite close to that of the sum of the individual volumes. Thus, when exact concentrations are not critical, you may find someone mixing 10 mL of solute and 90 mL of solvent to produce approximately 100 mL of a 10%-by-volume solution. The *correct* procedure for preparing such a solution is as follows. The measured amount of solute is placed in a container. Then solvent is added until the container is filled to the specified volume for the total solution (Figure 7.12). Usually, the container is a piece of glassware designed to hold a given volume of liquid (for example, a volumetric flask, Figure 7.13).

7.6 Quantitative Aspects of Solubility

EXAMPLE 7.1

What is the concentration in percent by volume of an alcohol–water solution containing 100 mL of alcohol in a total of 400 mL of solution?

$$\% \text{ by volume} = \frac{\text{volume of solute}}{\text{volume of solution}} \times 100$$

$$= \frac{100 \text{ mL}}{400 \text{ mL}} \times 100 = 25.0\%$$

EXAMPLE 7.2

How would you prepare 200 mL of a 40%-by-volume solution of alcohol in water?

$$40\% \text{ by volume} = \frac{\text{mL of alcohol}}{200 \text{ mL of total volume}} \times 100$$

Rearranging this equation, we get

$$\text{mL of alcohol} = \frac{40}{100} \times 200 = 80 \text{ mL of alcohol}$$

To prepare the solution, 80 mL of alcohol should be placed in a volumetric flask. Then water should be added to the flask until the flask is filled to the 200-mL mark.

EXAMPLE 7.3

If great precision is not required, what volume of alcohol and water should be mixed to yield about 50 mL of a 20%-by-volume solution of alcohol in water?

$$20\% = \frac{\text{volume of alcohol}}{50 \text{ mL}} \times 100$$

Rearranging to find the volume of alcohol required, we get

$$\text{Volume of alcohol} = \frac{20 \times 50}{100} = 10 \text{ mL}$$

> The total volume of the solution will approximately equal the sum of the volumes of alcohol and water.
>
> Volume of solution ≈ volume of alcohol + volume of water
>
> To find the amount of water required, subtract the volume of alcohol from the total volume.
>
> Volume of water = volume of solution − volume of alcohol
> = 50 mL − 10 mL = 40 mL of water
>
> When 10 mL of alcohol is mixed with 40 mL of water, the total volume may not be exactly 50 mL, but it will be close.

For solutions involving solutes that are solids, **percent-by-mass** concentrations are more commonly used.

$$\% \text{ by mass} = \frac{\text{mass of solute}}{\text{mass of solution}} \times 100$$

$$= \frac{\text{mass of solute}}{\text{mass of solute} + \text{mass of solvent}} \times 100$$

The mass of the solution *always* equals the mass of solute (or solutes) plus the mass of solvent. In this case, then, one can measure out separately the solute and solvent and then add the two together to get precisely the correct concentration (Figure 7.14).

FIGURE 7.14
Correct procedure for preparing percent-by-mass solution.

1. Weigh out solute.

2. Weigh out solvent.

3. Mix solute with solvent.

7.6 Quantitative Aspects of Solubility

EXAMPLE 7.4

What is the concentration in percent by mass of a solution of 5 g of NaCl dissolved in 495 g of water?

$$\% \text{ by mass} = \frac{5 \text{ g}}{5 \text{ g} + 495 \text{ g}} \times 100 = \frac{5 \text{ g}}{500 \text{ g}} \times 100 = 1\%$$

One advantage of dealing with aqueous solutions is that the density of the solvent, water, is 1 g/mL. For liquids, it is frequently easier to measure a specified volume rather than a given mass. If 25 g of water is required, 25 mL can be measured out.

EXAMPLE 7.5

Describe the preparation of 100 g of 4%-by-mass aqueous NaCl solution.

$$4\% = \frac{\text{mass of NaCl}}{100 \text{ g}} \times 100$$

$$\text{Mass of NaCl} = \frac{4}{100} \times 100 = 4 \text{ g}$$

$$\text{Amount of water} = 100 \text{ g} - 4 \text{ g} = 96 \text{ g} = 96 \text{ mL}$$

To prepare the solution, dissolve 4 g of NaCl in 96 mL of water.

EXAMPLE 7.6

How would you prepare about 50 mL of 2%-by-mass NaOH solution?

  Notice that you are being asked to prepare a given *volume* of a solution whose concentration is given in percent by *mass*. For dilute aqueous solutions, the density of the solution is roughly equal to the density of pure water. Therefore, we can make an approximation and say that 50 mL of the solution will have a mass of about 50 g.

$$2\% = \frac{\text{grams of NaOH}}{50 \text{ g solution}} \times 100$$

Rearranging, we get

$$\text{Grams of NaOH} = \frac{2 \times 50}{100} = 1 \text{ g}$$

To prepare the solution, measure out 1 g of NaOH. Add 49 g (49 mL) of water. The final volume will not be exactly 50 mL, but it will be close to that value.

Measurement of Very Low Concentrations

For very low concentrations of solute, such as those reported in clinical analyses of urine or blood samples, the unit **milligram percent** is often used. Milligram percent is defined as milligrams of solute per 100 mL of solution.

$$\text{Milligram percent} = \frac{\text{milligrams of solute}}{100 \text{ mL of solution}}$$

Normally there is about 320–350 mg of sodium ion per 100 mL of blood plasma. Thus, the normal concentration of sodium ion in blood plasma ranges from 320 to 350 mg %. The use of milligram percent avoids the sometimes awkward use of decimal numbers. For example, the above concentrations of sodium ion would be expressed as 0.320–0.350% by mass. (A 1 mg % solution is one-thousandth the concentration of a 1% solution.)

EXAMPLE 7.7

In the fluid surrounding the cells of the body, the concentration of cholesterol is 0.150% by mass and the concentration of lactic acid is 10 mg %. Which solute is present in higher concentration?

To answer the question, you need to have the concentrations in identical units in order to compare them. To convert the concentration of cholesterol to mg %, simply move the decimal point three places to the right.

$$0.150\% \text{ by mass} = 150 \text{ mg \%}$$

The concentration of the two substances can now be compared.

7.6 Quantitative Aspects of Solubility

> 10 mg % lactic acid
>
> 150 mg % cholesterol
>
> Now it is easy to see that the cholesterol concentration is higher.*

For even more dilute solutions, concentrations can be expressed in parts per million (**ppm**) or even parts per billion (**ppb**). A concentration in ppm is the same as one in milligrams per kilogram (mg/kg or, for aqueous solutions, mg/L), and ppb is identical to micrograms per kilogram (μg/kg or, for aqueous solutions, μg/L). These units are used to measure extremely low levels of toxic materials. For example, benzene is a compound that has been shown to produce leukemia-like symptoms in laboratory animals and humans. The Supreme Court has dealt with the question of whether the concentration of benzene in the air breathed by workers should be limited to 10 ppm or 1 ppm. The court decided that industries could not be required to lower the concentration from 10 to 1 ppm unless the higher concentration was *proved* to be dangerous.

ppm: parts per million

ppb: parts per billion

Percent (%): parts per hundred, pph

EXAMPLE 7.8

What is the concentration in ppm of an aqueous solution that contains 1.52 g of sodium cyanide in 250 L of solution?

The given concentration is 1.52 g/250 L. A concentration in ppm is equivalent to a concentration in mg/L. Therefore, use a metric conversion factor to convert g/L to mg/L.

$$\frac{1.52 \text{ g}}{250 \text{ L}} \times \frac{1000 \text{ mg}}{1 \text{ g}} = \frac{1520 \text{ mg}}{250 \text{ L}}$$

$$= 6.08 \text{ mg/L}$$

$$= 6.08 \text{ ppm}$$

As our technology becomes more sophisticated, our ability to detect minute quantities of materials increases. This increase in the sensitivity of analytical techniques raises questions. When a substance is first detected in the ppm or ppb range, is it a new contaminant in our environment, or has it been there all along at levels that were previously undetectable?

---

*Today mg % has been largely replaced by mg/dL, where dL is deciliter (= 10 mL).

Ratio Concentrations

The strength of a solution is sometimes presented as a ratio, for example, 1:1000 or 1:2000. Unfortunately, there is no universal agreement as to the meaning of the ratio. The ratio 1:1000 may be used to indicate that 1 g of pure solute (perhaps a drug) is to be dissolved in 1000 mL of solution. In the case of a liquid solute, the ratio may mean 1 mL of solute in 1000 mL of solution. Sometimes the ratio is presented as a dilution factor. Thus, 1:1000 may specify that one part of a sample is to be diluted with solvent until the volume of the solution is 1000 times the original volume. In specific instances, the meaning of the ratio may be clear, but because of the possibility for confusion, the use of ratio concentrations is being discouraged. Nonetheless, let us look at one example of the use of a ratio as a dilution factor.

EXAMPLE 7.9

The directions for use of a liquid drug indicate that it should be diluted 1:2000. How would 500 mL of the solution be prepared?

The ratio should be read as

$$\frac{1 \text{ part solute}}{2000 \text{ parts solution}}$$

The final volume of solution required is 500 mL (500 "parts").

Volume of solute

$$= 500 \text{ mL (parts) of solution} \times \frac{1 \text{ part solute}}{2000 \text{ parts solution}}$$

$$= 0.25 \text{ part (mL) solute}$$

To prepare the solution, 0.25 mL of the drug (as supplied) should be diluted to a total volume of 500 mL by addition of the appropriate solvent. This gives exactly the same final concentration as the dilution of 1 mL of drug to a total volume of 2000 mL.

Molarity

In Chapter 6 we learned that chemicals react in mole ratios specified by balanced chemical equations. It is often convenient, therefore, to

7.6 Quantitative Aspects of Solubility

know concentrations of solutions in terms of moles rather than mass or volume of solute. A solution that contains 1 mol of solute per liter of solution is called a 1 *molar* (or 1 M) solution. **Molarity**, then, is defined as the number of moles of solute divided by the number of liters of solution.

$$\text{Molarity (M)} = \frac{\text{moles of solute}}{\text{liters of solution}}$$

EXAMPLE 7.10

What is the molarity of a solution made by dissolving 3 mol of NaCl in enough water to make 6 L of solution?

$$\text{Molarity} = \frac{3 \text{ mol of solute}}{6 \text{ L of solution}} = 0.5 \text{ M}$$

EXAMPLE 7.11

What is the molarity of a solution of 120 g of NaOH in enough water to make 10 L of solution?

First we must determine the number of moles of NaOH. The formula weight of NaOH is 40.

$$1 \text{ mol NaOH} = 40 \text{ g NaOH}$$

Therefore,

$$120 \text{ g NaOH} \times \frac{1 \text{ mol NaOH}}{40 \text{ g NaOH}} = 3.0 \text{ mol NaOH}$$

The concentration of the solution is

$$\frac{3.0 \text{ mol}}{10 \text{ L}} = 0.30 \text{ M}$$

Frequently one needs to know how much solute is required for the preparation of a specified amount of solution of a given molarity. If

$$\text{Molarity} = \frac{\text{moles of solute}}{\text{liters of solution}}$$

then

$$\text{Moles of solute} = \text{molarity} \times \text{liters of solution}$$

or

$$\text{Millimoles of solute} = \text{molarity} \times \text{milliliters of solution}$$

EXAMPLE 7.12

How many grams of NaOH are required for the preparation of 500 mL (0.5 L) of 0.4 M solution?

We can use the rearranged equation to determine the number of moles of NaOH required.

$$\text{Moles of NaOH} = 0.4 \text{ M} \times 0.5 \text{ L}$$

$$= \frac{0.4 \text{ mol}}{1 \text{ L}} \times 0.5 \text{ L} = 0.2 \text{ mol}$$

But we were asked for the number of grams and not moles, so we must now calculate the number of grams in 0.2 mol.

$$1 \text{ mol NaOH} = 40 \text{ g NaOH}$$

Using this equality as a conversion factor, we find

$$0.2 \text{ mol} \times \frac{40 \text{ g}}{1 \text{ mol}} = 8 \text{ g}$$

FIGURE 7.15
Correct procedure for preparing a molar solution.

1. Calculate grams of solute needed for desired molarity.

2. Weigh out needed mass of solute.

3. Place solute into volumetric flask.

4. Add water (solvent) while stirring, until the mark is reached.

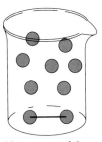 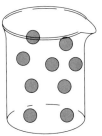 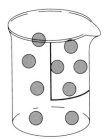

M, measure of the distance between solute particles.

Moles, count of all particles present.

M is the same in both segments; the number of moles is different.

FIGURE 7.16 Distinguishing between moles and molarity.

These solutions are prepared by weighing out the required amount of solute, placing it in a volumetric flask, and adding solvent to the mark (Figure 7.15).

It is very important that you understand the difference between moles and molarity. Moles is a count, in large groups, of the number of particles. Molarity is a measure of how crowded these particles are in the solution (see Figure 7.16). So to find the number of moles we must know the total mass and volume, but to find the molarity all we need to do is sample a portion of the solution. If you assume that an audience is evenly spread throughout an auditorium, then to find the number of persons in the auditorium you need to consider the entire auditorium. However, if you are interested in the crowdedness, or persons per available seats, all you need to do is check the number of persons in a given section and compare it to the number of seats available in that section.

Equivalents

When salts are placed in water, the crystals do not just dissolve into formula units but actually dissociate into ions. Thus, when an aqueous solution of $CaCl_2$ is prepared, the result is not a solution containing $CaCl_2$ systems. It is a solution containing Ca^{2+} ions and Cl^- ions (twice as many Cl^- as Ca^{2+} ions) all moving about rather independently. Such ionic solutes are referred to as **electrolytes** and are very important components of body fluids. They participate in transport across cell membranes, establish osmotic pressures (Section 7.9), and are conduits of nerve impulses that control muscle movement.

A unit related to the mole and sometimes used to report concentrations of some solutes in body fluids is the *equivalent*. An **equivalent** (equiv) is equal to the number of grams of a substance that will produce 1 mol of ionic charge in solution. By 1 mol of charge we mean 6.02×10^{23} positive or negative charges. One mole of sodium ions (Na^+) contains 1 mol of positive charge, but 1 mol of calcium ions

Formula unit: the smallest arrangement of ions that is electrically neutral

Electrolyte: an ionic solute

1 equivalent will produce 1 mol of ionic charge in solution.

(Ca^{2+}) contains 2 mol of positive charge. Therefore, 0.5 mol of Ca^{2+} is *equivalent* to 1 mol of Na^+ because both contain 1 mol of charge. The mass of 0.50 mol of calcium ions is 20 g, which is 1.0 equiv of Ca^{2+}, and the mass of 1.0 mol of sodium ions is 23 g, which is 1.0 equiv of Na^+.*

EXAMPLE 7.13

How many equivalents are in 1 mol of each of the following: K^+, Mg^{2+}, HCO_3^-, PO_4^{3-}?

This is easy. The number of equivalents per mole is equal to the charge on the ion (ignoring the sign of the charge).

The absolute value of the charge on K^+ is 1. Therefore, 1 mol of K^+ = 1 equiv.

$$\text{For } Mg^{2+}, \quad 1 \text{ mol} = 2 \text{ equiv}$$
$$\text{For } HCO_3^-, \quad 1 \text{ mol} = 1 \text{ equiv}$$
$$\text{For } PO_4^{3-}, \quad 1 \text{ mol} = 3 \text{ equiv}$$

EXAMPLE 7.14

How many equivalents are in 0.65 mol of SO_4^{2-}?

Since the charge on the ion is 2−,

$$1 \text{ mol} = 2 \text{ equiv}$$

Therefore,

$$0.65 \text{ mol} \times \frac{2 \text{ equiv}}{1 \text{ mol}} = 1.3 \text{ equiv}$$

EXAMPLE 7.15

How many equivalents are in 9.6 g of HPO_4^{2-}?

First calculate the number of moles in 9.6 g of HPO_4^{2-}. The formula weight of the ion is 96.

$$9.6 \text{ g} \times \frac{1 \text{ mol}}{96 \text{ g}} = 0.10 \text{ mol}$$

---

* A mole of an ion has the same mass as a mole of an equivalent uncharged formula unit. The charge on an ion results from the loss or gain of electrons. Electrons contribute so little to the mass of an atom that their contribution, or lack of it, is ignored.

7.6 Quantitative Aspects of Solubility

The hydrogen phosphate ion has a charge of 2−. Therefore,

$$1 \text{ mol} = 2 \text{ equiv}$$

Use this equality as a conversion factor.

$$0.10 \text{ mol} \times \frac{2 \text{ equiv}}{1 \text{ mol}} = 0.20 \text{ equiv}$$

The concentration of ionic solutes in body fluids is traditionally reported in milliequivalents (meq or sometimes mEq) per liter (Table 7.1). The concentration in units of equivalents per liter is also known as **normality** (N).

EXAMPLE 7.16

How many milliequivalents are in 0.125 equiv of a substance?

Milli- has the same meaning here that it generally has in the metric system.

$$1 \text{ equiv} = 1000 \text{ meq}$$

Thus,

$$0.125 \text{ equiv} \times \frac{1000 \text{ meq}}{1 \text{ equiv}} = 125 \text{ meq}$$

TABLE 7.1 Concentration of Electrolytes in Some Body Fluids

Electrolyte	Concentration in Body Fluids		
	Blood Plasma (meq/L)	Urine (meq/24 hr)	Sweat (meq/L)
Bicarbonate (HCO_3^-)	24–30		
Calcium (Ca^{2+})	2.1–2.6		
Chloride (Cl^-)	100–106	110–250	4–60
Magnesium (Mg^{2+})	1.6–2.4	6.0–8.5	
Phosphate (HPO_4^{2-}, $H_2PO_4^-$)	1–1.5		
Potassium (K^+)	4–5	40–80	
Sodium (Na^+)	136–145	80–180	10–80
Sulfate (SO_4^{2-})	0.5–1.5		

EXAMPLE 7.17

The concentration of calcium ion is reported as 210 meq/L in a sample of blood. How many equivalents per liter is this? How many grams of calcium are in 1 L of blood?

To convert meq to equivalents, just use the appropriate metric conversion factor.

$$\frac{210 \text{ meq}}{L} \times \frac{1 \text{ equiv}}{1000 \text{ meq}} = \frac{0.210 \text{ equiv}}{L}$$

To determine the number of grams of calcium ion in 1 L of blood, first convert equivalents to moles. For calcium ion (Ca^{2+}),

$$1 \text{ mol} = 2 \text{ equiv}$$

Thus,

$$0.210 \text{ equiv} = \frac{1 \text{ mol}}{2 \text{ equiv}} = 0.105 \text{ mol}$$

The gram formula weight of the calcium ion is 40.1 g/mol (the same as that of the calcium atom).

$$0.105 \text{ mol} \times \frac{40.1 \text{ g}}{1 \text{ mol}} = 4.21 \text{ g of calcium ion}$$

An important difference exists between concentrations expressed in molarity (or equivalents per liter) and those expressed in percent (or mg % or ppm, and so on). If you are trying to figure out how much solute to measure out for a particular percent concentration, the amount you measure is not affected by the identity of the solute. A 10% solution of NaOH contains 10 g of NaOH per 100 g of total solution. Similarly, 10% HCl and 10% $(NH_4)_2SO_4$ and 10% $C_{110}H_{190}N_3O_2Br$ solutions each contain 10 g of the specified solute per 100 g of solution. For molar solutions, however, the mass of solute in a solution of specified molarity is different for different solutes. A liter of 0.10 M solution requires 4.0 g (0.10 mol) of NaOH, 3.7 g (0.10 mol) of HCl, 13 g (0.10 mol) of $(NH_4)_2SO_4$, or 170 g (0.10 mol) of $C_{110}H_{190}N_3O_2Br$.

EXAMPLE 7.18

What mass of solute is required to prepare 1.0 L of each of the following solutions?

a. 1.0%-by-mass aqueous solution of NaOH
b. 1.0%-by-mass aqueous solution of $(NH_4)_2SO_4$
c. 1.0 M aqueous solution of NaOH
d. 1.0 M aqueous solution of $(NH_4)_2SO_4$

a.
$$\% \text{ by mass} = \frac{\text{mass of solute}}{\text{mass of solution}} \times 100$$

$$1.0\% = \frac{\text{g of solute}}{1000 \text{ g}} \times 1000$$

Rearranging, we get

$$\text{g of solute} = \frac{1.0\% \times 1000 \text{ g}}{100} = 10 \text{ g}$$

The 10 g of solute should be dissolved in 990 g of water to give a total of 1000 g of solution, which is approximately 1.0 L of aqueous solution.

b. The calculation is exactly the same as shown in (a).

c. One liter of a 1.0 M solution requires 1.0 mol of solute. The formula weight for NaOH is 40; therefore,

$$1.0 \text{ mol} = 40 \text{ g}$$

Dissolve 40 g of NaOH in water to give a total of 1.0 L of solution.

d. The formula weight of $(NH_4)_2SO_4$ is 132. Dissolve 132 g of $(NH_4)_2SO_4$ in water to give a total of 1.0 L of 1.0 M solution.

7.7 STRENGTHS OF DRUGS

Many drug manufacturers market their products for use in strengths that are not specified by concentration units discussed above. Instead they provide explicit instructions on how to prepare the drug for administration by professionals (referred to as *reconstitution* of the drug).

EXAMPLE 7.19

A drug label reads "Add 5.0 mL of sterile water; each mL of solution will contain 0.50 g." If the dose prescribed for a patient is

750 mg, how many milliliters of the reconstituted drug should be administered?

First, the drug must be reconstituted by addition of 5.0 mL of water to the container. After thorough mixing, the drug is ready for use. To calculate the dose in milliliters, the conversion factor given on the label (1.0 mL = 0.50 g) is used. Before that can be done, the specified dose (750 mg) must be converted to grams.

$$\text{Dose in mL} = 750 \text{ mg} \times \frac{1.0 \text{ g}}{1000 \text{ mg}} \times \frac{1.0 \text{ mL}}{0.50 \text{ g}} = 1.5 \text{ mL}$$

For some medications, the strength of the medication is reported literally in *units* (U). A **unit** of the substance is defined as an amount that elicits a particular effect. USP units, for example, meet standards set by the United States Pharmacopeia. The strength of some vitamins and antibotics, among other substances, is reported in this way.

EXAMPLE 7.20

The label on a container of crystalline penicillin G states, "RECONSTITUTION: add 9.6 mL diluent to provide 100 000 U per mL." If the prescribed dosage is 500 000 U, how much of the reconstituted drug should be administered?

Let us assume that you have correctly reconstituted the drug. The rest is easy.

$$\text{Dose in mL} = 500\,000 \text{ U} \times \frac{1 \text{ mL}}{100\,000 \text{ U}} = 5 \text{ mL}$$

7.8 Dilution of Stock Solutions

In scientific and clinical laboratories, stock solutions of known concentrations are often available. The stock solution can be diluted by addition of solvent to provide a whole range of concentrations. The key point is that addition of water (solvent) to a stock solution *does not affect the number of moles of solute.* It only dilutes the solution—lowers the concentration! (See Figure 7.17.) *Any* volume or concentration units can be used as long as you are consistent.

7.8 Dilution of Stock Solutions

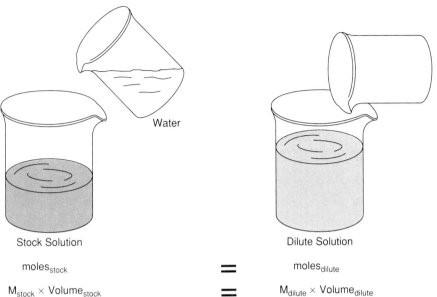

FIGURE 7.17
Dilution affects the molarity but not the moles.

EXAMPLE 7.21

A stock solution of 50% NaOH is available. How would 2 L of 10% solution be prepared?

$$\text{vol}_{stock} \times \text{conc}_{stock} = \text{vol}_{desired} \times \text{conc}_{desired}$$

Volume of stock solution $\times$ 50% = 2 L $\times$ 10%

Rearrangement gives

$$\text{Volume of stock solution} = \frac{2 \text{ L} \times 10\%}{50\%} = 0.4 \text{ L}$$

Take 0.4 L (400 mL) of the stock solution and add water to a total volume of 2 L.

EXAMPLE 7.22

If 5.0 mL of 4.0 M stock solution is diluted to 80 mL, what is the concentration of the final solution?

5.0 mL $\times$ 4.0 M = 80 mL $\times$ concentration of final solution

$$\text{Concentration of final solution} = \frac{5.0 \text{ mL} \times 4.0 \text{ M}}{80 \text{ mL}} = 0.25 \text{ M}$$

FIGURE 7.18
Application of calcium chloride salt cause snow to melt at temperatures below the normal freezing point of 0 °C. [Miro Vintoniv/Stock Boston.]

7.9 COLLIGATIVE PROPERTIES OF SOLUTIONS

Solutions have higher boiling points and lower freezing points than the corresponding pure solvent. The antifreeze in automobile cooling systems is there precisely because of these effects. If water alone were used as the engine coolant, it would boil away in the heat of summer and freeze in the depths of northern winters. Addition of antifreeze to the water raises the boiling point of the coolant and also prevents the coolant from freezing solid when the temperature drops below 0 °C. Salt is thrown on icy sidewalks and streets because the salt dissolves in the ice and lowers its freezing point (Figure 7.18). Therefore, the ice melts because the outdoor temperature is no longer low enough to maintain it as a solid.

To understand the reason for this, consider Figure 7.19. Beaker A contains a pure solvent. Some of the liquid molecules escape into the

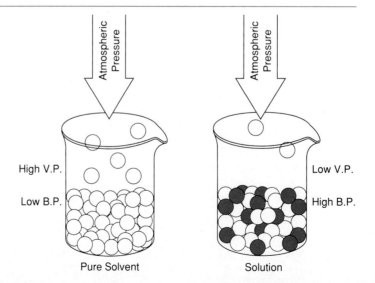

FIGURE 7.19
Relative vapor pressure and boiling points of a solvent and a solution.

7.9 Colligative Properties of Solutions

vapor phase, causing a given vapor pressure (vp). If we want to boil that solvent, we have to add enough heat to increase the vapor pressure until it is equal to the atmospheric pressure above the beaker. Beaker B contains a solution prepared from the same solvent as in A plus a solute. Its vapor pressure is *lower* than it was in A—the number of solvent molecules that have a chance to escape from the surface is not as large because solute molecules block their passage. If we want to boil this low vapor pressure solution, we need to add more heat (bring it to a higher temperature) than the solvent alone in beaker A to attain a vapor pressure equal to the atmospheric pressure. The boiling point (bp) of a solution is higher than that of the pure solvent.

In order to freeze a pure solvent the temperature must be lowered until the solvent molecules arrange in a crystalline form. If a solute is present, such as in a solution, it will act as a hindrance to the crystallization process. Therefore it requires even lower temperatures to force crystallization—the freezing point (fp) of a solution is lower than that of a pure solvent.

The extent to which freezing points and boiling points are affected by solutes is related to the number of solute particles present in solution. The higher the concentration of solute particles, the more pronounced the effect.

$$\Delta fp \propto \text{solute particles}$$

$$\Delta bp \propto \text{solute particles}$$

where Δ means a change in the property and $\propto$ means "is proportional to." **Colligative properties** of solutions are those properties, like vapor pressure depression, boiling point elevation, and freezing point depression, that depend on the number of solute particles present in solution. For living systems, perhaps the most important colligative property is osmotic pressure resulting from *osmosis*, a phenomenon we shall discuss in detail in the next section.

Colligative properties depend on the number of solute particles present.

Before we do that we must first consider a rather subtle aspect of solute concentration. See if you can answer the following questions. How many solute particles are in 1 L of a 1 M glucose, $C_6H_{12}O_6$, solution? How many in 1 L of 1 M sodium chloride. NaCl, solution? In 1 M calcium chloride, $CaCl_2$, solution? The answer "should" be 6×10^{23}, right? All the solutions contain 1 mol of their respective solute compounds. The question did *not* ask for the number of *formula units*, however; it asked for the number of *solute particles*. Glucose is a covalent compound; its atoms are all firmly tied together in molecules. In the glucose solution, each solute particle is a glucose molecule, and there *are* 6×10^{23} of these. But in the sodium chloride solution, each formula unit of NaCl consists of a separate sodium ion (Na^+) and chloride ion (Cl^-) in solution. In Section 3.14 we described the dissolution of sodium chloride in water. Individual ions are carried off into solution by solvent molecules. So 6×10^{23} formula units of NaCl produce 12×10^{23} particles in solution—6×10^{23} sodium ions plus 6×10^{23} chloride ions. Each calcium chloride unit produces three particles in solution—one calcium ion plus two chloride ions. Thus, the effect of a 1 M NaCl solution on colligative properties is twice that of a 1 M glucose solution. A calcium chloride solution has about three times the effect of a glucose solution of the same molarity. When colligative properties (specifically osmotic pressure, Section 7.10) are being discussed, concentration may be reported in terms of osmoles per liter or **osmolarity** (osmol/L). An **osmol** is a mole of solute particles. A 1 M NaCl solution contains 2 osmol of solute per liter of solution; a 1 M CaCl$_2$ solution contains 3 osmol/L. The osmolarity of a 1 M glucose solution is 1 osmol/L. The concentration of body fluids is typically reported in milliosmoles per liter (mosmol/L).

1 osmol = 1 mol of solute particles

7.10 Osmosis

In Section 7.3 we saw that solutions—both solute and solvent—pass through filter paper. The filter paper is **permeable** to all components of a true solution. Everyday experience tells us that other materials are **impermeable**. Neither solvent nor solute passes through the metal walls of cans or the glass walls of bottles and jars. Are there, perhaps, materials with intermediate properties, materials that will allow solvent molecules but not solute molecules to pass through, materials that are permeable to some solutes but not others? The answer, as you might suspect, is an emphatic *yes.*

Materials that permit the passage of some substances but not others are said to be **semipermeable**. Cell membranes, the lining of the digestive tract, and the walls of blood vessels are all semipermeable. The mechanism by which these naturally functioning semipermeable membranes control the flow of materials is exceedingly complex. One aspect of that control, however, is easy to understand.

The sieve model of osmosis pictures semipermeable membranes as having extremely small pores. The size of these pores is such that tiny molecules, such as water, can pass through but larger particles, such as sugar molecules, cannot. If a membrane with these characteristics separates a compartment containing pure water from one containing a sugar solution, an interesting thing happens. The volume of liquid in the compartment containing sugar increases while the volume in the pure water compartment decreases.

Use Figure 7.20 as a reference. On both sides of the membrane all molecules are moving about at random, occasionally bumping against the membrane. If a water molecule happens to hit one of the pores, it

7.10 Osmosis

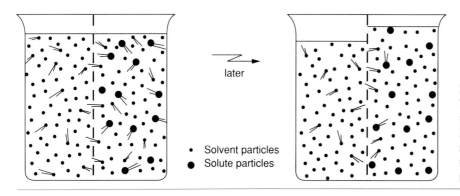

FIGURE 7.20
The sieve model of osmosis holds that the semipermeable membrane has pores large enough to permit the passage of small particles (water molecules) but too small to permit the passage of larger molecules.

passes through the membrane into the other compartment. When the much larger sugar molecule strikes a pore, it bounces back instead of passing through. The more sugar molecules there are in solution (i.e., the more concentrated the solution), the smaller the chance that a water molecule will strike a pore. Thus, in our example, there will be a *net* flow of water from the left compartment into the right compartment. This net diffusion of solvent through a semipermeable membrane is called **osmosis**. During osmosis, there is always a *net* flow of solvent from the more dilute (or pure solvent) area to the compartment in which the solution is more concentrated (Figure 7.21). The net diffusion of water would be *from* a 5% sugar solution *into* a 10% sugar solution.

> Osmosis: the net diffusion of water or other small molecules through a semipermeable membrane

As the liquid level in the right compartment builds up and that in the left compartment drops, the increased weight of fluid in the right compartment exerts pressure, which makes it more difficult for water molecules to enter that compartment. (Think of how a barometer works—Section 5.3.) Eventually the buildup of pressure is sufficient to

FIGURE 7.21
Net solvent flow through a semipermeable membrane occurs spontaneously in only one direction, from the compartment containing dilute solution (or pure solvent) into the compartment of concentrated solution. Remember: ordinarily the terms *dilute* and *concentrated* are used to describe the concentration of *solute*. The net flow of solvent is from where the solvent is more concentrated to where the solvent is less concentrated.

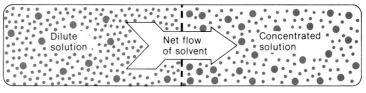

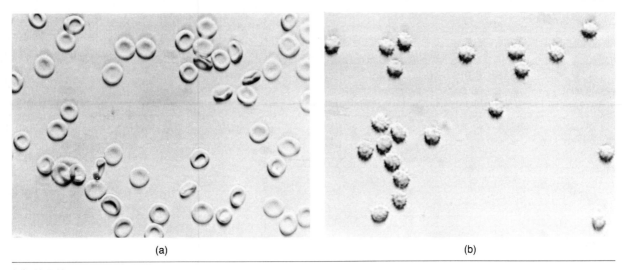

FIGURE 7.22
(a) Normal human red blood cells.
(b) After exposure to a hypertonic solution the cells are wrinkled and chriveled. [Courtesy of A. M. Winchester, Ph.D., Greeley, CO.]

Osmotic pressure: the pressure needed to prevent the net flow of solvent

Plasmolysis or hemolysis: the swelling of cells

Crenation: the wrinkling or shriveling of cells

prevent further net flow of water into that compartment. Things have not really come to a standstill; the rates at which water molecules move back and forth are just equal.

Instead of waiting for the liquid level to build up and stop the net flow of water, we can apply an external pressure to the compartment containing the more concentrated solution and accomplish the same thing. The precise amount of pressure needed to prevent the net flow of solvent from the dilute solution to the concentrated one is called the **osmotic pressure**. The magnitude of the osmotic pressure depends only on the concentration of solute particles, that is, on the osmolarity of the solution. The more particles (the higher the osmolarity), the greater the osmotic pressure. You can think of osmotic pressure and osmolarity as measures of the tendency of a solution to draw solvent into itself.

Living cells can be regarded as selectively permeable bags filled with solutions of ions, small and large molecules, and still larger cell components. Normally, the fluid surrounding a cell has the same osmolarity as the fluid within the cell. Flow in and flow out are about equal. If a cell is surrounded by a solution of lower osmolarity (a **hypotonic solution**), there is a net flow of water into the cell. The cell swells; it may even burst. The rupture of a cell by a hypotonic solution is called **plasmolysis**. If the cell is a red blood cell, the more specific term **hemolysis** is used. What happens if a cell is placed in a solution of higher osmolarity (a **hypertonic solution**)? The net flow of water is then out of the cell; the cell wrinkles and shrinks (Figure 7.22). The shriveling is called **crenation**, a process that can also lead to the death of the cell.

Solutions that exhibit the same osmotic pressure as that of the fluid inside the cell are said to be **isotonic**. Ordinarily, in replacing body fluids intravenously, the fluid must be isotonic. Otherwise, hemolysis or crenation results, and the patient's well-being is seriously jeopardized. A 0.92% sodium chloride solution, called physiological saline, and a 5.5% glucose (also called dextrose or blood sugar) solution are isotonic with the fluid inside red blood cells. The "D5W" so often referred to by television's doctors and paramedics is this approximately 5% solution of dextrose (the D) in water (the W). A 0.92% sodium chloride solution is about 0.16 M. A 5.5% glucose solution is approximately 0.31 M.

Isotonic solution: one with the same osmotic pressure as the fluid inside the cell

EXAMPLE 7.23

What are the osmolarities of these two isotonic solutions—0.92% sodium chloride and 5.5% glucose solutions?

The concentration of the glucose (dextrose) solution is 0.31 M. Each glucose molecule represents one particle in solution; therefore, the osmolarity of the solution is 0.31 osmol/L or 310 mosmol/L. Each formula unit of NaCl provides two particles in solution. Therefore, the osmolarity of the 0.16 M solution is 2 × 16 or 0.32 osmol/L or 320 mosmol/L.

Isotonic solutions have their limitations. Consider the case of a patient who is to be fed intravenously. Such a patient can handle only about 3.0 L of water in a day. If an isotonic solution of 5.5% glucose were to be used, 3.0 L of the solution would supply about 160 g of glucose. This would yield about 4.0 kcal/g to the patient, or a total of about 640 kcal, an amount woefully inadequate. Even a resting patient requires about 1400 kcal/day. For a person with serious burns, for example, requirements as high as 10 000 kcal/day have been recorded. We have oversimplified the situation. Such patients require other vital nutrients, and, in fact, a person can normally be given up to 1200 kcal through carefully formulated solutions. This still falls short of the requirements of many seriously ill people. One answer to the problem is to use very concentrated solutions (about six times as concentrated as isotonic solutions). Instead of being administered through the vein of an arm or a leg, this solution is infused directly through a tube into the superior vena cava, a large blood vessel leading to the heart (Figure 7.23). The large volume of blood flowing through this vein quickly dilutes the solution to levels that will not damage the blood. With this technique, patients have been given 5000 kcal/day and have even gained weight.

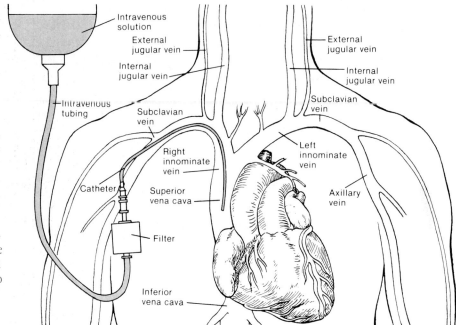

FIGURE 7.23
Concentrated nutrient solution may be infused directly into the superior vena cava. The solution is quickly diluted to near-isotonic strength by the large volume of blood flowing through the vein.

7.11 Colloids: The In-Between State

If sugar is dissolved in water, the sugar molecules and the water molecules become intimately mixed. The solution is homogeneous; that is, it has the same properties throughout. The sugar cannot be filtered out by ordinary filter paper, nor does it settle out on standing. On the other hand, if one tries to dissolve sand in water, the two substances may momentarily mix, but the sand rapidly settles to the bottom. The temporary dispersion of sand in water is called a **suspension**. By allowing the water to pass through a filter paper, one can trap the sand. The mixture is obviously heterogeneous, for part of it is clearly sand with one set of properties and part is water with another set of properties.

Is there nothing in between the true solution, with particles the size of ordinary molecules and ions, and the suspension with gross chunks of insoluble matter? Yes, there is something else, and it is a highly important arrangement of matter at that—the **colloidal state**.

Colloids are defined not by the kind of matter they contain but by the *size of the particles* involved. True solutions have particles of about 0.05 to about 0.25 nanometer (nm) in diameter (1 nm = 10^{-9} m). Suspensions have particles with diameters of 100 nm or more. Particles intermediate between these are said to be colloidal.

The properties of a **colloidal dispersion** are different from those of a true solution and also different from those of a suspension (Table 7.2).

7.11 Colloids: The In-Between State

TABLE 7.2 Properties of Solutions, Colloids, and Suspensions

Property	Solution	Colloid	Suspension
Particle size	0.1–1.0 nm	1–100 nm	>100 nm
Settles on standing?	No	No	Yes
Filter with paper?	No	No	Yes
Separate by dialysis?	No	Yes	Yes
Homogeneous?	Yes	Borderline	No

Colloidal particles cannot be filtered by filter paper, nor do they settle on standing. Colloidal dispersions usually appear milky or cloudy. Even those that appear clear will reveal a beam of light passing through the dispersion (Figure 7.24). This phenomenon, called the **Tyndall effect**, is not observed in true solutions. Colloidal particles, unlike tiny molecules, are large enough to scatter and reflect light off to the side. You have probably observed the Tyndall effect in a movie theater. The shaft of light that originates in the projection booth and ends at the movie screen is brought to you through the courtesy of colloidal dust particles in the air.

Scientists have defined eight different kinds of colloids, based on the physical state of the particles themselves (the dispersed phase) and that of the "solvent" (the dispersing phase). These are listed, with examples of each, in Table 7.3. The most important colloids in biological systems are emulsions—either liquids or solids dispersed in water. Cytoplasm is such a colloidal dispersion. Large biomolecules, such as starches and proteins, often form colloidal dispersions rather than true solutions in water.

In most colloidal dispersions, the particles are charged. This charge is

Tyndall effect—colloidal particles scatter and reflect light.

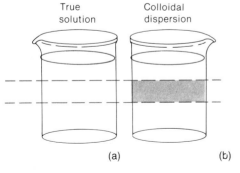

FIGURE 7.24
(a) A beam of light passes through a true solution without noticeable effect. The same beam, passing through a colloidal solution, is clearly visible because the light is scattered by the colloidal particles. This phenomenon is called the Tyndall effect. (b) Searchlight beams through fog, which is a dispersion of water in air, show the Tyndall effect.

TABLE 7.3 Types of Colloidal Dispersions

Type	Particle Phase	Medium Phase	Example
Foam	Gas	Liquid[a]	Whipped cream
Solid foam	Gas	Solid	Floating soap
Aerosol	Liquid	Gas	Fog, hair sprays
Liquid emulsion	Liquid	Liquid	Milk, mayonnaise
Solid emulsion	Liquid	Solid	Butter
Smoke	Solid	Gas	Fine dust or soot in air
Sol[b]	Solid	Liquid	Starch solutions, jellies
Solid sol	Solid	Solid	Pearl

[a] By their very nature, gas mixtures always qualify as solutions. The size of particles in gas mixtures and their homogeneity fulfill the requirements of solutions.
[b] Sols that set up in semisolid, jellylike form are called gels.

often due to the adsorption* of ions on the surface of a particle. A given colloid will preferentially adsorb only one kind of ion (either positive or negative) on its surface; thus, all the particles of a given colloidal dispersion bear like charges. Since like charges repel, the particles tend to stay away from one another. They cannot come together to form larger particles, which would settle out. Oil is ordinarily insoluble in water but can be emulsified by soap. The soap molecules form a negatively charged layer about the surface of each tiny oil droplet. These negative charges keep the oil particles from coming together and separating out. In a similar manner, bile salts emulsify the fats we eat, keeping them dispersed as tiny particles, which can be more efficiently digested. Milk is an emulsion in which fat droplets are stabilized by a coating of casein, a protein. Casein, soap, and bile salts are examples of **emulsifying agents**, substances that stabilize emulsions.

Colloids that are stabilized by surface charges on the colloidal particles can be made to coalesce and separate out by the addition of ions of opposite charge to the colloidal dispersion. Doubly or triply charged ions are particularly effective. Aluminium chloride ($AlCl_3$), which contains Al^{3+} ions, does an excellent job of breaking up colloids in which the particles are negatively charged.

7.12 Dialysis

In order to live, an organism must take in food and get rid of toxic wastes. The nutrients necessary to life must get into cells, and the wastes must get out. Generally, then, cell membranes must permit the

---

*In *ad*sorption, particles (such as ions) adhere to the surface of some substances. *Ab*sorbed particles would be taken into the body of the substances.

7.12 Dialysis

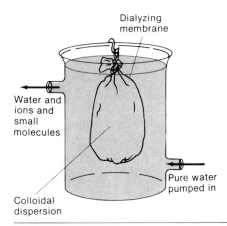

FIGURE 7.25
A simple apparatus for dialysis.

passage not only of water molecules but also of other small molecules and ions. At the same time, it is important that large molecules and colloidal particles not be lost from the cell. Membranes that pass small molecules and ions while holding back large molecules and colloidal particles are called **dialyzing membranes**. The process is called **dialysis**. It differs from osmosis in that osmotic membranes pass only solvent molecules.

Dialyzing membranes can be used to purify colloidal dispersions. The mixture to be purified is placed in a bag made of a dialyzing membrane, and the bag is placed in a container of pure water (Figure 7.25). Ions and small molecules pass out through the membrane, leaving the colloidal particles behind. Pure water is continually pumped past the bag, and the unwanted small particles are carried away. The dialyzing membrane may be an animal bladder, or it may be made from cellophane or other synthetic material.

The kidneys incorporate a complex dialyzing system responsible for the removal of certain potentially toxic waste products from the blood. By first gaining an understanding of the function of living kidneys, scientists have been able to construct artificial ones (Figure 7.26). Arti-

In dialysis, membranes pass small molecules and ions and hold back large molecules and colloidal particles.

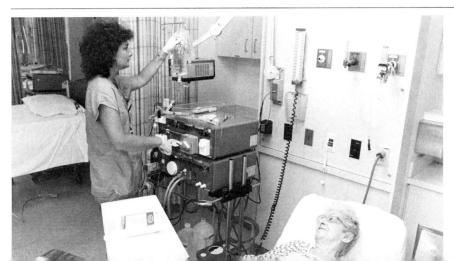

FIGURE 7.26
In renal dialysis elaborate machinery substitutes for human kidneys that are no longer capable of cleansing the blood. [Courtesy of the National Kidney Foundation.]

ficial kidneys are more elaborate in structure than the simple apparatus diagrammed in Figure 7.25, but their principle of operation is the same.

7.13 Solutions of the Body

Throughout this chapter we have mentioned solutions found in the body. Let us end the chapter with an overview of the complex interrelationships among these solutions.

The total volume of body fluids in the average 70-kg person is about 40 L. This fluid is divided into two major categories: **intracellular** fluid (the approximately 25 L held within the body's hundred trillion cells) and **extracellular** fluid (approximately 15 L located outside the cells). Included in the extracellular fluid are blood plasma (but not the fluid inside the red blood cells), interstitial fluid (the fluid between cells), lymph (interstitial fluid that is being transported from the spaces between the cells back to the blood circulatory system), cerebrospinal fluid (surrounding the brain and spinal cord), fluid in the gastrointestinal tract, urine, and a variety of other solutions.

Nutrients that make their way from the gastrointestinal tract into the blood must be carried and released to cells where the nutrients can be metabolized. Waste products from metabolism must be transported from the cells via the blood to the kidneys, where the waste products can be dumped in the urine, or to the lungs, where they can be exhausted to the atmosphere. That means that solutes must move from one solution in the body to another, from the blood to the interstitial fluid, to cells, and the reverse. We shall focus on this aspect of the complex interaction among the solutions of the body.

Figure 7.27 diagrams a small section of tissue, showing a capillary and nearby cells. The intracellular fluid, the interstital fluid, and blood all contain a variety of solutes and colloidal particles. All solutions contain relatively high levels of electrolytes, that is, charged particles such as

FIGURE 7.27
Fluid filters out of the arterial end of the capillary and back into the venous end. An imbalance in these processes results in a net loss of fluid to the interstitial space.

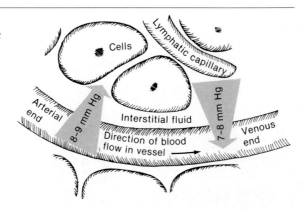

7.13 Solutions of the Body

TABLE 7.4 Osmolarity of Body Fluids

	Extracellular		
Solute	Plasma (mosmol/L)	Interstitial (mosmol/L)	Intracellular (mosmol/L)
Na^+	144	137	10
K^+	5	4.7	141
Ca^{2+}	2.5	2.4	
Mg^{2+}	1.5	1.4	31
Cl^-	107	112.7	4
HCO_3^-	27	28.3	10
HPO_4^{2-}, $H_2PO_4^-$	2	2	11
SO_4^{2-}	0.5	0.5	1
Protein	1.2	0.2	4
Other solutes	13	13	90.2
Total	303.7	302.2	302.2

potassium, sodium, chloride, and bicarbonate ions (Table 7.4). The electrolyte makeups of the *extracellular* fluids in the table (plasma and interstitial fluid) are very similar. The osmotic pressures due to these electrolytes are balanced. Compared to the electrolyte concentrations, the colloidal protein concentrations are very low. Yet it is the difference in the protein concentration that gives blood plasma a higher osmotic pressure than that exhibited by the interstitial fluid. The osmotic pressure associated with the excess of protein in plasma is referred to as **colloid osmotic pressure** or **oncotic pressure**.

As blood flows from the arterial end of the capillary to the venous end, water and smaller solute particles diffuse in and out of the capillary at an almost unbelievable rate. It is estimated that water diffuses in both directions throughout the body at the rate of 120 000 mL/min. Even more remarkable is the fact that in the healthy individual there is no net diffusion of water in either direction. Both incoming diffusion and outgoing diffusion are in perfect balance.

On the other hand, there is a net filtration of fluid out of the capillaries. **Filtration** refers to the transfer of fluid into and out of the capillary because of oncotic pressure (the osmotic pressure due to protein imbalance) and pressure differences due to the pumping action of the heart. At the arterial end of the capillary, oncotic pressure tends to pull water into the capillary, but the internal capillary pressure (imparted by the heart pump) is relatively high and tends to push fluid out. The balance of these pressures is about 8–9 mm Hg outward from the capillary. At the arterial end, fluid filters from the capillary into the interstitial space. At the venous end of the capillary, the oncotic pres-

Colloid osmotic pressure or oncotic pressure: osmotic pressure associated with excess protein in plasma

sure is the same, but the internal capillary pressure is lower because it is farther downstream from the heart. Now the balance of these pressures is about 7–8 mm Hg inward. About 90% of the fluid that filters out ultimately filters back in. That leaves about a 10% net loss to the interstitial space. For the body as a whole, this net transfer of fluid is estimated at slightly less than 2 mL/min.

Remember that *diffusion* of fluid occurs at the rate of roughly 120 000 mL/min. *But* diffusion is balanced and results in no net transfer of fluid volume (although it does result in exchange of nutrients, waste products, and so on). The tiny net amount of fluid transferred by filtration could, over time, result in a fatal buildup of fluid in the tissues. However, a second, specialized circulatory system—the lymphatic system (Figure 7.28)—is responsible for carrying the excess fluid in the interstitial spaces back to the main (blood) circulatory system. At the same time, the interstitial fluid and the intracellular fluid diffuse across the cell membrane to bring necessary nutrients into and waste products out of the cell. Note from Table 7.4 that even though the solute distributions in these two fluids are very different, the osmolarity of the solutions is the same. There is no push or pull of fluid due to unbalanced colloid osmotic pressure here.

If the overall balance of fluid transfer is disrupted, a condition called **edema** may result. The tissues become waterlogged and swollen as fluid accumulates in the interstitial spaces. The medical condition called **shock** can also result from a loss of fluid from the vascular (blood-carrying) system. The reduced blood volume causes a dramatic decrease in blood pressure. As a consequence, the blood's oxygen-transporting capability decreases, with potentially fatal results. When someone goes into shock, it means that the delicately balanced transport system in the body has gone awry and could fail completely. Treatment of the condition involves bringing the blood volume back to normal levels by transfusions.

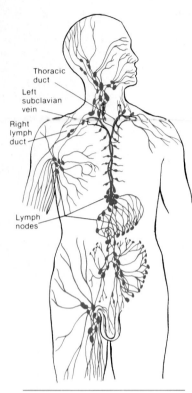

FIGURE 7.28
The lymphatic system.

Edema results from a disruption of the overall balance of fluid.

CHAPTER SUMMARY

Solutions are homogeneous mixtures of two or more components. The components are classified as solutes (the substances dissolved) or solvent (the component present in greatest amount). Aqueous solutions are solutions in which the solvent is water. In true solutions, solute cannot be filtered from solvent.

The concentration of a solution describes the relative amounts of solute and solvent, with the terms *dilute* and *concentrated* indicating low and high amounts of dissolved solute, respectively. Substances described as miscible form solutions when mixed in any proportions, but insoluble substances do not interact to form solutions. Most materials fall between the two extremes in solubility.

Temperature and pressure affect the solubility of materials, and these effects can be explained by the kinetic molecular theory. At a given temperature, an equilibrium can be established between undissolved and dissolved solute particles. A system in equilibrium is said to be saturated. When more solute is dissolved than is possible under equilibrium conditions, the solution is described as supersaturated.

Concentration can be reported in a variety of units. Percent concentration gives the amount of solute as a percentage of the solution, and the amount of solute and solution can be in volume units (percent by volume) or mass units (percent by mass). Concentration in milligram percent equals the number of milligrams of solute in 100 mL of solvent. Low concentrations are reported in ppm (mg/L) or ppb (μg/L). Concentrations may also be presented as ratios of solute to solution (solute:solution), but the units in ratio concentrations are not always clearly defined. Molarity is the concentration in moles of solute per liter of solution.

An equivalent is the number of grams of a substance that carries 1 mol of ionic charge. The equivalent (or the derived unit, milliequivalent) can be used to report concentrations of electrolytes (charged particles) in solution, particularly in body fluids. The term *unit* is used literally in describing the concentration of certain drugs and is a measure of the activity of the drug.

The dilution of a standard solution to obtain a solution of lower concentration can be accomplished using the following equation to determine the relative amounts required.

$$vol_{standard} \times conc_{standard} = vol_{diluted} \times conc_{diluted}$$

Colligative properties of solutions are those properties that depend only on the number of solute particles. Included among colligative properties are freezing point depression, boiling point elevation, and osmotic pressure due to osmosis. Osmosis is the net diffusion of water through a semipermeable membrane between two solutions of unequal osmolarity. The osmolarity (osmol/L) of a solution is its concentration in moles of particles per liter. Solutions of equal osmolarity have equal osmotic pressure; that is, the pressure required to prevent the net diffusion of water into the solutions is the same. Such solutions are said to be isotonic. A solution of lower osmolarity (lower osmotic pressure) is a hypotonic solution. Cells placed in hypotonic solutions undergo plasmolysis (termed hemolysis for a red blood cell). Solutions of higher osmotic pressure and osmolarity are hypertonic solutions. Cells placed in hypertonic solutions shrivel in a process termed crenation.

Mixtures exist in forms other than true solutions. Suspensions are temporary intermixtures that settle into separated components on standing. Colloidal dispersions are mixtures of colloidal particles in solvents. Colloidal particles are defined by their size, which is intermediate between that of solutes in true solutions and that of particles in suspensions. Whereas suspensions can be filtered apart, colloidal dispersions cannot. Colloidal dispersions also exhibit the Tyndall effect (light scattering), which true solutions do not. Colloidal dispersions are maintained by emulsifying agents, which are substances that provide charged particles that adsorb on the surface of the colloidal particles.

Dialysis is a phenomenon related to osmosis. In dialysis, solvent molecules and small solute particles pass through a semipermeable membrane that retards the passage of particles of colloidal size.

Exchange of solutes among the fluids of the body is largely accomplished by diffusion of smaller particles through semipermeable membranes separating the various solutions. In addition, a small net filtration of fluid from blood capillaries to the interstitial space occurs. The filtration is caused by an imbalance in pressures across the capillary. The fluid that is lost to the interstitial space is eventually returned to the blood via the lymphatic circulatory system.

CHAPTER QUIZ

1. For most systems with liquid or solid solutes, an increase in temperature causes solubility to
    a. increase then level off
    b. decrease
    c. increase
    d. remain unchanged

2. Which statement is correct for saturated solutions?
    a. Saturation levels are the same for all systems.
    b. Saturation levels are affected by temperature.
    c. Saturation and supersaturation are synonymous.
    d. Solute particles are in a static state.

3. Percent-by-mass solutions compare the mass of
    a. solute to solution   b. solute to solvent
    c. solvent to solute   d. solution to solute

4. Five hundred mL of a 2.5% mass/volume solution (another way to express percent concentration) contains how many grams of solute?
   a. 1250 g
   b. 500 g
   c. 12.5 g
   d. 2.5 g

5. How many grams of NaCl are contained in 500 mL of a 1.5 M NaCl solution?
   a. 0.013 g
   b. 19.5 g
   c. 39 g
   d. 44 g

6. A mole of Na_3PO_4 has how many equivalents?
   a. 1
   b. 2
   c. 3
   d. 4

7. How much water must be added to 300 mL of a 0.60 M solution to make it 0.4 M?
   a. 450 mL
   b. 300 mL
   c. 150 mL
   d. 100 mL

8. An increase in concentration of solute causes a solution to have
   a. lower boiling point
   b. higher melting point
   c. increased vapor pressure
   d. increased osmotic pressure

9. A hypertonic solution causes red cells to undergo
   a. crenation
   b. plasmolysis
   c. hemolysis
   d. swelling

10. The Tyndall effect is observed in
    a. solutions
    b. colloids
    c. suspensions
    d. solvents

PROBLEMS

7.1 Definitions

1. Define or explain and, where possible, illustrate these terms.
   a. solution
   b. solvent
   c. solute
   d. aqueous
   e. soluble
   f. insoluble
   g. miscible
   h. dilute
   i. concentrated
   j. saturated
   k. unsaturated
   l. supersaturated
   m. percent by volume
   n. percent by mass
   o. percent by weight
   p. milligram percent
   q. ppm
   r. ppb
   s. molarity
   t. equivalent
   u. electrolyte
   v. USP unit
   w. colligative property
   x. permeable
   y. semipermeable
   z. impermeable
   aa. osmosis
   bb. osmolarity
   cc. osmotic pressure
   dd. isotonic solution
   ee. hypotonic solution
   ff. hypertonic solution
   gg. crenation
   hh. plasmolysis
   ii. hemolysis
   jj. suspension
   kk. colloidal dispersion
   ll. Tyndall effect
   mm. emulsifying agent
   nn. dialysis
   oo. oncotic pressure
   pp. filtration

7.3 Some Properties of Solutions

2. Describe the properties of a true solution with respect to the following.
   a. size of solute particles
   b. distribution of solute and solvent particles
   c. filtration
   d. color and clarity

7.5 Dynamic Equilibria and Saturated Solutions

3. Use the kinetic molecular theory to explain why most solid solutes become more soluble with increasing temperature but gases become less soluble.

4. Fish live on oxygen dissolved in water. Would it be a good idea to thoroughly boil, then cool the water you place in a fish bowl? Explain.

5. In a dynamic equilibrium, two processes are occurring at the same rate. In the equilibrium involving a saturated solution, for which processes are the rates equal?

6. A supersaturated solution is maintained at constant temperature, and precipitation is induced by the addition of a seed crystal. When no more solid appears to precipitate, is the solution saturated, unsaturated, or supersaturated?

8
ACIDS AND BASES

DID YOU KNOW THAT...

- acids neutralize bases, and bases neutralize acids?
- acid–base reactions produce salts?
- in water solutions, acidity is always due to H_3O^+ ions?
- water solutions of acids and bases conduct electricity?
- some salts can change the acidity of water?
- carbonic acid, bicarbonates, and carbonates react with acids to form a gas?
- acids can destroy rocks?
- antacids are neutralizing agents?
- water can be an acid or a base?
- the higher the pH, the lower the acidity?
- buffers are acid–base shock absorbers?
- shock can cause alkalosis, and pneumonia can cause acidosis?

If our bodies are largely solutions—and they surely are—they are very special solutions. Delicate balances must be maintained among the many solutes in our blood and other body fluids. No balance is more important than that between compounds known as acids and bases. If the acidity of the blood changes very much, the blood loses its capacity to carry oxygen. Since many bodily processes produce acids, the control of acidity is, quite literally, a matter of life or death.

In this chapter, we shall look at some qualitative aspects of acids and bases—what they are and how they react. We shall also develop some more quantitative aspects of acid–base chemistry—how exact concentrations of acids and bases are determined and expressed and how the pH scale is defined. With this background we will be able to see how, through substances called buffers, the level of acidity can be controlled.

8.1 Acids and Bases: Some Definitions

Acid: a proton donor
Base: a proton acceptor

Just as there can be no oxidation without reduction, there can be no acid without a base. Like oxidation and reduction, *acid* and *base* can be defined in several ways. The definition of an **acid** as a *proton donor* and a **base** as a *proton acceptor* is most useful for our purposes. (We shall also explain what it means in a moment.) Before anyone knew of protons, however, they knew of acids and bases. Acids were substances that tasted sour and caused the color of litmus, a dye isolated from certain lichens, to turn red. Bases were compounds that tasted bitter and caused the color of litmus to turn blue. It was well known that if the right amounts of acid and base were mixed together, then the properties associated with acids and bases were lost. The mixture, for example, would taste neither sour not bitter, but salty.

As chemists became more sophisticated and began to think in terms of the structure of molecules, they discovered that the substances they regarded as acids were compounds capable of donating a proton to other compounds. The "other" compounds were bases. What does "donating a proton" mean? It sounds almost like nuclear chemistry, but it is not.

Consider the electron-dot structure of hydrogen chloride.

$$H\!:\!\ddot{\underset{..}{Cl}}\!:$$

Hydrogen shares a pair of electrons with chlorine. Now, what if we were to pull the hydrogen off and leave the electrons behind?

$$H\!:\!\ddot{\underset{..}{Cl}}\!: \longrightarrow H^+ + :\!\ddot{\underset{..}{Cl}}\!:^-$$

A hydrogen atom without an electron (the H^+ in the above equation) is a proton, $^1_1p^+$ or $^1_1H^+$. Thus when we say that an acid is a compound that

8.1 Acids and Bases: Some Definitions

can transfer a proton, we mean that it is a compound that can lose the hydrogen but keep the electrons. The only way a proton would leave a pair of electrons behind is if it were being offered a share in another pair of electrons. That is what a base does; consider the **hydroxide ion**.

Hydroxide ion: OH^-

$$[:\ddot{O}:H]^-$$

The oxygen in this ion has three unshared electron pairs, and it reacts with an acid.

$$H:\ddot{\underset{..}{Cl}}: + [:\ddot{O}:H]^- \longrightarrow H:\ddot{O}:H + :\ddot{\underset{..}{Cl}}:^-$$

An acid reacts with a base; a proton donor reacts with a proton acceptor.

You cannot buy hydroxide ion alone. To get hydroxide ion, you would have to buy a compound that contains this ion plus a cation—like sodium hydroxide, NaOH, or potassium hydroxide, KOH. When you mix NaOH with an acid like HCl, the sodium ion does not do anything in the reaction. It is often referred to as a spectator ion.

$$HCl + Na^+ + OH^- \longrightarrow Na^+ + Cl^- + H_2O$$

The sodium ion appears on both the left and right sides of the equation. In the reactant mixture, the positive sodium ion is partnered with the negative hydroxide ion. Among the products, its partner is the negative chloride ion. The sodium ion–chloride ion partnership is, in fact, a compound, sodium chloride. When an acid and a base react with each other, the mixture tastes salty because the reaction produces salt (and water). A more condensed form of the complete equation is

$$NaOH + HCl \quad NaCl + H_2O$$

This reaction of an acid with a base is referred to as **neutralization**. Here are some other acid–base reactions:

Neutralization: reaction of an acid with a base

$$HNO_3 + KOH \longrightarrow KNO_3 + H_2O$$
$$H_2SO_4 + 2\,NaOH \longrightarrow Na_2SO_4 + 2\,H_2O$$
$$CH_3COOH + NaOH \longrightarrow CH_3COONa + H_2O$$
$$HCl + NH_3 \longrightarrow NH_4Cl$$

The first reaction is completely analogous to our original example. We have just used a different acid and base. Here nitric acid, HNO_3, reacts with the base potassium hydroxide, KOH, to produce water and the salt potassium nitrate, KNO_3.

The second reaction illustrates a new point. Some acids can donate more than one proton. Sulfuric acid, H_2SO_4, can donate two and is therefore called a **diprotic** acid. There are also **triprotic** acids; an example is phosphoric acid, H_3PO_4.

The third equation in the list illustrates two points: (1) not all protons in an acid molecule are necessarily "acid"; (2) the presence of an "OH" does not necessarily mean that a compound is a base. Here is a more detailed structural formula for CH_3COOH, *acetic acid*.

$$\text{H}-\underset{\underset{\text{H}}{|}}{\overset{\overset{\text{H}}{|}}{\text{C}}}-\text{C}\underset{\text{O}-\text{H}}{\overset{\text{O}}{\diagup\!\!\!\diagdown}}$$

Only the hydrogen that is part of the OH group is acidic. The OH group is covalently bonded to the rest of the molecule; it is not a hydroxide ion. Remember, the variety of acidic compounds is enormous. All the acids we shall encounter, however, have one thing in common—they donate protons.

Finally, the fourth equation illustrates the fact that the base need not be a hydroxide. Let us rewrite this equation to show the ionic structure of the product.

$$HCl + NH_3 \longrightarrow NH_4^+ + Cl^-$$

First of all, this makes the acid–base nature of the reaction clearer. Hydrogen chloride, HCl, transfers a proton to ammonia, NH_3. Anyone who has been in a chemistry laboratory has probably seen this reaction. Ammonia, a gas at room temperature, is usually purchased and used in an aqueous solution. So, too, is hydrogen chloride, in which case it is referred to as *hydrochloric acid.* When bottles of aqueous ammonia and hydrochloric acid are opened, some NH_3 gas and HCl gas escape. Where they meet they form ammonium chloride, NH_4Cl, which appears as a white smoke. It is not unusual to see glassware coated with this fine white powder in laboratories where the two reagents are used frequently.

8.2 Acids and Bases in Aqueous Solutions

The acid–base reactions we are most interested in take place in water solutions. When an acid is dissolved in water, some of the water molecules act as bases.

$$HCl + H_2O \longrightarrow H_3O^+ + Cl^-$$
$$HNO_3 + H_2O \longrightarrow H_3O^+ + NO_3^-$$

8.2 Acids and Bases in Aqueous Solutions

The H_3O^+ species is called a **hydronium ion**. Thus, in aqueous solution, acids produce hydronium ions. The acids shown above are almost completely **ionized** in aqueous solution; that is, the originally covalent molecules react with water to form ions.

Hydronium ion: H_3O^+

Some acids are not as efficient at transferring their protons to water.

$$CH_3COOH + H_2O \rightleftharpoons H_3O^+ + CH_3COO^-$$

$$HCN + H_2O \rightleftharpoons H_3O^+ + CN^-$$

The double arrows indicate that in these reversible reactions the nonionized forms predominate. Because hydrogen chloride ionizes almost completely in water, hydrochloric acid is referred to as a **strong acid**. Because acetic acid (CH_3COOH) and hydrogen cyanide (HCN, called hydrocyanic acid in aqueous solution) ionize very little in water, they are called **weak acids**.

Sulfuric acid can transfer two protons.

$$H_2SO_4 + H_2O \longrightarrow H_3O^+ + HSO_4^-$$

$$HSO_4^- + H_2O \rightleftharpoons H_3O^+ + SO_4^{2-}$$

The arrow indicates that the first transfer reaction is essentially complete, which is why sulfuric acid is classified as a strong acid. The equilibrium involved in the second transfer favors the reactants to the left of the arrows; the hydrogen sulfate ion, HSO_4^-, is an acid but not a strong one. Table 8.1 lists some common acids with their classification according to strength.

TABLE 8.1 Some Representative Acids

Name	Formula	Classification	Number of Hydronium Ions from 1 000 000 Molecules in 0.01 M Solution
Sulfuric acid	H_2SO_4	Strong	1 220 000
Nitric acid	HNO_3	Strong	920 000
Hydrochloric acid	HCl	Strong	920 000
Phosphoric acid	H_3PO_4	Moderate	270 000
Lactic acid	$CH_3CHOHCOOH$	Weak	87 000
Acetic acid	CH_3COOH	Weak	13 000
Boric acid	H_3BO_3	Weak	1
Hydrocyanic acid	HCN	Weak	1

Common inorganic acids and bases (those without carbon) tend to be strong; that is, when placed in water, they dissociate completely into ions.

$$HCl + H_2O \xrightarrow[\text{in } H_2O]{100\%} H_3O^+ + Cl^-$$

$$NaOH \xrightarrow[\text{in } H_2O]{100\%} Na^+ + OH^-$$

Organic acids and bases (those containing carbon), on the other hand, tend to be weak—they dissociate only partially.

$$CH_3COOH \underset{\text{in } H_2O}{\overset{\ll 100\%}{\rightleftharpoons}} CH_3COO^- + H^+$$

What happens when a base is dissolved in water? Bases like sodium hydroxide are ionic compounds in the solid state. When they dissolve in water, the ions simply **dissociate**; that is, they separate from one another and wander about at random in the solution.

As we noted previously, ammonia, NH_3, is also a base. It, too, produces hydroxide ions in aqueous solution. Because ammonia is a covalent molecule, however, to do this it must ionize, and it ionizes by reacting with a water molecule.

$$NH_3 + H_2O \rightleftharpoons NH_4^+ + OH^-$$

In this instance, water acts like an acid and donates a proton to ammonia. Note that the arrows show that ammonia is a *weak* base. It does not produce hydroxide ions very efficiently in aqueous solution.

Strong acids and bases cause damage on contact with living cells. The action is nonspecific. All cells, regardless of type, are damaged more or less equally. These **corrosive** poisons produce what are known as chemical burns. Once the offending agent is neutralized or removed, the injuries are similar to burns from heat. They are often treated the same way.

Sulfuric acid is found in automobile batteries and in some drain cleaners (presumably because it dissolves drain-clogging hair). Hydrochloric acid (also called muriatic acid) is used in homes to clean calcium carbonate deposits from toilet bowls (Figure 8.1). Ingestion of concentrated solutions of sulfuric acid, hydrochloric acid, or any other strong acid causes corrosive damage to the digestive tract. As little as 10 mL of concentrated sulfuric acid, taken internally, may be fatal. Even diluted solutions of these strong acids should be handled carefully.

Sodium hydroxide (lye) is by far the most common of the strong bases and is found in household items like drain and oven cleaners. As

(a) (b)

FIGURE 8.1
Acids (a) and bases (b) found in the home. [Photos © The Terry Wild Studio.]

their labels suggest, these highly corrosive products should not be allowed to come into contact with tissue. Some basic detergent additives, such as carbonates, silicates, and borates, can cause corrosive damage to tissues, particularly those of the digestive tract and the eyes.

Even low concentrations of weak acids and bases are not necessarily safe. Ammonia is a weak base but a strong irritant. Bronchial spasms caused by inhalation of too much of this gas can lead to death. Fortunately, the odor of ammonia is so pungent that it usually drives a person from the area before it reaches toxic levels. Hydrogen cyanide is a weak acid and also is extremely toxic because the CN^- blocks the action of critical enzymes required for energy-producing oxidation–reduction reactions in cells.

8.3 NEUTRALIZATION IN AQUEOUS SOLUTION

When a strong acid reacts with a strong base in aqueous solution, the neutralization reaction can be written

$$H_3O^+ + OH^- \longrightarrow 2\,H_2O$$

because acid and base produce hydronium ions and hydroxide ions, respectively, in aqueous solutions. This equation is the **net ionic equation** for neutralization in aqueous solution.

An ionic equation that describes the reaction of a specific acid and base in aqueous solution must take into account the particular anion supplied by the acid and the cation provided by the base. When HCl ionizes to give a hydronium ion, it also produces a chloride ion. When

NaOH dissociates to give the hydroxide ion in aqueous solution, it also yields a sodium ion.

$$H_3O^+ + Cl^- + Na^+ + OH^- \longrightarrow Na^+ + Cl^- + 2\,H_2O$$

The sodium and chloride ions appear among the reactants and in unchanged form, among the products. In a *net* ionic equation, any substance that appears in identical form on both sides of the equation is not written. In this more complete form of ionic equation they are included. If the solution containing the products of this neutralization reaction were evaporated (driving off all the water, including the water formed in the reaction), only a crystalline ionic solid—a salt—would remain. **Salt** is a general term describing the ionic products of neutralization reactions. Sodium chloride is a salt, but not the only one; it is usually referred to as "table salt."

Salt: the ionic product of a neutralization reaction

While 1 mol of HCl neutralizes only 1 mol of NaOH, 1 mol of H_2SO_4 neutralizes 2 mol of NaOH.

$$\underset{1\text{ mol}\quad 1\text{ mol}}{HCl + NaOH} \longrightarrow NaCl + H_2O$$

$$\underset{1\text{ mol}\quad\quad 2\text{ mol}}{H_2SO_4 + 2\,NaOH} \longrightarrow Na_2SO_4 + 2\,H_2O$$

In neutralization reactions, then, 1 mol of H_2SO_4 is *equivalent* to 2 mol of HCl, that is, does the work of 2 mol of HCl. Alternatively, we may say that 0.5 mol of H_2SO_4 is equivalent to 1 mol of HCl. This is much the same situation encountered in Chapter 7 when we discussed equivalents of electrolytes in body fluids (Section 7.6). There we described an equivalent of electrolyte as the amount that contains a mole of charge. For acids and bases in aqueous solutions, an equivalent may be defined in terms of a mole of hydronium ions. An **equivalent of acid** is the amount that provides 1 mol of hydronium ions in solution. An **equivalent of base** is the amount that will react with 1 mol of hydronium ions. These definitions are perhaps best understood through examples.

EXAMPLE 8.1

How many equivalents of acid are there in 2 mol of each of the following: (a) H_3PO_4, (b) $H_2C_2O_4$ (oxalic acid), (c) HNO_3, (d) CH_3COOH?

a. The number of acidic hydrogens in the compound gives the number of equivalents per mole. Phosphoric acid, H_3PO_4, is a triprotic acid; it can transfer three protons to water molecules to form hydronium ions. Therefore,

8.3 Neutralization in Aqueous Solution

$$3 \text{ equiv of } H_3PO_4 = 1 \text{ mol of } H_3PO_4$$

This equality can be used as a conversion factor to transform the given value, 2 mol, into equivalents. Thus, for phosphoric acid,

$$2 \text{ mol} \times \frac{3 \text{ equiv}}{1 \text{ mol}} = 6 \text{ equiv}$$

b. Oxalic acid has two *ionizable* protons. For oxalic acid, then,

$$2 \text{ equiv of } H_2C_2O_4 = 1 \text{ mol of } H_2C_2O_4$$

Using this equality as a conversion factor, we get

$$2 \text{ mol} \times \frac{2 \text{ equiv}}{1 \text{ mol}} = 4 \text{ equiv}$$

c. Nitric acid has only one ionizable proton.

$$1 \text{ equiv of } HNO_3 = 1 \text{ mol of } HNO_3$$

$$2 \text{ mol} \times \frac{1 \text{ equiv}}{1 \text{ mol}} = 2 \text{ equiv}$$

d. Acetic acid also has only one ionizable proton (Section 8.1). Therefore, just as for nitric acid, 2 mol = 2 equiv.

EXAMPLE 8.2

To obtain 10 equiv, how many moles of each of the following must be used: (a) H_3BO_3, (b) HCN, (c) H_2CO_3?

a. For boric acid, H_3BO_3:

$$3 \text{ equiv} = 1 \text{ mol}$$

$$10 \text{ equiv} \times \frac{1 \text{ mol}}{3 \text{ equiv}} = 3.3 \text{ mol}$$

b. For hydrocyanic acid, HCN:

$$1 \text{ equiv} = 1 \text{ mol}$$

$$10 \text{ equiv} \times \frac{1 \text{ mol}}{1 \text{ equiv}} = 10 \text{ mol}$$

c. For carbonic acid, H_2CO_3:

$$2 \text{ equiv} = 1 \text{ mol}$$

$$10 \text{ equiv} \times \frac{1 \text{ mol}}{2 \text{ equiv}} = 5 \text{ mol}$$

EXAMPLE 8.3

How many equivalents of sulfuric acid are there in 4.9 g of the acid? The formula weight of sulfuric acid is 98.

We first determine the number of moles in 4.9 g.

$$4.9 \text{ g} \times \frac{1 \text{ mol}}{98 \text{ g}} = 0.050 \text{ mol}$$

Sulfuric acid has two ionizable hydrogen atoms. Therefore, 2 equiv = 1 mol.

$$0.050 \text{ mol} = \frac{2 \text{ equiv}}{1 \text{ mol}} = 0.10 \text{ equiv}$$

EXAMPLE 8.4

How many equivalents are there in 39 g of $Al(OH)_3$? The formula weight of aluminum hydroxide is 78.

$$39 \text{ g} \times \frac{1 \text{ mol}}{78 \text{ g}} = 0.50 \text{ mol}$$

Each formula unit of aluminum hydroxide can react with three hydronium ions.

$$Al(OH)_3 + 3 \text{ } H_3O^+ \longrightarrow Al^{3+} + 6 \text{ } H_2O$$

Therefore, 1 mol = 3 equiv, and

$$0.50 \text{ mol} \times \frac{3 \text{ equiv}}{1 \text{ mol}} = 1.5 \text{ equiv}$$

8.3 Neutralization in Aqueous Solution

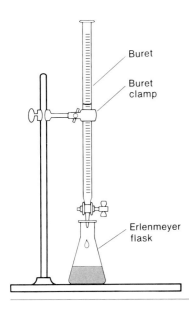

FIGURE 8.2
An apparatus for titration. A sample of unknown acid is measured into the flask. Base is added from a buret until an indicator changes color.

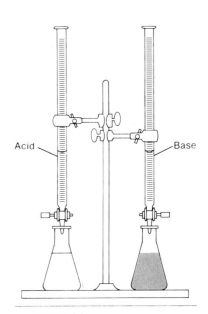

FIGURE 8.3
The indicator dye phenolphthalein is colorless in acidic solutions and red in basic solutions.

A technique called **titration** is often used to determine the amount of acid or base present in a sample. Assume that the unknown is an acid. The unknown (usually in aqueous solution) is placed in a container. Then a solution of base of known concentration (called a *standard base*) is added dropwise from a *buret*. (A buret is a piece of laboratory glassware designed to deliver known amounts of liquid into another container; see Figure 8.2.) As it is added, the base neutralizes the acid. Finally, one additional drop of base neutralizes the last bit of acid with a little bit of base left over. The addition of this last drop of base, therefore, causes the solution suddenly to swing from acidic to basic. How can one tell when this change occurs? One way is to use a dye, called an **indicator**, the color of which depends on the acidity of a solution. Litmus is such a dye, and a compound called phenolphthalein is another commonly used indicator (Figure 8.3). The indicator is added to the solution at the beginning of the titration, and, if everything is done correctly, the color change (**endpoint**) occurs when the number of equivalents of base present just equals the number of equivalents of acid (the **equivalence point**). The number of equivalents of base (and the number of equivalents of acid) can be calculated from the amount of base added and the concentration of the base.

Indicator: a dye whose color depends on the acidity of the solution

The endpoint of a titration occurs when the number of equivalents of base in the solution just equals the number of equivalents of acid in the solution.

Titration is a powerful technique for the determination of the concentration of many kinds of chemicals, not just acids and bases. All that is required is a detectable change at the equivalence point. Titration can be used to determine the concentration of oxidizing agents and reducing agents, the amount of silver in an ore sample, or the activity of an enzyme.

EXAMPLE 8.5

A 25.0-mL sample of an NaOH solution is titrated with 0.10 M HCl. The endpoint is reached when 30.0 mL of the acid has been added. What is the molarity of the base?

$$Na^+OH^- + H_3O^+Cl^- \longrightarrow 2\,H_2O + Na^+Cl^-$$

Volume:	25.0 mL	30.0 mL
Molarity:	?	0.10 M
Millimoles:		3.0 mmol

Assuming that at the endpoint, the equivalent point was also reached, the moles of H_3O^+ added equal the moles of OH^- initially present. Thus, if it took 3.0 mmol of hydronium ion, then there should also have been 3 mmol of hydroxide ion. Therefore, the concentration of the base is calculated as follows.

$$\text{Molarity of NaOH} = \frac{3.0 \text{ mmol}}{25.0 \text{ mL}} = 0.12 \text{ mol/L}$$

8.4 Salts in Water: Acidic, Basic, or Neutral

When an acid reacts with a base, the products are a salt and water. The process is called neutralization, but is the solution neutral? What if we simply take a salt and dissolve it in water? Would the solution be neutral? We might expect all salt solutions to be neutral (and many are), but some are acidic and others are basic. We can *predict* whether a solution of a salt will be acidic, basic, or neutral by considering the relative strengths of the acid and base from which the salt was formed. The following rules apply.

1. The salt of a strong acid and a strong base forms a neutral solution.
2. The salt of a strong acid and a weak base forms an acidic solution.
3. The salt of a weak acid and a strong base forms a basic solution.
4. The salt of a weak acid and a weak base may form an acidic, a basic, or (by chance) a neutral solution.

To apply these rules, you must be able to look at the formula of a salt and recognize the acid and base from which it was formed. You must also know which acids and bases are classified as strong and which are classified as weak (Table 8.1). The common water-soluble bases include the strong bases sodium hydroxide (NaOH) and potassium

8.4 Salts in Water: Acidic, Basic, or Neutral

hydroxide (KOH) and the weak base aqueous ammonia (frequently called ammonium hydroxide, NH_4OH). The cation of a salt originates in the base. The anion of the salt comes from the acid.

EXAMPLE 8.6

From what acid and base is the salt NaCl formed? Is a solution of NaCl acidic, basic, or neutral?

The cation in NaCl is Na^+; the base from which this cation was derived is NaOH, a strong base. The anion in NaCl is Cl^- from the strong acid HCl. According to rule 1, a solution of NaCl is neutral.

EXAMPLE 8.7

Is a solution of NH_4Cl acidic, basic, or neutral?

The base is NH_4OH (really aqueous NH_3) and the acid is HCl. Since aqueous NH_3 is a weak base and HCl is a strong acid, the solution is acidic (rule 2).

EXAMPLE 8.8

Is a solution of CH_3COOK acidic, basic, or neutral?

The formulas for most salts are written with the cation first and the anion second. Salts of acetic acid, however, may be written in the reverse order: $CH_3COO^-K^+$. (There are alternative formulas for acetic acid and its salts that use the more common order: $HC_2H_3O_2$ and $KC_2H_3O_2$.)

No matter how it is written, the salt is formed from the base, KOH, and acetic acid. The base is strong and the acid is weak. Therefore, the solution of the salt is basic (rule 3).

EXAMPLE 8.9

Is a solution of CH_3COONH_4 acidic, basic, or neutral?

Both the base, NH_4OH (aqueous NH_3), and the acid, CH_3COOH, are weak. Whether the solution would be acidic, basic, or neutral cannot be determined from the rules (rule 4).

Why would a salt be acidic or basic? Although the rules for predicting acidity or basicity suggest that the "stronger" partner dominates (for example, the salt of a *strong acid* and weak base gives an *acidic* solution), it is to the "weaker" partner that we must look for an answer.

Consider sodium acetate, the salt of a strong base and a weak acid. When this salt dissolves in water it produces sodium ions and acetate ions surrounded by water molecules. Acetate ions can react with water molecules to produce acetic acid.

$$CH_3COO^- + H_2O \longrightarrow CH_3COOH + OH^-$$

Acetic acid is a weak acid and quite happy to have its proton reattached. The water molecule, however, an even weaker acid, is not all that thrilled at having its proton removed and fights effectively to keep it.

$$CH_3COO^- + H_2O \rightleftharpoons CH_3COOH + OH^-$$

This just means that an equilibrium is established in which the two acids, water and acetic acid, compete for the proton. The equilibrium is shifted strongly to the left. (The stronger acid, CH_3COOH, releases the proton more readily, and the weaker acid, H_2O, holds on to its proton more strongly.) Nonetheless, an equilibrium exists, and some undissociated acetic acid and some excess hydroxide ions are present in solution. These hydroxide ions are what makes the solution basic.

It is important to note that there is no analogous equilibrium involving the other half of the original salt, the sodium ion. We could write an equation to suggest the possibility.

$$Na^+ + 2\,H_2O \not\longrightarrow NaOH + H_3O^+$$

Because sodium hydroxide is a *strong* base, however, it is completely ionized in solution. To be correct we would have to write the equation as follows.

$$Na^+ + 2\,H_2O \longrightarrow Na^+ + OH^- + H_3O^+$$

Since sodium ion appears on both sides of the equation, this reaction would boil down to

$$2\,H_2O \longrightarrow H_3O^+ + OH^-$$

which is simply a restatement of the fact that water itself ionizes slightly. The slight ionization of water gives equal amounts of hydronium ion and hydroxide ion, whereas the equilibrium involving the acetate ion produces a small excess of hydroxide ions in solution. The salt of a strong base and a weak acid gives a slightly basic solution.

8.4 Salts in Water: Acidic, Basic, or Neutral

For a solution of the salt of a weak base and a strong acid, precisely the opposite happens. Thus, for ammonium chloride (NH$_4$Cl), the ammonium ion exists in equilibrium with the base ammonia.

$$NH_4^+ + H_2O \rightleftharpoons H_3O^+ + NH_3$$

Ammonia, although a weak base, is a stronger base than water, so the equilibrium is shifted to the left. Some excess hydronium ions, however, are present in the solution because of this equilibrium. The chloride ion from ammonium chloride has no tendency to pick up a proton from water to form undissociated HCl because hydrochloric acid is a strong acid and the chloride ion simply will not hold on to the

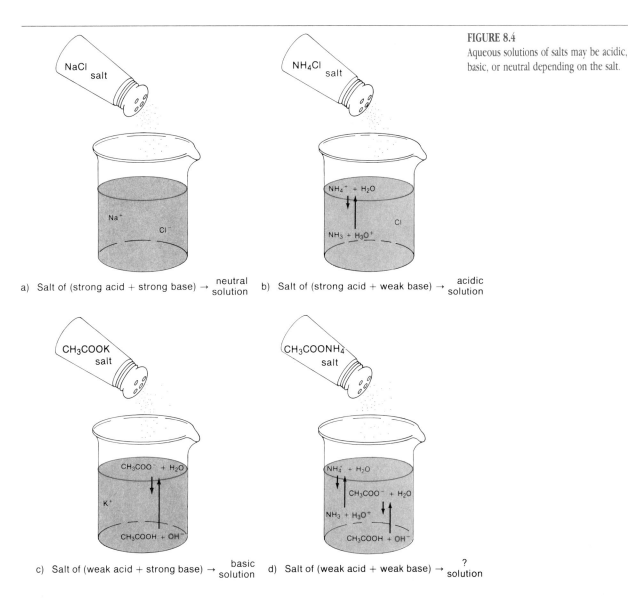

FIGURE 8.4
Aqueous solutions of salts may be acidic, basic, or neutral depending on the salt.

proton. The solution of a salt of a weak base and a strong acid contains excess hydronium ions and is slightly acidic.

In a solution of sodium chloride, neither ion has a tendency to enter into equilibrium with water molecules, and neither excess hydroxide ions nor excess hydronium ions are present. The solution is neutral.

Finally, both ions from a salt of a weak acid and a weak base enter equilibria. The solution of such a salt can be acidic, basic, or neutral, depending on which equilibrium predominates, that is, how much to the left or right each equilibrium is shifted (Figure 8.4).

8.5 Reactions of Acids with Carbonates and Bicarbonates

Add vinegar to baking soda and a vigorous fizzing action occurs. Some antacid preparations are designed to effervesce. What causes the fizz? Generally, it is the reaction of a carbonate or bicarbonate salt with an acid.

Carbonic acid, H_2CO_3, is a diprotic acid, and it can form two different salts. **Bicarbonates** are formed when carbonic acid transfers only one of its protons in a neutralization reaction.

$$H_2CO_3 + NaOH \longrightarrow \underset{\substack{\text{Sodium} \\ \text{bicarbonate}}}{NaHCO_3} + H_2O$$

Carbonates are the products formed when both protons are transferred.

$$H_2CO_3 + 2\,NaOH \longrightarrow \underset{\substack{\text{Sodium} \\ \text{carbonate}}}{Na_2CO_3} + 2\,H_2O$$

Because it is a weak acid, carbonic acid is not highly ionized in aqueous solution. To put it another way, carbonic acid holds on to its protons quite tightly. (Notice the apparent contradiction: weak acids hold tightly to their protons and strong acids do not.) Conversely, carbonate ions and bicarbonate ions pick up protons readily to form carbonic acid. The following equations summarize these facts.

$$H_2CO_3 + H_2O \rightleftharpoons H_3O^+ + HCO_3^-$$
$$HCO_3^- + H_2O \rightleftharpoons H_3O^+ + CO_3^{2-}$$

Carbonic acid is quite unstable. It decomposes to carbon dioxide gas and water.

8.5 Reactions of Acids with Carbonates and Bicarbonates

$$H_2CO_3 \longrightarrow H_2O + CO_2$$

The reaction is not related to the acidity of carbonic acid. It simply indicates that the compound, which happens to be an acid, is not stable. One cannot purchase a bottle of carbonic acid; it is too unstable to be isolated and bottled. It can be formed in solution, but as soon as it is formed, it begins to decompose.

The salts of carbonic acid are stable. You can buy sodium bicarbonate or sodium carbonate. If sodium bicarbonate is dissolved in water, a solution of sodium ions and stable bicarbonate ions is formed. If a strong acid is then added to the solution, the bicarbonate ions come into contact with the hydronium ions from the acid and immediately grab a proton to form carbonic acid.

$$HCO_3^- + H_3O^+ \longrightarrow H_2CO_3 + H_2O$$

Bicarbonate ion acts like a base; it accepts a proton. The carbonic acid that is formed is unstable and quickly falls apart, forming carbon dioxide and another molecule of water. Thus, the reaction of a strong acid with bicarbonate ion produces, ultimately, carbon dioxide and water.

$$HCO_3^- + H_3O^+ \longrightarrow \begin{array}{c} H_2CO_3 \\ + \\ H_2O \end{array} \longrightarrow H_2O + CO_2(g)$$

A similar reaction occurs with carbonate ions.

$$CO_3^{2-} + 2\,H_3O^+ \longrightarrow \begin{array}{c} H_2CO_3 \\ + \\ 2\,H_2O \end{array} \longrightarrow H_2O + CO_2(g)$$

Vinegar contains an acid (acetic acid, CH_3COOH) that is somewhat stronger than carbonic acid. Therefore, when vinegar is poured onto baking soda ($NaHCO_3$), the acetic acid transfers its proton to the bicarbonate ion to form carbonic acid, which immediately decomposes to produce carbon dioxide gas (the fizz).

$$CH_3COOH + Na^+ + HCO_3^- \longrightarrow CH_3COO^- + Na^+ + CO_2 + H_2O$$

Limestone and marble are mainly calcium carbonate ($CaCO_3$). Both are important building stones. Marble is also used in statues, monuments, and sculptures. The calcium carbonate in these objects is readily attacked by acids in the atmosphere or in rain. In some areas, the atmosphere has been made increasingly acidic in recent years, mainly

FIGURE 8.5
The effect of acid rain on the Black Forest in West Germany is shown dramatically by these two photographs taken 15 years apart. Sulfur, the agent held principally responsible for this catastrophe, imperils the natural balance in many countries. [Bossu/Sygma.]

by the burning of sulfur-containing fuels. The sulfur combines with oxygen to form sulfur dioxide gas.

$$S + O_2(g) \longrightarrow SO_2(g)$$

Some of the sulfur dioxide reacts further with oxygen to form sulfur trioxide gas.

$$2\,SO_2(g) + O_2(g) \longrightarrow 2\,SO_3(g)$$

The sulfur trioxide then reacts with water to form liquid sulfuric acid.

$$SO_3(g) + H_2O \longrightarrow H_2SO_4$$

The sulfuric acid, in the form of an aerosol mist or diluted by rainwater, furnishes the hydronium ions that dissolve the marble and limestone.

$$CaCO_3 + 2\,H_3O^+ \longrightarrow Ca^{2+} + CO_2(g) + 3\,H_2O$$

Acid mists and acidic rainwater also attack and dissolve many metals. Damage from air pollution to buildings, automobiles, and other structures and machines amounts to billions of dollars per year. These acid pollutants are also damaging to crops, to forests, and to human health (Figure 8.5).

8.6 ANTACIDS

The stomach secretes hydrochloric acid (HCl) as an aid in the digestion of food. Sometimes overindulgence or emotional stress leads to a condition called hyperacidity (too much acid). A number of antacids are available to treat this condition, many of them aggressively advertised. Indeed, sales of antacids in the United States are estimated to be about $500 million each year. Let us look at some popular antacids from the standpoint of acid–base chemistry.

Sodium bicarbonate, or baking soda ($NaHCO_3$), is an old standby antacid. It is probably safe and effective for occasional use by most people. Overuse will make the blood too alkaline, a condition called **alkalosis** (see Section 8.10). Sodium bicarbonate is not recommended for those with hypertension (high blood pressure) because high concentrations of sodium ion tend to aggravate the condition. The antacid in Alka-Seltzer is sodium bicarbonate. This popular remedy also contains citric acid and aspirin. When Alka-Seltzer is placed in water, the familiar fizz occurs because of the reaction of bicarbonate ions with hydronium ions from the acid.

$$HCO_3^- + H_3O^+ \longrightarrow CO_2 + 2\,H_2O$$

Alka-Seltzer came under attack because the aspirin in it might be harmful to people with ulcers and other stomach disorders. To offset that criticism, the makers introduced Alka-Seltzer Gold, which does not contain aspirin.

Calcium carbonate, frequently called precipitated chalk ($CaCO_3$), is another common antacid ingredient. It is fast acting and safe in small amounts, but regular use in large amounts can cause constipation. It also appears that calcium carbonate can cause *increased* acid secretion after a few hours. Temporary relief may be achieved at the expense of a worse problem later. Tums is simply flavored calcium carbonate. Alka 2 and Di-Gel liquid suspension also contain calcium carbonate as the antacid ingredient. The carbonate ion neutralizes acid by the reaction

$$CO_3^{2-} + 2\ H_3O^+ \longrightarrow CO_2 + 3\ H_2O$$

Aluminum hydroxide [$Al(OH)_3$] is another popular antacid ingredient. Like calcium carbonate, it can cause constipation in large doses. The hydroxide ions neutralize acid.

$$OH^- + H_3O^+ \longrightarrow 2\ H_2O$$

There is concern that antacids containing aluminum ions deplete the body of essential phosphate ions. The aluminum phosphate formed is insoluble and is excreted from the body.

$$Al^{3+} + PO_4^{3-} \longrightarrow AlPO_4$$

Aluminum hydroxide is the only antacid in Amphojel. It occurs in combination in many popular products.

Magnesium compounds constitute the fourth category of antacids. These include magnesium carbonate ($MgCO_3$) and magnesium hydroxide [$Mg(OH)_2$]. Milk of magnesia is a suspension of magnesium hydroxide in water. It is sold under a variety of brand names, the best known of which is probably Phillips. In small doses, magnesium compounds act as antacids. In large doses, they act as laxatives. Magnesium ions are poorly absorbed in the digestive tract. Rather, these small, dipositive ions draw water into the colon (large intestine), causing the laxative effect.

A variety of popular antacids combine aluminum hydroxide, with its tendency to cause constipation, and a magnesium compound, which acts as a laxative. These tend to counteract one another. Maalox and Mylanta are familiar brands.

Rolaids, a highly advertised antacid, contains the complex substance aluminum sodium dihydroxy carbonate [$AlNa(OH)_2CO_3$]. Both the hydroxide ion and the carbonate ion consume acid.

8.7 THE pH SCALE

Water can act as an acid (e.g., with ammonia), and it can also act as a base (e.g., with hydrogen chloride). Just what is water? Even the purest water is not all H_2O. In pure water, about 1 molecule in 500 million transfers a proton to another molecule, giving a hydronium ion and a hydroxide ion.

$$H_2O + H_2O \longrightarrow H_3O^+ + OH^-$$

The concentration of hydronium ion in pure water is 0.000 000 1, or 1×10^{-7}, mol/L (M). The concentration of hydroxide ion is also 1×10^{-7} M. Pure water, then, is a neutral solution, having an *excess* of neither hydronium ions nor hydroxide ions.

The product of the hydronium ion concentration and the hydroxide ion concentration is

$$(1 \times 10^{-7})(1 \times 10^{-7}) = 1 \times 10^{-14}$$

This product, called the **ion product of water**, is sometimes represented by the symbol K_w. K_w is a constant. If you add acid to pure water, thereby increasing the concentration of hydronium ions, the concentration of hydroxide ions falls until the product of the concentrations of the two ions equals 1×10^{-14}. Addition of base to pure water results in an analogous adjustment in ion concentrations (Figure 8.6). These facts can be summarized in a single equation.

$$K_w = [H_3O^+][OH^-] = 1 \times 10^{-14}$$

The bracketed symbols are a shorthand for the concentration of the enclosed ions in moles per liter. Thus $[H_3O^+]$ should be read as the "molarity of hydronium ion." This relationship between the concentrations of hydronium ions and hydroxide ions is valid for all aqueous solutions. If the concentration of one ion is known, the concentration of the other can be calculated.

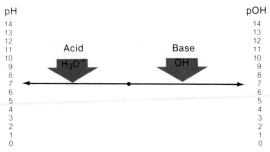
(a) When acid and base are in balance, the solution is neutral.

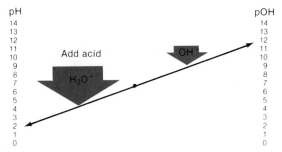
(b) When excess acid is added, pH decreases, and pOH increases.

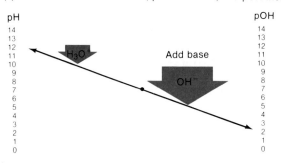
(c) When excess base is added, pH increases, and pOH decreases.

FIGURE 8.6
Effect of adding acid or base on the pH and pOH of a solution.

EXAMPLE 8.10

Lemon juice has a $[H_3O^+]$ of 0.01 M. What is the $[OH^-]$?

In exponential form, 0.01 M is written as 1×10^{-2} M. Since $[H_3O^+][OH^-] = 1 \times 10^{-14}$,

$$[OH^-] = \frac{1 \times 10^{-14}}{1 \times 10^{-2}}$$

To divide exponential numbers, one subtracts the exponent in the denominator from that in the numerator, so $-14 - (-2) = -14 + 2 = -12$.

$$[OH^-] = 1 \times 10^{-12}$$

8.7 The pH Scale

EXAMPLE 8.11

A sample of bile has an [OH$^-$] of 1×10^{-6} M. What is the [H$_3$O$^+$]?

$$[H_3O^+] = \frac{1 \times 10^{-14}}{1 \times 10^{-6}} = 1 \times 10^{-8}$$

The use of exponential numbers to express the often minute concentrations of hydronium and hydroxide ion is inconvenient. In 1909, S. P. L. Sorensen, working in the Carlsberg Laboratory in Denmark, proposed the convenient **pH** scale, where

$$pH = \log \frac{1}{[H_3O^+]} = -\log[H_3O^+]$$

The reaction of most people on first encountering this definition of pH is: "This is more convenient?" Yes, it really is. To determine the pH of a solution, you need only take the exponent of the hydronium ion concentration and reverse its sign. The pH of a solution whose hydronium ion concentration is 1×10^{-4} M is 4. The point is this: it is easier to say "The pH is 4," than to say, "The hydronium ion concentration is 1×10^{-4} M." They mean exactly the same thing.

Hydroxide ion concentration can be similarly expressed in terms of pOH.

$$pOH = -\log[OH^-]$$

If the concentration of hydroxide ion in an aqueous solution is 1×10^{-8} M, the pOH of the solution is 8.

The pH scale has been universally adopted. pH values are even quoted in advertisements for shampoos and cosmetics. The relationship between pH and [H$_3$O$^+$] is given in Table 8.2 and Figure 8.7. Note from the table that

$$pH + pOH = 14$$

TABLE 8.2 The Relationship Between pH and [H$_3$O$^+$] and Between pOH and [OH$^-$] (at 20 °C)

[H$_3$O$^+$]	pH	[OH$^-$]	pOH
1×10^{0}	0	1×10^{-14}	14
1×10^{-1}	1	1×10^{-13}	13
1×10^{-2}	2	1×10^{-12}	12
1×10^{-3}	3	1×10^{-11}	11
1×10^{-4}	4	1×10^{-10}	10
1×10^{-5}	5	1×10^{-9}	9
1×10^{-6}	6	1×10^{-8}	8
1×10^{-7}	7	1×10^{-7}	7
1×10^{-8}	8	1×10^{-6}	6
1×10^{-9}	9	1×10^{-5}	5
1×10^{-10}	10	1×10^{-4}	4
1×10^{-11}	11	1×10^{-3}	3
1×10^{-12}	12	1×10^{-2}	2
1×10^{-13}	13	1×10^{-1}	1
1×10^{-14}	14	1×10^{0}	0

EXAMPLE 8.12

The hydronium ion concentration is 0.000 01 M. What is the pH?

$$[H_3O^+] = 0.000\ 01\ M = 1 \times 10^{-5}\ M$$

The pH is the negative of the exponent of 10.

$$pH = -(-5) = 5$$

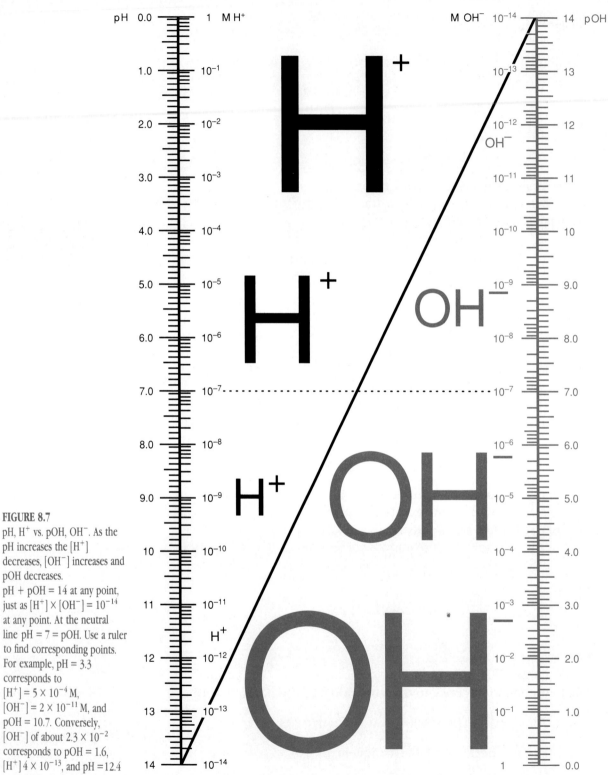

FIGURE 8.7
pH, H^+ vs. pOH, OH^-. As the pH increases the $[H^+]$ decreases, $[OH^-]$ increases and pOH decreases.
pH + pOH = 14 at any point, just as $[H^+] \times [OH^-] = 10^{-14}$ at any point. At the neutral line pH = 7 = pOH. Use a ruler to find corresponding points. For example, pH = 3.3 corresponds to $[H^+] = 5 \times 10^{-4}$ M, $[OH^-] = 2 \times 10^{-11}$ M, and pOH = 10.7. Conversely, $[OH^-]$ of about 2.3×10^{-2} corresponds to pOH = 1.6, $[H^+]$ 4×10^{-13}, and pH = 12.4

8.7 The pH Scale

EXAMPLE 8.13

The pH is 9. What is the hydronium ion concentration?
If the pH is 9, then the hydronium ion concentration in molarity can be expressed as the following power of 10.

$$[H_3O^+] = 10^{-pH} = 10^{-9} \text{ M}$$

In true scientific notation, this number would be expressed as 1×10^{-9}.

EXAMPLE 8.14

What is the hydroxide ion concentration for the solution described in Example 8.13?
Remember the ion product of water: $[H_3O^+][OH^-] = 1 \times 10^{-14}$. If $[H_3O^+] = 1 \times 10^{-9}$, then

$$[1 \times 10^{-9}][OH^-] = 1 \times 10^{-14}$$

$$[OH^-] = \frac{1 \times 10^{-14}}{1 \times 10^{-9}} = 1 \times 10^{-5} \text{ M}$$

EXAMPLE 8.15

What is the pOH of the solution described in Example 8.13?
In Example 8.14, we determined that the concentration of hydroxide ion is 1×10^{-5} M. The pOH is the negative of the exponent of 10.

$$pOH = -(-5) = 5$$

You could also determine the pOH without first calculating the concentration of hydroxide ion. Remember that pH + pOH = 14. From Example 8.13 we know that the pH is 9. Therefore,

$$9 + pOH = 14 \quad \text{and} \quad pOH = 14 - 9 = 5$$

TABLE 8.3 The Approximate pH of Some Solutions	
Solution	pH
0.1 M HCl	1.0
Gastric juices	1.6–1.8
Lemon juice	2.3
Vinegar	2.4–3.4
Soft drinks	2.0–4.0
Milk	6.3–6.6
Urine	5.5–7.0
Rainwater (unpolluted)	5.6
Saliva	6.2–7.4
Pure water	7.0
Blood	7.35–7.45
Egg white (fresh)	7.6–8.0
Bile	7.8–8.6
Milk of magnesia	10.5
Washing soda	12
0.1 M NaOH	13

EXAMPLE 8.16

What is the hydronium ion concentration of a solution whose pH is 3.6?

The hydronium ion concentration is $1 \times 10^{-3.6}$ M. In standard scientific notation, whole number powers of 10 are used. That means that the above number would be expressed as 2.5×10^{-4}. To perform this numerical conversion, logarithms are used. We will not ask you to do such conversions in this text, and therefore we shall ordinarily use only integer values for pH and pOH in problems. You should know, however, that $[H_3O^+]$ for a solution with a pH of 3.6 is between 1×10^{-3} and 1×10^{-4} M.

A pH of 7 represents a neutral solution. A pH below 7 represents an acidic solution, and one above 7 represents a basic solution. The lower the pH, the more acidic is the solution; the higher the pH, the more basic the solution (Table 8.3).

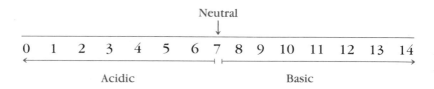

8.8 BUFFERS: THE CONTROL OF pH

Our bodies are acid factories. Our stomachs produce hydrochloric acid. Our muscles produce lactic acid. Starches and sugars produce pyruvic acid when they metabolize. Carbon dioxide from respiration produces carbonic acid in the blood. Our bodies must eliminate or neutralize these acids because excess acidity in the wrong place would kill us rather quickly. A **buffer solution** is one in which the pH remains relatively constant even if acid or base is added. Chemically, a buffer solution usually contains approximately equal concentrations of a weak acid and one of its salts or a weak base and one of its salts.

The pH of a buffer solution changes very little even with the addition of acid or base.

The first thing to realize is that a buffer does not necessarily maintain a pH of 7, that is, neutrality. Buffer solutions can be prepared to maintain almost any pH. For example, one can prepare a solution of acetic acid and sodium acetate that acts as a buffer at a pH value of about 5. This buffer can absorb significant amounts of additional acid or base without appreciable change in pH. An amount of acid that would change the pH of pure water by 4 full pH units would cause the buffer's

A buffer solution is composed of a weak acid and a salt of that weak acid.

8.8 Buffers: The Control of pH

pH to change by only 0.01 pH unit. The point is that a buffer is not supposed to maintain neutrality (pH 7) but whatever pH (including pH 7) it is set for.

How does a buffer work? The explanation is fairly simple: the buffer contains species that will tie up added hydronium ions or hydroxide ions. Consider a buffer solution consisting of acetic acid and sodium acetate (see Figure 8.8). If a strong acid is added to this solution, the hydronium ions produced by the added acid donate protons to the acetate ion in solution.

$$H_3O^+ + CH_3COO^- \rightleftharpoons CH_3COOH + H_2O$$

Although the reaction is reversible to a slight extent, most of the protons remain attached to the acetic acid product. Remember, acetic acid is a *weak* acid, and it cannot transfer the proton back to the water molecule efficiently. Since most of the protons are tied up in this way, the pH of the solution changes very little.

When a strong base is added, the hydroxide ions react with hydronium ions already in the buffer because of the presence of acetic acid.

$$OH^- + H_3O^+ \longrightarrow 2\ H_2O$$

Acetic acid is a *weak* acid, but it is an acid, and that means that it produces some hydronium ions in solution. The added hydroxide ions are tied up. Some of the hydronium ions originally present in the buffer are used up in this process. They are immediately replaced by further ionization of the acetic acid in the buffer.

$$CH_3COOH + H_2O \rightleftharpoons CH_3COO^- + H_3O^+$$

The concentration of hydronium ions returns to approximately the original value, and the pH is only slightly changed.

There are, then, two consequences of buffer action. First, the pH is maintained relatively constant. Neither the hydrogen ion nor the hydroxide ion concentration is allowed to rise or fall much. Second, the by-products of any buffer action are not new to the system. Thus, when excess acid reacts with the acetate anion, water and acetic acid are formed—and these molecules are already present in the system. The same can be said for the reaction of excess base. So the products of one reaction can then always be used as reagents for the opposite reaction.

There are many important buffer solutions. Most biochemical reactions, whether they occur in a laboratory or in our bodies, are carried out in buffered solutions. The buffers that control the pH of our blood will be discussed in the next section. Table 8.4 lists some buffers of interest and the pH range in which they operate.

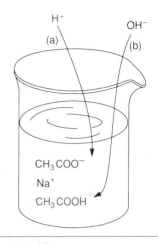

FIGURE 8.8
Acetic acid–sodium acetate buffer. (a) If an acid invades the system, it reacts with the acetate anion. (b) If a base invades the system, it reacts with the hydronium ion from acetic acid.

TABLE 8.4 Some Important Buffers

Buffer Components	Buffer System Names	pH[a]
$CH_3CHOHCOOH/$ $CH_3CHOHCOO^-$	Lactic acid–lactate ion	3.86
CH_3COOH/CH_3COO^-	Acetic acid–acetate ion	4.76
$H_2PO_4^-/HPO_4^{2-}$	Dihydrogen phosphate ion–monohydrogen phosphate ion	7.20
H_2CO_3/HCO_3^-	(Carbon dioxide) carbonic acid–bicarbonate ion	6.46[b]
NH_4^+/NH_3	Ammonium ion–ammonia	9.25

[a] The values listed are for solutions that are 0.1 M in each compound at 25 °C.
[b] This value includes dissolved CO_2 molecules as undissociated H_2CO_3. The value for H_2CO_3 alone is about 3.8.

8.9 Buffers in the Blood

The pH of the blood of higher animals is held remarkably constant. In humans, blood plasma normally varies between 7.35 and 7.45 in pH. Should the pH rise above 7.8 or fall below 6.8, a person may suffer irreversible damage to the brain or even die. Fortunately, human blood has not one, but at least three, buffering systems. Of these, the bicarbonate–carbonic acid buffering system is the most important.

When acids enter the blood, the resulting hydronium ions are taken up by bicarbonate ions to form undissociated carbonic acid and water. Most of the carbonic acid, in turn, decomposes to give carbon dioxide and water.

$$HCO_3^- + H_3O^+ \longrightarrow H_2CO_3 \longrightarrow H_2O + CO_2$$
$$+$$
$$H_2O$$

As long as there is sufficient bicarbonate ion to take up the added acid, the pH changes little.

If bases enter the bloodstream, the resulting hydroxide ions react with hydronium ions present to form water. The hydronium ions are present because of the slight ionization of the carbonic acid portion of the buffer. As hydronium ions are used up, more carbonic acid ionizes. As carbonic acid is used up, more can be formed from the large reservoir of dissolved carbon dioxide in the blood. Here, then, is the sequence of reactions involved in the neutralization of a base by the buffer.

1. Remove hydroxide ions: $\quad H_3O^+ + OH^- \longrightarrow 2\,H_2O$
2. Replace depleted hydronium ions: $\quad H_2CO_3 + H_2O \longrightarrow H_3O^+ + HCO_3^-$
3. Replace depleted carbonic acid: $\quad CO_2 + H_2O \longrightarrow H_2CO_3$

8.10 Acidosis and Alkalosis

If you add these equations together, canceling those species that appear as both reactant and product, you come up with the overall result:

$$CO_2 + OH^- \longrightarrow HCO_3^-$$

This is simply another way of saying that the added hydroxide ion is tied up, and that is why the pH does not change much.

Another blood buffer is the dihydrogen phosphate–monohydrogen phosphate ($H_2PO_4^-/HPO_4^{2-}$) system. Any acid reacts with monohydrogen phosphate ions to form dihydrogen phosphate ions.

$$HPO_4^{2-} + H_3O^+ \longrightarrow H_2PO_4^- + H_2O$$

The dihydrogen phosphate ion is a weak acid and exists in equilibrium with hydronium ion and monohydrogen phosphate ion.

$$H_2PO_4^- + H_2O \rightleftharpoons H_3O^+ + HPO_4^{2-}$$

Any base that comes into the blood would react with hydronium ions to form water. But more dihydrogen phosphate ions would ionize to replace these hydronium ions, leaving the pH essentially unchanged.

Proteins act as a third type of blood buffer. These complex molecules (Chapter 15) contain —COO⁻ groups, which, like acetate ions (CH_3COO^-), can act as proton acceptors. Proteins also contain —NH_3^+ groups, which, like ammonium ions (NH_4^+), can donate protons. If acid comes into the blood, hydronium ions can be neutralized by the —COO⁻ groups.

$$-COO^- + H_3O^+ \longrightarrow -COOH + H_2O$$

If base is added, it can be neutralized by the —NH_3^+ groups.

$$-NH_3^+ + OH^- \longrightarrow -NH_2 + H_2O$$

These three buffers (and perhaps others) act to keep the pH of the blood constant. Buffers can be overridden by large amounts of acid or base; their capacity is not infinite. The blood buffers can be overwhelmed if the body's metabolism goes awry.

8.10 ACIDOSIS AND ALKALOSIS

Have you ever had your muscles hurt after prolonged physical activity? If so, you have had your blood buffers overloaded. Muscle contraction produces lactic acid. This acid ionizes somewhat more strongly than carbonic acid and thus tends to lower the pH of the blood (it tends to release more hydronium ions into the blood). Moderate amounts of lactic acid can be handled by the blood buffers. For bicarbonate, the reaction is

$$\underset{\text{Lactic acid}}{CH_3\overset{\overset{\displaystyle OH}{|}}{C}HCOOH} + HCO_3^- \longrightarrow \underset{\text{Lactate ion}}{CH_3\overset{\overset{\displaystyle OH}{|}}{C}HCOO^-} + H_2CO_3$$

Excessive amounts of lactic acid overload the buffers, however, and the pH is lowered. Nerve cells respond to the increased acidity by sending a message of pain to the brain.

If the pH of the blood falls below 7.35, the condition is called **acidosis**. If the pH of the blood rises above 7.45, **alkalosis** sets in. Faulty respiration or metabolic problems can cause these pathological conditions.

In severe cases of starvation, the body gets its energy from the oxidation of stored fats. The acidic products of fat metabolism can build up in the blood, leading to acidosis. Fad diets, such as those that severely limit the intake of carbohydrates, can also lead to acidosis. Other causes of acidosis include kidney failure and diarrhea (which cause excretion of bicarbonate), diabetes, slow breathing due to emphysema, pneumonia, poliomyelitis, heart failure, or holding the breath—all of which cause the carbon dioxide level to increase.

Respiratory alkalosis may occur when a person hyperventilates. This pattern of rapid breathing vents too much carbon dioxide to the atmosphere. As the CO_2 level in the blood falls, bicarbonate ions react with hydronium ions to produce more CO_2. We have seen this equation before.

$$HCO_3^- + H_3O^+ \longrightarrow H_2CO_3 \longrightarrow H_2O + CO_2$$
$$+$$
$$H_2O$$

Hydronium ions are used up, the pH of the blood increases, and the condition called alkalosis exists. Other causes of alkalosis include vomiting, fever, hysteria, shock, and, of course, ingestion of antacids.

Normally, the respiratory system functions to maintain the proper blood pH. Remember that the preceding equation is reversible. When the blood buffers are unable to compensate for a change in pH, the body immediately reacts by changing the rate of respiration. Carbon dioxide in the blood is either conserved or vented, shifting the HCO_3^-/CO_2 equilibrium to the left or right, depending on whether the blood has become alkaline or acidic. One important function of the kidneys is the control of blood pH by the selective excretion and retention of certain ions. The kidneys react more slowly than the respiratory center, but the kidneys have a much greater capacity to control blood pH.

Acidosis occurs if the pH of the blood falls below 7.35.

Alkalosis occurs if the pH of the blood rises above 7.45.

CHAPTER SUMMARY

Acids are proton donors, and bases are proton acceptors. In aqueous solutions acids transfer protons to water molecules to produce hydronium ions (H_3O^+) and bases yield hydroxide ions (OH^-). Acids that yield two hydronium ions per formula unit are called diprotic acids, and those that yield three are described as triprotic.

The process by which an originally covalent acid or base produces hydronium ions or hydroxide ions, respectively, in solution is termed *ionization*. When an ionic compound, such as sodium hydroxide, separates into independent cations and anions in solution, the term *dissociation* is used. Strong acids and bases are those that ionize or dissociate completely in solution. For weak acids and bases, the equilibrium involving the covalent and ionized forms strongly favors the nonionized forms. A list of common acids is given in Table 8.1. Sodium and potassium hydroxide are strong bases, and ammonia is a weak base.

Many acids and bases are found in common household products. Several of these (H_2SO_4, HCl, NaOH) are corrosive poisons. Others (HCN, NH_3), although classified as weak in terms of acidity or basicity, are nonetheless quite toxic.

Neutralization is the reaction of an acid and a base to produce a salt and water. In aqueous solutions the net ionic equation for the neutralization reaction involving a strong acid and a strong base is

$$H_3O^+ + OH^- \rightarrow 2\,H_2O$$

An equivalent of acid is the amount of acid that produces 1 mol of hydronium ions in solution, and an equivalent of base is the amount that combines with 1 mol of hydronium ions. The number of equivalents per mole of acid or base is equal to the number of hydronium ions generated or captured by one formula unit of the substance.

In a titration, the amount of acid (or base) present in an unknown sample is determined by neutralization of the unknown with a known amount of base (or acid). The endpoint of the titration can be determined by use of an indicator, a dye that changes color when the acidity of a solution changes. Ideally the endpoint coincides with the equivalence point, that is, the point at which the same numbers of equivalents of acid and base have been added.

Depending on the makeup of the salt, an aqueous solution of a salt may be acidic, basic, or neutral. The salt of a strong acid and a strong base gives a neutral solution. A solution of the salt of a weak acid and a strong base is basic, and that of a strong acid and weak base is acidic. The character of a solution of the salt of a weak acid and a weak base is difficult to predict. In all cases, the nature of the solution is determined by an equilibrium involving the ion from a weak "parent" and water.

Carbonic acid (H_2CO_3) is a weak, unstable diprotic acid. Its salts, the bicarbonates (HCO_3^-) and carbonates (CO_3^{2-}), react with strong acids to produce ultimately carbon dioxide and water. Much of the damage inflicted by acid rain on buildings and other structures results from the reaction of this pollutant with $CaCO_3$ in limestone and marble. Bicarbonates and carbonates are also found as ingredients in antacid tablets, which are used to neutralize stomach acid, HCl. Other bases, such as $Mg(OH)_2$ and $Al(OH)_3$, are also used in antacid preparations.

In any aqueous solution, the product of the hydronium ion concentration and the hydroxide ion concentration is a constant.

$$[H_3O^+][OH^-] = 1 \times 10^{-14} = K_w$$

K_w is called the ion product of water. Since the ion product of water is a constant, in aqueous solution an increase in hydronium ion concentration will be accompanied by a decrease in hydroxide ion concentration and vice versa. pH is a measure of the hydronium ion concentration in aqueous solution.

$$\text{pH} = -\log[H_3O^+]$$

pOH is an analogous measure of the hydroxide ion concentration in aqueous solution, and

$$\text{pH} + \text{pOH} = 14$$

A solution with a pH of 7 is neutral. If the pH is below 7, the solution is acidic; if the pH is above 7, the solution is basic.

Buffers are substances that maintain the pH of a solution in the presence of added acid or base. Most buffers consist of a weak acid and its salt or a weak base and its salt. Buffers maintain the pH of a solution by tying up added hydronium or hydroxide ions. Because the maintenance of proper pH is so important to life processes, human blood carries at least three different buffer systems. These include the bicarbonate–carbonic acid–carbon dioxide system, the dihydrogen phosphate–monohydrogen phosphate system, and blood proteins. If the blood buffers are overwhelmed, the pH of the blood changes. If the pH rises above 7.45, the condition is called alkalosis. If pH falls below 7.35, the condition is referred to as acidosis. A variety of respiratory and metabolic problems may lead to these pathological conditions, but ordinarily the respiratory system and the kidneys back up the blood buffers in maintaining the optimum pH of the blood.

CHAPTER QUIZ

1. The hydrogen ion is
   a. H^+ b. H_3O^+ c. OH^- d. HOH^-
2. H_3PO_4 is a(n)
   a. monoprotic organic acid
   b. polyprotic inorganic base
   c. triprotic inorganic acid
   d. aprotic organic base
3. The strongest acid in the following list is
   a. HCl   b. HCN
   c. CH_3COOH d. H_3BO_3
4. In neutralization, 0.60 mol of H_3PO_4 is equivalent to _____ mol HCl.
   a. 0.2   b. 0.6   c. 1.8
   d. none of these is right
5. In neutralization, 0.10 mol of NaOH is equivalent to _____ mol of $Ca(OH)_2$.
   a. 0.05   b. 0.10   c. 0.20   d. 0.30
6. It takes 22.0 mL of 0.1 M NaOH to neutralize 100 mL of an H_2SO_4 solution. The molarity of the acid is
   a. 0.011   b. 0.022   c. 1.1   d. 2.2
7. When placed in water, NH_4Cl produces a solution that is
   a. acidic   b. basic
   c. neutral   d. unpredictable
8. The pH of a solution that is 1.0×10^{-3} M NaOH is
   a. 3   b. 13   c. 11   d. -3
9. The combination NaCl/HCl is not a good buffer because
   a. it needs water
   b. no Na^+ ions are allowed
   c. it has a strong acid
   d. it must be organic.

10. Alkalosis can be caused by all but which one of the following?
    a. diarrhea   b. vomiting
    c. fever   d. shock

PROBLEMS

8.1 Definitions

1. Define and, where possible, illustrate the following terms.
   a. acid   b. base
   c. hydroxide ion   d. neutralization
2. Contrast acids and bases with respect to taste and effect on litmus.
3. Identify the first compound in each equation as an acid or a base.
   a. $C_5H_5N + H_2O \rightarrow C_5H_5NH^+ + OH^-$
   b. $C_6H_5OH + H_2O \rightarrow C_6H_5O^- + H_3O^+$
   c. $CH_3COCOOH + H_2O \rightarrow$
      $CH_3COCOO^- + H_3O^+$
   d. $C_6H_5SH + H_2O \rightarrow C_6H_5S^- + H_3O^+$
   e. $CH_3NH_2 + H_2O \rightarrow CH_3NH_3^+ + OH^-$
   f. $C_6H_5SO_2NH_2 + H_2O \rightarrow$
      $C_6H_5SO_2NH^- + H_3O^+$
4. Give examples of a monoprotic, a diprotic, and a triprotic acid.
5. Indicate whether the acid is monoprotic, diprotic, or triprotic.
   a. $H_2SO_3 + 2 H_2O \rightarrow SO_3^{2-} + 2 H_3O^+$
   b. $CH_3CHOHCOOH + H_2O \rightarrow$
      $CH_3CHOHCOO^- + H_3O^+$
   c. $CH_2(COOH)_2 + 2 H_2O \rightarrow$
      $CH_2(COO^-)_2 + 2 H_3O^+$
   d. $H_3AsO_4 + 3 H_2O \rightarrow AsO_4^{3-} + 3 H_3O^+$
   e. $H_4P_2O_7 + 4 H_2O \rightarrow P_2O_7^{4-} + 4 H_3O^+$

8.2 Acids and Bases in Aqueous Solutions

6. Define:
   a. hydronium ion
   b. ionization
   c. dissociation
7. Give formulas and names of three strong acids and three weak acids.
8. Give formulas and names for a strong base and a weak base.

Problems

9. Each of the following equations presents an equilibrium between two acids. In each case, identify the acids and indicate which is stronger.
   a. $HBr + F^- \rightleftharpoons HF + Br^-$
   b. $HCN + F^- \rightleftharpoons HF + CN^-$
   c. $HIO_3 + H_2PO_4^- \rightleftharpoons H_3PO_4 + IO_3^-$
   d. $HCOOH + CCl_3COO^- \rightleftharpoons HCOO^- + CCl_3COOH$
   e. $HClO_3 + NO_3^- \rightleftharpoons HNO_3 + ClO_3^-$

10. Bases are proton acceptors. Each equilibrium in Problem 9 also includes a pair of bases. Identify the bases in each equation, and indicate which is stronger.

11. Complete the following by writing formulas for the expected products and balancing the equations.
    a. $HCl + LiOH \rightarrow$
    b. $Al(OH)_3 + HCl \rightarrow$
    c. $H_2SO_4 + Mg(OH)_2 \rightarrow$

12. Complete the following by writing formulas for the expected products and balancing the equations.
    a. $NaHCO_3 + HNO_3 \rightarrow$
    b. $CaCO_3 + HBr \rightarrow$
    c. $H_2SO_4 + K_2CO_3 \rightarrow$

13. Complete the equation: $NH_3 + HI \rightarrow$

8.3 Neutralization in Aqueous Solution

14. Define:
    a. salt
    b. titration
    c. equivalent
    d. indicator
    e. endpoint
    f. equivalence point

15. Write the *net ionic equation* for a neutralization reaction involving a strong acid and a strong base in aqueous solution.

16. Write the *net ionic equation* for the reaction of a bicarbonate salt with a strong acid in aqueous solution.

17. Calculate the concentration of an HCl solution if neutralization of 20 mL of it requires
    a. 40 mL of 0.25 M NaOH
    b. 10 mL of 0.50 M KOH
    c. 10 mL of 0.50 M $Ca(OH)_2$

18. A 20-mL sample of gastric fluid (assume all acidity is due to HCl) is neutralized by 25 mL of 0.10 M NaOH. What is the concentration of the HCl solution?

19. Write the *net ionic equation* for the reaction of a carbonate salt with a strong acid in aqueous solution.

20. How many equivalents are there in
    a. 1 mol of H_2SO_4?
    b. 1 mol of $Al(OH)_3$?

21. How many equivalents are there in
    a. 2 mol of H_3PO_4?
    b. 0.5 mol of NH_3?
    c. 10 mol of CH_3COOH?

22. How many moles are required to provide 1 equiv of each of the following?
    a. HNO_3 b. $Mg(OH)_2$

23. How many moles are required to provide the indicated number of equivalents of each of the following?
    a. 6 equiv of H_3PO_4
    b. 0.2 equiv of H_2SO_4
    c. 1.5 equiv KOH

24. How many equivalents are there in each of the following?
    a. 162 g of HBr
    b. 6.0 g of LiOH
    c. 3.4 g of H_2S

25. How many equivalents are there in each of the following?
    a. 1.71 g of $Ba(OH)_2$
    b. 49 g of H_3PO_4
    c. 15 g of CH_3COOH

26. Would 1 mol of NaOH neutralize
    a. 1 mol of HNO_3?
    b. 1 mol of CH_3COOH?
    c. 3 mol of H_3PO_4?
    d. 0.5 mol of H_2SO_4?

27. Would 0.50 equiv of $Ca(OH)_2$ neutralize all of the acid in
    a. 0.50 equiv of H_3PO_4?
    b. 1.0 equiv of HCl?
    c. 0.25 equiv of H_3BO_3?

28. In an ideal titration, the endpoint occurs at the equivalence point. What is the difference between an endpoint and an equivalence point?

8.4 Salts in Water: Acidic, Basic, or Neutral

29. Write an equation for the equilibrium established when CH_3COO^- is placed in water.
30. Write an equation for the equilibrium established when NH_4^+ is placed in water.
31. Classify the aqueous solution of each of these salts as acidic, basic, or neutral.
    a. KCl
    b. NaCN
    c. NH_4CN
    d. CH_3COOK
    e. $(NH_4)_2SO_4$
    f. Na_2SO_4
32. A weak acid is titrated with a strong base. Would the solution at the equivalence point be acidic, basic, or neutral? Explain.

8.5 Reactions of Acids with Carbonates and Bicarbonates

33. Write an equation that describes the effect of acid rain on marble.

8.6 Antacids

34. Name some of the active ingredients in antacids.
35. Should a person who has hypertension be advised to use baking soda or milk of magnesia as an antacid? Why?

8.7 The pH Scale

36. Define:
    a. pH
    b. pOH
    c. K_w
37. What is the pH of each of the following solutions?
    a. 1×10^{-2} M HCl
    b. 1×10^{-3} M HNO_3
38. What is the pH of each of the following solutions?
    a. 0.0001 M HCl
    b. 0.000 01 M HNO_3
    c. 0.1 M HBr
39. What is the pOH of each of the following solutions?
    a. 1×10^{-1} M NaOH
    b. 1×10^{-5} M KOH
40. What is the pOH of each of the following solutions?
    a. 0.001 M NaOH
    b. 0.01 M KOH
41. What is the pOH of each of the solutions in Problems 37 and 38?
42. What is the pH of each of the solutions in Problems 39 and 40?
43. Indicate whether each of the following pH values represents an acidic, a basic, or a neutral solution.
    a. 11
    b. 4
    c. 7
    d. 3.4
44. Answer Problem 43 assuming that the values given are for pOH.

8.8 Buffers: The Control of pH

45. Use acetic acid and acetate ion to explain how a buffer controls pH.
46. Use ammonia and ammonium chloride to explain how a buffer controls pH.

8.9 Buffers in the Blood

47. Write the equation for the *decomposition* of carbonic acid.
48. Identify three buffer systems operating in the blood.
49. What groups on protein buffers react with added acid and base?
50. If acid is added to an unbuffered solution, will the pH increase or decrease?

8.10 Acidosis and Alkalosis

51. Define:
    a. alkalosis
    b. acidosis
    c. buffer
52. If someone has alkalosis, is the blood pH too high or too low?
53. When pneumonia blocks respiratory passageways and limits the transfer of carbon dioxide from blood to the atmosphere, blood CO_2 levels build up. Is the pH of the blood higher or lower than normal in this situation?
54. A common cause of death in young children is acidosis resulting from severe diarrhea. In severe diarrhea, large amounts of bicarbonate ion are excreted from the body. Why should this lead to acidosis?
55. A first aid procedure in some cases involves placing a paper bag over the patient's face. Explain.

9

HYDROCARBONS

DID YOU KNOW THAT...

▶ "organic" compounds are so named because the first ones were derived from living organisms?

▶ there are about 10 million known compounds and 95% of them are organic?

▶ organic compounds are generally low-boiling and insoluble in water?

▶ sodium bicarbonate can be used as a fire extinguisher?

▶ 10 carbon atoms and 22 hydrogen atoms can be linked to each other in 75 different ways?

▶ alkanes with 5–16 carbon atoms are liquids, those with less than 5 carbon atoms are gases, and those with more than 16 are solids?

▶ many plastics are the products of polymerization of alkenes?

▶ many dental sealants polymerize with teeth?

▶ benzene is a cyclic hydrocarbon with special properties?

▶ inducing vomiting is not the proper treatment for people who swallow gasoline?

Definitions from the 18th and 19th centuries—
  Inorganic: from a nonliving source
  Organic: from a living system

Scientists of the eighteenth and nineteenth centuries studied compounds isolated from rocks and ores and from the atmosphere and oceans, and they labeled them **inorganic compounds** because they were obtained from nonliving systems. Compounds obtained from plants and animals were called **organic compounds** because they were isolated from organized (living) systems. The early chemists believed that only living organisms, within their tissues, could synthesize organic compounds, that some **vital force** was required for such syntheses. By the middle of the nineteenth century, however, a number of "organic" compounds had been prepared using ordinary laboratory techniques. The vital-force theory had to be discarded, but the labels *organic* and *inorganic* remained.

Modern definition: Organic chemistry is the chemistry of the compounds of carbon.

Today, **organic chemistry** is defined simply as the chemistry of the compounds of carbon. It may seem strange that we divide chemistry into two branches, one of which considers the chemistry of one element while the other handles the chemistry of the more than 100 remaining elements. This division seems more reasonable when one discovers that, of the 10 million or more compounds that have been characterized, the overwhelming majority contain carbon. Carbon has a unique ability to form stable, covalent bonds with itself and with other elements in infinite variations. The molecules thus produced may contain only one or over a million carbon atoms. So complex is the chemistry of carbon that we shall approach its study by dividing its millions of compounds into families. We will study one family at a time and begin by concentrating on the simpler members of each family. Eventually we will move to a consideration of those molecules that deserve to be called organic in the old sense—complex, carbon-containing molecules that determine the form and function of living systems. We might pause to ponder how far science has come since the eighteenth century, when scientists believed it was impossible to synthesize even the simplest organic molecule. In 1980, the Supreme Court ruled that a new form of life, developed in a laboratory, could be patented.

9.1 A Comparison of Organic and Inorganic Compounds

Organic compounds, like inorganic ones, obey all the natural laws. Often there is no clear distinction in chemical or physical properties between organic and inorganic molecules. Nevertheless, it may be useful to compare and contrast typical members of each class. Table 9.1 lists a variety of properties of the inorganic compound sodium chloride (NaCl, common table salt) and the organic compound benzene (C_6H_6, a common solvent once widely used to strip furniture for refinishing).

Most organic compounds have relatively low melting points. Many,

9.2 The Hydrocarbons

TABLE 9.1 A Comparison of an Inorganic Compound and an Organic Compound

Name	Formula	Solubility in H_2O	Solubility in Gasoline	Flammable?	mp	bp	Density	Bonding
Benzene	C_6H_6	Insoluble	Soluble	Yes	5.5 °C	80 °C	0.88 g/cm^3	Covalent
Sodium chloride	NaCl	Soluble	Insoluble	No	801 °C	1413 °C	2.7 g/cm^3 (crystal)	Ionic

like benzene, are liquids at room temperature. The typical inorganic substance, like sodium chloride, is a crystalline solid with a high melting point.

The typical organic compound is insoluble in water and less dense than water. An attempt to mix an organic liquid with water usually produces two layers, with the organic layer on top. The typical inorganic compound is readily soluble in water.

The typical organic compound is highly flammable, an important point to remember when one is working with organic compounds. Both the anesthetic ether and gasoline are flammable organic substances that form explosive mixtures with air. Typical inorganic compounds are nonflammable. Some inorganic compounds, such as water, baking soda (sodium bicarbonate, $NaHCO_3$), and borax ($Na_2B_4O_7 \cdot 10H_2O$), are used in fighting fires. (Sodium bicarbonate is one of the few carbon-containing compounds classified as inorganic.)

The typical organic compound is covalently bonded, and the properties of organic compounds reflect this fact. The typical inorganic compound is ionic and exhibits the properties characteristic of ionic bonding. Do not forget, though, that there are covalent inorganic compounds (water, for example) and organic ions (acetate ion, for example).

9.2 THE HYDROCARBONS

We shall begin our study of organic compounds by considering those containing only two elements, carbon and hydrogen. These compounds are called **hydrocarbons** (a reasonable name, you will admit). The hydrocarbons are themselves subdivided into several families of compounds. Family membership is determined by the type of bonding found in a compound. **Saturated hydrocarbons**—hydrocarbons that contain only single bonds—are called **alkanes**. The families of **unsaturated hydrocarbons** include the **alkenes**, hydrocarbons containing a double bond; **alkynes**, hydrocarbons with a triple bond;

Hydrocarbons contain only hydrogen and carbon.

and an important class called the **aromatic hydrocarbons**, whose special bonding we will describe later.

Each of these families will be considered in turn. You will soon find that most of the principles developed in the discussion of one family are also valid for the next family. Thus, although organic chemistry includes a large amount of material, much of it can be summarized in a relatively few fundamental concepts.

9.3 THE ALKANES

Alkanes contain carbons and hydrogens connected only by single bonds.

The alkanes are compounds containing only carbon and hydrogen atoms connected by single bonds. Carbon is a Group IVA element; a carbon atom forms four bonds. The simplest alkane is methane, CH_4, whose electron-dot formula is

$$\begin{array}{c} H \\ H:\ddot{C}:H \\ H \end{array}$$

Most organic molecules, including methane, are not two-dimensional. Methane is tetrahedral (Figure 9.1). We will ordinarily represent this compound, as

$$\begin{array}{c} H \\ | \\ H-C-H \\ | \\ H \end{array}$$

but a more accurate model is pictured in Figure 9.2. The trouble is, the more accurate picture is more difficult to draw. That is why we ordinarily use **structural formulas** like the one given above for methane. These formulas show you in what order atoms are attached but do not attempt to portray accurately the three-dimensional geom-

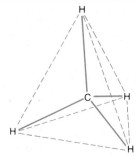

FIGURE 9.1
The tetrahedral methane molecule.

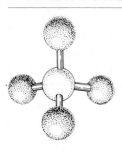

FIGURE 9.2
Ball-and-stick model of methane.

9.3 The Alkanes

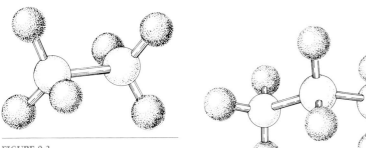

FIGURE 9.3 Ball-and-stick model of ethane.

FIGURE 9.4 Ball-and-stick model of propane.

etry of the molecule. To emphasize the relationship between structural formulas and the more accurate three-dimensional models, we will continue to show both representations while we discuss the smaller alkane molecules.

Ethane (C_2H_6) is the next smallest alkane (Figure 9.3).

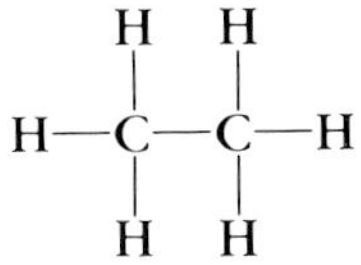

$$\begin{array}{c} H\;\;H \\ |\;\;\;| \\ H-C-C-H \\ |\;\;\;| \\ H\;\;H \end{array}\quad \text{ethane}$$

The three-carbon alkane C_3H_8 is called propane. A three-dimensional model of this compound is shown in Figure 9.4, and the two-dimensional structural formula is

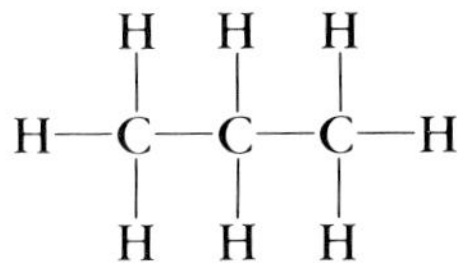

$$\begin{array}{c} H\;\;H\;\;H \\ |\;\;\;|\;\;\;| \\ H-C-C-C-H \\ |\;\;\;|\;\;\;| \\ H\;\;H\;\;H \end{array}\quad \text{propane}$$

A pattern is now becoming apparent. We can build alkanes of any length simply by tacking carbon atoms together in long chains and adding sufficient hydrogen atoms to give each of the carbon atoms a total of four bonds. Even the naming of these compounds follows a pattern. For compounds of five carbon atoms or more, the stems are derived from the Greek or Latin names for the number of carbon atoms in the molecule (Table 9.2). The compound names end in *-ane*, signifying that the compounds are alk*anes*. Table 9.3 gives structural formulas and names for continuous-chain alkanes up to 10 carbon atoms in length. We do not need to stop at 10 carbon atoms. One could hook together 100 or 1000 or 1 million. We can make an infinite number of alkanes simply by lengthening the chain. But lengthening the chain is not our (or nature's) only option. Chain-branching is also possible, and this possibility leads us to the topic of **isomerism**.

TABLE 9.2 Stems That Indicate the Number of Carbon Atoms

Stem	Number
Pent-	5
Hex-	6
Hept-	7
Oct-	8
Non-	9
Dec-	10

TABLE 9.3 The First 10 Normal Alkanes

Name	Molecular Formula	Structural Formula	No. of Possible Isomers
Methane	CH_4	H–C(H)(H)–H	1
Ethane	C_2H_6	H–C(H)(H)–C(H)(H)–H	1
Propane	C_3H_8	H–C(H)(H)–C(H)(H)–C(H)(H)–H	1
Butane	C_4H_{10}	H–[C(H)(H)]₄–H	2
Pentane	C_5H_{12}	H–[C(H)(H)]₅–H	3
Hexane	C_6H_{14}	H–[C(H)(H)]₆–H	5
Heptane	C_7H_{16}	H–[C(H)(H)]₇–H	9
Octane	C_8H_{18}	H–[C(H)(H)]₈–H	18
Nonane	C_9H_{20}	H–[C(H)(H)]₉–H	35
Decane	$C_{10}H_{22}$	H–[C(H)(H)]₁₀–H	75

9.4 ISOMERISM

Isomers are defined as different compounds with the same molecular formula. The compound butane (C_4H_{10}) is shown in Table 9.3. This compound boils at 0 °C. A second compound, whose boiling point is −12 °C, has the same molecular formula, C_4H_{10}. The *structural formula* of the second compound is *not* the same as butane's. Instead of having four carbon atoms connected in a continuous chain, this new compound has a continuous chain of only three carbon atoms. The fourth carbon is branched off the middle carbon of this three-carbon chain.

Isomers: different compounds with the same numbers of each type of atom

```
      H   H   H
      |   |   |
  H—C—C—C—H
      |   |   |
      H   |   H
          |
      H—C—H
          |
          H
```

Because it is an isomer of butane, this branched four-carbon alkane is called isobutane (Figure 9.5).

Note that the following drawings do *not* represent isomers of butane.

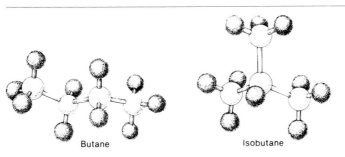

FIGURE 9.5
Ball-and-stick models of butane and isobutane.

Butane   Isobutane

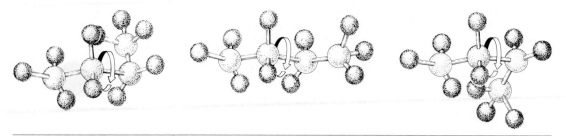

FIGURE 9.6
Rotation about single bonds. The right-hand portion of the molecule is spinning about the single bond between the second and third carbon atoms. The same molecule can thus assume different arrangements.

Each of these structural formulas shows a continuous four-carbon chain. The chain may be twisted or bent, but it is still one continuous chain. The different arrangements result when various parts of the molecule spin about the single bonds that join them to the rest of the molecule. Figure 9.6 shows a model of a butane molecule in which the right half is twisting about the bond between the second and third carbon atoms in the chain. Thus, the three models shown in Figure 9.6 and the preceding three structural formulas all represent butane. For our own convenience, we will usually represent butane as a straight chain of four carbon atoms stretched from left to right across the page. We will ordinarily draw the longest continuous chain in any molecule horizontally across the page and hang any branches above or below this main chain.

There are three isomeric five-carbon alkanes, called pentane (I), isopentane (II), and neopentane (III), respectively.

$$
\begin{array}{ccc}
\text{I} & \text{II} & \text{III}
\end{array}
$$

Normal alkanes have nonbranching chains of carbon atoms.

Those alkanes that have a continuous chain of carbon atoms with no branching are frequently referred to as the **normal alkanes**. Compound I is therefore sometimes called n-pentane, meaning normal or continuous-chain pentane. Table 9.3, which pictures the normal alkanes up to 10 carbon atoms long, also indicates the number of isomers possible for alkanes with 1–10 carbon atoms. Note how rapidly the number of isomers increases as more carbon atoms are incorporated. By the time there are 10 carbon atoms involved, we are faced with making up names for 75 different isomers.

9.5 IUPAC Nomenclature

To bring order to the chaotic naming of newly discovered compounds, the *International Union of Pure and Applied Chemistry* (IUPAC—sometimes pronounced "eye you pack") has established rules for organic nomenclature (the system of naming for organic compounds). The rules for alkanes are summarized here.

1. The names of alkanes end in *-ane*. The names of the unbranched alkanes having up to 10 carbon atoms are given in Table 9.3.
2. The names of branched-chain alkanes are made up of two parts. The latter part of the name is determined by the length of the longest continuous chain of carbon atoms in the compound. For example, the compounds

are considered derivatives of pentane because there are five carbon atoms in the longest continuous chain. Both compounds have names that end in *pentane*.

3. The first part of the name specifies which groups are attached to the "parent" chain. Like the parent chain, the branched groups are given names based on the number of carbon atoms they contain. The names for the branches end in *-yl*, however, instead of *-ane*. Thus, a one-carbon branch, such as those in the molecules above, is called *methyl*. The structures pictured in rule 2 are both methylpentanes. A two-carbon group hanging from the parent chain is called *ethyl*. There is only one alkane named propane, but a chain of three carbon atoms can be attached to a longer chain in two different ways. The attachment may be from an end carbon atom of the three-carbon chain or from the middle carbon atom.

These groups are called propyl and isopropyl, respectively.

A saturated hydrocarbon group is called an alkyl group.

The alkyl groups we are most likely to encounter are listed in Table 9.4. Just as saturated hydrocarbon molecules are called, as a class, alkanes, the saturated hydrocarbon groups are referred to as **alkyl groups**. Methyl, ethyl, and isopropyl are all alkyl groups. Sometimes the symbol "R—" is used to represent an alkyl group when one does not wish to specify a particular group.

4. Arabic numerals are used to indicate the positions at which the alkyl groups are attached on the parent chain. The numbering of the parent chain is always started at the end that will provide the lowest number for the position of the attached group (sometimes called the **substituent group**). The compounds shown in rule 2 are 2-methylpentane and 3-methylpentane, respectively.

5. If more than one group is attached to the parent chain, each group must be named and numbered. The official IUPAC rules state that the groups should be listed in alphabetical order, but many chemists still list them in order of increasing size. You may sometimes hear 4-ethyl-2-methylhexane called "2-methyl-4-ethylhexane." If two or more identical groups are attached to the main chain, another prefix is used to indicate whether there are two or three or more such identical groups. The prefixes are *di-* for two, *tri-* for three, and *tetra-* for four.

2,2-Dimethylhexane   2,2,4-Trimethylpentane

When two identical groups are located at the same position on the parent chain, the number must be repeated. Notice also, from the above examples, that commas are used to separate numbers from one another and hyphens are used to separate numbers from words.

All of this sounds very complicated. We can assure you it is much easier to do than to read about. Here are some examples.

TABLE 9.4 Common Alkyl Groups

Name	Structural Formula	Condensed Structural Formula
Methyl	H—CH₂— (with H above and below C)	CH_3-
Ethyl	H—CH₂—CH₂—	CH_3CH_2-
Derived from propane		
Propyl	H—CH₂—CH₂—CH₂—	$CH_3CH_2CH_2-$
Isopropyl	H—CH₂—CH—CH₂—H	CH_3CHCH_3 (with bond below middle C)
Derived from butane		
Butyl	H—CH₂—CH₂—CH₂—CH₂—	$CH_3CH_2CH_2CH_2-$
Secondary butyl (*sec*-butyl)	H—CH₂—CH—CH₂—CH₂—H	$CH_3CHCH_2CH_3$
Isobutyl	(CH₃ branch on middle C of propyl chain)	$\underset{\|}{CH_3}CHCH_2-$ above shown as CH_3 over CHCH₂—
Tertiary butyl (*tert*-butyl)	(CH₃ branch on central C with three CH₃ groups)	$CH_3-\underset{\|}{\overset{CH_3}{C}}-CH_3$

EXAMPLE 9.1

Name the compound

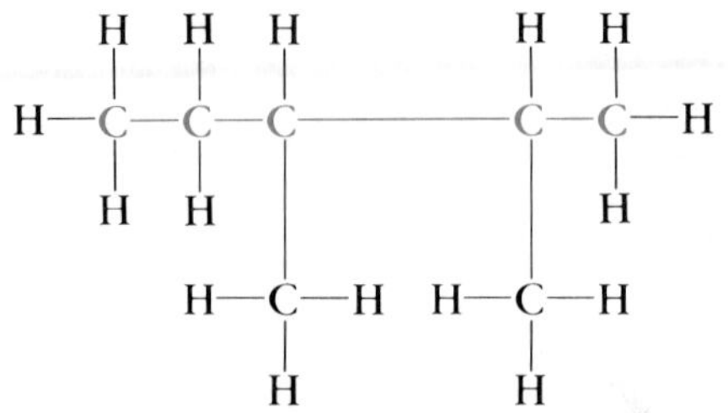

```
      H  H  H           H  H
      |  |  |           |  |
   H—C—C—C———————C—C—H
      |  |  |           |
      H  H  |           H
            |           |
         H—C—H     H—C—H
            |           |
            H           H
```

The longest continuous chain has five carbon atoms. Two methyl groups are attached to the second and third carbon atoms (not the third and fourth—remember to count from the end of the parent chain that gives the lowest numbers). The correct name is 2,3-dimethylpentane.

EXAMPLE 9.2

Name the compound

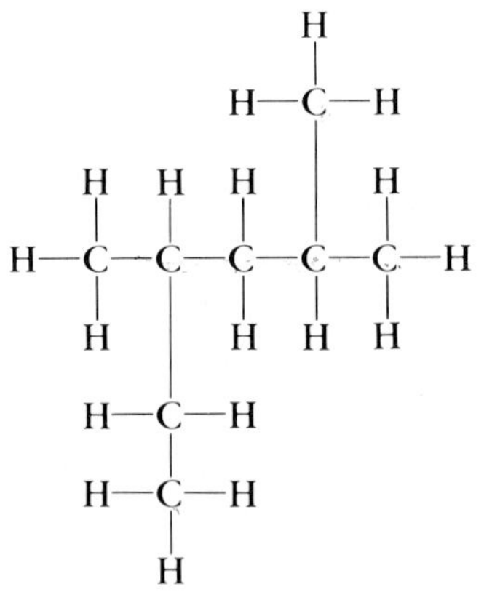

```
                         H
                         |
                      H—C—H
         H  H  H         |  H
         |  |  |         |  |
      H—C—C—C—C—C—H
         |  |  |  |  |
         H  |  H  H  H
            |
         H—C—H
            |
         H—C—H
            |
            H
```

The correct name is 2,4-dimethylhexane, not 2-ethyl-4-methylpentane. This is a fooler. The parent compound is the longest continuous chain. Although this is usually drawn horizontally across the page, it does not have to be. In this compound, the longest chain contains six, not five, carbon atoms.

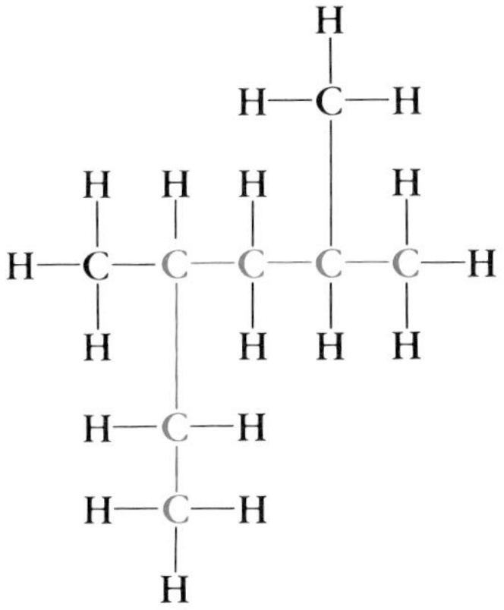

9.5 IUPAC Nomenclature

```
                    H
                    |
                H—C—H
    H   H       |       H
    |   |       |       |
H—C—C—C—C—C—H
    |   |   |   |   |
    H   |   H   H   H
        H—C—H
        |
        H—C—H
        |
        H
```

EXAMPLE 9.3

Draw the structural formula for 4-isopropyl-2-methylheptane.

In drawing compounds, always start with the parent chain, heptane in this case.

—C—C—C—C—C—C—C—

Then add the groups at their proper positions. You can number the parent chain from either direction as long as you are consistent (do not change directions in the middle of a problem).

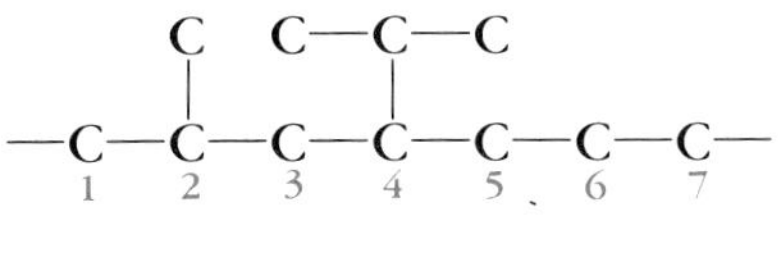

```
        C   C—C—C
        |   |
—C—C—C—C—C—C—C—
 1   2   3   4   5   6   7
```

Finally, fill in all the hydrogen atoms. (Each carbon atom must have four bonds.)

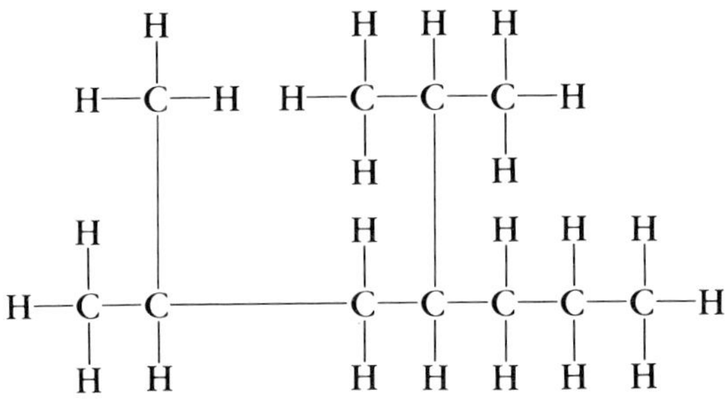

```
            H           H   H   H
            |           |   |   |
        H—C—H   H—C—C—C—H
            |           |   |
            |           H   H
            |
        H   |       H       H   H   H
        |   |       |       |   |   |
    H—C—C———————C—C—C—C—H
        |   |       |   |   |   |   |
        H   H       H   H   H   H   H
```

You can condense this formula by writing the hydrogens right next to the carbons to which they are attached.

$$\begin{array}{c} \qquad\qquad CH_3\quad\ CH_3 \\ \quad CH_3\qquad\ \ CH \\ CH_3CH-CH_2-CH-CH_2CH_2CH_3 \end{array}$$

9.6 Properties of Alkanes

Petroleum is a major source of alkanes (and of hydrocarbons in general). As it comes from the ground, petroleum is of limited use. It must be processed in various ways to provide the familiar fractions listed in Table 9.5. As can be seen from the table, alkanes having from 1 to 4 carbon atoms per molecule are gases at room temperature. The normal alkanes with 5 to about 16 carbon atoms per molecule are liquids, and those having more than 16 carbon atoms per molecule are solids (Table 9.6). Alkanes are nonpolar and are essentially insoluble in the polar solvent water. The densities of liquid alkanes are less than that of water. Therefore, when added to water, liquid alkanes float on top (think of oil slicks from tanker spills). Alkanes do dissolve many organic substances of low polarity, such as fats, oils, and waxes. Mixtures of alkanes are therefore frequently used as organic solvents.

Chemically, the alkanes are the least reactive of all organic compounds. In fact, they undergo so few reactions that they are sometimes called **paraffins**, from the Latin words meaning "little affinity." Paraffin

Alkanes are sometimes called paraffins.

TABLE 9.5 Typical Petroleum Fractions

Fraction	Typical Range of Hydrocarbons	Approximate Range of Boiling Point (°C)	Typical Uses
Natural gas	CH_4 to C_4H_{10}	Less than 40	Fuel, starting materials for plastics
Gasoline	C_5H_{12} to $C_{12}H_{26}$	40–200	Fuel, solvents
Kerosene	$C_{12}H_{26}$ to $C_{16}H_{34}$	175–275	Diesel fuel, jet fuel, home heating, cracking to gasoline
Heating oil	$C_{15}H_{32}$ to $C_{18}H_{38}$	250–400	Industrial heating, cracking to gasoline
Lubricating oil	$C_{17}H_{36}$ and up	Above 300	Lubricants
Residue	$C_{20}H_{42}$ and up	Above 350 (some decomposition)	Paraffin, asphalt

9.6 Properties of Alkanes

TABLE 9.6 Physical Properties of Selected Alkanes

Name	Molecular Formula	Melting Point (°C)	Boiling Point (°C)	Density at 20 °C (g/mL)
Methane	CH_4	−183	−162	(Gas)
Ethane	C_2H_6	−172	−89	(Gas)
Propane	C_3H_8	−187	−42	(Gas)
Butane	C_4H_{10}	−138	0	(Gas)
Pentane	C_5H_{12}	−130	36	0.626
Hexane	C_6H_{14}	−95	69	0.659
Heptane	C_7H_{16}	−91	98	0.684
Octane	C_8H_{18}	−57	126	0.703
Nonane	C_9H_{20}	−54	151	0.718
Decane	$C_{10}H_{22}$	−30	174	0.730
Undecane	$C_{11}H_{24}$	−26	196	0.740
Dodecane	$C_{12}H_{26}$	−10	216	0.749
Tridecane	$C_{13}H_{28}$	−6	235	0.757
Tetradecane	$C_{14}H_{30}$	6	254	0.763
Pentadecane	$C_{15}H_{32}$	10	271	0.769
Hexadecane	$C_{16}H_{34}$	18	280	0.775
Heptadecane	$C_{17}H_{36}$	22	302	(Solid)
Octadecane	$C_{18}H_{38}$	28	316	(Solid)
Nonadecane	$C_{19}H_{40}$	32	330	(Solid)
Eicosane	$C_{20}H_{42}$	37	343	(Solid)

wax is a mixture of solid alkanes frequently used as a seal for homemade preserves. The inertness of the wax makes it ideal for such use.

Alkanes do undergo a reaction of major commercial importance. They can be oxidized in a highly exothermic reaction referred to as **combustion**. We have seen the combustion reaction of methane frequently in this book.

Combustion (burning) of alkanes is a highly exothermic reaction.

$$CH_4 + 2\,O_2 \longrightarrow CO_2 + 2\,H_2O + \text{heat}$$

The most important product of this reaction is the heat of combustion, used for cooking food, heating homes, and drying clothes. Combustion of other alkanes provides the energy to power automobiles, trains, and planes; to produce electricity; and to run much of our industry.

Alkanes also undergo reactions with fluorine, chlorine, and bromine to produce halogenated hydrocarbons. Again let us illustrate the reaction with methane.

$$CH_4 + Cl_2 \xrightarrow{\text{heat}} CH_3Cl + HCl$$

The organic product we have shown is called chloromethane or methyl chloride. It is not the only possible organic product. The methane starting material may have more than one of its hydrogens replaced by chlorine. Other possible products are

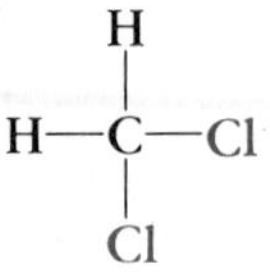

$$H-\underset{\underset{Cl}{|}}{\overset{\overset{H}{|}}{C}}-Cl \qquad Cl-\underset{\underset{Cl}{|}}{\overset{\overset{H}{|}}{C}}-Cl \qquad Cl-\underset{\underset{Cl}{|}}{\overset{\overset{Cl}{|}}{C}}-Cl$$

Dichloromethane (methylene chloride) 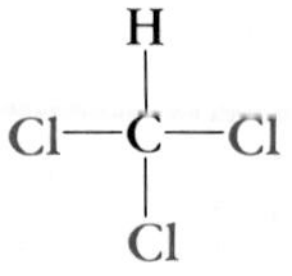   Trichloromethane (chloroform) 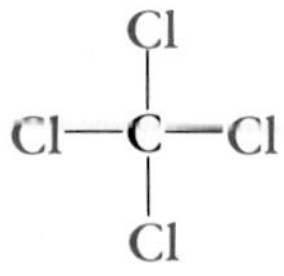   Tetrachloromethane (carbon tetrachloride)

Larger hydrocarbons can produce even more complex mixtures of products.

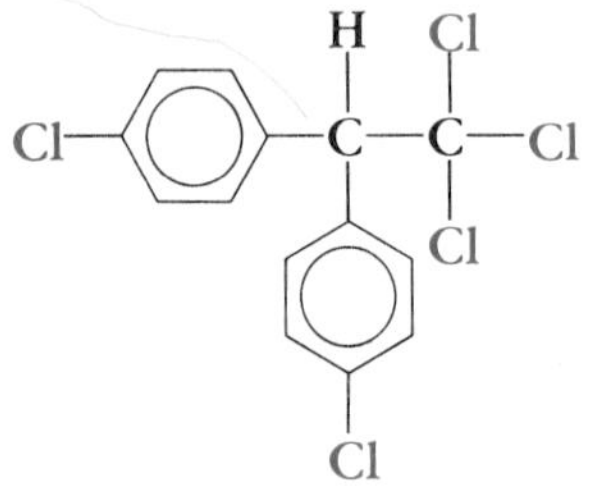

Halogenated hydrocarbons (some produced from hydrocarbons other than alkanes) have enjoyed widespread commercial use. They serve as a solvents, cleaning fluids, insecticides, refrigerator coolants, and in a host of other uses. They are also, unfortunately, a major source of pollution. Thus, while the use of the halogenated hydrocarbon insecticide DDT (*d*ichloro*d*iphenyl*t*richloroethane) (Figure 9.7) greatly reduced the incidence of malaria in various areas of the world, it has also been responsible for the near extinction of several species of birds, massive fish kills, and the eradication of many types of beneficial insects. Chloroform and carbon tetrachloride, both widely used at one time, are carcinogens (cancer-producing compounds). It can certainly be said that we have become much more cautious in our use of halogenated hydrocarbons as their negative effects on our environment have become evident.

FIGURE 9.7
The structure of DDT. The hexagonal rings in this structure are discussed in Section 9.12.

9.7 ALKENES: STRUCTURE AND NOMENCLATURE

Alkenes have a carbon–carbon double bond.

The next family of hydrocarbons that we shall consider is the **alkenes** (note the *-ene* ending). This family is characterized by the presence of a carbon–carbon double bond in its molecules. Names, structures, and physical properties of a few representative alkenes are given in Table 9.7.

We have used only condensed structural formulas in this table. Such formulas are much easier to set in type, but they are sometimes confusing to newcomers to organic chemistry. Thus, $CH_2=CH_2$ stands for

$$\begin{array}{c} H \quad H \\ \overset{..}{C}::\overset{..}{C} \\ H \quad H \end{array} \qquad \text{or} \qquad \overset{H}{\underset{H}{\diagdown}}C=C\overset{H}{\underset{H}{\diagup}}$$

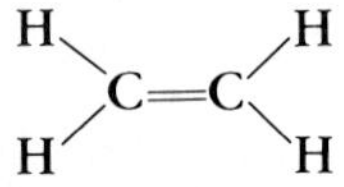

9.7 Alkenes: Structure and Nomenclature

TABLE 9.7 Physical Properties of Some Selected Alkenes

IUPAC Name	Molecular Formula	Condensed Structure	Melting Point (°C)	Boiling Point (°C)
Ethene	C_2H_4	$CH_2\!=\!CH_2$	−169	−104
Propene	C_3H_6	$CH_3CH\!=\!CH_2$	−185	−47
1-Butene	C_4H_8	$CH_3CH_2CH\!=\!CH_2$	−185	−6
1-Pentene	C_5H_{10}	$CH_3CH_2CH_2CH\!=\!CH_2$	−138	30
1-Hexene	C_6H_{12}	$CH_3(CH_2)_3CH\!=\!CH_2$	−140	63
1-Heptene	C_7H_{14}	$CH_3(CH_2)_4CH\!=\!CH_2$	−119	94
1-Octene	C_8H_{16}	$CH_3(CH_2)_5CH\!=\!CH_2$	−102	121

The double bond is shared by the two carbon atoms and does not involve the hydrogen atoms, although the condensed formula does not make this point obvious. Figure 9.8 shows a model of this compound, which is commonly called *ethylene*.

Before we consider the rest of the alkenes in Table 9.7, let us pause to discuss the nomenclature of alkenes. Just as some of the alkanes have common names (2-methylpropane is commonly called isobutane), so do some alkenes. As we indicated above, the two-carbon alkene is called ethylene. The only three-carbon alkene is frequently called propylene. Four alkene isomers have the molecular formula C_4H_8. Whenever large numbers of isomers occur, common names are more difficult to deal with. For the most part, the IUPAC system is used in naming the higher alkenes. Some of the IUPAC rules for alkenes follow.

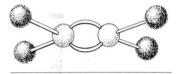

FIGURE 9.8
Ball-and-stick model of ethylene.

1. All alkenes have names ending in *-ene*.
2. The longest chain of atoms *containing the double bond* is the parent compound. The name has the same stem as the corresponding alkane, but the ending is changed from *-ane* to *-ene*. Thus, propylene ($CH_3CH\!=\!CH_2$) is named *propene* in the IUPAC system.
3. When necessary, the position of the double bond along the parent chain is specified by giving the number of the *first* carbon atom involved in the double bond. For example, the compound $CH_3CH\!=\!CHCH_2CH_3$ has the double bond between the second and third carbon atoms. Its name is 2-pentene.
4. Substituent groups are named and numbered as usual. Thus

$$CH_3-\underset{\underset{\displaystyle CH_3}{|}}{CH}-CH_2-CH\!=\!CH-CH_3$$

is 5-methyl-2-hexene. Note that the numbering of the parent

chain is always done in such a way as to give the double bond the lowest number, even if that forces a substituent group to have a higher number. We say that the double bond has *priority* in numbering.

EXAMPLE 9.4

Name the compound

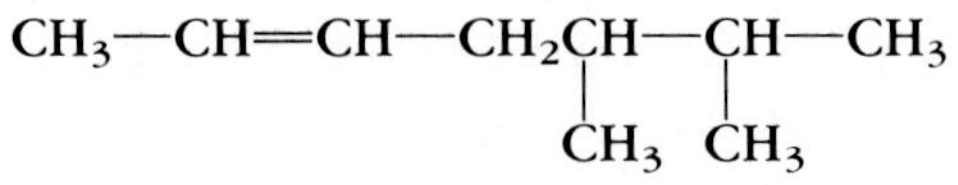

$$CH_3-CH=CH-CH_2CH-CH-CH_3$$
$$||$$
$$CH_3CH_3$$

The longest continuous chain has seven carbon atoms. To give the first carbon of the double bond the lowest number, we start numbering from the left.

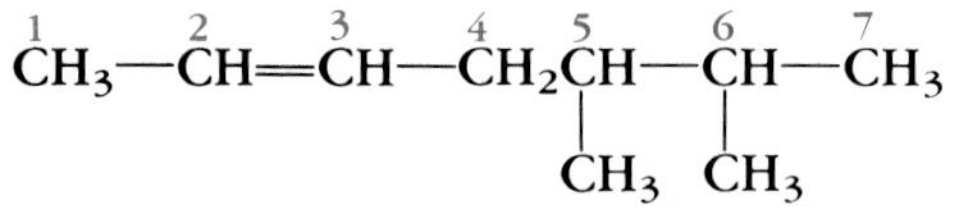

$$\overset{1}{C}H_3-\overset{2}{C}H=\overset{3}{C}H-\overset{4}{C}H_2\overset{5}{C}H-\overset{6}{C}H-\overset{7}{C}H_3$$
$$||$$
$$CH_3CH_3$$

The name of the compound is 5,6-dimethyl-2-heptene.

EXAMPLE 9.5

Draw the structural formula for 3,4-dimethyl-2-pentene.
To draw this compound, first write down the parent chain of five carbon atoms.

$$C-C-C-C-C$$

Then add the double bond between the second and third carbon atoms (this is a 2-pentene).

$$\overset{1}{C}-\overset{2}{C}=\overset{3}{C}-\overset{4}{C}-\overset{5}{C}$$

You could have decided to count from right to left instead of from left to right. It does not matter as long as you are consistent (count the same way when you attach substituents).
Now add the groups at their proper positions and fill in all the hydrogen atoms.

9.8 Geometric Isomerism in Alkenes

$$
\begin{array}{c}
\overset{\displaystyle H}{\underset{\displaystyle H}{|}}\overset{\displaystyle H}{\underset{\displaystyle H}{|}} \\
H\;\;H-\overset{|}{C}-H\;\;H-\overset{|}{C}-H\;\;H \\
H-\overset{|}{\underset{|}{C}}-\overset{|}{\underset{|}{C}}=\overset{}{C}\text{\textemdash}\overset{|}{\underset{|}{C}}\text{\textemdash}\overset{|}{\underset{|}{C}}-H \\
H\;\;H\;\;H\;\;H
\end{array}
$$

or

$$
\begin{array}{c}
CH_3\;\;CH_3 \\
\;|\;\;\;\;\;| \\
CH_3\text{\textemdash}CH\text{=}C\text{\textemdash}CH\text{\textemdash}CH_3
\end{array}
$$

9.8 GEOMETRIC ISOMERISM IN ALKENES

Four alkene isomers have the formula C_4H_8. The first, 1-butene, is easy.

$$CH_2\text{=}CHCH_2CH_3$$

We can also draw an isomer analogous to the alkane isobutane.

$$
\begin{array}{c}
CH_3 \\
\;| \\
CH_2\text{=}C\text{\textemdash}CH_3
\end{array}
$$

The common name of this compound is isobutylene. Its IUPAC name is 2-methyl-1-propene.

A third isomer structure comes to mind fairly easily.

$$CH_3CH\text{=}CHCH_3$$

The carbon skeleton is the same as that of 1-butene, but the double bond is shared by the second and third carbon atoms instead of the first and second. It turns out that there are two different compounds with a four-carbon chain and a double bond between the second and third carbon atoms. There are two different 2-butene isomers. This is a type of isomerism we have not encountered before, and it depends on restricted rotation within the molecule because of the double bond.

Remember that atoms and groups of atoms are free to spin about single bonds (Section 9.4). Two atoms connected by a double bond are not free to spin relative to each other, however. Although the models we present in Figure 9.9 do not "look" like the actual molecules, they do offer a mechanical equivalent to the structural feature that restricts rotation in the molecule. The two connectors between the doubly

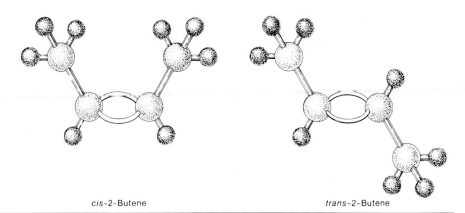

FIGURE 9.9
Models of the 2-butenes. The carbons joined by single bonds are free to spin about those bonds, but the doubly bonded carbons are restricted in this regard.

cis-2-Butene   trans-2-Butene

bonded carbon atoms prevent these atoms from spinning freely as singly bonded carbon atoms can. Thus, the two arrangements of 2-butene shown in Figure 9.9 really represent two different compounds. One cannot be converted to the other unless the double bond is broken first. To distinguish the two different isomers, one is called *cis*-2-butene and the other is *trans*-2-butene.

$$\begin{array}{cc} CH_3 \quad\quad CH_3 & CH_3 \quad\quad H \\ C=C & C=C \\ H \quad\quad\quad H & H \quad\quad\quad CH_3 \end{array}$$

cis-2-Butene   *trans*-2-Butene

Cis: same side
Trans: opposite side

The **cis** isomer is the one with both methyl groups on the same side of the molecule. In the **trans** isomer, the methyl groups are on opposite sides. (Draw a straight line passing through the two carbon atoms in the double bond. If the methyl groups fall on the same side of this line, the compound is *cis*. If they fall on opposite sides, the compound is *trans*.)

If either one of the carbon atoms in a double bond has two identical groups bonded to it, then **cis–trans isomerism** (called **geometric isomerism**) is not possible. In such a case, it does not matter which of the two identical groups is up or down, since they both look the same. Propene cannot exhibit cis–trans isomerism because it has two hydrogen atoms bonded to one of the carbon atoms in the double bond. We can draw two *seemingly* different propenes (structures IV and V).

$$\begin{array}{cc} H \quad\quad CH_3 & H \quad\quad H \\ C=C & C=C \\ H \quad\quad\quad H & H \quad\quad\quad CH_3 \end{array}$$

IV   V

The second structure is not really different from the first, however. You

9.8 Geometric Isomerism in Alkenes

need only pick it up from the page and flip it over to see that the two formulas are identical.

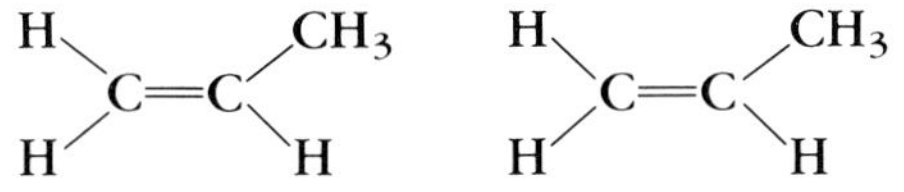

$$\underset{\text{IV}}{\overset{H}{\underset{H}{>}}C=C\overset{CH_3}{\underset{H}{<}}} \qquad \underset{\substack{V \\ \text{(flipped over)}}}{\overset{H}{\underset{H}{>}}C=C\overset{CH_3}{\underset{H}{<}}}$$

EXAMPLES 9.6

Which of the following compounds can exist as geometric isomers? Draw the cis and trans isomers where possible.

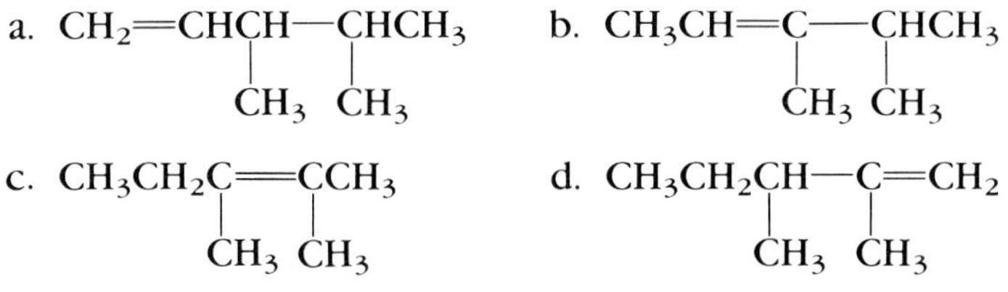

a. $CH_2\!=\!CHCH\!-\!CHCH_3$
   $\qquad\qquad\;\;\; | \quad\;\; |$
   $\qquad\qquad\; CH_3 \; CH_3$

b. $CH_3CH\!=\!C\!-\!CHCH_3$
   $\qquad\qquad\;\;\; | \quad\; |$
   $\qquad\qquad\; CH_3 \; CH_3$

c. $CH_3CH_2C\!=\!CCH_3$
   $\qquad\qquad | \;\; |$
   $\qquad\;\; CH_3 \; CH_3$

d. $CH_3CH_2CH\!-\!C\!=\!CH_2$
   $\qquad\qquad\;\;\; | \quad\; |$
   $\qquad\qquad\; CH_3 \; CH_3$

a. The carbon atom at the left side of the double bond has two hydrogen atoms bonded to it. Geometric isomerism is not possible for this compound.

b. This compound does exist in cis and trans forms.

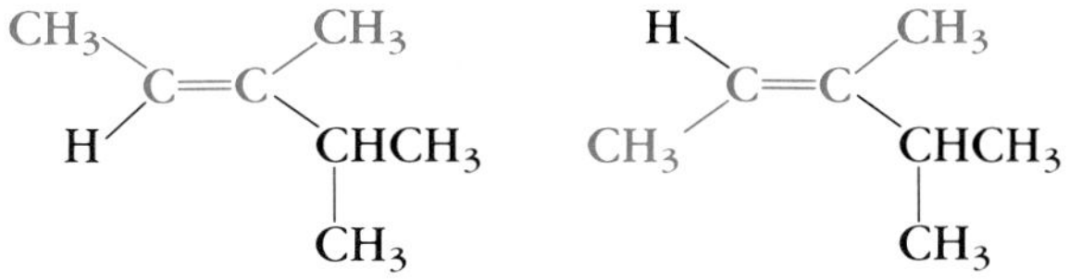

$$\underset{H}{\overset{CH_3}{>}}C=C\underset{\substack{CHCH_3 \\ | \\ CH_3}}{\overset{CH_3}{<}} \qquad \underset{CH_3}{\overset{H}{>}}C=C\underset{\substack{CHCH_3 \\ | \\ CH_3}}{\overset{CH_3}{<}}$$

c. The carbon atom at the right side of the double bond has two methyl groups bonded to it. Geometric isomerism is not possible for this compound.

d. This compound has two hydrogen atoms bonded to the carbon atom at the right of the double bond. No geometric isomerism is possible for this compound.

EXAMPLE 9.7

Draw all alkenes with the formula C_5H_{10} and indicate which exist as cis and trans isomers. Give the IUPAC names for the isomers.

First we will draw the various possible carbon skeletons incorporating a double bond.

$$CH_2{=}CHCH_2CH_2CH_3 \qquad CH_3CH{=}CHCH_2CH_3$$

1-Pentene   2-Pentene
VI   VII

$$\begin{array}{cc} \qquad CH_3 & \qquad CH_3 \\ CH_2{=}\overset{|}{C}{-}CH_2CH_3 & CH_3{-}\overset{|}{C}{=}CH{-}CH_3 \end{array}$$

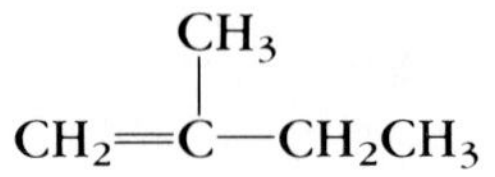

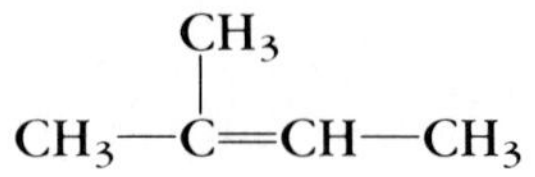

2-Methyl-1-butene   2-Methyl-2-butene
VIII   IX

$$\overset{CH_3}{\underset{|}{}}$$
$$CH_3CHCH{=}CH_2$$

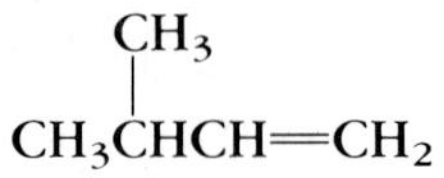

3-Methyl-1-butene
X

Of these, only VII exists as cis and trans isomers.

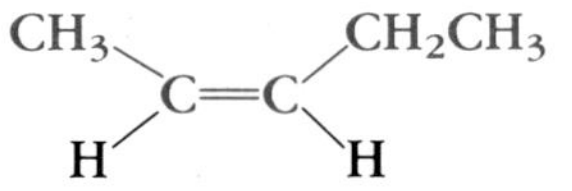

$$\begin{array}{cc} CH_3 \diagdown \quad \diagup CH_2CH_3 & CH_3 \diagdown \quad \diagup H \\ \quad C{=}C & \quad C{=}C \\ H \diagup \quad \diagdown H & H \diagup \quad \diagdown CH_2CH_3 \end{array}$$

cis-2-Pentene 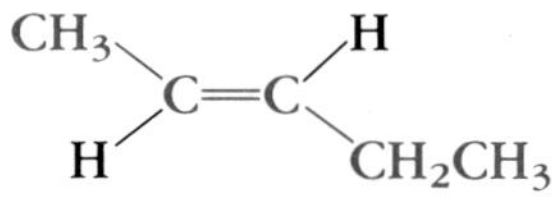   trans-2-Pentene

Structures VI, VIII, and X each have two hydrogen atoms on one of their doubly bonded carbon atoms, and structure IX has two methyl groups on one of its doubly bonded carbon atoms. Note that

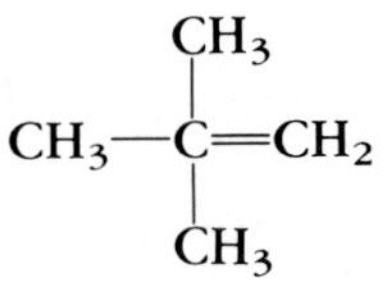

$$CH_3{-}\underset{\underset{CH_3}{|}}{\overset{\overset{CH_3}{|}}{C}}{=}CH_2 \qquad \text{(incorrect structure)}$$

is not a possible isomer. The central carbon atom has five bonds in this structure, and carbon atoms can form only four covalent bonds.

Geometric isomerism may seem to have little practical importance. Biochemical processes, however, such as those related to vision, are very sensitive to just this sort of molecular geometry. We can *see* because a cis molecule in the retina of our eyes can absorb light and change to a trans molecule. That change triggers an electric signal that our brains interpret as sight.

9.9 PROPERTIES OF ALKENES

The physical properties of alkenes are similar to those of alkanes of corresponding formula weight. Also, like alkanes (and all other hydrocarbons), the alkenes burn; that is, they undergo combustion. There is a major difference between the two families. The alkenes possess a double bond, and this structural feature makes the alkenes much more chemically reactive than the alkanes.

Alkenes have physical properties similar to those of alkanes.

The typical reactions of alkenes are **addition reactions**. A double bond involves *two* shared pairs of electrons. In addition reactions, one of these pairs is uncoupled and used to attach additional groups to the molecule. The organic product no longer has a double bond. Here is the reaction of chlorine with ethene.

Alkenes typically undergo addition reactions.

$$H_2C=CH_2 + Cl-Cl \longrightarrow H-\underset{\underset{Cl}{|}}{\overset{\overset{H}{|}}{C}}-\underset{\underset{Cl}{|}}{\overset{\overset{H}{|}}{C}}-H$$

Let us contrast this with the reaction of chlorine with eth*ane*.

$$H-\underset{\underset{H}{|}}{\overset{\overset{H}{|}}{C}}-\underset{\underset{H}{|}}{\overset{\overset{H}{|}}{C}}-H + Cl-Cl \xrightarrow{heat} H-\underset{\underset{H}{|}}{\overset{\overset{H}{|}}{C}}-\underset{\underset{H}{|}}{\overset{\overset{H}{|}}{C}}-Cl + H-Cl$$

Note that in the alkane reaction a hydrogen atom had to be removed from a carbon atom to make room for the incoming chlorine atom. The chlorine atom was *substituted* for the hydrogen atom on ethane. The alkane reaction is a **substitution** reaction, and a second product, HCl, is also formed. In the alkene reaction there is only one product. The two reactants, ethene and chlorine, simply add together to form a single new compound, 1, 2-dichloroethane. Note also that the product of this *addition* reaction is no longer an alkene but a halogenated alk*ane*.

Alkenes undergo many different addition reactions. We described above the reaction with chlorine. **Bromination** is also possible.

$$H_2C=CH_2 + Br-Br \longrightarrow H-\underset{\underset{Br}{|}}{\overset{\overset{H}{|}}{C}}-\underset{\underset{Br}{|}}{\overset{\overset{H}{|}}{C}}-H$$

Ethene   Bromine   1,2-Dibromoethane
          (brownish red)   (colorless)

Bromine, Br_2, forms brownish red solutions. When an alkene is added to such a solution, the color disappears because the alkene reacts with the bromine. The decolorization of bromine solutions is frequently used as a simple test for the presence of alkenes.

Halogenation—the addition of a halogen

Chlorination (the addition of chlorine) and **bromination** (the addition of bromine) are referred to more generally as **halogenation** reactions.

Hydrogenation—the addition of hydrogen

Another important addition reaction is **hydrogenation**, the addition of hydrogen, H_2. A catalyst is required for the reaction, usually nickel (Ni), platinum (Pt), or palladium (Pd).

$$\underset{\text{Ethene}}{\overset{H}{\underset{H}{\diagup}}C=C\overset{H}{\underset{H}{\diagdown}}} + \underset{\text{Hydrogen}}{H-H} \xrightarrow{Ni} \underset{\text{Ethane}}{H-\underset{\underset{H}{|}}{\overset{\overset{H}{|}}{C}}-\underset{\underset{H}{|}}{\overset{\overset{H}{|}}{C}}-H}$$

The product of this reaction is an alkane with the same carbon skeleton as the original alkene. An unsaturated molecule is converted to a saturated molecule. In Chapter 14 we will consider an industrial application of this reaction when we take up the topic of saturated and unsaturated fats.

Another important addition reaction of alkenes is that with water. This reaction, called **hydration**, also requires a catalyst, in this case a strong acid like sulfuric acid.

$$\underset{\text{Ethene}}{\overset{H}{\underset{H}{\diagup}}C=C\overset{H}{\underset{H}{\diagdown}}} + \underset{\text{Water}}{H-OH} \xrightarrow{H_2SO_4} \underset{\text{Ethyl alcohol}}{H-\underset{\underset{H}{|}}{\overset{\overset{H}{|}}{C}}-\underset{\underset{OH}{|}}{\overset{\overset{H}{|}}{C}}-H}$$

Vast quantities of ethyl alcohol, for use as an industrial solvent, are made this way. Although the alcohol is identical to that used in alcoholic beverages, federal law requires that all drinking alcohol be produced by the natural process called fermentation (which we will consider in the next chapter).

Polymerization—a process that links hundreds or thousands of alkene molecules

Perhaps the most important commercial reaction of alkenes is **polymerization**. In this reaction, alkene adds to alkene. The process does not stop with a one-to-one addition but involves hundreds or thousands of alkene molecules in each reaction.

$$\cdots + \overset{H}{\underset{H}{\diagup}}C=C\overset{H}{\underset{H}{\diagdown}} + \overset{H}{\underset{H}{\diagup}}C=C\overset{H}{\underset{H}{\diagdown}} + \overset{H}{\underset{H}{\diagup}}C=C\overset{H}{\underset{H}{\diagdown}} + \cdots \longrightarrow \cdots -\overset{H}{\underset{H}{|}}{C}-\overset{H}{\underset{H}{|}}{C}-\overset{H}{\underset{H}{|}}{C}-\overset{H}{\underset{H}{|}}{C}-\overset{H}{\underset{H}{|}}{C}-\overset{H}{\underset{H}{|}}{C}- \cdots$$

TABLE 9.8 A Selection of Addition Polymers

Monomer	Polymer Segment	Polymer Name	Typical Uses
$H_2C=CH_2$	$-[CH_2-CH_2]_n-$	Polyethylene	Plastic bags, bottles, toys, electrical insulation
$H_2C=CH-CH_3$	$-[CH_2-CH(CH_3)]_n-$	Polypropylene	Indoor–outdoor carpeting, bottles
$H_2C=CH-C_6H_5$	$-[CH_2-CH(C_6H_5)]_n-$	Polystyrene	Simulated wood furniture, styrofoam insulation and packing materials
$H_2C=CH-Cl$	$-[CH_2-CHCl]_n-$	Poly(vinyl chloride), PVC	Plastic wrap, simulated leather (Naugahyde), phonograph records, garden hoses
$H_2C=CCl_2$	$-[CH_2-CCl_2]_n-$	Poly(vinylidene chloride)	Food wrap (Saran)
$F_2C=CF_2$	$-[CF_2-CF_2]_n-$	Polytetrafluoroethylene, Teflon	Nonstick coating for cooking utensils, electrical insulation
$H_2C=CH-CN$	$-[CH_2-CH(CN)]_n-$	Polyacrylonitrile, Orlon, Acrilan, Creslan, Dynel	Yarns, wigs
$H_2C=CH-O-C(=O)-CH_3$	$-[CH_2-CH(O-C(=O)-CH_3)]_n-$	Poly(vinyl acetate), PVA	Adhesives, textile coatings, chewing gum resin, paints
$H_2C=C(CH_3)-C(=O)-O-CH_3$	$-[CH_2-C(CH_3)(C(=O)-O-CH_3)]_n-$	Poly(methyl methacrylate), Lucite, Plexiglas	Glass substitute, bowling balls

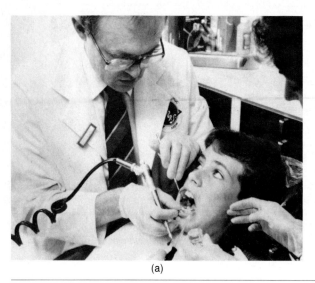 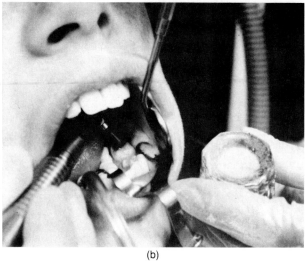

FIGURE 9.10
(a) A dental hygienist works chairside applying sealant material, which is then polymerized using visible light. (b) The sealant is brushed into the pits and fissures on the surface of molars. [Courtesy of U.S. Department of Health and Human Services.]

Many small alkene units (referred to as **monomers**) combine to produce a very large product molecule (the **polymer**). The reaction we have shown above illustrates the conversion of ethylene (ethene) to polyethylene. Remember, even though it is called poly*ethylene*, the product is *not* an alkene. This is an addition reaction, and in addition reactions the double bond of the alkene starting material is destroyed.

There are many different possible alkene monomers. Table 9.8 lists just a few.

One increasingly popular use of polymers is in dental sealants. These formulations commonly incorporate a compound (called bis-GMA) that is a distant, complex cousin of the Plexiglas or Lucite monomer (Table 9.8). The tooth surface is first etched with phosphoric acid, and then the sealant preparation is applied. The polymerization reaction occurs right on the tooth surface and is initiated by exposure of the sealant to heat or visible or ultraviolet light. Sealants effectively prevent the development of dental caries (cavities) in the pits and fissures that are characteristic of the chewing surfaces of molars. In 1984, a conference sponsored by the National Institute of Dental Research recommended the use of pit and fissure sealants as part of routine dental care (Figure 9.10).

We will encounter another type of polymerization when we study large biological molecules, such as carbohydrates and proteins.

9.10 Alkynes

In alkenes, carbon atoms in the double bond share two pairs of electrons. Carbon atoms can also share three pairs of electrons, forming triple

9.10 Alkynes

bonds. Compounds containing such bonds are called **alkynes**. The structural formula for the simplest alkyne is

$$H:C:::C:H \quad \text{or} \quad H-C\equiv C-H$$

The common name of this compound is *acetylene* (Figure 9.11). The IUPAC nomenclature parallels that for alkenes, except that the family ending is *-yne* rather than *-ene*. The official name for acetylene is ethyne. Notice that the common name, acetylene, is deceptive. It sounds very much like ethylene or propylene. Remember that acetylene is an alkyne and that ethylene and propylene are alkenes.

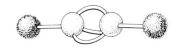

FIGURE 9.11
Ball-and-stick model of acetylene.

EXAMPLE 9.8

Give IUPAC names for (a) CH₃CH₂C≡CH and
(b) CH₃CHC≡CCH₃.
         |
         CH₃

a. The compound has a continuous chain of four carbon atoms, and the triple bond involves the first and second carbon atoms of the chain. The name is 1-butyne.
b. The longest continuous chain has five carbon atoms. This chain is numbered from right to left to give the triple bond a lower number. Therefore, the methyl substituent is attached at position 4. The name is 4-methyl-2-pentyne.

Because there is only one group attached to each carbon atom in a triple bond, there is no such thing as cis–trans isomerism for alkynes.

The alkynes are similar to alkenes in both physical and chemical properties. They undergo addition reactions, but with the possibility of adding twice as much of a reagent as an alkene can add. For example, in hydrogenation, acetylene can add one molecule of hydrogen.

$$H-C\equiv C-H + H-H \xrightarrow{Ni} \begin{array}{c} H \\ \diagdown \\ C=C \\ \diagup \\ H \end{array} \begin{array}{c} H \\ \diagup \\ \\ \diagdown \\ H \end{array}$$

Or acetylene can add two molecules of hydrogen.

$$H-C\equiv C-H + 2\,H-H \xrightarrow{Ni} H-\underset{\underset{H}{|}}{\overset{\overset{H}{|}}{C}}-\underset{\underset{H}{|}}{\overset{\overset{H}{|}}{C}}-H$$

One *or* two pairs of electrons in the triple bond can react.

9.11 Cyclic Hydrocarbons

The hydrocarbons we have encountered so far have been composed of open-ended chains of carbon atoms. Carbon and hydrogen atoms can also hook up in other arrangements in which closed rings are formed. The simplest possible ring-containing, or **cyclic**, hydrocarbon has the molecular formula C_3H_6.

Cyclic hydrocarbons: carbons connected in a ring system

$$
\begin{array}{c}
\text{H} \quad \text{H} \\
\diagdown \diagup \\
\text{C} \\
\diagup \quad \diagdown \\
\text{H}-\text{C}-\text{C}-\text{H} \\
| \quad | \\
\text{H} \quad \text{H}
\end{array}
$$

The compound is called cyclopropane (Figure 9.12).

Names of cycloalkanes (cyclic compounds containing only single bonds) are formed by addition of the prefix *cyclo-* to the name of the open-chain compound with the same number of carbon atoms as are in the ring. Names and structures of a number of cycloalkanes are given in Figure 9.13. The figure also includes a cycloalk*ene* and a cyclic compound carrying a group attached to the main ring.

Note that the carbon atoms in the rings of each compound form a regular geometric figure. For example, the three carbon atoms of cyclopropane form a triangle. Therefore, a triangle is frequently used to represent cyclopropane. Similarly, the five carbon atoms of cyclopentane form a pentagon, and we use a pentagon to represent cyclopentane. It is understood that the carbon atoms occur at the angles of the particular figure and that each carbon atom is attached to a sufficient number of hydrogen atoms to give the carbon atom four bonds.

The physical and chemical properties of the cyclic hydrocarbons generally correspond to those of their open-chain counterparts. This statement is *not* true for one very special group of cyclic hydrocarbons, of which a compound called benzene is the most prominent member.

FIGURE 9.12
Ball-and-stick model of cyclopropane.

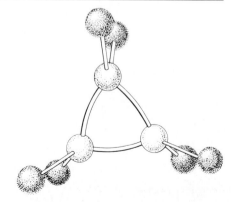

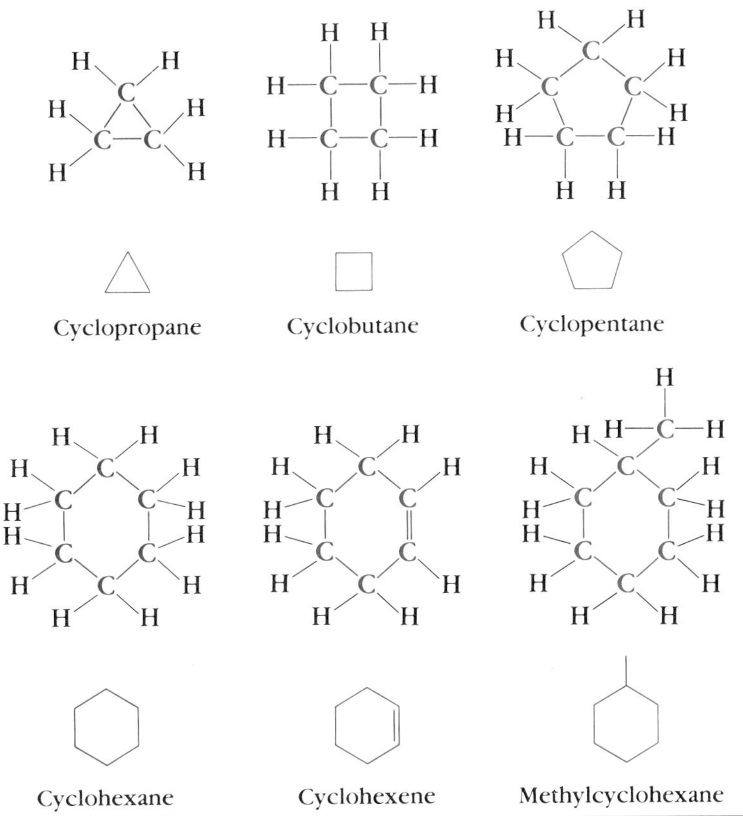

FIGURE 9.13 Some cyclic hydrocarbons.

9.12 BENZENE

The molecular formula of benzene is C_6H_6. We can write many structures that correspond to this formula. Three of these structures are

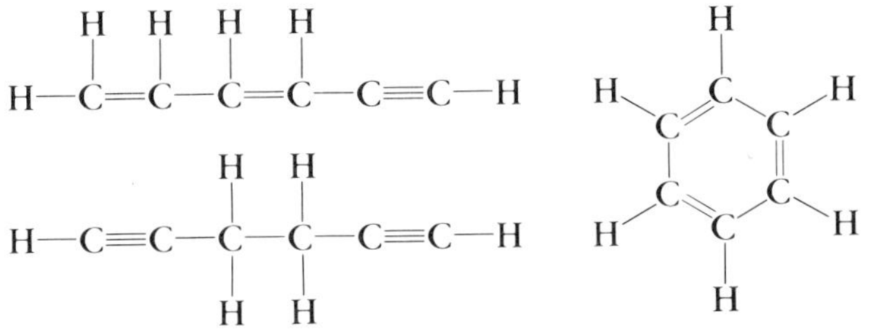

The real substance, benzene, does not have the properties that one would predict from these structures. For example, if benzene really contained double or triple bonds, it would be expected to undergo addition reactions readily. It does not.

Friedrich August Kekulé, a German chemist, proposed the structure on the right. It is accurate in some respects. Benzene does have a six-member ring structure. It is also an unsaturated molecule; that is, it contains something other than single bonds. To account for benzene's inertness toward the typical reactions of unsaturated compounds, Kekulé's structure has been modified. In order for benzene to behave as it does, the "extra" or double-bond electrons cannot be located between specific carbon atoms as they would be in ordinary double bonds. All six of these electrons are regarded as being equally shared by all six carbon atoms in the ring. Therefore, the structure of benzene is now drawn

where the circle represents the six "extra" electrons. No real double bonds exist, just a stable ring of electrons that resists being disrupted. Benzene does not enter into addition reactions because that would destroy the ring of electrons.

The abbreviated formula for benzene is

An abbreviated Kekulé structure is still used for benzene in some texts.

9.13 Aromatic Hydrocarbons

Many compounds are chemically similar to benzene. Several of those discovered in the early days had pleasant odors and were known as aromatic compounds. The label *aromatic* is also used today: an **aromatic compound** is simply a compound that has, like benzene, an unusually stable ring of electrons. All the nonaromatic hydrocarbons that we have considered—the alkanes, alkenes, etc.—are referred to collectively as **aliphatic compounds** to distinguish them from aromatic compounds. "Aliphatic" originally meant that the source of the compound was a fat. Today, however, it simply means "not aromatic."

A number of aromatic compounds can be derived from benzene by substitution of a variety of groups for one or more of the hydrogen atoms. Substitution of a methyl group for one hydrogen gives methyl-

Aromatic compound—an unusually stable compound containing a ring of electrons

Aliphatic = nonaromatic

9.13 Aromatic Hydrocarbons

benzene, more commonly known as toluene (pronounced to rhyme with "doll, you mean").

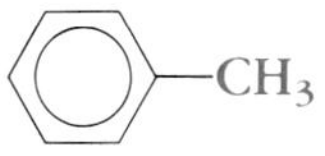

C₆H₅—CH₃

Remember that in this formula those points that are not shown attached to a substituent carry a hydrogen atom. The molecular formula for toluene is C_7H_8.

Substitution of an ethyl group for a hydrogen atom gives ethylbenzene:

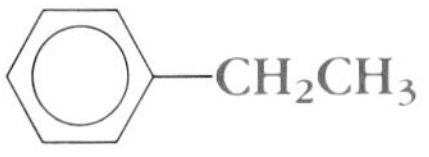

C₆H₅—CH₂CH₃

Occasionally, when the group attached to the benzene ring does not have a simple name, it is the benzene ring that is named as a substituent group. In these cases, the ring is referred to as the **phenyl group**. Phenyl is C_6H_5- or

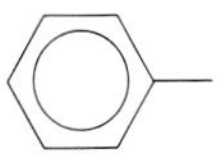

C₆H₅—

Thus, the following compound is named 2-phenylpentane.

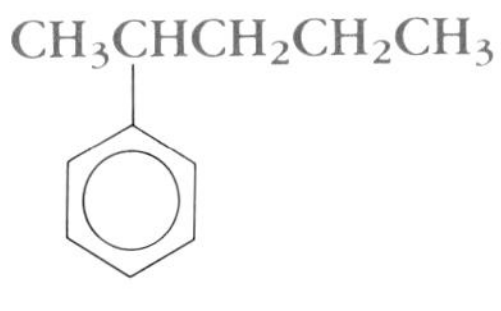

$$CH_3CHCH_2CH_2CH_3$$
        |
       C₆H₅

When two substituents are attached to the benzene ring, we must use some way of indicating their relative positions. One (familiar) method uses numbers. The other method uses the prefixes *ortho-*, *meta-*, and *para-* to indicate relative positions. The prefix *ortho-* is abbreviated *o-* and indicates substituents on adjacent carbon atoms. An ortho-substituted compound is a 1,2-disubstituted benzene. The prefix *meta-* (*m-*) is used for 1,3-disubstituted benzenes, and the prefix *para-* (*p-*) is used with 1,4-disubstituted benzenes. All four of the following structures represent the same compound, *m*-dinitrobenzene (NO_2 is called the *nitro* group).

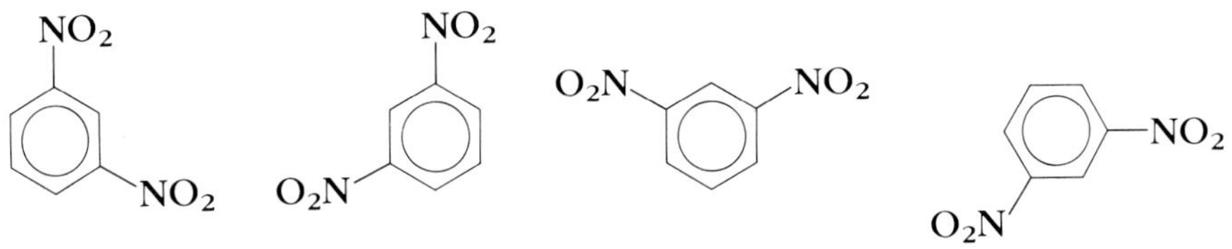

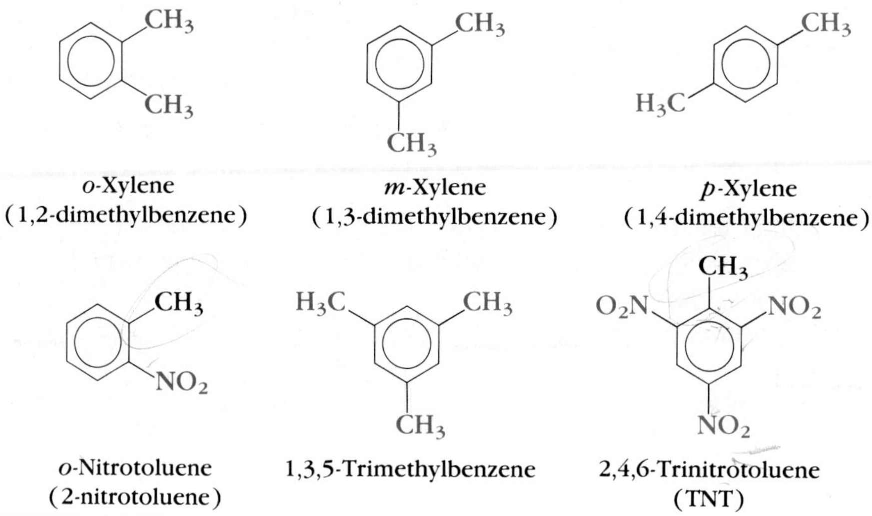

FIGURE 9.14
Some aromatic hydrocarbons and derivatives.

o-Xylene
(1,2-dimethylbenzene)

m-Xylene
(1,3-dimethylbenzene)

p-Xylene
(1,4-dimethylbenzene)

o-Nitrotoluene
(2-nitrotoluene)

1,3,5-Trimethylbenzene

2,4,6-Trinitrotoluene
(TNT)

For three or more substituents, the numbering system must be used. Examples are shown in Figure 9.14.

In another group of aromatic hydrocarbons, two or more benzene rings are "fused" together; that is, the rings have two or more carbon atoms in common. The simplest of these is naphthalene ($C_{10}H_8$).

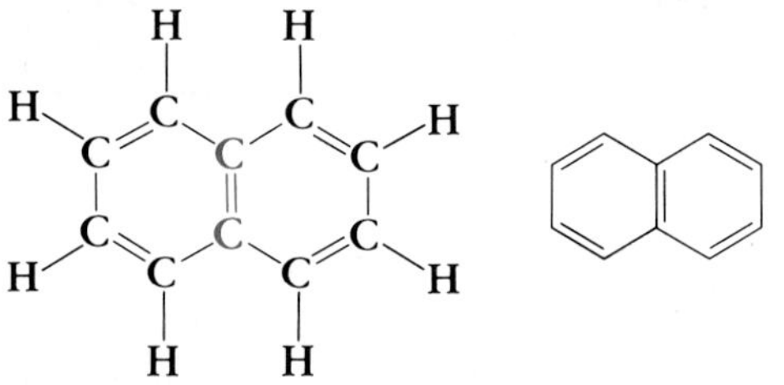

Benzene, toluene, and the xylenes (Figure 9.14) are liquids at room temperature. The compounds involving fused rings, the **polycyclic aromatic hydrocarbons**, are solids. All are insoluble in water, but most are soluble in organic solvents like hexane.

Like all hydrocarbons, aromatic hydrocarbons burn. In other chemical reactions, aromatic hydrocarbons do not behave like the unsaturated hydrocarbons we have previously encountered. Aromatic hydrocarbons resist addition reactions, but they do undergo a number of substitution reactions. Benzene, for example, can be chlorinated in the presence of a catalyst (iron, Fe, in this case).

$$C_6H_6 + Cl_2 \xrightarrow{Fe} C_6H_5Cl + HCl$$

We have written this equation using molecular formulas to emphasize that this is a substitution reaction. One of benzene's hydrogen atoms is replaced by a chlorine atom. Many organic chemists would write the equation as follows:

Because the hydrogen atoms are not shown explicitly in these structures, it is easy to forget that this is a substitution reaction. Remember—the chlorine atom *replaces* a hydrogen atom. Note also that only the organic starting material and product are emphasized. The inorganic product is ignored completely. (It is still formed; it is simply not shown.) Not only the catalyst, but the inorganic reagent, too, is written by the arrow. You will encounter this form of chemical equation again as we proceed through organic chemistry.

9.14 Hydrocarbons and Health

The range of the physiological effects of the hydrocarbons is broad. The liquids are excellent solvents for nonpolar materials. If aspirated into the lungs (as can happen if one attempts to siphon gasoline from a gas tank by mouth), these compounds cause "chemical pneumonia." They dissolve fatlike molecules from the cell membranes in the alveoli. The cells become less flexible, and the alveoli are no longer able to expel fluids. The buildup of fluids is similar to that which occurs in bacterial or viral pneumonia. People who swallow gasoline, petroleum distillates, or similar substances should not be made to vomit. This only increases the chance of their getting the hydrocarbons into their lungs.

The effects of gaseous hydrocarbons vary. Methane appears to be physiologically inert; that is, it does not react at all in the body. It can still kill you. One common method of suicide is to breathe the methane that fuels a gas stove. In this case, methane does not react chemically, but it excludes oxygen from the lungs, thus causing asphyxiation. Other gaseous hydrocarbons act as anesthetics in high concentration. Ethene and, particularly, cyclopropane have been used medically as inhalation anesthetics.

Most aromatic hydrocarbons present toxic hazards. Benzene, whether inhaled or ingested, may cause convulsions or even death by respiratory failure. Chronic exposure to lower levels of benzene can depress the formation of blood cells by bone marrow. Prolonged chronic exposure can cause death. Many polycyclic aromatic hydrocarbons are carcinogenic. Typical of these compounds is 3,4-benzpyrene (benz[*a*]pyrene).

Benzpyrene is known to induce cancer in laboratory animals. It is found in cigarette smoke and automobile exhaust fumes.

Not all hydrocarbons are bad. Heavier liquid alkanes, when applied to the skin, act as emollients (skin softeners). Such alkane mixtures as mineral oil can be used to replace natural skin oils washed away by frequent bathing and swimming. Petroleum jelly (Vaseline is one brand) is a semisolid mixture of hydrocarbons that can be applied as an emollient or simply as a protective film. Water and water solutions (e.g., urine) will not dissolve such a film, which explains why petroleum jelly protects a baby's tender skin from diaper rash.

Nonetheless, if there is any doubt as to the physiological effect of a hydrocarbon, it is always best to assume the worst and protect oneself accordingly.

CHAPTER SUMMARY

Organic chemistry is the study of the compounds of carbon. In contrast to the typical inorganic compound, the typical organic compound is flammable, low-melting, water-insoluble, and covalently bonded.

The millions of organic compounds are organized into families that share common structural and chemical characteristics. Hydrocarbons (compounds incorporating only carbon and hydrogen) are divided into two general categories, the saturated compounds containing only single bonds and the unsaturated compounds with multiple bonding. Alkanes are the only family of saturated hydrocarbons, but unsaturated hydrocarbons include alkenes (incorporating double bonds), alkynes (incorporating triple bonds), and aromatic hydrocarbons (whose structure is characterized by a stable ring of electrons). The term *aliphatic* is used for all compounds that are not classified as aromatic. Petroleum is a major source of both aromatic and aliphatic hydrocarbons.

One reason for the large number of organic compounds is carbon's ability to bond with itself to form long chains, both branched and unbranched. In addition, compounds of carbon exhibit isomerism, a phenomenon in which different compounds possess the same molecular formula. Alkenes in which two different groups are attached to the carbon atoms in the double bond exhibit a special kind of isomerism referred to as geometric isomerism.

$$CH_3 \diagdown C=C \diagup CH_3 \qquad CH_3 \diagdown C=C \diagup H$$
$$H \diagup \diagdown H \qquad H \diagup \diagdown CH_3$$

cis trans

For simple hydrocarbons common names may be used, but the large number of isomers and the complexity of some compounds require the more organized nomenclature developed under the auspices of the International Union of Pure and Applied Chemistry (IUPAC). Abbreviated nomenclature rules can be summarized as follows.

1. The longest continuous carbon chain in a compound is named as the parent by using a stem that indicates the number of carbon atoms (see Table 9.3) and an ending that indicates the family (-ane for alkanes, -ene for alkenes, -yne for alkynes).
2. The position of a double or triple bond in the chain is specified by the number of the first carbon atom in the multiple bond (counting from the nearest chain end).
3. Branches hanging from the parent chain are named as indicated in Table 9.4. The branch (substituent) names are given first, followed by the name of the parent compound. The position of attachment of a branch is given by citing the number of the carbon atoms in the parent chain to which it is connected. If identical groups appear on the parent chain, that fact is

Problems

indicated by a prefix attached to the name of the group (di- for two, tri- for three, tetra- for four).

4. In those compounds in which a ring of carbon atoms is present (unless the compound is aromatic), the number of carbon atoms in the ring determines the name of the parent. The parent name is prefixed with the term cyclo-.

5. The simplest aromatic compound is named benzene. The nomenclature of aromatic compounds either uses benzene as the parent name (as in ethylbenzene, $C_6H_5CH_2CH_3$) or gives a special name to the compound as a whole [for example, toluene ($C_6H_5CH_3$) or the xylenes ($C_6H_4(CH_3)_2$)]. The position of substitution on a disubstituted benzene ring can be specified by numbers or by the prefixes ortho- (1,2-), meta- (1,3-), and para- (1,4-). Numbers are used if more than two positions are substituted. When the benzene ring itself is treated as a substituent, it is referred to as the phenyl group.

Naphthalene

is an example of a fused-ring polycyclic aromatic hydrocarbon.

CHAPTER QUIZ

1. Compared to KCl, CCl₄ would be
   a. high-boiling
   b. water-soluble
   c. covalent
   d. high-melting

2. "Unsaturated" means
   a. organic
   b. covalent
   c. multiple bonding
   d. fatty

3. The stem "hept-" refers to which number?
   a. 6
   b. 7
   c. 8
   d. 9

4. The hydrocarbon with skeletal structure

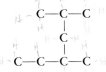

   is called

   a. 2-tert-butylbutane
   b. 2,4-dimethylhexane
   c. 1-isopropyl-2-methylbutane
   d. 2-methyl-4-ethylpentane

5. Which of the following could be isomers?
   a. C_4H_8 and C_4H_8
   b. C_4H_8 and C_5H_8
   c. C_4H_8 and C_4H_{10}
   d. C_4H_8 and C_5H_9

6. Which is *not* a typical reaction of alkenes?
   a. halogenation
   b. substitution
   c. bromination
   d. hydrogenation

7. Name the compound

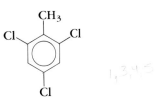

   a. trichloromethylbenzene
   b. 2,4,6-trichlorotoluene
   c. 1-methyl-2-chloro-4-chloro-6-chlorobenzene
   d. benzotrichlorotoluene

8. Hydration of alkenes yields as one of its products
   a. an alkane
   b. a water alkene mixture
   c. an alcohol
   d. an alkyne

9. Which of the following can exhibit cis–trans isomerism?
   a. alkanes
   b. alkenes
   c. alkynes
   d. polymers

10. Which is the symbol for cyclopropane?

    a. ⬡   b. ▢   c. △   d. ⬠

PROBLEMS

9.1 A Comparison of Organic and Inorganic Compounds

1. List three ways in which a "typical" organic compound differs from a "typical" inorganic compound.

9.2 The Hydrocarbons

2. Define, illustrate, or give an example for each of the following terms.
   a. hydrocarbon   b. alkane
   c. paraffin   d. saturated
   e. unsaturated   f. alkene
   g. alkyne

9.3 The Alkanes

3. How many carbon atoms are there in each of the following?
   a. ethane   b. heptane
   c. butane   d. nonane

4. Write the structural formulas for the following.
   a. heptane
   b. isopentane
   c. 2,2,5-trimethylhexane
   d. 4-ethyl-3-methyloctane

9.4 Isomerism

5. Define:
   a. *normal* alkanes
   b. isomers
   c. geometric isomers

6. Indicate whether the structures in each set represent the same compound or isomers.

   a. CH_3CH_3 and $\begin{array}{c}CH_3\\|\\CH_3\end{array}$

   b. $\begin{array}{c}CH_3\\|\\CH_3CH_2\end{array}$ and $CH_3CH_2CH_3$

   c. $\begin{array}{c}CH_3\\|\\CH_3CH_2CHCH_2CH_3\end{array}$ and $\begin{array}{c}CH_3\\|\\CH_3CHCH_2CH_2CH_3\end{array}$

   d. $\begin{array}{c}CH_3\\|\\CH_3CHCH_2CH_3\end{array}$ and $\begin{array}{c}CH_3\\|\\CH_3CH_2CH\\|\\CH_3\end{array}$

   e. $\begin{array}{c}CH_3\ \ CH_3\\|\ \ \ \ \ |\\CH_3CH_2CH-CH_2\end{array}$ and $\begin{array}{c}CH_3\ \ CH_3\\|\ \ \ \ \ |\\CH_2CH_2CHCH_3\end{array}$

9.5 IUPAC Nomenclature

7. Define:
   a. R—   b. substituent
   c. alkyl group

8. Draw the structural formulas of the four-carbon alkanes (C_4H_{10}). Identify butane and isobutane and give the IUPAC names for the compounds.

9. Write structures for the five isomeric hexanes (C_6H_{14}). Name each by the IUPAC system.

10. Draw the following alkyl groups.
    a. ethyl   b. isopropyl
    c. *tert*-butyl   d. isobutyl

11. Name the following compounds by the IUPAC system.
    a. $\begin{array}{c}CH_3CH_2CHCH_2CH_2CH_3\\|\\CH_3\end{array}$

    b. $\begin{array}{c}CH_3CH_2CH_2CHCH_2CH_2CH_3\\|\\CH\\/\ \ \backslash\\CH_3\ \ CH_3\end{array}$

9.6 Properties of Alkanes

12. Write an equation for the combustion of propane.

13. Write an equation for the bromination of ethane.

9.7 Alkenes: Structure and Nomenclature

14. Write structural formulas for the following.
    a. propylene
    b. 3-ethyl-2-pentene
    c. 3-*tert*-butyl-1-hexene
    d. 2,3-dimethyl-2-butene

15. Write structural formulas for the following.
    a. *cis*-2-hexene   b. *trans*-3-hexene

16. Write structural formulas for the following.
    a. chloroform
    b. 1,1,2,2-tetrachloroethene

17. Name the following compounds by the IUPAC system.
    a. $\begin{array}{c}CH_2{=}CCH_2CH_2CH_3\\|\\CH_3\end{array}$

Problems

b. CH₃C=CHCH₂CHCH₃
      | |
      CH₃   CH₃

c.
$$\begin{array}{c} CH_3CH_2 \\ \diagdown \\ C=C \\ \diagup \\ H \end{array} \begin{array}{c} \\ \diagup \\ \diagdown \\ H \end{array} CH_2CH_3$$

9.8 Geometric Isomerization in Alkenes

18. Indicate whether the structures in each set represent the same compound or isomers.

a. CH₃CH=CHCH₃ and
CH₃CH₂CH=CH₂

b.
CH₃ CH₃
 | |
CH₃C=CCH₃   and   CH₃C=CCH₃
                   | |
                  CH₃ CH₃

c. CH₃CH₂CH₂CH=CCH₃ and
                        |
                       CH₃

CH₃CHCH=CHCH₂CH₃
 |
CH₃

d. CH₂=CCH₂CH₃   and   CH₂=CCH₃
        | |
       CH₃   CH₂CH₃

e.
$$\begin{array}{c} CH_3 \\ \diagdown \\ C=C \\ \diagup \\ CH_3CH_2 \end{array} \begin{array}{c} CH_2CH_3 \\ \diagup \\ \diagdown \\ CH_2CH_2CH_3 \end{array}$$ and
$$\begin{array}{c} CH_3 \\ \diagdown \\ C=C \\ \diagup \\ CH_3CH_2 \end{array} \begin{array}{c} CH_2CH_2CH_3 \\ \diagup \\ \diagdown \\ CH_2CH_3 \end{array}$$

19. Including geometric isomers, there are five isomeric alkenes with the following basic carbon skeleton and the molecular formula C₆H₁₂.

       C
       |
C—C—C—C—C

Draw all five isomers and name them by IUPAC rules.

20. Draw the three alkyne isomers with the formula C₅H₈. Give IUPAC names for the compounds.

9.9 Properties of Alkenes

21. Define:
a. monomer   **b.** polymer

22. Give the reagents required for the following transformations.

a. (cyclopentane) → (chlorocyclopentane)

b. CH₃CH=CHCH₃ → CH₃CHCH₂CH₃
                            |
                           OH

c. CH₂=CHCH₃ → CH₂CHCH₃
                    | |
                   Cl Cl

23. What starting materials are required to complete the transformations shown?

a. ? $\xrightarrow{H_2}{Ni}$ (cyclohexane)

b. ? $\xrightarrow{2\,Cl_2}$ H—C—C—H
                             | |
                            Cl Cl (with Cl Cl on top)

c. ? $\xrightarrow[H_2SO_4]{H_2O}$ CH₃CHCH₃
                                        |
                                       OH

24. Which reactions in Problem 23 are addition reactions?

25. Which reactions in Problem 23 are substitution reactions?

26. Write an equation for the hydrogenation of methylpropene (isobutylene).

27. Write an equation for the bromination of ethene.

28. In working this problem, refer to Table 9.8. Draw a section of polymer chain that is a minimum of six carbon atoms long for each of the following.
a. PVC   **b.** Teflon   **c.** Saran

29. What would the polymer formed from 1,2-dichloroethene look like?

9.10 Alkynes

30. Write structural formulas for the following.
a. acetylene   **b.** 1-butyne

31. Name the following compounds by the IUPAC system.
a. CH₃CH₂CH₂C≡CH
b. CH₃C≡CCH₃

32. Write an equation for the bromination of ethyne (with 2 mol of bromine).
33. Give the reagents required for the following transformation.

$$HC\equiv CCH_3 \rightarrow CH_3CH_2CH_3$$

9.11 Cyclic Hydrocarbons

34. Write structural formulas for the following.
    a. cyclohexane   b. cyclopentene
35. Name the following compounds by the IUPAC system.

    a. 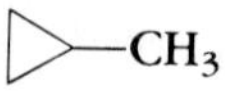 ▷—CH₃   b. ▢

36. Write an equation for the hydration of cyclohexene.

9.12 Benzene

37. Write an equation for the bromination of benzene in the presence of iron.

9.13 Aromatic Hydrocarbons

38. Define:
    a. aromatic compound
    b. aliphatic compound
    c. phenyl group
39. Indicate whether the compound is aromatic or aliphatic.

    a. 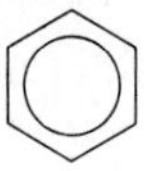   b. 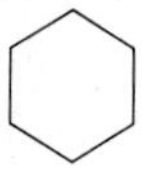   c.    d.

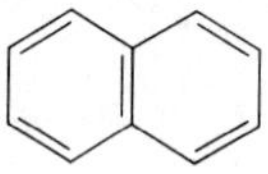

40. Identify the substitution pattern as meta, ortho, or para.

    a. 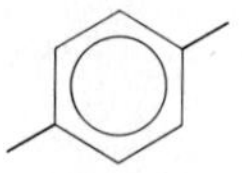   b. 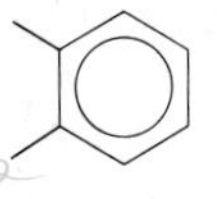   c.

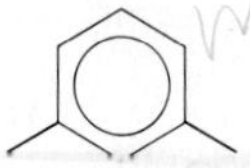

41. Write structural formulas for the following.
    a. toluene
    b. *m*-diethylbenzene
    c. naphthalene
    d. *p*-dichlorobenzene
    e. 2,4-dinitrotoluene
    f. 1,2,4-trimethylbenzene
42. Name the following compounds by the IUPAC system.

    a. 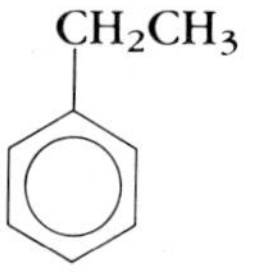 (ethylbenzene)   b. 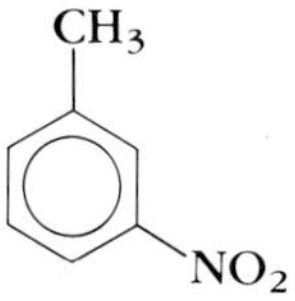 (methyl, NO₂)   c. 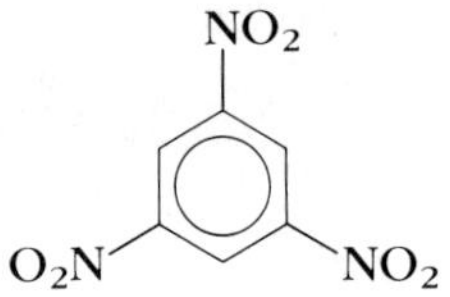 (1,3,5-trinitro substituted)

43. What starting material is required to complete the following transformation?

    $? \xrightarrow{Cl_2, Fe}$ (1-chloronaphthalene)

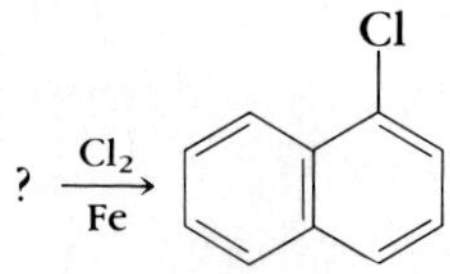

44. What is the danger in swallowing liquid alkanes?
45. What physiological effect do ethene and cyclopropane share?
46. Name a physiological effect of some polycyclic aromatic hydrocarbons.

10

SOME OXYGEN-CONTAINING ORGANIC COMPOUNDS

DID YOU KNOW THAT...

▶ alcohols were among the earliest known compounds?
▶ alcohols, ethers, and phenols can be considered to be derivatives of water?
▶ alcohols have strong hydrogen bonding?
▶ alcohols can be dehydrated to alkenes *or* ethers?
▶ the chemistry of a functional group allows us to know the chemistry of all the compounds within that group?
▶ ethanol is an antidote for methanol poisoning?
▶ "proof" is twice the percentage of alcohol by volume?
▶ antifreeze contains ethylene glycol?
▶ phenols are antiseptics?
▶ there are many thousands of ethers?
▶ formalin is a formaldehyde solution used as a biological preservative?

The earliest written histories record the isolation and use by primitive peoples of the compound known as alcohol. According to Genesis, Noah planted a vineyard after the flood, drank wine from its grapes, and became drunk. Human ingenuity has reached some sort of peak in finding sources of alcohol. Noah may have preferred grapes, but alcohol has been obtained from the fermentation of other fruits, grains, potatoes, and even cactus. Actually, the compound we call alcohol is only one member of a family of organic compounds known as the alcohols. Other members of this family are such familiar substances as cholesterol and sugars.

The name of another family of organic compounds to be considered in this chapter, the ethers, has become almost synonymous with anesthesia. Anyone who has ever been in a hospital would recognize the pungent, antiseptic odor of phenol, the simplest member of the third family discussed in this chapter.

These three organic families can be regarded as derivatives of water. Water is by far the most important inorganic compound we have studied. It should not be surprising, then, that organic compounds derived from water are also of critical importance to health and life.

Aldehydes and ketones are the oxidation products of alcohols. These two families of compounds constitute a most diverse range of familiar substances—from sex hormones to the aroma of green leaves. They are responsible for both the flavors of cinnamon, vanilla, and butter and the sickeningly sweet smell of some rancid foods.

In this chapter we shall examine these five closely related families of oxygen-containing organic compounds. The next chapter will consider a sixth such family, which is extensive and requires a chapter of its own.

10.1 Functional Groups

Consider the water molecule.

$$\underset{H\qquad H}{O}$$

It is a bent molecule with the central oxygen atom attached to two hydrogen atoms. If one of these hydrogen atoms is removed and replaced with an alkyl group (R—), we have

$$\underset{R\qquad H}{O}$$

An alcohol contains a hydroxyl (—OH) group.

This is the general formula for the **alcohol** family. The alkyl group may be methyl, ethyl, isopropyl, or an aliphatic group too complicated to have a simple name. As long as the **hydroxyl group** (—OH) is attached to an aliphatic group, the compound is an alcohol.

10.1 Functional Groups

If the hydroxyl group is attached directly to an aromatic ring, a different family of compounds is produced. Compounds in which an aromatic group (Ar—) is attached to a hydroxyl group are called **phenols**.

A phenol has a hydroxyl group attached to an aromatic ring.

$$Ar-O-H$$

The chemistry of the phenols is sufficiently different from that of the alcohols to justify treatment of the two classes of compounds as separate, if closely related, families. Nonetheless, for both families the chemistry is largely determined by the hydroxyl group. Even the physical properties are determined to a large extent by that group. A group of atoms that confers characteristic chemical and physical properties on a family of organic compounds is called a **functional group**. The hydroxyl group is only one functional group. We have already encountered others. The carbon–carbon double bond in alkenes and the

A functional group confers characteristic properties on a family of organic compounds.

TABLE 10.1 Selected Organic Functional Groups

Name of Class	Functional Group	General Formula of Class	Name of Class	Functional Group	General Formula of Class
Alkane	None	R—H			R—N—R (with R above)
Alkene	—C=C—	R—C=C—R (with R's)			
Alkyne	—C≡C—	R—C≡C—R	Carboxylic acid	—C(=O)—O—H	R—C(=O)—O—H
Alcohol	—C—O—H	R—O—H			
			Ester	—C(=O)—O—C—	R—C(=O)—O—R
Ether	—C—O—C—	R—O—R			
			Amide	—C(=O)—N—	R—C(=O)—N—H (with H)
Aldehyde	—C(=O)—H	R—C(=O)—H			
Ketone	—C(=O)—	R—C(=O)—R			R—C(=O)—N—R (with H)
Amine	—C—N—	R—N—H (with H)			R—C(=O)—N—R (with R)
		R—N—R (with H)			

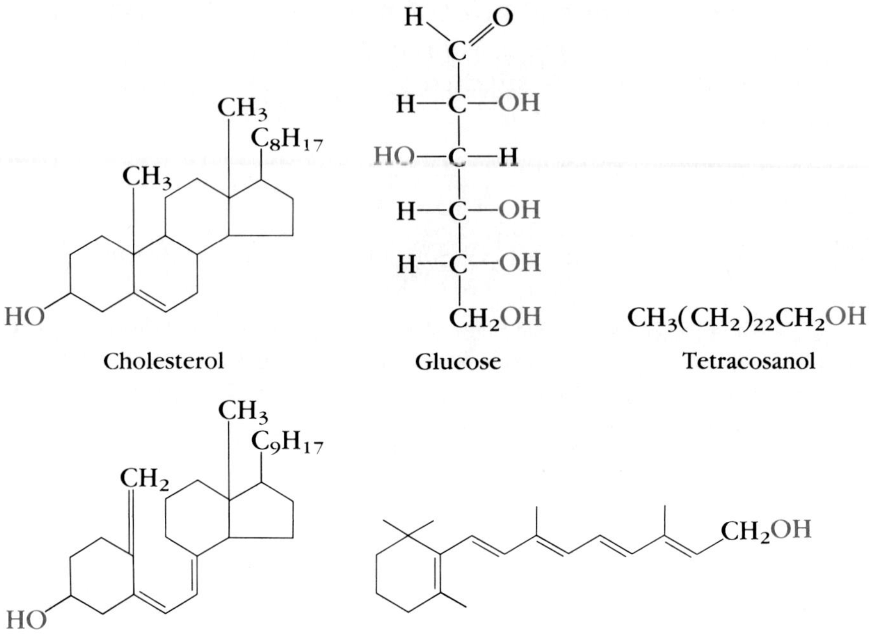

FIGURE 10.1
Some compounds containing the alcohol functional group. Cholesterol is found in all body tissues and is the main constituent of gallstones. Glucose is a simple sugar. Tetracosanol is an ingredient of beeswax. Vitamin D_2 is essential for proper formation of bones, and vitamin A is needed for good vision.

Cholesterol

Glucose

Tetracosanol: $CH_3(CH_2)_{22}CH_2OH$

Vitamin D_2 (calciferol)

Vitamin A (retinol)

carbon–carbon triple bond in alkynes are functional groups. In both instances, these structural features confer on the members of the families a particular chemical reactivity. For example, alkenes and alkynes tend to undergo addition reactions. The halogens in halogenated hydrocarbons are functional groups, although we did not consider in detail the particular reactions associated with these groups. The alkanes are characterized by their *lack* of a distinct functional group. Other functional groups will serve as unifying concepts for the following material. Some of the more important functional groups are listed in Table 10.1. For ready reference, this table is reproduced on the inside back cover.

A "compound containing a hydroxyl group" takes in a lot of territory, as Figure 10.1 illustrates. The concept of a functional group, however, permits us to study simple molecules containing the functional group and then generalize our findings to cover even the most complex members of the group.

10.2 The Alcohols: Nomenclature

The simplest alcohol has the molecular formula CH_4O. Its structural formula is

10.2 The Alcohols: Nomenclature

$$H-\underset{\underset{H}{|}}{\overset{\overset{H}{|}}{C}}-O-H \quad \text{or} \quad CH_3OH$$

Notice that the compound can be thought of as a methyl group (CH_3-) joined to a hydroxyl group ($-OH$). For simple alcohols like this, common names are often used. We name the alkyl group and then add the word *alcohol* to indicate the presence of the hydroxyl group. The simplest alcohol is *methyl alcohol*.

Ethyl alcohol is

$$H-\underset{\underset{H}{|}}{\overset{\overset{H}{|}}{C}}-\underset{\underset{H}{|}}{\overset{\overset{H}{|}}{C}}-O-H \quad \text{or} \quad CH_3CH_2OH$$

As there are two propyl groups (Table 9.4), there are two propyl alcohols.

$$CH_3CH_2CH_2OH \qquad CH_3\underset{\underset{OH}{|}}{CH}CH_3$$

Propyl alcohol Isopropyl alcohol

There are four butyl groups and four butyl alcohols. The names of the butyl groups were given in Table 9.4. Alcohols are subdivided into three classes, called primary (1°), secondary (2°), and tertiary (3°). These classes are based on the number of carbon atoms attached to the carbon atom that bears the *hydroxyl* group. Figure 10.2 illustrates the classification system using the isomeric butyl alcohols.

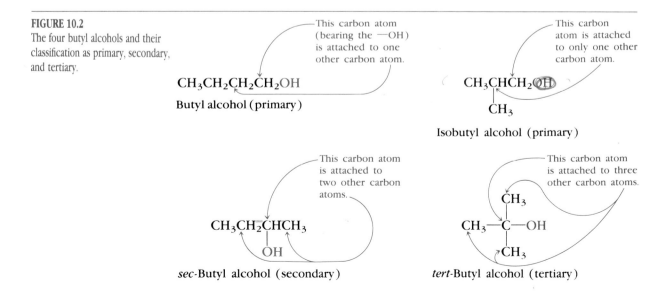

FIGURE 10.2
The four butyl alcohols and their classification as primary, secondary, and tertiary.

EXAMPLE 10.1

Classify each of the following alcohols as primary, secondary, or tertiary: (a) ethyl alcohol, (b) isopropyl alcohol, (c) propyl alcohol.

a. Ethyl alcohol (CH_3CH_2OH) has two hydrogen atoms and only one carbon atom bonded to the carbon atom holding the —OH group. It is a primary alcohol.
b. In isopropyl alcohol,

$$\underset{\underset{OH}{|}}{CH_3CHCH_3}$$

the carbon atom holding the —OH group is also directly bonded to one hydrogen atom and two carbon atoms. Isopropyl alcohol is a secondary alcohol.
c. Propyl alcohol ($CH_3CH_2CH_2OH$), like ethyl alcohol, is a primary alcohol.

EXAMPLE 10.2

How would you classify methyl alcohol?
Methyl alcohol cannot be classified under this system. There is only one carbon atom in methyl alcohol; therefore, it is attached to no other carbon atoms and is neither primary, secondary, nor tertiary.

More complex alcohols are generally given IUPAC names. These systematic names are often used even for the simplest members of the family. In the IUPAC system, alcohols are named for the alkane corresponding to the longest continuous chain of carbon atoms. The final -e of the alkane name is dropped and replaced by the ending -ol. If necessary, the position of the hydroxyl group is indicated by a number placed immediately in front of the name of the longest, or parent, chain. The molecule is always numbered from the end nearer the functional group, giving the hydroxyl group the lowest possible number.

EXAMPLE 10.3

Give the IUPAC names for methyl alcohol and ethyl alcohol.
Methyl alcohol (CH_3OH) is named as a derivative of methane

10.2 The Alcohols: Nomenclature   331

(CH_4). Drop the -e, add -ol, and you have it: methanol. To name CH_3CH_2OH, drop the final -e from the name of the corresponding alkane, ethane (CH_3CH_3), and add -ol. The IUPAC name is ethanol.

EXAMPLE 10.4

Give IUPAC names for the two propyl alcohols.

$$\overset{3}{C}H_3\overset{2}{C}H_2\overset{1}{C}H_2OH \quad\quad CH_3\underset{OH}{\overset{|}{C}H}CH_3$$

The names are, respectively, 1-propanol (not 3-propanol) and 2-propanol.

EXAMPLE 10.5

Give the IUPAC name for *tert*-butyl alcohol.

$$CH_3-\underset{OH}{\overset{CH_3}{\underset{|}{\overset{|}{C}}}}-CH_3$$

The longest continuous chain contains three carbon atoms, and the hydroxyl group is on the second carbon atom of this chain, yielding, for the moment, 2-propanol. In addition, a methyl group is attached to the second carbon atom of the parent chain, giving 2-methyl-2-propanol.

EXAMPLE 10.6

Give the IUPAC name for

$$\overset{7}{C}H_3\underset{CH_3}{\overset{6}{\overset{|}{C}}H}\overset{5}{C}H_2\overset{4}{C}H_2\underset{OH}{\overset{3}{\overset{|}{C}}H}\overset{2}{C}H_2\overset{1}{C}H_3$$

The name is 6-methyl-3-heptanol (*not* 2-methyl-5-heptanol).

10 SOME OXYGEN-CONTAINING ORGANIC COMPOUNDS

TABLE 10.2 Physical Properties of Some Common Alcohols

Name	Formula	Boiling Point (°C)	Solubility (g/100 g H$_2$O)
Methyl alcohol	CH$_3$OH	64	∞
Ethyl alcohol	CH$_3$CH$_2$OH	78	∞
Propyl alcohol	CH$_3$CH$_2$CH$_2$OH	97	∞
Butyl alcohol	CH$_3$CH$_2$CH$_2$CH$_2$OH	118	7.9
Pentyl alcohol	CH$_3$(CH$_2$)$_3$CH$_2$OH	138	2.3
Hexyl alcohol	CH$_3$(CH$_2$)$_4$CH$_2$OH	156	0.6
Octyl alcohol	CH$_3$(CH$_2$)$_6$CH$_2$OH	195	0.05
Decyl alcohol	CH$_3$(CH$_2$)$_8$CH$_2$OH	228	Insoluble
Isopropyl alcohol	CH$_3$CHOHCH$_3$	82	∞
Isobutyl alcohol	(CH$_3$)$_2$CHCH$_2$OH	108	10.2
sec-Butyl alcohol	CH$_3$CHOHCH$_2$CH$_3$	99	12.5
tert-Butyl alcohol	(CH$_3$)$_3$COH	83	∞

10.3 Physical Properties of Alcohols

Most of the common alcohols are liquids at room temperature (Table 10.2). The simplest one, methanol, boils at 65 °C. The hydrocarbon ethane, which has nearly the same formula weight as methanol, boils at −89 °C and is a gas at room temperature. Here we see the pronounced effect of hydrogen bonding (Section 3.12) on boiling points. Figure 10.3 shows how alcohol molecules may be associated through hydrogen bonding. Such association is less extensive in alcohols than in water because an alkyl group (R—) has replaced one of the hydrogen atoms. No hydrogen bond can occur through that alkyl group. Because of this, methanol has a lower boiling point than water even though its formula weight is about twice that of water.

Alcohols of low formula weight are soluble in water. Indeed, methyl, ethyl, and the two propyl alcohols can be mixed with water in all proportions; that is, they are completely miscible with water. As the

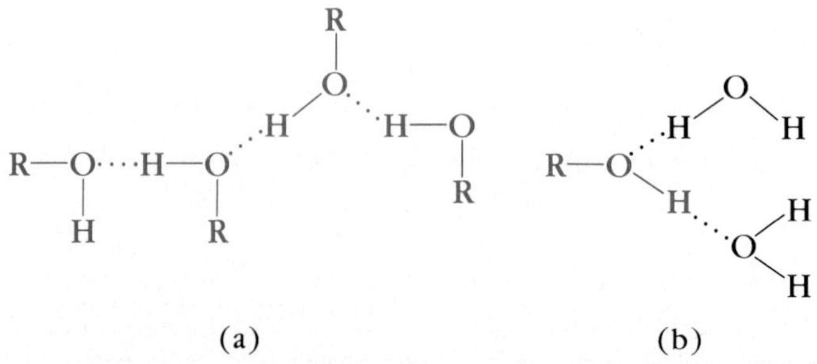

FIGURE 10.3
(a) Alcohols have high boiling points relative to the hydrocarbons because of strong intermolecular hydrogen bonding. (b) Alcohols of low molecular weight are soluble in water because the molecules can hydrogen-bond to water molecules.

10.4 CHEMICAL PROPERTIES OF ALCOHOLS

The reactions of alcohols are the reactions of the hydroxyl functional group. We shall limit our discussion here to two reactions that are of great importance in biological systems: dehydration and oxidation.

Dehydration (removal of water) is usually accomplished by heating the alcohol in the presence of concentrated sulfuric acid. The hydroxyl group is removed from the alcohol carbon, and a hydrogen atom is removed from an adjacent carbon atom, giving an alkene.

Dehydration: the removal of water

$$CH_3CH_2OH \xrightarrow[180\ °C]{\text{conc. } H_2SO_4} CH_2=CH_2 + H_2O$$

Ethanol → Ethene

Under proper conditions it is possible to carry out a dehydration involving two molecules of alcohol. In this case, the hydroxyl group of one alcohol is removed and only the hydrogen atom of the hydroxyl group of the second alcohol molecule is removed. The two organic groups remaining combine to form an ether molecule.

$$CH_3CH_2-OH + H-O-CH_2CH_3 \xrightarrow[140\ °C]{\text{conc. } H_2SO_4} CH_3CH_2-O-CH_2CH_3 + H_2O$$

Two molecules of ethanol → Diethyl ether

Thus, depending on conditions, one can prepare either alkenes or ethers by dehydration of alcohols. A 180 °C, dehydration of ethanol gives ethene. At 140 °C, the main product of dehydration of ethanol is diethyl ether. (Ethers will be discussed in more detail later in this chapter.)

Dehydration reactions also take place in living cells. One cannot have concentrated sulfuric acid or temperatures of 180 °C in living cells and expect the cells to survive. So our bodies use enzymes, such as aconitase, as catalysts for reactions like the following.

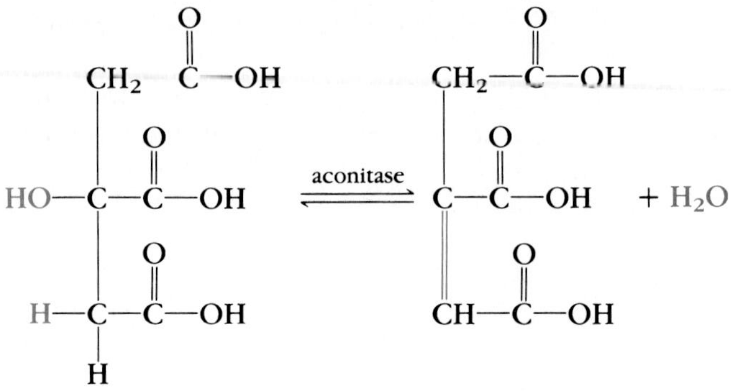

$$\begin{array}{c}\text{CH}_2\text{—C(=O)—OH} \\ | \\ \text{HO—C(—C(=O)—OH)} \\ | \\ \text{H—C(H)—C(=O)—OH}\end{array} \xrightleftharpoons{\text{aconitase}} \begin{array}{c}\text{CH}_2\text{—C(=O)—OH} \\ \| \\ \text{C—C(=O)—OH} \\ \| \\ \text{CH—C(=O)—OH}\end{array} + \text{H}_2\text{O}$$

Citric acid   *cis*-Aconitic acid

The compounds involved are more complex than the ethanol and ethene we used in our previous example, but the reaction is not. Ignore all the parts hanging off to the right of the molecules; they remain unchanged (in this reaction) in the product. The only thing that has happened is that a hydrogen atom and a hydroxyl group have been eliminated from the starting material, and the product contains a double bond. The point to be made is this: If you know the chemistry of a particular functional group, you know the chemistry of a thousand or a hundred thousand different individual compounds. Alcohols have a potential for undergoing dehydration, and you will find that big ones, little ones, and ones that incorporate other functional groups all dehydrate if conditions are right.

The second type of reaction that we want to consider for alcohols is **oxidation**. Primary and secondary alcohols (and methanol) are readily oxidized. Tertiary alcohols are not. Potassium dichromate ($K_2Cr_2O_7$) is typically used as the oxidizing agent in the laboratory. For example, in acid solution, potassium dichromate oxidizes 1-propanol to a compound called propionaldehyde.

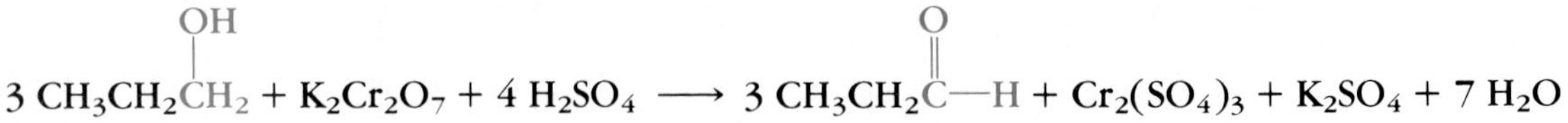

$$3\ CH_3CH_2\overset{OH}{\underset{|}{C}}H_2 + K_2Cr_2O_7 + 4\ H_2SO_4 \longrightarrow 3\ CH_3CH_2\overset{O}{\overset{\|}{C}}\!-\!H + Cr_2(SO_4)_3 + K_2SO_4 + 7\ H_2O$$

Organic chemists regularly simplify equations to emphasize only the change involving the organic molecules. Thus, the above reaction can be written

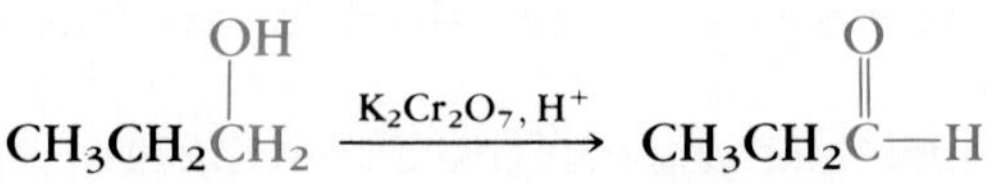

$$CH_3CH_2\overset{OH}{\underset{|}{C}}H_2 \xrightarrow{K_2Cr_2O_7,\ H^+} CH_3CH_2\overset{O}{\overset{\|}{C}}\!-\!H$$

1-Propanol   Propionaldehyde

The required inorganic reagents are written above the arrow (a place reserved for catalysts by inorganic chemists). The inorganic by-products are ignored. In this way, all attention is focused on the organic starting material and product, and less time is spent writing frequently complicated equations.

We have shown a primary alcohol being oxidized to an aldehyde. Aldehydes themselves (as we shall see in Section 10.14) are even more easily oxidized than alcohols. Therefore, the aldehyde is removed from the reaction mixture as soon as it is formed to protect it from further oxidation.

Secondary alcohols are oxidized to compounds called *ketones*. Oxidation of 2-propanol by potassium dichromate gives acetone.

$$CH_3-\underset{\underset{\text{2-Propanol}}{|}}{\overset{OH}{\underset{}{CH}}}-CH_3 \xrightarrow{K_2Cr_2O_7,\ H^+} CH_3-\overset{O}{\underset{}{\overset{\|}{C}}}-CH_3$$

2-Propanol (a secondary alcohol) → Acetone (a ketone)

Unlike aldehydes, these ketone products are relatively resistant to further oxidation.

Tertiary alcohols are also resistant to oxidation. Ordinary oxidizing agents like potassium dichromate bring about no change in this class of alcohols.

$$CH_3-\underset{\underset{CH_3}{|}}{\overset{\overset{OH}{|}}{C}}-CH_3 \xrightarrow{K_2Cr_2O_7,\ H^+} \text{no reaction}$$

2-Methyl-2-propanol (*tert*-butyl alcohol)

Oxidation of an alcohol involves the removal of two hydrogen atoms.

$$-\underset{|}{\overset{\overset{O-H}{|}}{C}}-H \longrightarrow -\underset{|}{\overset{\overset{O}{\|}}{C}}$$

A tertiary alcohol has no hydrogen atom on the carbon atom holding the hydroxyl group and thus is not oxidized.

10.5 PHYSIOLOGICAL PROPERTIES OF ALCOHOLS

The simple alcohols are generally poisonous to some degree, as Table 10.3 indicates. Note that no lethal dose is given for methanol

Simple alcohols are generally poisonous.

TABLE 10.3 Lethal Doses (Orally) of Some Alcohols in Rats

Alcohol	Structure	LD_{50} (g/kg body wt)[a]
Methanol	CH_3OH	b
Ethanol	CH_3CH_2OH	10.6
1-Propanol	$CH_3CH_2CH_2OH$	1.87
2-Propanol	$CH_3CHOHCH_3$	5.8
1-Butanol	$CH_3CH_2CH_2CH_2OH$	4.36
1-Hexanol	$CH_3(CH_2)_4CH_2OH$	4.59
Ethylene glycol	$HOCH_2CH_2OH$	8.54
Glycerol	$HOCH_2CHOHCH_2OH$	>25

[a] LD_{50} = lethal dose for 50% of the population of test animals.

[b] No LD_{50} is given for methanol. Its acute (short-term) toxicity is not terribly high. However, it is metabolized to formaldehyde (HCHO) in the body, so that the chronic (long-term) toxicity is quite high. The LD_{50} for formaldehyde administered orally to rats is 0.070 g/kg of body weight. The LD_{50} for acetaldehyde, the metabolite of ethanol, is 1.9 g/kg.

Data from Susan Budavari (Ed.), *The Merck Index*, 11th ed., Rahway, NJ: Merck, 1989.

(wood alcohol*). While its acute (short-term) toxicity is not terribly high, it can cause permanent blindness or death, even in small doses. Each year many accidental injuries are attributed to this alcohol, which is frequently mistaken for its less harmful relative, ethanol.

The reason methanol is so dangerous is that humans and other primates have liver enzymes that oxidize primary alcohols (and methanol) to aldehydes. Methanol is oxidized to formaldehyde.

$$CH_3OH \xrightarrow{\text{liver enzymes}} \underset{\text{Formaldehyde}}{H-\overset{\overset{\displaystyle O}{\|}}{C}-H}$$

Methanol

Formaldehyde reacts rapidly with the components of cells. It causes the coagulation of proteins in much the same way that cooking causes an egg to coagulate. This property of formaldehyde accounts for the great toxicity of methanol.

Ethanol is also toxic, but much less so than methanol. When most people say "alcohol," meaning liquor or that which is imbibed, they are referring to ethanol. Most alcoholic beverages are made by the fermentation of starches or sugars by yeast (Figure 10.4). On an industrial scale, molasses from sugarcane or starches from various types of grain (hence the term *grain alcohol*) are fermented to give aqueous solutions of ethanol. Yeast contains enzymes (biological catalysts) re-

* Methanol is also called *wood alcohol* because it can be condensed from the vapors that form when wood is heated in the absence of air, a process called **destructive distillation**.

quired for this conversion. The fermentation process has been known for a long time, and enzymes from all kinds of sources, including saliva, have been used. Explorer Captain James Cook reported during his voyages on "kava," a heady drink made by Tongans, who chewed bits of root, deposited these in a wooden bowl, mixed them with water, and served the brew. In reactions controlled by these enzymes, carbohydrates are broken down to ethanol and carbon dioxide.

$$(C_6H_{10}O_5)_n + n\,H_2O \xrightarrow{\text{yeast}} 2n\,C_2H_5OH + 2n\,CO_2$$

The n is used in this equation to indicate that carbohydrates come in all sizes. In the production of champagne or beer, part of the carbon dioxide is retained, making a carbonated beverage.

The fermentation process is capable of yielding aqueous solutions containing up to 18% ethanol. As the ethanol concentration builds to these levels, it inhibits the catalytic activity of the yeast, and the system shuts itself down. To produce a more concentrated solution, the solution isolated directly from fermentation is distilled. Such concentrated solutions (usually in the range of 40–50% ethanol) are called *distilled spirits* and are prepared in distilleries (legally) or backwoods stills (illegally). The fermentation product can be concentrated to as much as 95% ethanol by distillation. This material is frequently used as a solvent for drugs meant for internal consumption.

Concentrated solutions of ethanol are called distilled spirits.

In the body, ethanol is oxidized by the same liver enzymes that operate on methanol. The product obtained from ethanol is acetaldehyde.

$$\underset{\text{Ethanol}}{CH_3CH_2{-}OH} \xrightarrow{\text{liver enzymes}} \underset{\text{Acetaldehyde}}{CH_3\overset{O}{\underset{\|}{C}}{-}H}$$

FIGURE 10.4
Alcohol can be made by the fermentation of nearly any type of starchy or sugary material. (Photo © The Terry Wild Studio.)

TABLE 10.4 Relationship Between Drinks Consumed, Blood-Alcohol Level, and Behavior[a]

Number of Drinks[b]	Blood-Alcohol Level (% by volume)	Behavior
2	0.05	Mild sedation; tranquility
4	0.10	Lack of coordination
6	0.15	Obvious intoxication
10	0.30	Unconsciousness
20	0.50	Possible death

[a] Data would be approximate for 70-kg (154-lb) person.
[b] Based on 30-mL (1-oz) "shots" of 90-proof whiskey or 360-mL (12-oz) bottles of beer, consumed rapidly.

This product is oxidized, in turn, to acetic acid (a normal constituent of cells) and ultimately to carbon dioxide and water. Ethanol, administered intravenously, has long been used as an antidote for methanol poisoning. The ethanol preferentially loads up the liver enzymes in humans. The enzymes are tied up oxidizing ethanol to acetaldehyde and cannot convert methanol to the dangerously toxic fomaldehyde. Thus, the unoxidized methanol is gradually excreted from the body.

We noted previously that ethanol itself is toxic. A pint of pure ethanol, rapidly ingested, would kill most people. Even the strongest alcoholic beverages, however, are seldom more than 90 proof (45% ethanol).* Excessive ingestion over a long period of time leads to deterioration of the liver and loss of memory—and possibly to strong physiological addiction. Ethanol acts as a mild hypnotic (sleep inducer). Perhaps this is fortunate, for a heavy drinker usually "passes out" before ingesting a lethal dose (Table 10.4). Although it generally acts as a depressant (i.e., it reduces the level of consciousness and the intensity of our reactions to environmental stimuli), ethyl alcohol in small amounts seems sometimes to act as a stimulant. Any such effect, however, is probably due to the alcohol's action in relaxing tensions and relieving inhibitions.

Alcoholic beverages are highly taxed. To discourage the diversion of untaxed industrial ethanol to use in beverages, the industrial material is **denatured** by addition of other chemicals that make the alcohol unfit to drink. Among the common denaturants are gasoline and methanol.

Denatured alcohol—added chemicals make the alcohol unfit to drink

---

* The "proof" is twice the percentage of alcohol by volume. The term has its origin in a seventeenth-century method for testing whiskey. Dealers were perhaps too often tempted to increase profits by adding water to the booze. A qualitative method for testing the whiskey was to pour some of it on gunpowder and ignite it. If the gunpowder ignited after the alcohol had burned away, that was considered "proof" that the whiskey did not contain too much water.

2-Propanol (isopropyl alcohol), an ingredient in rubbing alcohol and other products for external use, is sometimes mistaken for ethanol and is ingested. Although more toxic than ethanol, it seldom causes fatalities. Instead it induces vomiting. It does not stay down long enough to kill you (though perhaps long enough to make you wish you were dead). Other higher alcohols behave in a similar manner.

The lower alcohols have a mild antiseptic action. Aqueous solutions of 60–70% ethanol are often used to clean the skin before an injection, the drawing of blood, or minor surgery.

10.6 MULTIFUNCTIONAL ALCOHOLS

The simple alcohols we have met thus far contain only one hydroxyl group each. Several important compounds that are frequently encountered contain more than one hydroxyl group per molecule. The **glycols** contain two hydroxyl groups per molecule, and the most important of these is ethylene glycol.

Glycols contain two hydroxyl groups per molecule.

$$\begin{array}{cc} CH_2\!\!-\!\!CH_2 \\ | & | \\ OH & OH \end{array}$$

This compound is the main ingredient in permanent antifreeze mixtures for automobile radiators. Ethylene glycol is a sweet, somewhat viscous liquid. With two hydroxyl groups, there is extensive intermolecular hydrogen bonding. Thus, ethylene glycol has a high boiling point and does not boil away when used as antifreeze. It is also completely miscible with water.

Ethylene glycol is quite toxic. As with methanol, its toxicity is due to an oxidation product. Liver enzymes oxidize ethylene glycol to oxalic acid (an aldehyde is produced as an intermediate in this reaction).

$$\underset{\text{Ethylene glycol}}{\begin{array}{cc} OH & OH \\ | & | \\ CH_2\!\!-\!\!CH_2 \end{array}} \xrightarrow{\text{liver enzymes}} \underset{\text{Oxalic acid}}{\begin{array}{cc} O & O \\ \| & \| \\ HO\!\!-\!\!C\!\!-\!\!C\!\!-\!\!OH \end{array}}$$

This compound crystallizes as its calcium salt, calcium oxalate (CaC_2O_4), in the kidneys, leading to renal damage. Such injury can lead to kidney failure and death. As with methanol poisoning, the usual treatment for ethylene glycol poisoning is ethanol, administered to load up the liver enzymes and thus block them from catalyzing the conversion of ethylene glycol to oxalic acid.

The physical properties of propylene glycol

$$\begin{array}{cc} CH_3CH\!\!-\!\!CH_2 \\ | & | \\ OH & OH \end{array}$$

TABLE 10.5 Properties of Some Polyhydric Alcohols

Name	Formula	Boiling Point (°C)	Solubility in Water
Ethylene glycol	CH_2OHCH_2OH	198	∞
Propylene glycol	$CH_3CHOHCH_2OH$	188	∞
Glycerol	$CH_2OHCHOHCH_2OH$	290 (decomposes)	∞
Sorbitol	$CH_2OH(CHOH)_4CH_2OH$	(Solid)	∞

are quite similar to those of ethylene glycol (Table 10.5). Its physiological properties, however, are quite different. Propylene glycol is essentially nontoxic, and it can be used as a solvent for drugs. It is also used as a moisturizing agent for foods. Like other alcohols, propylene glycol can be oxidized by liver enzymes.

$$CH_3-\underset{\underset{\text{Propylene glycol}}{}}{\overset{OH}{\underset{|}{C}H}}-\underset{}{\overset{OH}{\underset{|}{C}H_2}} \xrightarrow{\text{liver enzymes}} CH_3-\underset{\underset{\text{Pyruvic acid}}{}}{\overset{O}{\underset{\|}{C}}}-\overset{O}{\underset{\|}{C}}-OH$$

In this case, however, the product is pyruvic acid, a normal intermediate in carbohydrate metabolism in the body.

Glycerol (glycerin) contains three hydroxyl groups per molecule.

$$\underset{OH}{\overset{|}{C}H_2}-\underset{OH}{\overset{|}{C}H}-\underset{OH}{\overset{|}{C}H_2}$$

This sweet, syrupy liquid is essentially nontoxic and is a normal product of the hydrolysis of fats within our bodies.

Carbohydrates are the most important polyfunctional alcohols. They constitute one of the major classes of biological molecules and will be taken up, as a class, later in this book.

10.7 The Phenols

Compounds with a hydroxyl group attached directly to an aromatic ring are called **phenols**. The parent compound, C_6H_5OH, is itself called *phenol*. Other compounds of interest to us are best known by special, nonsystematic names. A variety of phenols are shown in Figure 10.5.

By their very nature, phenols never contain fewer than six carbon atoms. Phenols are generally solids with low melting points or oily

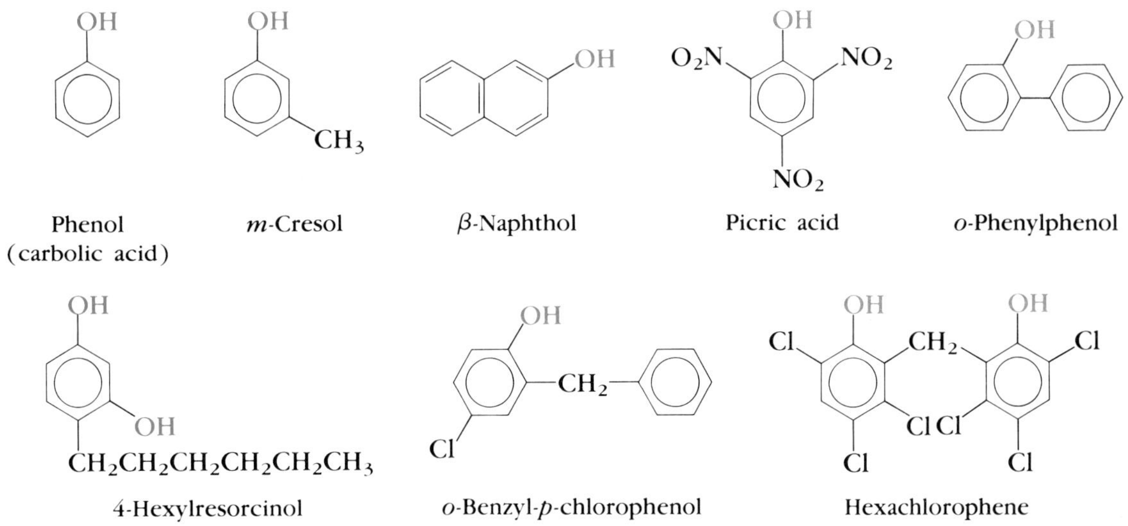

10.7 The Phenols

FIGURE 10.5 Some phenols of interest. Cresols are used as wood preservatives, β-naphthol is a dye intermediate, picric acid is an explosive, and the remaining phenols are antiseptics or disinfectants.

liquids at room temperature. Most are only sparingly soluble in water. They have found wide use as germicides or antiseptics (substances that kill microorganisms on living tissue) and as disinfectants (substances intended to kill micoorganisms on furniture, fixtures, floors, and so on).

The first widely used antiseptic was phenol itself, which is also called *carbolic acid* (not the same as *carbonic acid*, H_2CO_3). Joseph Lister used it for antiseptic surgery in 1867. Unfortunately, phenol kills not only undesirable microorganisms but all types of cells. Applied to the skin, it can cause severe burns. In the bloodstream it is a systemic poison, that is, one that is carried to and affects all parts of the body. Its severe side effects led to searches for safer antiseptics.

One of the most active phenolic antiseptics is 4-hexylresorcinol. Much more powerful than phenol as a germicide, it has fewer undesirable side effects and is safe enough to be used as the active ingredient in some mouthwashes.

Prominent among disinfectants are the compounds o-phenylphenol and o-benzyl-p-chlorophenol. These compounds are the main active ingredients in preparations such as Lysol.

Hexachlorophene was once widely used in germicidal cleaning solutions (pHisohex) and as an ingredient in deodorant soaps and other cosmetics. In the United States, products contained at most 3% hexachlorophene. The compound was generally considered a safe and effective antibacterial agent. Then, in 1972, an outbreak of neurological disease among infants in France was traced to a baby powder called Bébé that contained over 20% hexachlorophene. More than 30 infants died. In the United States hexachlorophene was banned from all products intended for over-the-counter sales. Although it is still available by prescription and for use in hospitals, concentrations may not exceed 3%.

10.8 The Ethers

Ethers have two carbon atoms attached to an oxygen atom.

We can start with a water molecule and derive a structure for the third family of compounds to be discussed in this chapter, the **ethers**. To accomplish this we must substitute carbon groups for both of water's hydrogen atoms. There are three possible ways of doing this.

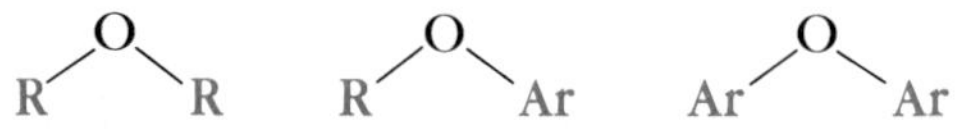

$$R-O-R \quad R-O-Ar \quad Ar-O-Ar$$

All the above formulas represent ethers. A compound is an ether as long as there are two carbon groups attached to the oxygen atom, whether the groups are aliphatic or aromatic.

Simple ethers are simply named. Just name the groups attached to the oxygen atom and then add the generic name *ether* (Figure 10.6). For symmetrical ethers (those in which both organic groups are identical), the group name should be preceded by the prefix *di-*, although the prefix is sometimes dropped in common usage. The names *methyl ether* and *dimethyl ether* refer to the same compound, but the latter is preferred.

Ethers have no hydrogen atoms bonded to the oxygen atom; hence, molecules in the pure liquid are incapable of intermolecular hydrogen bonding. Ethers have boiling points about the same as those of alkanes of comparable formula weight. For example, diethyl ether (formula weight = 74) boils at 35 °C, and pentane (formula weight = 72) boils at 36 °C. Ether molecules do have an oxygen atom, however. They can participate in hydrogen bonding if some other kind of molecule, like water, supplies the appropriate hydrogen.

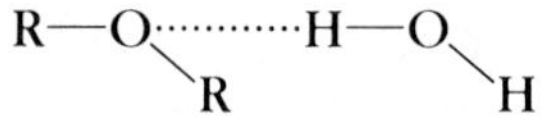

$$R-O\cdots H-O$$
$$RH$$

Consequently, ethers have about the same water solubilities as their isomeric alcohols. Both 1-butanol and diethyl ether have the molecular

FIGURE 10.6
Some representative ethers. Notice that in some ethers (ethylene oxide, dioxane, and tetrahydrofuran) the oxygen is incorporated as part of a ring.

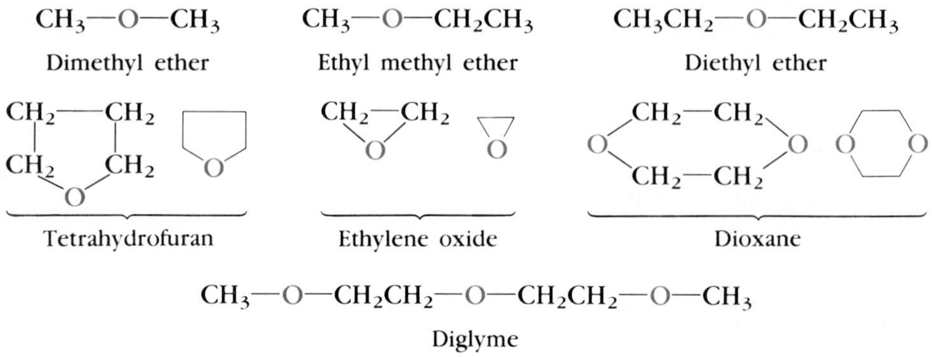

CH_3-O-CH_3 $CH_3-O-CH_2CH_3$ $CH_3CH_2-O-CH_2CH_3$
Dimethyl ether   Ethyl methyl ether   Diethyl ether

Tetrahydrofuran   Ethylene oxide   Dioxane

$CH_3-O-CH_2CH_2-O-CH_2CH_2-O-CH_3$
Diglyme

formula $C_4H_{10}O$ (i.e., they are isomers). Each is soluble to the extent of about 8 g in 100 g of water.

The ether functional group is quite unreactive chemically. The inertness of ethers makes them excellent solvents for organic materials. Diethyl ether is often used in the extraction of organic compounds from plant and animal sources or from mixtures of organic and inorganic substances. After the extraction, the volatile ether is easily removed by evaporation, leaving the desired organic components as a residue.

10.9 GENERAL ANESTHETICS

Anesthetics are depressants. Ethanol, a depressant, has been used as an anesthetic in some circumstances. A **general anesthetic** acts on the brain to produce unconsciousness as well as insensitivity to pain (Figure 10.7).

Diethyl ether was the first general anesthetic. A Boston dentist, William Morton, introduced it into surgical practice in 1846. Inhalation of diethyl ether vapor produces unconsciousness by depressing the activity of the central nervous system. Diethyl ether is relatively safe because there is a fairly wide gap between the effective level for anesthesia and the lethal dose. The disadvantages are its high flammability and its side effect, nausea.

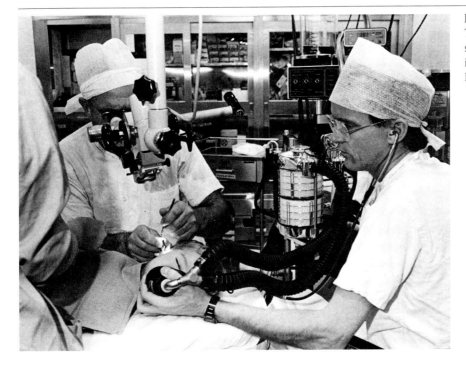

FIGURE 10.7
The use of effective anesthetics during surgery opened the way for a revolution in medicine. [Ulrike Welsch/Photo Researchers, Inc.]

Morton tried nitrous oxide (N_2O), or laughing gas, without success before he tried diethyl ether. Joseph Priestley discovered nitrous oxide in 1772. Its narcotic effect was noted, and it soon came to be used widely at laughing-gas parties among the nobility. Nitrous oxide, mixed with oxygen, finds some use in modern anesthesia; it is quick-acting but not very potent. Concentrations of 50% or greater must be used to be effective. When nitrous oxide is mixed with ordinary air instead of oxygen, not enough oxygen gets into the patient's blood, and permanent brain damage can result.

Chloroform ($CHCl_3$) was introduced as a general anesthetic in 1847. Its use quickly became popular after Queen Victoria gave birth to her eighth child while anesthetized by chloroform in 1853. It has also been used in veterinary medicine. Chloroform is nonflammable and produces effective anesthesia, but it has a number of serious drawbacks: it has a narrow safety margin (the effective dose is close to the lethal dose), and it also causes liver damage. Chloroform has been listed as a carcinogen by the Environmental Protection Agency.

In Chapter 9 we encountered the most potent anesthetic gas, cyclopropane. Small amounts rapidly produce insensitivity to pain without rendering the patient unconscious. The great drawback is that cyclopropane forms explosive mixtures with air throughout its effective range of concentrations. Special equipment and an experienced anesthetist are required for its use.

Modern anesthetics include fluorine-containing compounds such as halothane, enflurane, and methoxyflurane.

$$\underset{\text{Halothane}}{\overset{\overset{F}{|}\;\;\overset{Br}{|}}{\underset{\underset{F}{|}\;\;\underset{Cl}{|}}{F-C-C-H}}} \qquad \underset{\text{Enflurane}}{\overset{\overset{H}{|}\;\;\overset{F}{|}\;\;\overset{H}{|}}{\underset{\underset{F}{|}\;\;\underset{F}{|}\;\;\underset{}{}}{F-C-O-C-C-Cl}}} \qquad \underset{\text{Methoxyflurane}}{\overset{\overset{H}{|}\;\;\overset{F}{|}\;\;\overset{H}{|}}{\underset{\underset{H}{|}\;\;\underset{F}{|}\;\;\underset{Cl}{|}}{H-C-O-C-C-Cl}}}$$

Notice that the first compound is not an ether, but the last two are. These compounds are nonflammable and are relatively safe for the patient. Their safety for operating room personnel, however, has been questioned, particularly in the case of halothane. For example, female medical personnel exposed over extended periods incur a higher rate of miscarriages than the general population.

Modern surgical practice has moved away from the use of a single anesthetic. Generally, a patient is given a strong sedative (for example, a barbiturate) by injection to produce unconsciousness. The gaseous anesthetic is then administered to provide insensitivity to pain and to keep the patient unconscious. A relaxant, such as curare,* also may be

---

*Curare is the arrow poison used by South American Indian tribes. Large doses of curare kill by causing a complete relaxation of all muscles. Death occurs because of respiratory failure.

employed. Curare produces muscle relaxation; thus, only light anesthesia is required. This practice avoids the hazards of deep anesthesia.

The potency of an anesthetic is related to its solubility in fat (a relatively nonpolar substance). General anesthetics seem to work by dissolving in the fatlike membranes of nerve cells. This changes the permeability of the membranes, and the conductivity of the nerve cells is depressed.

10.10 THE CARBONYL GROUP: A CARBON-OXYGEN DOUBLE BOND

The first functional group we encountered was the carbon–carbon double bond in alkenes. In the alcohols, we saw a functional group in which an oxygen atom was attached to a carbon atom. The **carbonyl** (pronounced "carbon-eel") group incorporates a feature of each of these other functional groups. It involves a carbon–oxygen double bond.

Carbonyl groups have a carbon–oxygen double bond.

The bonds to a carbonyl group are actually arranged as we have drawn them, pointing to the corners of a triangle.

Like other structural formulas, however, the formulas for carbonyl-containing compounds are frequently represented with distorted bond angles to conserve space. Therefore, we shall often represent the carbonyl group in the following way.

$$-\overset{\overset{\displaystyle O}{\|}}{C}- \quad \text{or} \quad -\overset{\displaystyle |}{C}=O$$

All three formulas show that the carbon atom in the carbonyl group can bond to two other groups.

Aldehydes and *ketones* both contain carbonyl groups (Figure 10.8). These two families of compounds offer us an opportunity to study the carbonyl group in its simplest surroundings. In the next chapter we

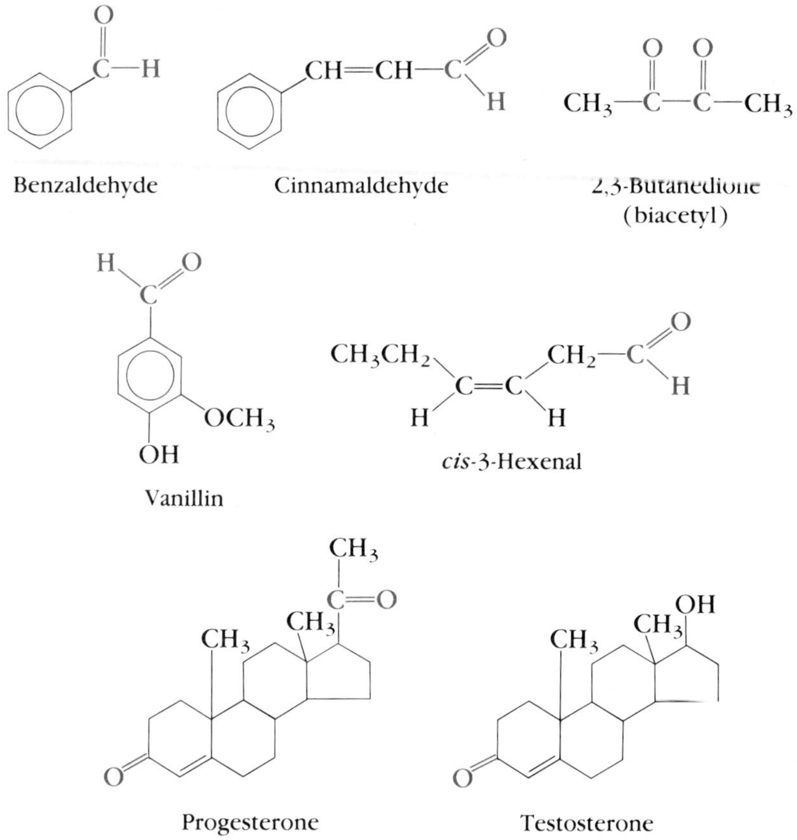

FIGURE 10.8
Some interesting aldehydes and ketones. Benzaldehyde is an oil found in almonds. Cinnamaldehyde is oil of cinnamon. 2,3-Butanedione is responsible for the flavor of butter, and vanillin, of course, gives vanilla its flavor. *cis*-3-Hexenal provides an herbal odor. Progesterone is a female sex hormone, and testosterone is a male sex hormone.

shall consider somewhat more complicated functional groups that incorporate a carbon–oxygen double bond. We shall ultimately find ourselves running into this ubiquitous grouping of atoms in carbohydrates, fats, proteins, nucleic acids, hormones, vitamins, and the host of organic compounds critical to the functioning of living systems.

What is the difference between an aldehyde and a ketone? It appears to be rather trivial at first. In **ketones**, two carbon groups are attached to the carbonyl carbon. These general formulas all represent ketones.

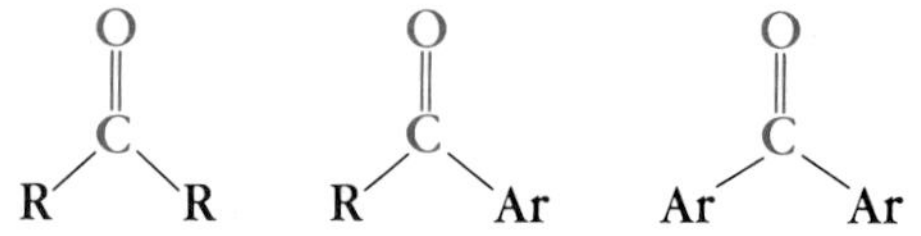

Ketones have two carbon groups attached to the carbonyl carbon.

In **aldehydes**, at least one of the attached groups must be a hydrogen atom. These compounds are all aldehydes.

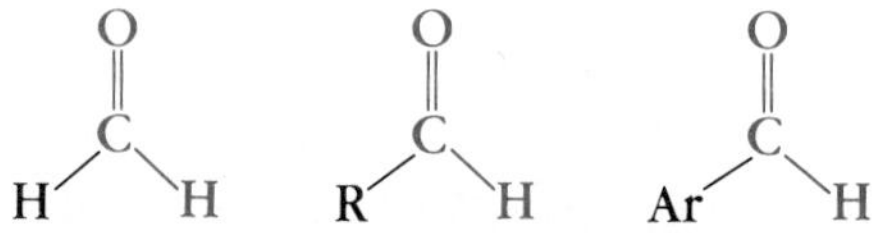

Aldehydes have at least one hydrogen attached to the carbonyl carbon.

10.11 Naming Common Aldehydes

Aldehydes and ketones share many common properties, as one might expect for compounds with the same functional group. But they are different in other respects, different enough to warrant their classification into two families.

10.11 NAMING COMMON ALDEHYDES

As with most compounds isolated from natural sources, aldehydes came to be known by common names long before IUPAC rules were established. The simplest aldehyde, and the only one with two hydrogen atoms attached to the carbonyl group, contains only one carbon atom and is known as formaldehyde. The aldehyde with two carbon atoms is called acetaldehyde, and that with three carbons is propionaldehyde. All three of these aldehydes have been mentioned previously as products of the oxidation of methanol, ethanol, and 1-propanol, respectively. These aldehydes and a few others of interest are shown in Figure 10.9.

The IUPAC names of aldehydes are derived from those of the corresponding alkanes. Select the longest continuous chain of carbon atoms that contains the functional group. Take the name of the alkane with that number of carbon atoms, drop the -e, and add the ending -al. Because the IUPAC ending for alcohols is -ol, there is occasionally some confusion unless great care is exercised in writing and pronouncing the IUPAC names of these two families. The one-carbon alcohol is methanol, with the -ol ending pronounced "all."* The one-carbon aldehyde is methanal, with the ending pronounced like the man's name Al. IUPAC names of the first five aldehydes are given in Figure 10.9.

In the IUPAC system, the carbonyl carbon atom is *always* taken as the first carbon of the parent chain.

*Some chemists pronounce the -ol as in "old" to further emphazise the difference.

FIGURE 10.9
Some aldehydes of interest.

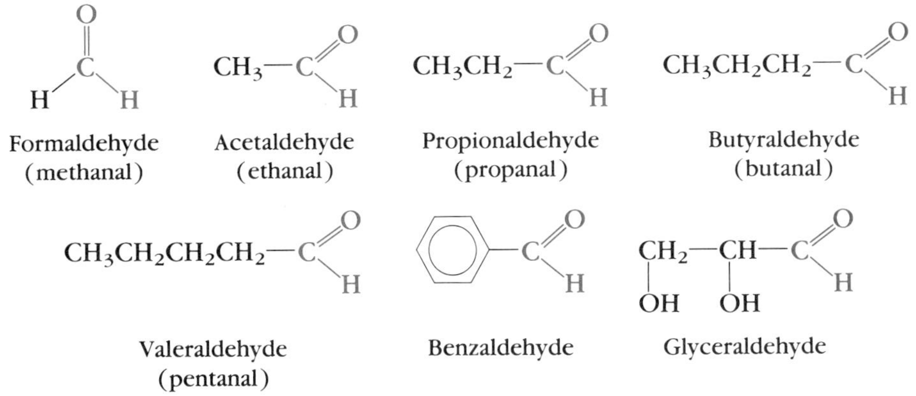

Formaldehyde (methanal)

Acetaldehyde (ethanal) — $CH_3-C(=O)H$

Propionaldehyde (propanal) — $CH_3CH_2-C(=O)H$

Butyraldehyde (butanal) — $CH_3CH_2CH_2-C(=O)H$

Valeraldehyde (pentanal) — $CH_3CH_2CH_2CH_2-C(=O)H$

Benzaldehyde

Glyceraldehyde — $CH_2(OH)-CH(OH)-C(=O)H$

EXAMPLE 10.7

Give the IUPAC name for

$$CH_3CH_2CH_2CH_2CH_2\overset{\overset{O}{\|}}{C}-H$$

The parent chain contains six carbon atoms (including the carbonyl carbon). Hexane is the name for a six-carbon alkane; therefore, this aldehyde is hexanal.

EXAMPLE 10.8

Give the IUPAC name for

$$CH_3CH_2\underset{\underset{CH_3}{|}}{CH}-\overset{\overset{O}{\|}}{C}-H$$

Note the four carbon atoms in the longest continuous chain (including the carbonyl group) and the methyl group on the second carbon atom of this chain. (Remember: The carbonyl carbon is **numbered 1**, so we count from the right here.) The compound is 2-methylbutanal.

10.12 Naming Common Ketones

The carbonyl group in a ketone must be attached to two carbon groups. Therefore, the simplest ketone has three carbon atoms and is known far and wide as **acetone** (Figure 10.10). The name is unique and does not correspond to the first in a series of similar common names. Generally, ketones are given common names consisting of the names of the groups attached to the carbonyl group, followed by the word *ketone*. (Note the similarity to the naming of ethers.) Another name for acetone, then, is dimethyl ketone. With four carbon atoms, we have ethyl methyl ketone (Figure 10.10). If names for the groups attached to the carbonyl group are known, this common naming system can be applied.

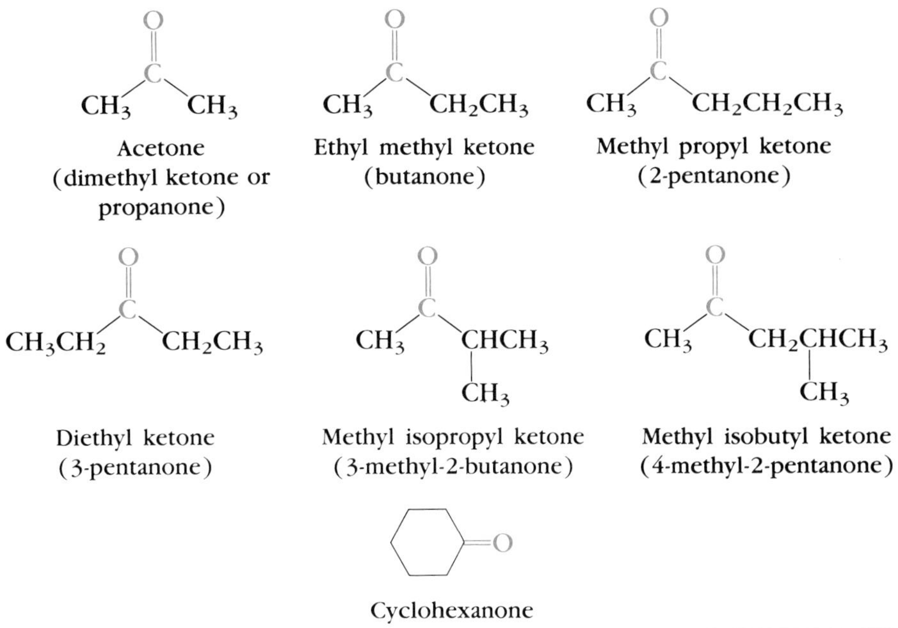

10.12 Naming Common Ketones

FIGURE 10.10 Some common ketones.

Acetone (dimethyl ketone or propanone): CH₃—CO—CH₃

Ethyl methyl ketone (butanone): CH₃—CO—CH₂CH₃

Methyl propyl ketone (2-pentanone): CH₃—CO—CH₂CH₂CH₃

Diethyl ketone (3-pentanone): CH₃CH₂—CO—CH₂CH₃

Methyl isopropyl ketone (3-methyl-2-butanone): CH₃—CO—CH(CH₃)CH₃

Methyl isobutyl ketone (4-methyl-2-pentanone): CH₃—CO—CH₂CH(CH₃)CH₃

Cyclohexanone

In the IUPAC system, the longest continuous chain to which the oxygen atom is doubly bonded is selected as the parent chain. The *-e* ending of the corresponding alkane name is dropped and replaced with *-one*. Acetone thus becomes propanone, and ethyl methyl ketone is called butanone. In higher ketones, a number indicates the position of the doubly bonded oxygen as well as any substituents. Figure 10.10 shows an assortment of ketones and lists both IUPAC and common names.

EXAMPLE 10.9

Write the structural formula for 3-hexanone, and give its common name.

The longest chain has six carbon atoms, and the doubly bonded oxygen atom is located at position 3 along this chain.

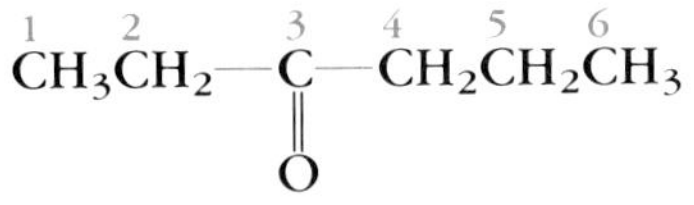

$$\overset{1}{C}H_3\overset{2}{C}H_2-\overset{3}{C}(=O)-\overset{4}{C}H_2\overset{5}{C}H_2\overset{6}{C}H_3$$

The two alkyl groups attached to the carbonyl group are ethyl and propyl. The common name for this compound is ethyl propyl ketone.

> **EXAMPLE 10.10**
>
> Write the structural formula for diisopropyl ketone, and give its IUPAC name.
>
> Two isopropyl groups must be attached to a carbonyl group.
>
> $$CH_3CH-\underset{\underset{CH_3}{|}}{\overset{\overset{O}{\|}}{C}}-\underset{\underset{CH_3}{|}}{CHCH_3}$$
>
> There are five carbon atoms in the longest continuous chain containing the carbonyl group, which is located at position 3 on this chain. Attached to this chain at the second and fourth positions are two methyl groups. The IUPAC name of this compound is 2,4-dimethyl-3-pentanone.

10.13 Physical Properties of Aldehydes and Ketones

The carbon and oxygen atoms of the carbonyl group share two pairs of electrons, but they do not share them equally. The electronegative oxygen atom has a much greater attraction for the bonding pairs. The carbon atom is left with a partial positive charge; the oxygen atom, with a partial negative charge. The bond is polar.

$$\overset{\delta+\delta-}{C=O}$$

The polarity of the carbon–oxygen double bond is greater than that of

TABLE 10.6 Boiling Points of Compounds with Similar Molecular Weights and Different Types of Intermolecular Forces

Compound	Molecular Weight	Intermolecular Force	Boiling Point (°C)
$CH_3CH_2CH_2CH_3$	58	Dispersion only	0
$CH_3OCH_2CH_3$	60	Weak dipole	6
$CH_3\overset{\overset{O}{\|}}{C}CH_3$	58	Strong dipole	56
$CH_3CH_2CH_2OH$	60	Hydrogen bonding	97

10.14 Chemical Properties of Aldehydes and Ketones

TABLE 10.7 Physical Properties of Selected Aldehydes and Ketones

Compound	Formula	Boiling Point (°C)	Solubility in Water (g/100 g H_2O)
Formaldehyde	HCHO	−21	Very soluble
Acetaldehyde	CH_3CHO	20	∞
Propionaldehyde	CH_3CH_2CHO	49	16
Butyraldehyde	$CH_3CH_2CH_2CHO$	76	7
Valeraldehyde	$CH_3CH_2CH_2CH_2CHO$	103	Slightly soluble
Benzaldehyde	C_6H_5CHO	178	0.3
Acetone	CH_3COCH_3	56	∞
Ethyl methyl ketone	$CH_3COCH_2CH_3$	80	26
Methyl propyl ketone	$CH_3COCH_2CH_2CH_3$	102	6.3
Diethyl ketone	$CH_3CH_2COCH_2CH_3$	101	5

the carbon–oxygen single bond and significantly affects the boiling points of aldehydes and ketones. The carbon–oxygen single bond in ethers has little effect on boiling points (Table 10.6). The effect of the carbonyl group, however, is still not comparable to that of the hydrogen bonding in alcohols. Pure aldehydes and ketones cannot form intermolecular hydrogen bonds. (Remember that the hydrogen attached to the carbonyl group in an aldehyde is attached at the *carbon* atom, not the *oxygen* atom, of the functional group.) Aldehydes and ketones, because they contain oxygen, can hydrogen-bond with water molecules. These families are about as soluble in water as alcohols of comparable weight.

Aldehydes and ketones cannot form intermolecular hydrogen bonds.

Physical properties of selected aldehydes and ketones are summarized in Table 10.7. Formaldehyde is a gas at room temperature and is generally available as a 40% aqueous solution called **formalin**. This solution was formerly used as a biological preservative.

10.14 Chemical Properties of Aldehydes and Ketones

For chemists, aldehydes and ketones are among the most interesting families of organic compounds because of the many reactions in which they participate. Nonetheless, we shall consider only two reactions that are especially important in the study of biochemical processes. The first of these is the oxidation of aldehydes. In a sense, this discussion is a continuation of Section 10.4, where we considered the oxidation of alcohols to aldehydes and ketones.

Aldehydes and ketones undergo many reactions.

Aldehydes have a hydrogen atom attached to the carbonyl group.

Aldehydes are readily oxidized and yield compounds called carboxylic acids. Ketones have no hydrogen atom attached to the carbonyl group and resist oxidation.

$$\underset{\text{An aldehyde}}{\text{R}-\overset{\overset{\displaystyle O}{\|}}{\text{C}}-\text{H}} \xrightarrow{[O]} \underset{\text{Carboxylic acid}}{\text{R}-\overset{\overset{\displaystyle O}{\|}}{\text{C}}-\text{OH}}$$

$$\underset{\text{A ketone}}{\text{R}-\overset{\overset{\displaystyle O}{\|}}{\text{C}}-\text{R}'} \xrightarrow{[O]} \text{no reaction}$$

The [O] in these equations is a general way of representing an oxidizing agent.

Aldehydes are among the most easily oxidized of organic compounds, and this fact enables chemists to identify them. By using very mild oxidizing agents, aldehydes can be distinguished not only from ketones, but also from alcohols and other families that require stronger oxidizing agents. Silver ion (Ag^+) and copper(II) ion (Cu^{2+}) are two such gentle reagents. For once, let us concentrate on the inorganic reagent in these reactions. When silver is used as the oxidizing agent, it is reduced to metallic silver.

$$Ag^+ + e^- \longrightarrow Ag$$

Under appropriate conditions and using a silver reagent developed by Professor Bernhard Tollens (and called, logically enough, **Tollens' reagent**), the metallic silver appears as a beautiful mirror on the inside of the reaction vessel (e.g., a test tube).

Copper(II) ion is blue in aqueous solution. When such a solution is used to oxidize aldehydes: the copper(II) ion is converted to copper(I) ion. The copper(I) ion precipitates from solution as copper(I) oxide, Cu_2O, which is a brick-red solid. Thus, if an aldehyde is present to reduce copper(II) to copper(I), the reagent changes from a clear blue solution to a red-orange precipitate. Two copper reagents have been developed for detecting the presence of the aldehyde functional group. One is called **Fehling's solution**, and the other, **Benedict's solution**. Both contain copper(II) ion and differ only in the anion used to keep the copper in solution until it is reduced.

Tollens', Benedict's, and Fehling's reagents are all used as simple chemical tests for aldehydes because, in each case, reaction is accompanied by a readily observed change in the appearance of the reagents. The test with copper is frequently used to detect the presence of the sugar glucose (which contains an aldehyde group) in urine, for

10.14 Chemical Properties of Aldehydes and Ketones

example, as a way of monitoring the condition of someone with diabetes.

Ketones as well as aldehydes can participate in the last reaction we shall discuss in this chapter. Alcohols add to the carbonyl double bond in these compounds. An understanding of this reaction is critical to an understanding of the structure of carbohydrates. Since carbohydrates are relatively complex molecules, we shall first examine the reaction using simpler compounds as models.

When we say that an alcohol adds to a carbonyl group, we mean that the hydrogen atom of the alcohol's hydroxyl group pulls loose and attaches to the oxygen atom of the carbonyl group. The remainder of the alcohol attaches to the carbon atom of the carbonyl through the hydroxyl's oxygen atom. That verbal description of the process should make you appreciate all the more the simple clarity of a chemical equation.

$$R-\underset{H}{\overset{O}{\overset{\|}{C}}} + H-O-R' \rightleftharpoons R-\underset{H}{\overset{OH}{\underset{|}{C}}}-OR'$$

The product of this reaction is called a **hemiacetal** and is generally quite unstable. It readily reverts to the aldehyde and alcohol.

Reaction between an aldehyde and an alcohol yields a hemiacetal.

Hemiacetals can be forced to react further with alcohols to form **acetals**. A second molecule of alcohol reacts with the hemiacetal in the presence of dry hydrogen chloride (which serves as an acid catalyst). The overall reaction for acetaldehyde (ethanal) and methanol is

$$CH_3-\overset{O}{\overset{\|}{C}}-H + H-O-CH_3 \rightleftharpoons CH_3-\underset{H}{\overset{OH}{\underset{|}{C}}}-OCH_3 \xrightarrow[\text{dry HCl}]{H-O-CH_3} CH_3-\underset{H}{\overset{OCH_3}{\underset{|}{C}}}-OCH_3 + H_2O$$

A hemiacetal   An acetal

Note that the first part of the reaction is a typical addition reaction; the double bond becomes a single bond. The second part is a typical dehydration reaction, an ether formation.

Unlike hemiacetals, acetals are quite stable and can be isolated in pure form. Acetals (and **ketals**, formed from ketones) can be converted back to the original starting materials by treatment with aqueous acids.

$$CH_3-\underset{H}{\overset{OCH_3}{\underset{|}{C}}}-OCH_3 + H_2O \xrightarrow{H^+} CH_3-\overset{O}{\overset{\|}{C}}-H + 2\ CH_3OH$$

The formation of hemiacetals involves the reaction of two different functional groups, a carbonyl group and a hydroxyl group. What if both are located on the same molecule? The same reaction takes place.

$$\underset{\text{CH}_2\text{CH}_2\text{CH}_2\overset{\text{OH}}{\text{C}}\text{H}}{} \longrightarrow \text{ring form with O, OH}$$

Perhaps the reaction is easier to visualize if the starting material is bent to resemble the arrangement of the product.

This is precisely the reaction that occurs in carbohydrates. Ribose is one of many carbohydrate molecules that form *intra*molecular (within a molecule) hemiacetals.

Ribose (open-chain form) → Ribose (ring form)

Once again, let us caution you not to be overwhelmed by complex structures. Always look where the action is. This is the same reaction we first illustrated with acetaldehyde and methanol. The hydroxyl hydrogen atom transfers to the carbonyl oxygen atom. The hydroxyl oxygen atom attaches to the carbonyl carbon atom. All the other parts of this molecule (including all the other hydroxyl groups) are merely spectators; they are not involved in the reaction. This is an important lesson to learn. When we do finally concentrate on biochemical reactions, you should recognize the same chemistry you have seen before. We will find more complex compounds but the same fundamental chemistry.

CHAPTER SUMMARY

Functional groups are groupings of atoms that confer characteristic chemical and physical properties on the organic compounds in which they are found. The carbon–carbon double bond (C=C), hydroxyl group (—OH), and carbonyl group (C=O) are three examples of functional groups.

Three families of oxygen-containing organic compounds can be considered as derivatives of water: alcohols, phenols, and ethers. Alcohols are those compounds in which one hydrogen atom of the water molecule is replaced by an aliphatic group. In phenols, one hydrogen atom of the water molecule has been replaced by an aromatic group. If both hydrogen atoms of water are replaced by aliphatic or aromatic groups, the compound is an ether.

The common names of alcohols consist of the name of the alkyl group attached to the oxygen followed by *alcohol*. In the IUPAC system, alcohols are named in the same way as alkanes except that the family ending is *-ol* (instead of *-e*) and the position of the hydroxyl group is specified by counting from the nearest end of the parent chain. Alcohols may be classified as primary, secondary, or tertiary according to the rules illustrated in Figure 10.3. Wood alcohol (methanol), grain alcohol (ethanol), and rubbing alcohol (2-propanol) are common names that indicate the source or use of the particular alcohol. The term *alcohol* is frequently used for the compound ethanol. The common polyfunctional alcohols include ethylene glycol, propylene glycol, and glycerol.

The names of ethers consist of the names of the two groups attached to the oxygen atom (given in alphabetical order) followed by *ether*. Phenols are named as derivatives of the parent compound *phenol* or possess unique names. Phenol is also called carbolic acid.

Because they can form hydrogen bonds, alcohols have higher boiling points than ethers, but the smaller members of both families show appreciable solubility in water, a hydrogen-bonding solvent. The hydroxyl group of alcohols also gives them more chemical reactivity than that possessed by ethers. Alcohols can dehydrate (lose a molecule of water) to form either alkenes or, if two alcohol molecules are involved, ethers. Primary alcohols can be oxidized to aldehydes, which can in turn be oxidized to carboxylic acids. Secondary alcohols are oxidized to ketones. Tertiary alcohols resist oxidation.

Ethanol is produced in the fermentation of carbohydrates. The product of fermentation never exceeds a concentration in water of 18%, but higher concentrations can be obtained by distillation. The concentration of alcohol may be reported in proof, which is twice the numerical value in percent. Ethanol is a depressant but is less toxic than methanol. Both compounds are oxidized by liver enzymes, first to aldehydes (which cause most of the physiological damage) and ultimately to carbon dioxide and water. Because ethanol intended for drinking is highly taxed, untaxed industrial ethanol is denatured by adding other chemicals.

Ethylene glycol, which is used in antifreeze, is also extremely toxic because it is oxidized in the liver to oxalic acid. This compound precipitates as its calcium salt in the kidneys, inflicting serious renal damage. Many phenols are used as antiseptics and disinfectants. Diethyl ether (sometimes referred to as ether) is an effective general anesthetic. In current use as anesthetics are several halogen-containing compounds, among them some ethers. Modern anesthesiology often combines a sedative and a muscle relaxant with a general anesthetic for the best results.

Both aldehydes and ketones incorporate a carbonyl group. A hydrogen atom must be attached to the carbonyl carbon atom in an aldehyde. In ketones, the carbonyl is attached to two carbon-containing groups. The common names of aldehydes incorporate special stems (shown in Figure 10.9) and end in *aldehyde*. The family ending in the IUPAC system is *-al*. The common names of ketones consist of the names of the two groups attached to the carbonyl group followed by *ketone*. The family ending for IUPAC names is *-one*. Acetone is another name for propanone.

The polar carbonyl group gives aldehydes and ketones boiling points intermediate between those of ethers and alcohols. As is true of other oxygen-containing families, the smaller aldehydes and ketones are appreciably soluble in water.

Aldehydes are the most easily oxidized organic compounds and will react with a number of weak oxidizing agents, such as Tollens' reagent (a form of

Ag$^+$) and Fehling's and Benedict's reagents (forms of Cu^{2+}). Both aldehydes and ketones can react with alcohols to form hemiacetals and acetals or hemiketals and ketals.

CHAPTER QUIZ

1. Which of the following is an alcohol?
   a. H—OH
   b. R—OH
   c. Ar—OH
   d. K—OH

2. In secondary alcohols the carbon in —COH is attached to how many other carbons?
   a. none
   b. one
   c. two
   d. three

3. Isobutyl alcohol belongs to which class of alcohols?
   a. linear
   b. primary
   c. secondary
   d. tertiary

4. Oxidation of secondary alcohols yields
   a. aldehydes
   b. ketones
   c. acids
   d. no new product

5. Which is *not* an ether?
   a. R—O—R
   b. R—O—Ar
   c. R—O—H
   d. Ar—O—Ar

6. Which of the following exhibit no hydrogen bonding to other molecules of the same compound?
   a. alcohols
   b. phenols
   c. ethers
   d. glycols

7. Which is a common modern hospital anesthetic?
   a. halothane
   b. chloroform
   c. nitrous oxide
   d. ether

8. Which is an aldehyde?
   a. R—C(=O)—R
   b. R—C(=O)—Ar
   c. H—C(=O)—R
   d. Ar—C(=O)—R

9. Which is another name for ethanal?
   a. formaldehyde
   b. acetaldehyde
   c. propionaldehyde
   d. butyraldehyde

10. For similar molecular weights, which have the highest boiling point?
    a. aldehydes
    b. ketones
    c. ethers
    d. alcohols

PROBLEMS

10.1 Functional Groups

1. What is a functional group?

2. Give the structure of and name the functional group in
   a. alkenes
   b. alcohols
   c. ketones

10.2 The Alcohols: Nomenclature

3. Write structural formulas for the eight isomeric pentyl alcohols (C$_5$H$_{12}$O). (*Hint*: Three are derived from pentane, four are derived from isopentane, and one is derived from neopentane.) Name the alcohols by the IUPAC system.

4. Classify the alcohols in Problem 3 as primary, secondary, or tertiary.

5. Name these compounds.
   a. CH$_3$CH$_2$CH$_2$CH$_2$CH(OH)CH$_3$
   b. CH$_3$CH(CH$_3$)CH$_2$OH
   c. CH$_3$CH(OH)CH$_3$
   d. CH$_3$CH$_2$CH$_2$CH$_2$CH$_2$CH$_2$OH

6. Write structural formulas for the following.
   a. *tert*-butyl alcohol
   b. 3-hexanol
   c. 3,3-dimethyl-2-butanol
   d. pentyl alcohol
   e. 4-methyl-2-hexanol

7. Give the IUPAC names for the compounds referred to by these names.
   a. grain alcohol
   b. wood alcohol
   c. rubbing alcohol

10.3 Physical Properties of Alcohols

8. Without consulting tables, arrange the compounds in order of increasing boiling points: ethanol, 1-propanol, methanol.

9. Without consulting tables, arrange the compounds in order of increasing solubility in water: methanol, 1-butanol, 1-octanol.

10.4 Chemical Properties of Alcohols

10. Classify each of these conversions as oxidation, dehydration, or hydration. (Only the organic starting material and product are shown.)

   a. $CH_3OH \rightarrow H-\overset{\overset{O}{\|}}{C}-H$

   b. $CH_3\overset{\overset{OH}{|}}{C}HCH_3 \rightarrow CH_3CH=CH_2$

   c. $CH_3\overset{\overset{OH}{|}}{C}HCH_3 \rightarrow CH_3\overset{\overset{O}{\|}}{C}CH_3$

   d. $HOOCCH=CHCOOH \rightarrow HOOCCH_2\overset{\overset{OH}{|}}{C}HCOOH$

   e. $2\ CH_3OH \rightarrow CH_3OCH_3$

11. Each of the butyl alcohols is treated with potassium dichromate in acid. Draw the product (if any) expected from each of the four isomeric alcohols.

12. Write an equation for the dehydration of 2-propanol to yield an alkene.

13. Write an equation for the dehydration of 2-propanol to yield an ether.

14. Draw the ether that would result from the *intra*molecular dehydration of $HOCH_2CH_2CH_2CH_2CH_2OH$.

15. Draw the alkene that would form from the dehydration of cyclohexanol.

10.5 Physiological Properties of Alcohols

16. Which of the compounds in Problem 7 is most toxic? Which is least toxic?

17. What is denatured alcohol? Why is some alcohol denatured?

18. Is 90-proof liquor the product of simple fermentation, or should it be called distilled spirits?

19. Why is methanol so much more toxic to humans than ethanol?

10.6 Multifunctional Alcohols

20. Write the structural formulas for the following.
    a. propylene glycol   b. glycerol

21. Without consulting tables, arrange the compounds in order of increasing boiling points: butane, ethylene glycol, 1-propanol.

22. Without consulting tables, arrange the compounds in order of increasing boiling points: diethyl ether, propylene glycol, 1-butanol.

23. Without consulting tables, arrange the compounds in order of increasing solubility in water: pentane, propylene glycol, diethyl ether.

24. Why is ethylene glycol so much more toxic to humans than propylene glycol?

25. What chemical compound is used in the treatment of acute methanol or ethylene glycol poisoning? How does it work?

10.7 The Phenols

26. Name these compounds.

    a. (phenol with Cl at meta position, OH)   b. (phenol with NO_2 at ortho position, OH)

27. Write structural formulas for the following.
    a. *m*-iodophenol
    b. *p*-methylphenol (*p*-cresol)
    c. 2,4,6-trinitrophenol (picric acid)

28. Draw the structural formula for carbolic acid, and give another name for this compound.

10.8 The Ethers

29. Draw and name the isomeric ethers with the formula $C_5H_{12}O$.

30. Name these compounds.
    a. $CH_3CH_2CH_2OCH_2CH_2CH_3$
    b. $CH_3CH_2OCHCH_2CH_3$
        $\qquad\qquad\quad |$
        $\qquad\qquad\ CH_3$

31. Write structural formulas for the following.
    a. ethyl methyl ether
    b. diisopropyl ether
    c. isopropyl phenyl ether

10.9 General Anesthetics

32. What is a general anesthetic?

33. Which of these anesthetics is dangerous because of its flammability?
    a. diethyl ether   b. halothane
    c. chloroform   d. cyclopropane

34. For each of the following anesthestics, describe a disadvantage associated with its use. Do not include flammability.
    a. nitrous oxide   b. halothane
    c. diethyl ether

35. What application does curare have in modern anesthesiology?

10.10 The Carbonyl Group

36. Draw structures of and name the four isomeric aldehydes having the formula $C_5H_{10}O$.

37. Draw structures and give common and IUPAC names for the three isomeric ketones having the formula $C_5H_{10}O$.

10.11 Naming Common Aldehydes

38. Give suitable names to the following.

    a. benzaldehyde structure (C6H5-CHO)   b. $CH_3CH_2\overset{O}{\underset{\|}{C}}-H$

    c. $CH_3\overset{CH_3}{\underset{\underset{CH_3}{|}}{C}}CH_2\overset{O}{\underset{\|}{C}}-H$

39. Write structural formulas for the following.
    a. valeraldehyde
    b. 3-methylheptanal
    c. p-nitrobenzaldehyde

10.12 Naming Common Ketones

40. Give suitable names to the following.

    a. $CH_3CH_2\overset{O}{\underset{\|}{C}}CH_2\overset{}{\underset{\underset{CH_3}{|}}{C}}HCH_3$

    b. cyclopentanone structure

41. Write structural formulas for the following.
    a. 4-methylcyclohexanone
    b. isobutyl *tert*-butyl ketone
    c. 2,4-dimethyl-3-pentanone

10.13 Physical Properties of Aldehydes and Ketones

42. Which compound has the higher boiling point: acetone or 2-propanol?

43. Which compound has the higher boiling point: butanal or 1-butanol?

44. Which compound has the higher boiling point: dimethyl ether or acetaldehyde?

10.14 Chemical Properties of Aldehydes and Ketones

45. Write the equation for the reaction of acetaldehyde with each of the following.
    a. 1 mol of CH_3OH
    b. 2 mol of CH_3OH, with dry HCl present
    c. 1 mol of $HOCH_2CH_2OH$, with dry HCl present
    d. Ag^+
    e. Cu^{2+}
    f. $K_2Cr_2O_7$

46. Write the equation for the reaction, if any, of acetone with the reagents in Problem 45.

47. Indicate whether Tollens' reagent could be used to distinguish between the compounds in each set. Explain your reasoning.
    a. 1-pentanol and pentanal
    b. 2-pentanol and 2-pentanone
    c. pentanal and 2-pentanone
    d. pentanal and pentane
    e. 2-pentanone and pentane

48. Assume that a *stronger* oxidizing agent, such as $K_2Cr_2O_7$, could be used as a test for distinguishing among compounds. For each set in Problem 47, indicate whether this reagent would distinguish between the two compounds. Explain your reasoning.

49. Draw the hemiacetal formed from the intramolecular reaction of

    $HOCH_2CH_2CH_2CH_2\overset{O}{\underset{\|}{C}}-H$

11

ORGANIC ACIDS AND DERIVATIVES

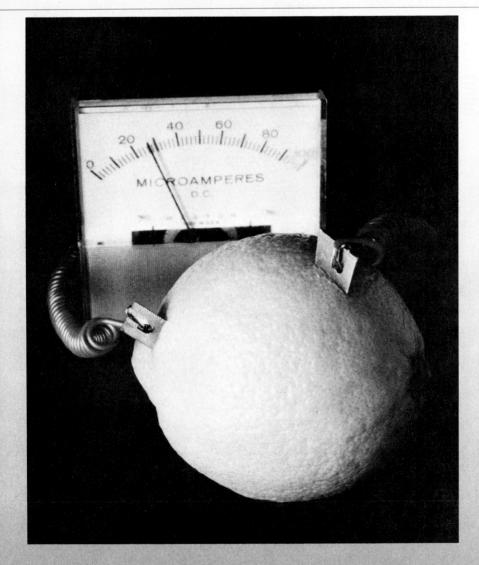

DID YOU KNOW THAT...

- a red ant's bite injects an acid?
- fermentation of cider and honey produces vinegar?
- esters and amides are derivatives of carboxylic acids?
- acids dimerize because of hydrogen bonding?
- soaps are salts of long-chain acids?
- aspirin was first produced in the United States in 1899?
- heroin is made from morphine?
- some hallucinogenic and psychedelic drugs are amides?

Organic acids were known long before inorganic acids, such as HCl and H₂SO₄, were isolated. When primitive people tried to produce alcohol by fermentation, they sometimes got vinegar (which contains acetic acid) instead. Naturalists of the seventeenth century knew that the sting of a red ant's bite was due to an organic acid that the pest injected into the wound. It has long been recognized that the crisp, tart flavor of citrus fruits was produced by an organic compound appropriately called citric acid. The acetic acid of vinegar, the formic acid of red ants, and the citric acid of fruits all belong to the same family of compounds, the carboxylic acids.

A number of derivatives of carboxylic acids are also important. Amides, of which proteins are perhaps the most spectacular example, and esters, which include fats, are two classes of acid derivatives that we shall consider in this chapter. As usual, we shall concentrate on relatively simple acids, esters, and amides, postponing our examination of the more complex worlds of proteins and fats (with one exception) until after we complete our survey of the families of organic compounds.

11.1 Acids and Their Derivatives: The Functional Groups

We spoke of the carbonyl group in the previous chapter, and we noted that it is this functional group that determines the chemistry of aldehydes and ketones. The carbonyl group is also incorporated in carboxylic acids and the derivatives of carboxylic acids. The carbonyl group in these compounds is only one part of the functional group that characterizes the families, however.

The functional group of the **carboxylic acids** is the **carboxyl** group.

$$-C\underset{OH}{\overset{O}{\lessgtr}} \quad \text{or} \quad -COOH$$

This group can be considered a combination of the *carb*onyl group ($>C=O$) and the hydr*oxyl* group ($-OH$), but it has characteristic properties of its own.

In the **amides**, the functional group is

$$-C\underset{N-}{\overset{O}{\lessgtr}}$$

Here we have the carbonyl group attached to a nitrogen atom. The properties of the amide functional group are different from those of the

11.2 Some Common Acids: Structures and Names

TABLE 11.1 Carboxylic Acid Derivatives

Family	Functional Group	Example	Common Name	IUPAC Name
Carboxylic acid	—C(=O)—OH	CH₃C(=O)—OH	Acetic acid	Ethanoic acid
Amide	—C(=O)—N—	CH₃C(=O)—NH₂	Acetamide	Ethanamide
Ester	—C(=O)—O—C—	CH₃C(=O)—OCH₃	Methyl acetate	Methyl ethanoate

simple carbonyl group and those of simple nitrogen-containing compounds, called amines (Chapter 12).

The functional group of the **esters** looks a little like that of an ether and a little like that of a carboxylic acid.

As you should now suspect, compounds incorporating this group do not behave like acids or ethers. Esters are a distinctive family of compounds.

All the families we shall discuss in this chapter, except for the carboxylic acids themselves, are regarded as derived from the acids. In each case, the hydroxyl group of the acid functional group is replaced with some other group in the derivative. Table 11.1 gathers all these functional groups in one location to permit you to compare and contrast the various groups more easily. The table also offers an example (with names) for each type of compound. We shall consider nomenclature in more detail as we take up each of these families separately.

11.2 SOME COMMON ACIDS: STRUCTURES AND NAMES

Most of the organic acids that we shall consider are derived from natural sources and have common names associated with these sources. We have already encountered a variation of these common names in our study of aldehydes. The names of acid derivatives are also based on the common names of the acids. In many ways, the common nomenclature of carboxylic acids will serve the same basic function as the IUPAC rules for naming alkanes. Once you have learned the common names for the acids, the naming of derivatives will involve only slight modifications of the original acid names.

The simplest organic acid is formic acid (from the Latin *formica*, "ant"). It was first obtained by the destructive distillation of ants. The structure of formic acid is

$$H-\overset{\overset{O}{\|}}{C}-OH$$

Aerobic: with air

Acetic acid is the two-carbon acid. It can be made by the **aerobic** ("with air") fermentation of a mixture of cider and honey. This reaction produces a solution of vinegar that contains about 4–10% acetic acid plus a number of other compounds that give the vinegar flavor. The structure of acetic acid is

$$H-\overset{\overset{H}{|}}{\underset{\underset{H}{|}}{C}}-\overset{\overset{O}{\|}}{C}-OH \quad \text{or} \quad CH_3COOH$$

It is probably the most familiar *weak* acid used in educational and industrial chemistry laboratories.

Table 11.2 lists several members of the family of carboxylic acids and the derivation of their common names.

Although IUPAC names are seldom encountered in everyday use, they are readily derived from the names of the alkanes with the same number of carbon atoms. Just drop the *-e* from the alkane name and add *-oic acid*. Thus, the IUPAC name for formic acid (HCOOH) is methanoic acid, and that for acetic acid (CH_3COOH) is ethanoic acid. Table 11.2 also presents IUPAC names for other simple (though not necessarily small) members of the family.

Nature prefers an even number of carbon atoms in its acids. Therefore, beyond six carbon atoms, we have listed only the commonly isolated, even-numbered acids.

EXAMPLE 11.1

Give the IUPAC name for

$$CH_3CH_2\underset{\underset{CH_3}{|}}{CH}COOH$$

11.2 Some Common Acids: Structures and Names

TABLE 11.2 Some Common Aliphatic Acids

Condensed Formula	IUPAC Name	Common Name	Derivation of Common Name
HCOOH	Methanoic acid	Formic acid	Latin *formica*, "ant"
CH$_3$COOH	Ethanoic acid	Acetic acid	Latin *acetum*, "vinegar"
CH$_3$CH$_2$COOH	Propanoic acid	Propionic acid	Greek *protos*, "first," and *pion*, "fat"
CH$_3$CH$_2$CH$_2$COOH	Butanoic acid	Butyric acid	Latin *butyrum*, "butter"
CH$_3$(CH$_2$)$_3$COOH	Pentanoic acid	Valeric acid	Named for its occurrence in the root of valerian, a medicinal herb
CH$_3$(CH$_2$)$_4$COOH	Hexanoic acid	Caproic acid	
CH$_3$(CH$_2$)$_6$COOH	Octanoic acid	Caprylic acid	Latin *caper*, "goat"
CH$_3$(CH$_2$)$_8$COOH	Decanoic acid	Capric acid	
CH$_3$(CH$_2$)$_{10}$COOH	Dodecanoic acid	Lauric acid	Laurel tree
CH$_3$(CH$_2$)$_{12}$COOH	Tetradecanoic acid	Myristic acid	*Myristica fragrans* (nutmeg)
CH$_3$(CH$_2$)$_{14}$COOH	Hexadecanoic acid	Palmitic acid	Palm tree
CH$_3$(CH$_2$)$_{16}$COOH	Octadecanoic acid	Stearic acid	Greek *stear*, "tallow"

The longest continuous chain contains four carbon atoms; the compound is therefore named as a substituted butanoic acid. The methyl substituent is at the second carbon atom. The compound is 2-methylbutanoic acid.

EXAMPLE 11.2

Draw chloroacetic acid.
 Acetic acid contains two carbon atoms.

$$-\overset{|}{\underset{|}{C}}-\overset{O}{\underset{}{\overset{\|}{C}}}-OH$$

The carboxyl group already has four bonds. Therefore, the chlorine atom must be on the remaining carbon atom.

$$H-\overset{H}{\underset{Cl}{\overset{|}{C}}}-\overset{O}{\overset{\|}{C}}-OH$$

The simplest aromatic acid is called benzoic acid.

C$_6$H$_5$—COOH

TABLE 11.3 Some Common Dicarboxylic Acids

Structure	IUPAC Name	Common Name	Derivation of Common Name
HOOC—COOH	Ethanedioic acid	Oxalic acid	Greek *oxys*, "sharp" or "acid"
HOOC—CH_2—COOH	Propanedioic acid	Malonic acid	From malic acid (Latin *malum*, "apple")
HOOC$(CH_2)_2$COOH	Butanedioic acid	Succinic acid	Latin *succinum*, "amber"
HOOC$(CH_2)_3$COOH	Pentanedioic acid	Glutaric acid	From glutamic acid (Latin *gluten*, "glue")
HOOC$(CH_2)_4$COOH	Hexanedioic acid	Adipic acid	Latin *adeps*, "fat"
o-C₆H₄(COOH)₂	1,2-Benzenedicarboxylic acid	Phthalic acid	From na*phth*alene
p-HOOC—C₆H₄—COOH	1,4-Benzenedicarboxylic acid	Terephthalic acid	From na*phth*alene

Others are called by common names or may be named as derivatives of benzoic acid.

A number of common dicarboxylic acids, both aliphatic and aromatic, are presented in Table 11.3. Again, the common names are the names most often used for these compounds.

11.3 Physical Properties of Carboxylic Acids

Carboxylic acids show strong intermolecular hydrogen bonding.

Carboxylic acids are highly polar and exhibit strong intermolecular hydrogen bonding. Consequently, these compounds have higher boiling points than even the alcohols of comparable formula weights. Ethanol (with a formula weight of 46) boils at 78 °C, while formic acid (with the same formula weight) boils at 100 °C. Similarly, 1-propanol (with a formula weight of 60) boils at 97 °C, while acetic acid (with the same formula weight) boils at 118 °C.

Good evidence exists that, even in the vapor phase, some of the hydrogen bonds between acid molecules are not broken. The particular structure of the carboxyl group permits two molecules to hydrogen-bond strongly to each other.

11.4 Chemical Properties of Carboxylic Acids

$$R-C\overset{O\cdots\cdots H-O}{\underset{O-H\cdots\cdots O}{}}C-R$$

In many situations, the interaction is so strong that the **dimer** (the two-molecule unit) acts as a single particle. Solutions of carboxylic acids exhibit smaller freezing point depressions, lower osmotic pressures, and so on, than one would expect based on the concentration of the acid. That is because we are counting as two separate molecules a combination that is behaving as one piece.

The carboxyl groups readily hydrogen-bond to water molecules. The acids having one to four carbon atoms are colorless liquids that are completely miscible with water. Solubility decreases with an increasing number of carbon atoms. Hexanoic acid, $CH_3(CH_2)_4COOH$, is soluble only to the extent of 1.0 g per 100 g of water. Palmitic acid, $CH_3(CH_2)_{14}COOH$, is essentially insoluble.

Most of the carboxylic acids have irritating, obnoxious odors. Formic acid, HCOOH, has a very sharp and penetrating odor. The familiar odor of vinegar is a weakened version of the odor of acetic acid, CH_3COOH. Butyric acid, $CH_3(CH_2)_2COOH$, smells like rancid butter, and the aroma of valeric acid, $CH_3(CH_2)_3COOH$, has been descriptively pinpointed as "essence of old gym sneakers." With acids of higher formula weight, the odor weakens because the compounds are less volatile.

Pure acetic acid freezes at 16.6 °C. Since this is only slightly below normal room temperature (about 20 °C), acetic acid solidifies when cooled only slightly. In poorly heated laboratories of a century or so ago (and in some laboratories of our increasingly energy-conscious world), acetic acid was often found frozen on the reagent shelf. For that reason, pure acetic acid (sometimes referred to as concentrated acetic acid) came to be known as glacial acetic acid, a name that survives to this day.

11.4 CHEMICAL PROPERTIES OF CARBOXYLIC ACIDS

In Chapter 8, we noted that acids turn blue litmus red, neutralize bases, and taste sour. Of the three properties of acids listed, you are probably most familiar with the last—the sour taste. Vinegar is sour because it contains acetic acid. Grapefruits and lemons are sour because they contain citric acid. Sour milk contains lactic acid. The acids we eat are, for the most part, organic acids. Historically, the first organic acids came from plant or animal matter, that is, from organisms. The strong acids we encountered in Chapter 8, such as hydrochloric acid (HCl), nitric acid (HNO_3), and sulfuric acid (H_2SO_4), are made from minerals and hence are called **mineral acids**.

Acids:
▶ blue litmus → red
▶ neutralize bases
▶ taste sour

Most carboxylic acids are weak acids. They ionize only slightly in aqueous solution.

$$R-COOH + H_2O \rightleftharpoons R-COO^- + H_3O^+$$

Acetic acid was one of the weak acids listed in Chapter 8. Those carboxylic acids that are water-soluble (like acetic acid) form solutions that change the color of litmus from blue to red. Carboxylic acids react with bases such as sodium hydroxide, sodium carbonate, and sodium bicarbonate to form salts.

$$R-COOH + NaOH \longrightarrow R-COO^- Na^+ + H_2O$$
$$2\,R-COOH + Na_2CO_3 \longrightarrow 2\,R-COO^- Na^+ + H_2O + CO_2$$
$$R-COOH + NaHCO_3 \longrightarrow R-COO^- Na^+ + H_2O + CO_2$$

Carboxylic acids that are insoluble in water dissolve in aqueous solutions of hydroxide, carbonate, or bicarbonate ions because the insoluble acids react to form ionic salts that *are* water-soluble. The behavior of decanoic acid is illustrated in Figure 11.1. The reaction

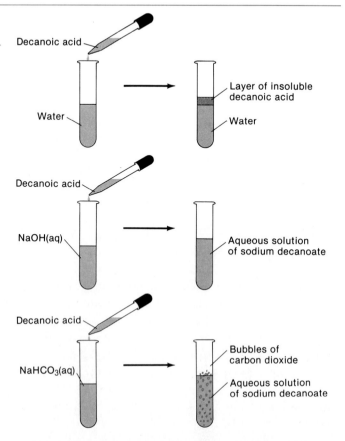

FIGURE 11.1
Decanoic acid is insoluble in water but soluble in aqueous sodium hydroxide or bicarbonate.

with sodium bicarbonate can be used as a simple chemical test for carboxylic acids. An organic compound of unknown identity is mixed with an aqueous solution of sodium bicarbonate. If the compound dissolves with the evolution of carbon dioxide gas, the unknown may be a carboxylic acid.

Carboxylic acids react with bases to form salts. We have encountered one of these salts, sodium acetate, in a number of previous chapters. If you remember the name of this salt, you will have a pattern for naming the other salts of carboxylic acids. As is usual with salts, the cation is named first. The anion is named by changing the *-ic* ending of the corresponding acid name to *-ate*. Either the common or the IUPAC name for the acid may be used.

$$CH_3-\overset{O}{\underset{\|}{C}}-O^-\ Na^+ \qquad \underset{}{\bigcirc}-\overset{O}{\underset{\|}{C}}-O^-\ NH_4^+ \qquad \left(CH_3CH_2-\overset{O}{\underset{\|}{C}}-O^-\right)_2 Ca^{2+}$$

Sodium acetate   Ammonium benzoate   Calcium propionate
(sodium ethanoate)   (calcium propanoate)

11.5 ESTERS: STRUCTURES AND NOMENCLATURE

Esters are the products of the dehydration of a carboxylic acid and an alcohol (or phenol).

Dehydration of a carboxylic acid and an alcohol results in an ester.

$$\underset{\text{Carboxylic acid}}{R-\overset{O}{\underset{\|}{C}}-OH} + \underset{\text{Alcohol}}{HO-R'} \xrightarrow{H^+} \underset{\text{Ester}}{R-\overset{O}{\underset{\|}{C}}-OR'} + H_2O$$

Ester formation (**esterification**) is very similar to the dehydration reaction described in Section 10.4, in which two alcohol molecules yielded an ether.

If you think of esters as derived from acids and alcohols, the naming of esters is logical. First, name the carbon group derived from the alcohol (or phenol), designated by R' in the above equation. Then name the part of the ester derived from the acid

$$R-\overset{O}{\underset{\|}{C}}-O-$$

as if it were an anion (Section 11.4). Figure 11.2 lists some esters of interest and their names.

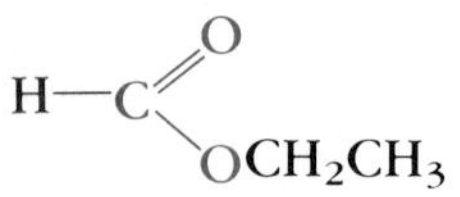

$$H-\underset{OCH_2CH_3}{\overset{O}{\overset{\|}{C}}}$$

Ethyl formate
(ethyl methanoate)

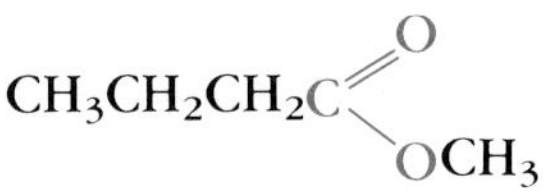

$$CH_3CH_2CH_2\underset{OCH_3}{\overset{O}{\overset{\|}{C}}}$$

Methyl butyrate
(methyl butanoate)

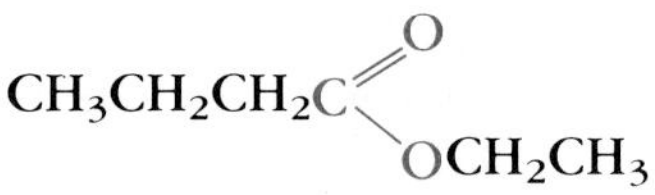

$$CH_3CH_2CH_2\underset{OCH_2CH_3}{\overset{O}{\overset{\|}{C}}}$$

Ethyl butyrate
(ethyl butanoate)

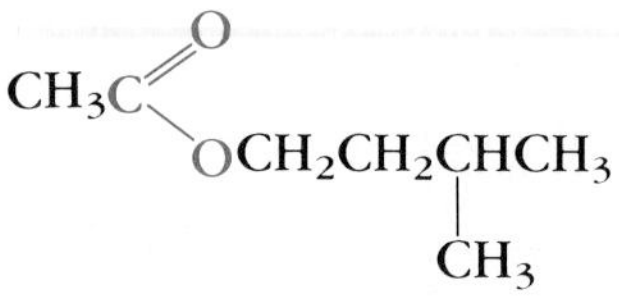

$$CH_3\underset{OCH_2CH_2CHCH_3}{\overset{O}{\overset{\|}{C}}}$$
$$CH_3$$

Isopentyl acetate
(isopentyl ethanoate)

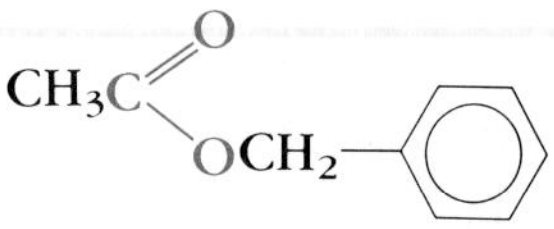

$$CH_3\underset{OCH_2-\text{C}_6H_5}{\overset{O}{\overset{\|}{C}}}$$

Benzyl acetate
(benzyl ethanoate)

$$CH_3\underset{OCH_2CH_2CH_2CH_2CH_2CH_2CH_2CH_3}{\overset{O}{\overset{\|}{C}}}$$

Octyl acetate
(octyl ethanoate)

FIGURE 11.2
Some esters of interest. Ethyl formate is an artificial rum flavor. Methyl butyrate occurs in apples, and ethyl butyrate occurs in pineapples. Isopentyl acetate is banana oil, used as a solvent and in flavoring, while benzyl acetate is oil of jasmine, used in perfumes. Octyl acetate occurs in oranges.

EXAMPLE 11.3

What are the common and IUPAC names for this ester?

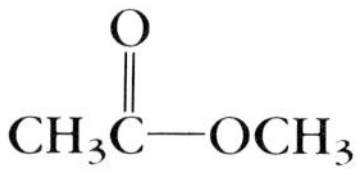

$$CH_3\overset{O}{\overset{\|}{C}}-OCH_3$$

The alkyl group attached to oxygen is methyl. The group

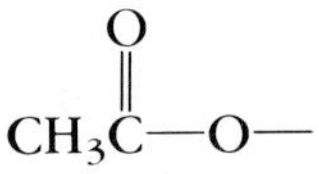

$$CH_3\overset{O}{\overset{\|}{C}}-O-$$

is derived from acetic acid (which has two carbon atoms). Its name is acetate. The compound is methyl acetate. The IUPAC name for the two-carbon acid unit is ethanoate. The IUPAC name for the ester is methyl ethanoate.

EXAMPLE 11.4

What is the name of this ester?

$$\underset{\text{phenyl benzoate}}{\text{C}_6\text{H}_5-\overset{\overset{\text{O}}{\|}}{\text{C}}-\text{O}-\text{C}_6\text{H}_5}$$

The group attached by a single bond to the oxygen atom of the carboxyl group is phenyl. The acid portion corresponds to the benzoate group (from benzoic acid). Therefore, the compound is phenyl benzoate.

EXAMPLE 11.5

Draw the structure for ethyl valerate.
  It is easier to start with the acid portion. Draw the valerate (five-carbon) group first.

$$\text{CH}_3\text{CH}_2\text{CH}_2\text{CH}_2\overset{\overset{\text{O}}{\|}}{\text{C}}-\text{O}-$$

Then attach the ethyl group to the bond that ordinarily holds the hydrogen atom in the carboxyl group.

$$\text{CH}_3\text{CH}_2\text{CH}_2\text{CH}_2\overset{\overset{\text{O}}{\|}}{\text{C}}-\text{O}-\text{CH}_2\text{CH}_3$$

11.6 Physical Properties of Esters

The molecules of an ester are polar but are incapable of hydrogen-bonding with one another. Esters thus have considerably lower boiling points than the isomeric carboxylic acids. As one might expect, the boiling points of esters are about intermediate between those of ketones and ethers of comparable formula weights.
  Ester molecules are capable of hydrogen-bonding to water molecules, rendering esters of low formula weight somewhat water-soluble. Borderline solubility occurs in those molecules that have three to five carbon atoms.
  The aromas of many esters are as pleasant as the odors of carboxylic acids are bad. The characteristic fragrances of many flowers and fruits are due to esters (Figure 11.2).

Esters are polar but cannot form hydrogen bonds.

11.7 Chemical Properties of Esters: Hydrolysis

Hydrolysis of an ester results in replacement of alkoxy (—OR) group with hydroxyl (—OH) group.

Unlike the acids from which they are formed, esters are neutral compounds that exhibit neither acidic nor basic properties. Esters typically undergo chemical reactions in which the **alkoxy** group (—OR′) is replaced by another group. One such reaction is **hydrolysis**, or splitting with water. Hydrolysis is catalyzed by either acid or base.

Acidic hydrolysis is the reverse of the esterification reaction.

$$R-\overset{O}{\underset{\|}{C}}-OR' + H_2O \xrightarrow{H^+} R-\overset{O}{\underset{\|}{C}}-OH + R'OH$$

In basic or alkaline hydrolysis,

$$R-\overset{O}{\underset{\|}{C}}-OR' + NaOH \longrightarrow R-\overset{O}{\underset{\|}{C}}-O^- Na^+ + R'OH$$

the salt of the acid is isolated. Acid formed in the splitting of the ester reacts immediately with base in the reaction mixture and produces the salt. The treatment of methyl benzoate with aqueous sodium hydroxide gives sodium benzoate and methyl alcohol.

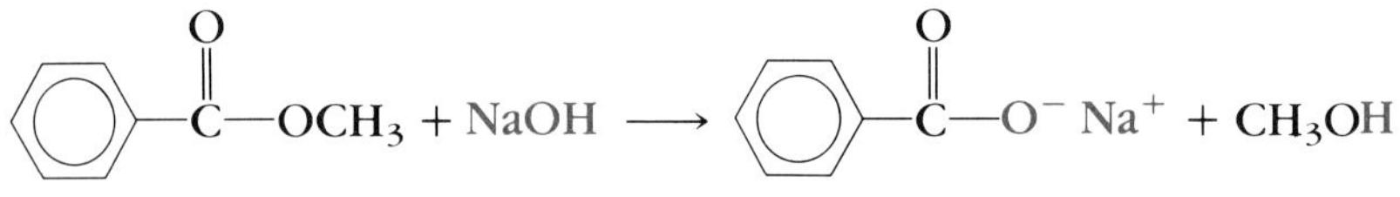

$$C_6H_5-\overset{O}{\underset{\|}{C}}-OCH_3 + NaOH \longrightarrow C_6H_5-\overset{O}{\underset{\|}{C}}-O^- Na^+ + CH_3OH$$

Saponification: the making of a soap (a salt of a fatty acid)

Alkaline hydrolysis of fats and oils (esters of glycerol and long-chain carboxylic acids) is called **saponification** (from the Latin *sapon-*, "soap"). Soaps are the salts of fatty acids (Figure 11.3).

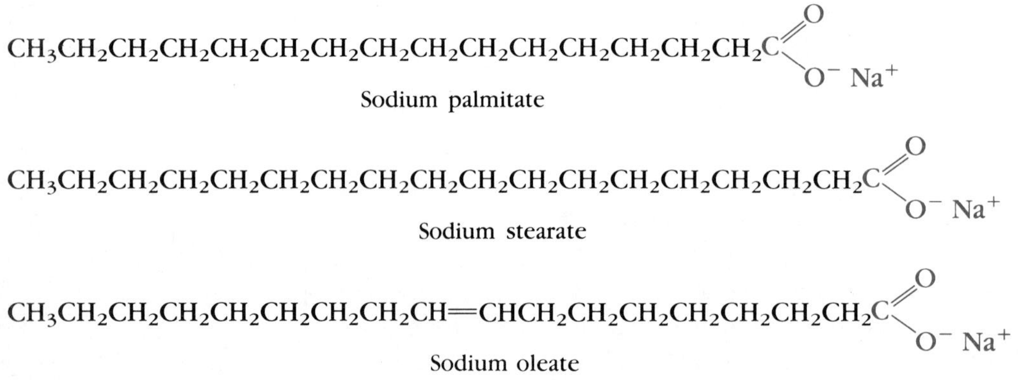

$$CH_3CH_2CH_2CH_2CH_2CH_2CH_2CH_2CH_2CH_2CH_2CH_2CH_2CH_2CH_2C\overset{O}{\underset{O^- Na^+}{\diagdown}}$$
Sodium palmitate

$$CH_3CH_2CH_2CH_2CH_2CH_2CH_2CH_2CH_2CH_2CH_2CH_2CH_2CH_2CH_2CH_2CH_2C\overset{O}{\underset{O^- Na^+}{\diagdown}}$$
Sodium stearate

$$CH_3CH_2CH_2CH_2CH_2CH_2CH_2CH_2CH=CHCH_2CH_2CH_2CH_2CH_2CH_2CH_2C\overset{O}{\underset{O^- Na^+}{\diagdown}}$$
Sodium oleate

FIGURE 11.3
Soap is a mixture of fatty acid salts such as these.

11.8 Physiological Properties of Selected Esters

People have long sought relief from pain. Alcohol, opium, cocaine, and Indian hemp (marijuana) were used as medicines for relief of pain in some early societies. The first successful synthetic pain relievers were derivatives of salicylic acid (Figure 11.4). Salicylic acid was first isolated from willow bark in 1860, although an English clergyman named Edward Stone had reported to the Royal Society as early as 1763 that an extract of willow bark was useful in reducing fever. Salicylic acid is itself a good **analgesic** (pain reliever) and **antipyretic** (fever reducer), but it is very sour and irritating when taken by mouth. Chemists sought to modify the structure of the molecule to remove this undesirable property while retaining (or even improving) the desirable properties.

Analgesic—pain reliever
Antipyretic—fever reducer

Sodium salicylate was first used in 1875. It was more pleasant than salicylic acid when taken by mouth but proved to be highly irritating to the lining of the stomach. Phenyl salicylate (salol) was introduced in 1886. It passes unchanged through the stomach. In the small intestine it is hydrolyzed to the desired salicylic acid, but phenol, which is quite toxic, is also formed. Acetylsalicylic acid (aspirin) was first introduced in 1899 and soon became the largest-selling drug in the world.

Aspirin is a registered trade name of the German Bayer Company. In Germany, Canada, and other countries, *aspirin* still means the Bayer brand. Other brands are sold as acetylsalicylic acid, or ASA. In the United States, Bayer has lost its rights to the trade name, and aspirin is now used as a generic name for acetylsalicylic acid. Aspirin is an analgesic and antipyretic. It reduces inflammation and hinders the

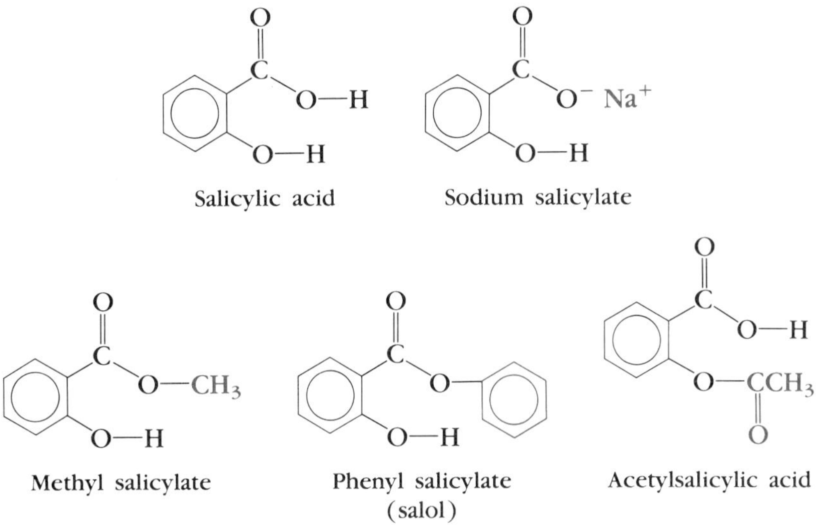

FIGURE 11.4
Salicylic acid and some of its derivatives.

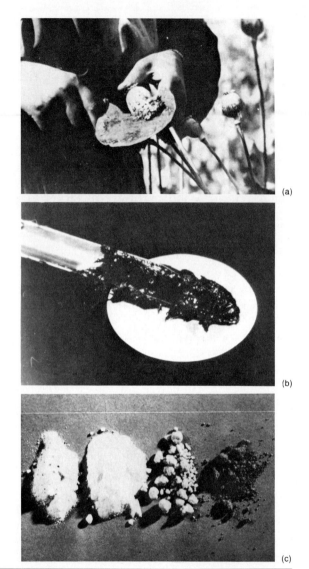

FIGURE 11.5
Morphine is the narcotic isolated from raw opium. Codeine and heroin are derivatives of morphine. (a) Scraping raw opium. (b) Opium gum. (c) Smoking opium, codeine, heroin, and morphine. [Courtesy of the U.S. Department of Justice, Drug Enforcement Administration, Washington, DC.]

clotting of blood. An aspirin *tablet* usually contains 325 mg (5 grains) of acetylsalicylic acid. The tablet is held together with some sort of inert binder, such as starch, clay, or sugar. Various brands of aspirin have been extensively tested. The conclusions of impartial studies are invariably the same: the only significant difference between brands is price.

Morphine, an opium alkaloid, is a **narcotic**, that is, a drug that produces both **sedation** (narcosis) and relief of pain (analgesia) (Figure 11.5). It is also strongly addictive. Chemists, therefore, also sought to modify the properties of morphine.

Morphine is both a phenol and an alcohol. When both of these groups are esterified, heroin (diacetylmorphine) is produced.

11.8 Physiological Properties of Selected Esters

Morphine

Heroin (diacetylmorphine)

Chemists at the Bayer Company first prepared this morphine derivative in 1874. Heroin received little attention until 1890, when it was proposed as an antidote for morphine addiction! Shortly thereafter, Bayer and others were advertising heroin as a sedative for coughs (Figure 11.6), often in the same advertisement with aspirin. It was soon found that heroin induced addiction more quickly than morphine, however, and that heroin addiction was harder to cure.

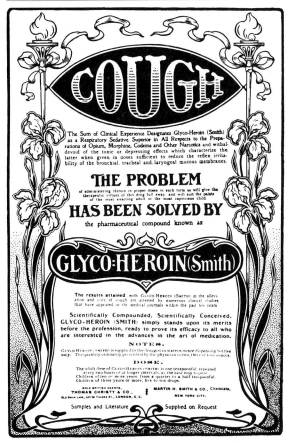

FIGURE 11.6
Heroin was advertised as a safe medicine in 1900. It was widely marketed as a sedative for coughs. [Courtesy of the U.S. Bureau of Narcotics and Dangerous Drugs, Washington, DC.]

The physiological action of heroin is similar to that of morphine, except that heroin produces a stronger feeling of euphoria for a longer period of time. Heroin is so strongly addictive that it seems that one or two injections are sufficient to induce dependence in some individuals. Heroin is not legal in the United States, even by prescription. Some individuals are lobbying for legalization of the drug for use with terminally ill people who have great pain.

Many thousands of morphine derivatives have been synthesized over the years. Only a few show significant analgesic activity. Most are strongly addictive. In fact, it now seems likely that the same molecular features that provide relief of pain also create chemical dependence.

11.9 Amides: Structures and Nomenclature

In the **amide** functional group, a nitrogen atom is attached to a carbonyl group.

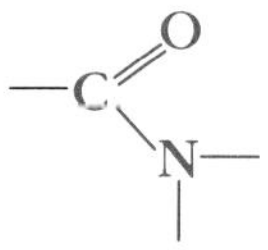

$$-C\underset{N-}{\overset{\displaystyle\parallel O}{}}$$

If the two remaining bonds to the nitrogen atom are attached to hydrogen atoms, the compound is called a **simple amide**. If one or both of the two remaining bonds to the nitrogen atom are attached to alkyl or aromatic groups, the compound is called a **substituted amide**.

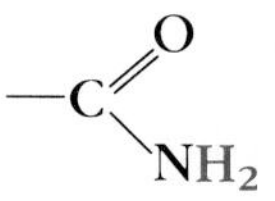

$$-C\overset{\displaystyle O}{\underset{NH_2}{}} \qquad -C\overset{\displaystyle O}{\underset{NHR}{}}$$

A simple amide 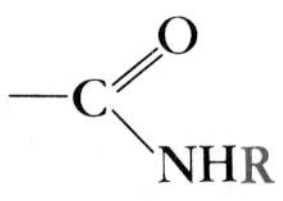   A substituted amide

The names of simple amides are formed by dropping the ending *-ic* (or *-oic*) from the name of the acid and adding the suffix *-amide*.

EXAMPLE 11.6

Name the compound

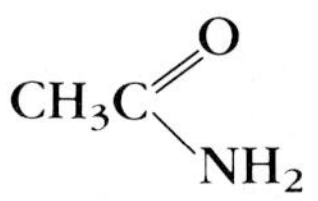

$$CH_3C\overset{\displaystyle O}{\underset{NH_2}{}}$$

11.9 Amides: Structures and Nomenclature

This amide is derived from acetic acid. Drop the *-ic* suffix, attach the ending *-amide*, and you have the name: acetamide (or ethanamide in the IUPAC system).

EXAMPLE 11.7

Name the compound

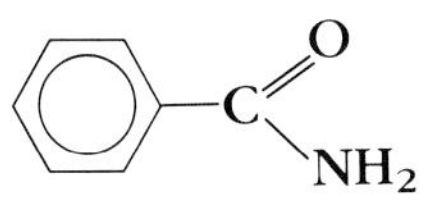

$$\text{C}_6\text{H}_5-\text{C}(=\text{O})-\text{NH}_2$$

This amide is derived from benzoic acid. Drop the *-oic*, add *-amide*, and you have it: benzamide.

In substituted amides, the alkyl groups attached to the nitrogen atom must also be identified. Instead of using a number to specify location, chemists indicate the group's attachment to nitrogen by a capital letter *N*.

EXAMPLE 11.8

Name the compound

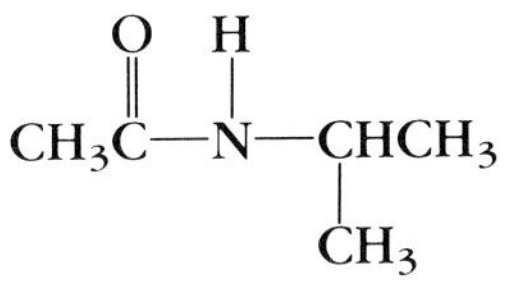

$$\text{CH}_3\text{C}(=\text{O})-\text{N}(\text{H})-\text{CH}(\text{CH}_3)_2$$

The acid portion of the molecule (the portion incorporating the carbonyl group and to one side of the nitrogen atom) contains two carbon atoms and is derived from acetic acid. The compound will therefore be named as a substituted acetamide. The substituent attached directly to the nitrogen atom is an isopropyl group. The name of the compound is *N*-isopropylacetamide. (The IUPAC name is *N*-isopropylethanamide.)

If the substituent on the nitrogen atom is phenyl, the compound is named as an **anilide**. The *-ic* or *-oic* ending of the acid name is replaced with *-anilide* instead of *-amide*.

> **EXAMPLE 11.9**
>
> Draw butyranilide.
>   The compound is a substituted four-carbon amide (derived from butyric acid).
>
> $$CH_3CH_2CH_2\overset{\overset{O}{\|}}{C}-N-$$
>
> The suffix *-anilide* indicates that there is a phenyl group substituted for one of the two hydrogen atoms on the nitrogen atom of the simple amide.
>
> $$CH_3CH_2CH_2\overset{\overset{O}{\|}}{C}-NH-\phi$$

11.10 Physical Properties of Amides

At room temperature, most simple amides are solids.

With the exception of the simplest amide (formamide, whose melting point is 3 °C), simple amides are solids at room temperature. These amides have both relatively high melting points and high boiling points because of strong intermolecular hydrogen bonding.

$$R-\overset{\overset{O}{\|}}{C}\underset{\underset{H}{|}}{-N}-H\cdots\cdots O=\overset{\underset{|}{N-H\cdots\cdots}}{\underset{H}{\overset{R}{C}}}$$

Similar hydrogen bonding plays a critical role in determining the structure and properties of proteins. This ability of amides to form hydrogen bonds also accounts for their solubility in water. The borderline for water solubility occurs at five or six carbon atoms.

11.11 Chemical Properties of Amides: Hydrolysis

Amides are neutral compounds.

Generally, the amides are neutral compounds, showing neither appreciable acidity nor significant basicity in water. The amides can be hydrolyzed in water, but at a slower rate than comparable esters. Like the ester hydrolysis, the reaction is catalyzed by acid or base. Acidic hydrolysis of a simple amide gives a carboxylic acid and an ammonium salt.

$$CH_3CH_2\overset{O}{\underset{\|}{C}}-NH_2 + HCl + H_2O \longrightarrow CH_3CH_2\overset{O}{\underset{\|}{C}}-OH + NH_4Cl$$

Basic hydrolysis gives a salt of the carboxylic acid and ammonia.

$$CH_3CH_2\overset{O}{\underset{\|}{C}}-NH_2 + NaOH \longrightarrow CH_3CH_2\overset{O}{\underset{\|}{C}}-O^- \, Na^+ + NH_3$$

It may be easier to see why the products of the two reactions differ if we consider the hydrolysis products that would form if the reaction could be carried out in the absence of added acid or base.

$$CH_3CH_2\overset{O}{\underset{\|}{C}}-NH_2 + H_2O \longrightarrow CH_3CH_2\overset{O}{\underset{\|}{C}}-OH + NH_3$$

The products of this reaction are an acid (the carboxylic acid) and a base (ammonia). If we actually carry out the hydrolysis in the presence of hydrochloric acid, then some of the hydrochloric acid reacts with the basic product ammonia to form ammonium chloride. If, instead, we use sodium hydroxide to speed the reaction, then some of this base reacts with the carboxylic acid product to form the sodium salt of the carboxylic acid. Remember: Chemical principles are not confined to the chapters in which they are introduced. If acids react with bases to form salts in Chapter 8, they also react in Chapter 11. If a reaction under consideration happens to produce an acidic product, that product always exhibits the properties we have previously described for acids.

We will encounter amide hydrolysis again when we consider the digestion of proteins. As usual, the body uses enzymes rather than the strong acid or base catalysts employed in the laboratory.

11.12 PHYSIOLOGICAL PROPERTIES OF SELECTED AMIDES

Physiologically, the most important group of amides are certainly the proteins. These compounds are so important that we will delay our consideration of them until we can devote an entire chapter to the subject.

Many drugs contain the amide function. Acetanilide was once used as an antipyretic. It is quite toxic, however, and has been largely

Many drugs contain the amide group.

replaced by aspirin and by two of its derivatives, phenacetin and acetaminophen, both of which are derived from acetanilide by substitution on the benzene ring.

$$CH_3CH_2O-\langle\bigcirc\rangle-NH-\overset{O}{\underset{\|}{C}}-CH_3 \qquad HO-\langle\bigcirc\rangle-NH-\overset{O}{\underset{\|}{C}}-CH_3$$

Phenacetin
(*p*-ethoxyacetanilide)

Acetaminophen
(*p*-hydroxyacetanilide)

Phenacetin has about the same effectiveness as aspirin in reducing fever and relieving minor aches and pains. It has been implicated in damage to the kidneys and in blood abnormalities. Phenacetin was banned from use in the United States in 1983. Acetaminophen (Tylenol, Datril, and so on) is also comparable to aspirin in the relief of pain and reduction of fever (but generally not in price). it is frequently recommended for people who are allergic to aspirin. Acetaminophen does not reduce inflammation nor does it interfere with the clotting of blood.

One of the most interesting amides of all is the *N,N*-diethylamide of lysergic acid, better known as LSD (from the German *Lysergsäure diethylamid*).

FIGURE 11.7
Illicit dosage forms of lysergic acid diethylamide (LSD). [Courtesy of the U.S. Bureau of Narcotics and Dangerous Drugs, Washington, DC]

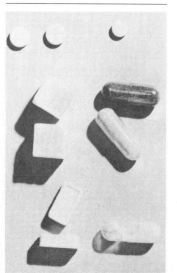

LSD

LSD is a member of a class of drugs that qualitatively change the way in which we perceive things (Figure 11.7). These drugs are called **hallucinogenic**, **psychotomimetic**, or **psychedelic** drugs, because they induce hallucinations, psychoses, and colorful visions. LSD is probably the most powerful hallucinogen of them all. The usual dosage is about 10–100 μg. To give you an idea of how small 10 μg is: one aspirin tablet contains 300 000 μg of aspirin. "Mind-expanding" qualities have been attributed to LSD, but there is no sound evidence confirming this attribution.

Is LSD a dangerous drug? A few facts are known, but most are disputed. For example, in one study, LSD administered to pregnant hamsters caused gross fetal deformities. Other studies, however, seemed to suggest that LSD does not cause chromosomal damage.

The great concern over the problem is due, in part, to the thalidomide tragedy of the late 1950s and early 1960s. Thalidomide was a completely legal, amidelike drug used as a tranquilizer.

Thalidomide

It was considered so safe, based on laboratory studies, that it was widely prescribed for pregnant women and, in Germany, was available without a prescription. The human population took several years to provide information that laboratory animals had not provided. The drug had a disastrous effect on developing human embryos. Children born to women who had taken the drug during the first 12 weeks of pregnancy had phocomelia, a condition characterized by the shortening or absence of arms or legs and other physical defects. The drug was widely used in Germany and Great Britain, and these two countries bore the brunt of the tragedy. The United States escaped relatively unscathed because an official of the Food and Drug Administration had believed that there was evidence to doubt the drug's safety and therefore did not approve it for use in the United States (Figure 11.8).

FIGURE 11.8
Frances Kelsey of the U.S. Food and Drug Administration withheld approval for the tranquilizer thalidomide to be distributed in the United States. Because of this, few "thalidomide babies" were born in this country. [Courtesy of the U.S. Food and Drug Administration, Washington, DC.]

CHAPTER SUMMARY

The functional groups of the carboxylic acids, esters, and amides are, respectively,

—C(=O)—OH   —C(=O)—OR   —C(=O)—N—

The latter two families are considered derivatives of the carboxylic acids, and the nomenclature for these compounds reflects that fact.

The common and IUPAC names for simple aliphatic acids are given in Table 11.2. Table 11.3 presents the names of the common dicarboxylic acids. The simplest aromatic carboxylic acid is benzoic acid, and more complex aromatic acids either have special common names or are named as derivatives of benzoic acid. The naming of esters and amides is demonstrated below using representative examples.

Structure	Common Name	IUPAC Name
CH_3COH	Acetic acid	Ethanoic acid
$CH_3COCH_2CH_3$	Ethyl acetate	Ethyl ethanoate
CH_3CNH_2	Acetamide	Ethanamide
$CH_3CNHC_6H_5$	Acetanilide	Ethananilide

Simple amides are those in which the nitrogen atom is bonded to two hydrogen atoms. In substituted amides, the nitrogen atom is bonded to one or two alkyl or aryl groups. The attachment of these groups to the nitrogen atom is specified by the use of the letter N in place of a number to specify position.

The molecules of all three families of compounds are polar, but, unlike carboxylic acids and amides, ester molecules cannot form hydrogen bonds with one another. Therefore, esters generally exhibit lower boiling points than acids or amides of comparable formula weight. The smaller members of all three families show appreciable water solubility because of hydrogen bonding between water molecules and the organic molecules.

The smaller carboxylic acids have irritating, unpleasant odors, whereas esters of moderate size have characteristically pleasant aromas. Amides have no distinctive odor.

Carboxylic acids are weak acids and form salts when treated with aqueous bases, such as sodium hydroxide or sodium bicarbonate. As is true of other salts, the names of these salts first list the cation and then the anion formed from the carboxylic acid. The name of the anion is derived from the acid name by dropping the -ic acid ending and appending -ate. Esters are formed by the combination of a carboxylic acid and an alcohol.

$$R-\overset{O}{\underset{\|}{C}}-OH + R'OH \overset{H^+}{\rightleftharpoons} R-\overset{O}{\underset{\|}{C}}-OR' + H_2O$$

The reverse reaction, in which an ester is converted to a carboxylic acid and an alcohol, is called hydrolysis. In the alkaline hydrolysis of an ester, the carboxylic acid product is obtained in the form of its salt. This reaction is termed saponification.

Hydrolysis of a simple amide produces a carboxylic acid and ammonia. The form in which these products are isolated depends on the conditions (acidic or basic) under which the reaction is carried out.

$$R-\overset{O}{\underset{\|}{C}}-NH_2 + HCl + H_2O \longrightarrow$$

$$R-\overset{O}{\underset{\|}{C}}-OH + NH_4Cl$$
$$\text{Salt}$$

$$R-\overset{O}{\underset{\|}{C}}-NH_2 + NaOH \longrightarrow R-\overset{O}{\underset{\|}{C}}-O^-\ Na^+ + NH_3$$
$$\text{Salt}$$

Some esters (e.g., aspirin) and amides (e.g., phenacetin and acetaminophen) are analgesics (pain relievers) or antipyretics (fever reducers), or both. Heroin, an ester of morphine, is a strongly addictive narcotic, and LSD (N,N-diethyllysergamide) is a powerful hallucinogenic drug.

CHAPTER QUIZ

1. The carboxyl group is
   a. —COH
   b. —COOH
   c. —CHO
   d. —CHOH

2. Methyl ethanoate originates by reaction of methanol with which of the following?
   a. ethane
   b. ethene
   c. acetic acid
   d. dimethyl ether

3. The IUPAC name for acetic acid is
   a. methanoic acid
   b. ethanoic acid
   c. propanoic acid
   d. butanoic acid

4. The following molecules all have approximately the same molecular weight. Which would have the highest boiling point?
   a. ethanol
   b. formic acid
   c. propane
   d. ethanal

5. Which is most soluble in water?
   a. CH_3COOH
   b. $CH_3(CH_2)_4COOH$
   c. $CH_3(CH_2)_8COOH$
   d. $CH_3(CH_2)_{10}COOH$

6. $CH_3CH_2CH_2COOCH_3$ is called
   a. butyl methylate
   b. butyl formate
   c. methyl butyrate
   d. methyl acetate

7. Which are incapable of hydrogen bonding between molecules of the same compound?
   a. esters
   b. acids
   c. alcohols
   d. amides

8. Hydrolysis of which molecules will produce an alcohol?
   a. acid
   b. ester
   c. amide
   d. substituted amide

Problems

9. The amide functional group is
   a. —CONH$_2$
   b. —CH$_2$NH$_2$
   c. —CH$_2$NHCH$_2$
   d. —COONH$_2$

10. An analgesic is a
    a. fever reducer
    b. sedative
    c. pain reliever
    d. narcotic

PROBLEMS

11.1 Acids and Their Derivatives: The Functional Groups

1. Draw the functional groups in
   a. aldehydes
   b. ketones
   c. carboxylic acids
   d. esters
   e. ethers
   f. amides

11.2 Some Common Acids: Structures and Names

2. Give common names for the straight-chain carboxylic acids containing the given number of carbon atoms.
   a. 4
   b. 6
   c. 12
   d. 18

3. Draw structural formulas for the following compounds.
   a. heptanoic acid
   b. 3-methylbutanoic acid
   c. 3-chloropentanoic acid

4. Draw structural formulas for the following compounds.
   a. o-nitrobenzoic acid
   b. p-chlorobenzoic acid
   c. 2,3-dibromobenzoic acid

5. Draw structural formulas for the following compounds.
   a. oxalic acid
   b. phthalic acid

6. Give the common and the IUPAC name for each of the following.
   a. HCOOH
   b. CH$_3$CH$_2$COOH
   c. CH$_3$(CH$_2$)$_8$COOH
   d. HOOCCH$_2$CH$_2$CH$_2$COOH

11.3 Physical Properties of Carboxylic Acids

7. Which compound would have the higher boiling point?

CH$_3$CH$_2$CH$_2$—O—CH$_2$CH$_3$

$$CH_3CH_2CH_2\overset{O}{\underset{\|}{C}}-OH$$

8. Which compound would have the higher boiling point?

CH$_3$CH$_2$CH$_2$CH$_2$CH$_2$OH

$$CH_3CH_2CH_2\overset{O}{\underset{\|}{C}}-OH$$

9. Which compound would have the higher boiling point?

$$CH_3CH_2CH_2\overset{O}{\underset{\|}{C}}-OH$$

$$CH_3CH_2\overset{O}{\underset{\|}{C}}-O-CH_3$$

10. Without consulting tables, arrange these in order of increasing boiling point.
    a. butyl alcohol
    b. propionic acid
    c. pentane
    d. methyl acetate

11. Which compound is more soluble in water?

$$CH_3\overset{O}{\underset{\|}{C}}-OH \quad CH_3CH_2CH_2CH_3$$

11.4 Chemical Properties of Carboxylic Acids

12. Draw the structural formulas for the following salts.
    a. potassium acetate
    b. calcium oxalate

13. Name each of the following salts.

    a.
    $$\text{C}_6\text{H}_5\overset{O}{\underset{\|}{C}}-O^-\,Na^+$$

    b. $(CH_3CH_2\overset{O}{\underset{\|}{C}}-O^-)_2\,Ca^{2+}$

c. $CH_3\overset{O}{\underset{\|}{C}}-O^-\ NH_4^+$

d. $(CH_3CH_2CH_2\overset{O}{\underset{\|}{C}}-O^-)_2\ Zn^{2+}$

e. benzene-1,2-dicarboxylate Ca^{2+} (phthalate calcium salt)

14. Which compound is more soluble in water?

   $C_6H_5-\overset{O}{\underset{\|}{C}}-OH$ $C_6H_5-\overset{O}{\underset{\|}{C}}-O^-\ Na^+$

15. Write an equation for the reaction of decanoic acid
    a. with aqueous NaOH
    b. with aqueous $NaHCO_3$.

16. Write an equation for the reaction of benzoic acid
    a. with aqueous NaOH
    b. with aqueous $NaHCO_3$

17. Benzoic acid is insoluble in water. If the reactions described in Problem 16 were carried out in test tubes, what would you observe?

18. Complete these equations.

    a. $CH_3CH_2\overset{O}{\underset{\|}{C}}-OH \xrightarrow{NaOH}$

    b. $C_6H_5-COOH \xrightarrow{KOH}$

    c. $HOOC-COOH \xrightarrow{excess\ NaOH}$

    d. phthalic acid (1,2-benzenedicarboxylic acid) $\xrightarrow{excess\ NaHCO_3}$

11.5 Esters: Structures and Nomenclature

19. Draw structural formulas for the following esters.
    a. methyl acetate
    b. heptyl acetate
    c. phenyl acetate

20. Draw structural formulas for the following esters.
    a. ethyl pentanoate
    b. ethyl 3-methylhexanoate

21. Draw structural formulas for the following esters.
    a. ethyl benzoate
    b. isopropyl benzoate
    c. phenyl benzoate

22. Draw structural formulas for the following esters.
    a. ethyl butyrate
    b. isopropyl propionate
    c. diethyl oxalate

23. Name the following esters.

    a. $C_6H_5-\overset{O}{\underset{\|}{C}}-O-CH_3$

    b. $CH_3-O-\overset{O}{\underset{\|}{C}}-H$

    c. $CH_3CH_2\overset{O}{\underset{\|}{C}}-O-CH_2CH_3$

    d. $CH_3CH_2CH_2O-\overset{O}{\underset{\|}{C}}-CH_3$

    e. $CH_3CH_2\overset{O}{\underset{\|}{C}}-OCH_2CH_2CH_2$ (cyclic)

    f. $CH_3CH_2CH_2\overset{O}{\underset{\|}{C}}-O-C_6H_5$

24. Complete these equations.

    a. $CH_3\overset{O}{\underset{\|}{C}}-OH + CH_3CH_2CH_2OH \xrightarrow{H^+}$

    b. $HO-\overset{O}{\underset{\|}{C}}CH_2\overset{O}{\underset{\|}{C}}-OH + 2\ CH_3OH \xrightarrow{H^+}$

11.6 Physical Properties of Esters

25. Which compound is more soluble in water?

   $CH_3\overset{O}{\underset{\|}{C}}-O-CH_3$

$$CH_3CH_2CH_2CH_2\overset{\overset{O}{\|}}{C}-O-CH_2CH_3$$

26. Of the families of compounds discussed in this chapter,
    a. which has members with characteristically unpleasant odors?
    b. which group has characteristically pleasant aromas?

11.7 Chemical Properties of Esters: Hydrolysis

27. Write an equation for the acid-catalyzed hydrolysis of ethyl acetate.
28. Write an equation for the base-catalyzed hydrolysis of ethyl acetate.
29. Complete these equations.

    a. $CH_3CH_2CH_2O-\overset{\overset{O}{\|}}{C}-\text{C}_6\text{H}_5 + H_2O \xrightarrow{H^+}$

    b. $CH_3\underset{\underset{CH_3}{|}}{CH}\overset{\overset{O}{\|}}{C}-O-CH_2CH_3 + H_2O \xrightarrow{H^+}$

11.8 Physiological Properties of Selected Esters

30. Define and give an example for each of the following
    a. analgesic   b. antipyretic
31. Examine the labels of at least five "combination" pain relievers (e.g., Excedrin, Empirin, Anacin). Make a list of the ingredients in each. Look up the properties (medical use, dosage, side effects, toxicity, and so on) in a reference work like *The Merck Index*.
32. Aspirin is a chemical compound.
    a. What is its structure and chemical name?
    b. In what ways may one brand of aspirin differ from another?
    c. In what ways must brands of aspirin be the same?
33. Do a cost analysis on at least five brands of plain aspirin, calculating the cost per grain. Also calculate the cost per gram (1.0 g = 15 grains).

11.9 Amides: Structures and Nomenclature

34. Draw structural formulas for the following amides.
    a. butanamide
    b. hexanamide
35. Draw structural formulas for the following amides.
    a. valeramide
    b. propionamide
36. Draw structural formulas for the following amides.
    a. *N*-methylacetamide
    b. *N*-ethylbenzamide
    c. *N,N*-dimethylbenzamide
37. Draw structural formulas for the following amides.
    a. acetanilide
    b. benzanilide
38. Name the following amides.

    a. $\text{C}_6\text{H}_5-\overset{\overset{O}{\|}}{C}-NH_2$

    b. $CH_3\underset{\underset{CH_3}{|}}{CH}NH-\overset{\overset{O}{\|}}{C}CH_3$

    c. $CH_3\overset{\overset{O}{\|}}{C}-NH_2$

    d. $Cl-\text{C}_6\text{H}_4-\overset{\overset{O}{\|}}{C}-NH_2$

    e. $H_2N-\overset{\overset{O}{\|}}{C}-CH_2-\overset{\overset{O}{\|}}{C}-NH_2$

    f. $CH_3CH_2\overset{\overset{O}{\|}}{C}-\underset{\underset{CH_3}{|}}{N}-CH_3$

    g. $CH_3CH_2CH_2\overset{\overset{O}{\|}}{C}-NH-\text{C}_6\text{H}_5$

11.10 Physical Properties of Amides

39. Which compound would have the higher boiling point?

$$CH_3CH_2CH_2\overset{O}{\underset{\|}{C}}-NH_2$$

$$CH_3\overset{O}{\underset{\|}{C}}-O-CH_2CH_3$$

40. Which compound is more soluble in water?

$$CH_3C{\equiv}CCH_3 \qquad CH_3\overset{O}{\underset{\|}{C}}-NH_2$$

11.11 Chemical Properties of Amides: Hydrolysis

41. Write an equation for the base-catalyzed hydrolysis of benzamide.

42. Write an equation for the acid-catalyzed hydrolysis of benzamide.

43. Complete these equations.

a. $CH_3\overset{O}{\underset{\|}{C}}-NH_2 + HCl + H_2O \longrightarrow$

b. Ph−C(=O)−N(CH$_3$)−CH$_3$ + NaOH ⟶

44. Pellagra is a vitamin-deficiency disease. Corn contains the antipellagra factor nicotinamide, which is not readily absorbed in the digestive tract. When corn is treated with quicklime (calcium hydroxide) to make hominy, the vitamin is rendered more available. What sort of reaction takes place? Write the equation. Nicotinamide is

(pyridine-3-carboxamide structure: pyridine ring with −C(=O)−NH$_2$)

11.12 Physiological Properties of Selected Amides

45. Define and give an example for each of the following.
   a. narcotic   **b.** hallucinogenic

46. How does heroin differ from morphine in structure? In physiological properties?

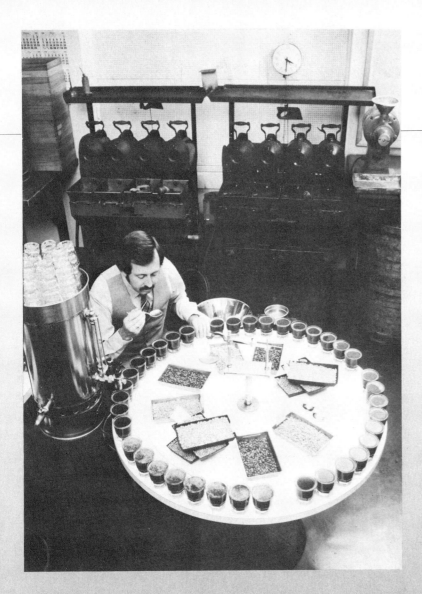

12
AMINES

DID YOU KNOW THAT...

- amines are chemicals with baselike properties?
- amines can be considered derivatives of ammonia?
- alkaloids are naturally occurring amines?
- morphine has been used for thousands of years?
- cadaverine is the amine produced by decaying flesh?
- novocaine and lidocaine are local anesthetics?
- proteins contain both amine and acid groups?
- amines play a major role in mental health?
- adrenaline (epinephrine) increases blood pressure?
- lithium carbonate has been used to control manic depression?
- reserpine is a most effective sedative?
- slight structural changes may result in vast changes in physiological activity?
- barbital is used to "put dogs to sleep"?
- thiopental has been labeled a "truth drug"?
- after 20 years of use, Valium was found to be addictive?

If carbon compounds are the basis of life, nitrogen compounds are the bases of life (pun intended). Ammonia is a weak base. Amines, which are organic derivatives of ammonia, are also weak bases. These organic bases occur in living and, especially, in once-living but now decaying organisms.

Nitrogen is an essential constituent of many physiologically active compounds. All enzymes—indeed, all proteins—contain nitrogen. Many vitamins and hormones contain nitrogen. Most drugs incorporate nitrogen atoms. And nitrogenous bases are part of the complex structure of the compounds that carry our genetic heritage, the nucleic acids DNA and RNA (bases in acids!). In this chapter we shall discuss the amines generally, as well as a number of related nitrogen-containing compounds that exhibit interesting physiological effects. We shall save the discussion of proteins and nucleic acids for later chapters.

12.1 Structure and Classification of Amines

Amines are organic derivatives of ammonia.

In Chapter 10, we saw that alcohols and ethers can be considered derivatives of water. Similarly, **amines** are organic derivatives of ammonia. Amines are classified according to the number of carbon atoms bonded directly to the nitrogen atom. A primary amine has one carbon group attached to the nitrogen, a secondary amine has two, and a tertiary amine has three.

$$\begin{array}{cccc} H-N-H & R-N-H & R-N-H & R-N-R \\ | & | & | & | \\ H & H & R & R \end{array}$$

Ammonia   A primary amine   A secondary amine   A tertiary amine

The use of the terms *primary, secondary,* and *tertiary* must be distinguished from our previous use of these terms in connection with the alcohols (Section 10.2). For example, consider structures I and II.

$$\begin{array}{cc} NH_2 & OH \\ | & | \\ CH_3-CH-CH_3 & CH_3-CH-CH_3 \\ I & II \end{array}$$

Compound I is a primary amine because only one of nitrogen's bonds is attached to a carbon atom. Compound II is a *secondary* alcohol. Alcohols are designated primary, secondary, or tertiary on the basis of the number of carbon atoms bonded *not* to oxygen but to the carbon

12.2 Simple Amines

atom *attached* to oxygen. Another significant difference can be seen in structures III and IV.

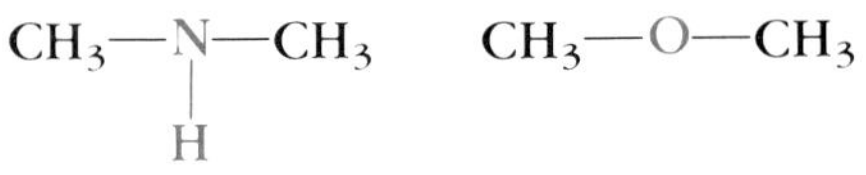

$$CH_3-\underset{\underset{H}{|}}{N}-CH_3 \qquad CH_3-O-CH_3$$

$$\text{III} \qquad\qquad\qquad \text{IV}$$

Compound III is a secondary amine, but compound IV is an ether (*not* an alcohol, secondary or otherwise). When there is only one carbon group attached to oxygen, the compound is an alcohol. But if there are two, the compound is an ether. In contrast, whether there are one or two (or three) alkyl or aromatic groups attached to nitrogen, the compounds are all classed as amines.

12.2 SIMPLE AMINES

To name simple aliphatic amines, simply specify the alkyl groups attached to the nitrogen atom and add the suffix *-amine*.

EXAMPLE 12.1

Draw structural formulas for, give the names of, and classify the amines with the formula C_3H_9N.

We are asked to draw all the isomers of the three-carbon amines. There are two three-carbon alkyl groups, propyl and isopropyl, and two amines contain these groups.

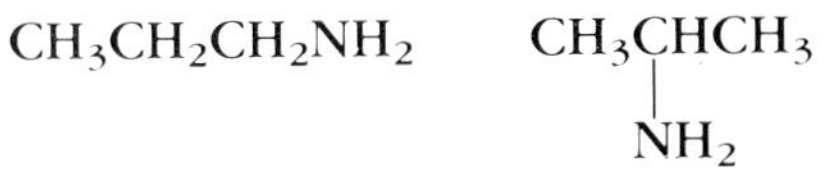

$$CH_3CH_2CH_2NH_2 \qquad CH_3\underset{\underset{NH_2}{|}}{C}HCH_3$$

The compounds are propylamine and isopropylamine, respectively. Both are primary amines because, in each compound, only one of the nitrogen bonds is connected to a carbon atom.

We do not need to limit ourselves to compounds in which all carbon atoms appear in a single alkyl group. We can divide the three carbon atoms in the molecule between two groups, one a methyl group and the other an ethyl group.

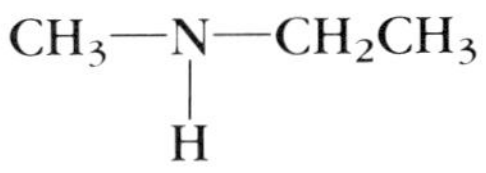

$$CH_3-\underset{\underset{H}{|}}{N}-CH_2CH_3$$

This compound is a secondary amine named ethylmethylamine.

Finally, we can divide the carbon atoms among three separate groups, in this case, three methyl groups.

$$CH_3-\underset{\underset{CH_3}{|}}{N}-CH_3$$

This is trimethylamine, a tertiary amine.

EXAMPLE 12.2

Draw and classify *tert*-butylamine.
  The structural formula is

$$CH_3-\underset{\underset{CH_3}{|}}{\overset{\overset{CH_3}{|}}{C}}-NH_2$$

tert-Butylamine is a primary (not a tertiary) amine.

The primary amine in which the nitrogen atom is attached directly to a benzene ring is called **aniline**. Aromatic amines are either named as derivatives of this parent compound or have acquired unique common names (Figure 12.1).

Compounds in which the nitrogen atom is attached to both a benzene ring and an alkyl group are also named as derivatives of aniline. The alkyl groups are named first, and their position of attachment (i.e., at the nitrogen atom) is indicated by the capital letter *N*.

N-Methylaniline   *N,N*-Dimethylaniline

EXAMPLE 12.3

Draw the structures for *p*-ethylaniline and *N*-ethylaniline.
  Both compounds are derivatives of aniline. The first compound

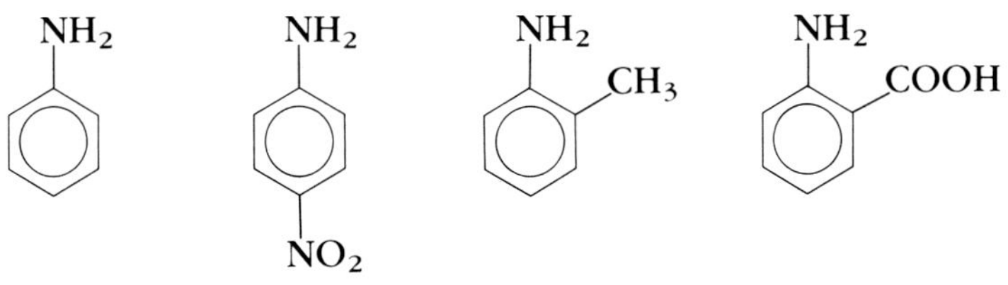

12.3 Heterocyclic Amines

Aniline — NH$_2$ on benzene ring
p-Nitroaniline — NH$_2$ and NO$_2$ para on benzene ring
o-Toluidine — NH$_2$ and CH$_3$ ortho on benzene ring
Anthranilic acid — NH$_2$ and COOH ortho on benzene ring

FIGURE 12.1 Selected aromatic amines.

is a primary amine with an ethyl group located *para* to the *amino* (—NH$_2$) group.

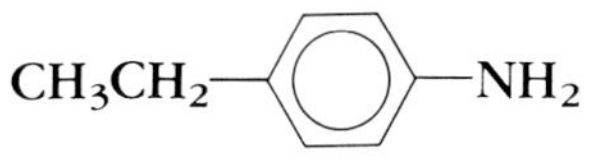

$$CH_3CH_2-\text{(benzene ring)}-NH_2$$

The second compound is a secondary amine with the ethyl group attached to the *nitrogen*.

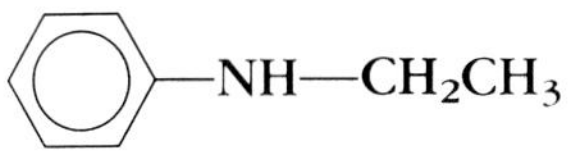

$$\text{(benzene ring)}-NH-CH_2CH_3$$

For amines that incorporate other functional groups or those in which alkyl groups cannot be simply named, the amino group is named as a substituent. We have referred to amino acids before; these are the compounds from which proteins are made. An amino acid contains a carboxylic acid group and an amino group. All have distinctive common names. The compound

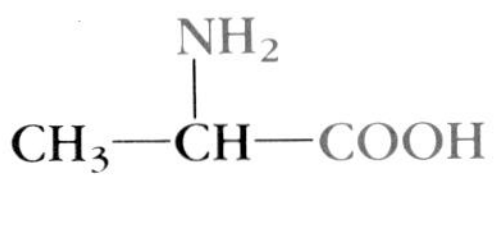

$$CH_3-\underset{\underset{NH_2}{|}}{CH}-COOH$$

is called alanine. Other rarely used names for it are 2-aminopropanoic acid (IUPAC) and α-aminopropionic acid. [Alpha (α) refers to the carbon atom next to the carbonyl group.]

12.3 Heterocyclic Amines

In Chapter 9 we encountered a variety of cyclic hydrocarbons. All the atoms in the rings of these compounds were carbon atoms. In Figure 10.6 we pictured several cyclic ethers in which the oxygen atom of the ether functional group was one of the atoms in the ring. Those compounds with rings incorporating only carbon atoms are called

Heterocyclic rings contain an element other than carbon in the ring.

carbocyclic compounds. Compounds like the cyclic ethers, which incorporate not only carbon but also other elements in the ring, are called **heterocyclic compounds**. A large number of naturally occurring amines are heterocyclic. Morphine (Section 11.8) is just such an amine. So is lysergic acid, from which LSD is made (Section 11.12). Unlike most of the carbocyclic compounds we have studied, heterocyclic amines have common names that are not simply adaptations of the names of the corresponding open-chain compounds. Figure 12.2 pictures a selection of the more important, simple heterocyclic amines. (If you regard some of these as *not* simple, then we urge you to look again at the structures of morphine or LSD.)

The compounds pyrrole and pyrrolidine each have four carbon atoms and one nitrogen atom in a ring. Pyrrole is an aromatic compound with a stable ring of electrons similar to that of benzene. (The ordinarily nonbonding pair of electrons that sits on a nitrogen atom is part of this ring of electrons. We will be looking at the nitrogen atom's "nonbonding" pair of electrons in Section 12.5 when we consider the chemical properties of amines.) Pyrrolidine, with four more hydrogen atoms than pyrrole, behaves like an aliphatic amine. Imidazole also has a five-membered ring, but it contains two nitrogen atoms and only three carbon atoms. Like pyrrole, imidazole has aromatic properties.

Pyridine and piperidine each have five carbon atoms and one ni-

---

FIGURE 12.2
A selection of heterocyclic amines.

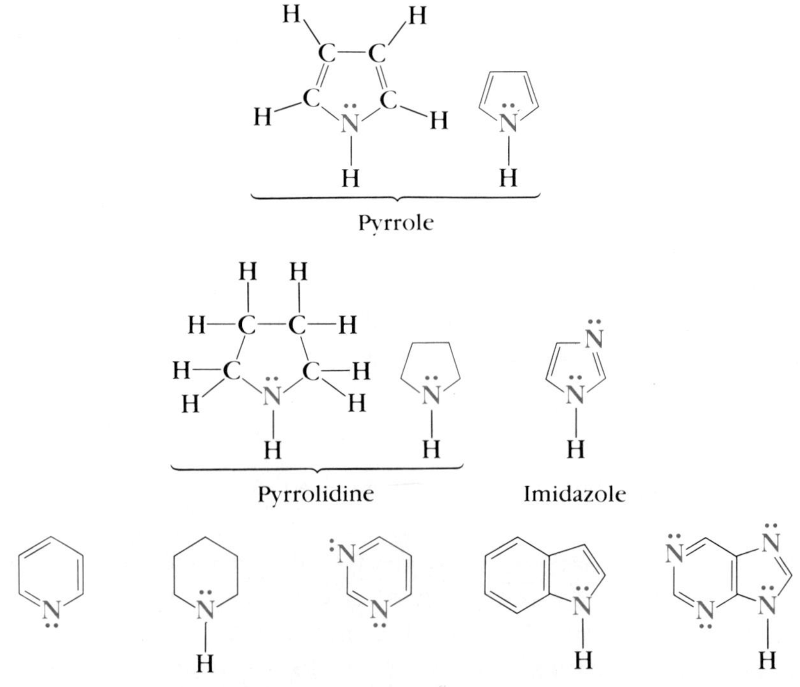

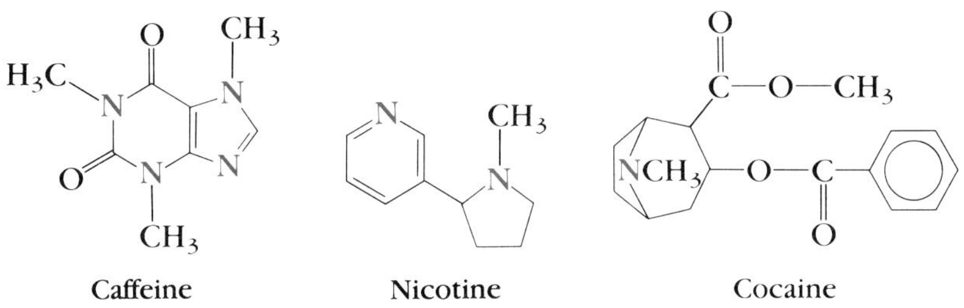

12.4 Physical Properties of Amines

FIGURE 12.3 Some familiar alkaloids.

trogen atom. Pyridine is aromatic; piperidine is aliphatic. Another six-membered heterocycle is pyrimidine, which has two nitrogen atoms and four carbon atoms in its ring. Pyrimidine is another aromatic compound.

There are other heterocyclic compounds that have two rings sharing a common side (a situation we encountered with naphthalene among the carbocyclic compounds). Indole has a benzene ring fused with a pyrrole ring. Purine has a pyrimidine ring sharing a side with an imidazole structure. Compounds related to purine and pyrimidine make up a part of the structure of nucleic acids, the genetic material of cells. When we finally take up the topic of nucleic acids, you will see one of the truly outstanding examples of molecular design. The heterocyclic amines play a major role in that design.

As we shall see in Section 12.5, amines are basic compounds. Many amines, particularly heterocyclic ones, occur naturally in plants and are called **alkaloids**, which means "like alkalies." Knowledge of many of these, at least in crude form, dates back to antiquity. Morphine (Section 11.8), an alkaloid isolated from opium, has been used for thousands of years. A few other well-known alkaloids are shown in Figure 12.3.

12.4 PHYSICAL PROPERTIES OF AMINES

Primary and secondary amines have hydrogen atoms attached to nitrogen atoms; thus, they are capable of intermolecular hydrogen bonding. These forces are not as strong as those between alcohol molecules (which have hydrogen bonded to oxygen, a more electronegative element than nitrogen). Amines boil at higher temperatures than alkanes but at lower temperatures than alcohols of comparable formula weight. Tertiary amines have no hydrogen bonded to nitrogen and cannot form intermolecular hydrogen bonds. They have boiling points comparable to those of the ethers (Table 12.1).

All three classes of amines can hydrogen-bond to water. Amines of low molecular weight are quite soluble in water; the borderline of water solubility occurs at five or six carbon atoms.

Primary and secondary amines are capable of intermolecular hydrogen bonding.

TABLE 12.1 Physical Properties of Some Amines and Comparable Oxygen-Containing Compounds

Compound	Class	Molecular Weight	Boiling Point (°C)
Butylamine	1°	73	78
Diethylamine	2°	73	55
Butyl alcohol	—	74	118
Propylamine	1°	59	49
Trimethylamine	3°	59	3
Ethyl methyl ether	—	60	6

Amines have strong odors.

Amines have interesting (!) odors. The simple ones smell very much like ammonia. Higher aliphatic amines smell like decaying fish. Or perhaps we should put it the other way around: decaying fish give off odorous amines. The stench of rotting flesh is due in part to putrescine and cadaverine, two compounds that are diamines.

$$H_2NCH_2CH_2CH_2CH_2NH_2 \qquad H_2NCH_2CH_2CH_2CH_2CH_2NH_2$$

1,4-Diaminobutane (putrescine)   1,5-Diaminopentane (cadaverine)

12.5 Chemical Properties of Amines

Amines as Bases

Let us look at the structure of ammonia and the amines a little more closely. In ammonia, three hydrogen atoms are bonded to the nitrogen atom. The nitrogen atom also has an unshared pair of electrons.

$$H:\overset{..}{\underset{H}{N}}:H \quad \text{or} \quad H-\overset{..}{\underset{|\atop H}{N}}-H$$

Ammonia can undergo a reaction in which it shares its normally nonbonding electrons. It can accept a proton and form the ammonium ion.

$$H:\overset{..}{\underset{H}{N}}:H + H_3O^+ \longrightarrow \left[H:\overset{H}{\underset{H}{\overset{..}{N}}}:H \right]^+ + H_2O$$

This, of course, is the reaction of ammonia as a base. An amine also has an unshared pair of electrons, and it too can act as a base.

12.5 Chemical Properties of Amines

$$R\!:\!\underset{H}{\overset{..}{N}}\!:\!H + H_3O^+ \longrightarrow \left[R\!:\!\underset{H}{\overset{H}{\overset{..}{N}}}\!:\!H\right]^+ + H_2O$$

Just as amines are regarded as organic derivatives of ammonia, the ions formed in reactions such as the above are regarded as derivatives of the ammonium ion. Just as amines can come with one, two, or three organic groups attached to the nitrogen atom, any or all of the hydrogen atoms on the ammonium ion can be replaced by aliphatic or aromatic groups.

$$\left[\begin{array}{c}H\\|\\R-N-H\\|\\H\end{array}\right]^+ \left[\begin{array}{c}R\\|\\R-N-H\\|\\H\end{array}\right]^+ \left[\begin{array}{c}R\\|\\R-N-R\\|\\H\end{array}\right]^+ \left[\begin{array}{c}R\\|\\R-N-R\\|\\R\end{array}\right]^+$$

The ion in which all four hydrogen atoms have been replaced by alkyl groups is termed a *quaternary ammonium ion*. Compounds that incorporate this type of ion are called **quaternary ammonium salts** (see Table 12.2).

Ammonium ions in which one or more hydrogen atoms have been replaced with alkyl groups are named in a manner analogous to that used for simple amines. The alkyl groups are names as substituents, and the parent species is regarded as the ammonium ion.

TABLE 12.2 The Classification of Amines

Class	Symbol	General Formula	Examples	
Primary	1°	R—NH_2	$CH_3CH_2CH_2NH_2$	⌬—NH_2
Secondary	2°	R—NH—R	CH_3NHCH_3	⌬—$NHCH_3$
Tertiary	3°	R—NR—R	CH_3—N(CH_3)—CH_3	⌬—N(CH_3)—CH_3
Quaternary (salts)	4°	R—N$^+$R$_2$—R	CH_3—N$^+$(CH_3)$_2$—CH_3	⌬—CH_2N$^+$(CH_3)$_2$—CH_3

> **EXAMPLE 12.4**
>
> Name the following ions: (a) $CH_3\overset{+}{N}H_3$, (b) $(CH_3)_2\overset{+}{N}H_2$, (c) $(CH_3)_3\overset{+}{N}H$, (d) $(CH_3)_4\overset{+}{N}$.
> a. Methylammonium ion
> b. Dimethylammonium ion
> c. Trimethylammonium ion
> d. Tetramethylammonium ion.

When aniline behaves as a base and picks up a proton, the cation formed is called the anilinium ion.

$$C_6H_5-\overset{+}{N}H_3$$

Anilinium ion

Ammonia is a weak base.

$$:NH_3 + H_2O \rightleftharpoons NH_4^+ + OH^-$$

Simple amines are somewhat more basic than ammonia but are still much weaker bases than compounds like sodium hydroxide. Aromatic amines like aniline are much weaker bases than even ammonia. In general, then, when amines are placed in water, the equilibrium favors the nonionized forms.

$$R\ddot{N}H_2 + H_2O \rightleftharpoons RNH_3^+ + OH^-$$

Amines do react with acids, such as the mineral acids, to form salts.

$$\begin{array}{c} CH_3 \\ | \\ CH_3-N: \\ | \\ CH_3 \end{array} + HNO_3 \longrightarrow \left[\begin{array}{c} CH_3 \\ | \\ CH_3-N-H \\ | \\ CH_3 \end{array}\right]^+ NO_3^-$$

Trimethylamine → Trimethylammonium nitrate

Amine salts are named like other salts: the name of the cation is given first, followed by that of the anion. Remember that the ions formed from aliphatic amines are named as substituted ammonium ions.

12.5 Chemical Properties of Amines

EXAMPLE 12.5

Name the salt formed in the reaction of ethylamine with hydrochloric acid.

The reaction is

$$CH_3CH_2NH_2 + H_3O^+ + Cl^- \longrightarrow CH_3CH_2NH_3^+ \, Cl^- + H_2O$$

The salt is named ethylammonium chloride.

EXAMPLE 12.6

Draw the structure of ethylmethylammonium acetate.

When names seem to reach halfway across the page, always break them into parts. Consider the cation first. Ammonium is the parent, and that means that you have a nitrogen atom with four bonds to it. Two alkyl groups are named, ethyl and methyl. These two groups are attached to the nitrogen atom, which means you need two hydrogen atoms to complete the bonding. Because this is a cation, you also need a positive charge.

$$\left[CH_3CH_2 - \underset{H}{\overset{H}{N}} - CH_3 \right]^+$$

The anion is the acetate ion, CH_3COO^-. The salt is

$$\left[CH_3CH_2 - \underset{H}{\overset{H}{N}} - CH_3 \right]^+ CH_3COO^-$$

Salts of aniline are named as anilinium compounds. An older system, still in use for naming drugs, calls the salt of aniline and hydrochloric acid "aniline hydrochloride." By this older system, the formula of the compound is frequently drawn to correspond to the name.

C₆H₅—NH₃⁺ Cl⁻   C₆H₅—NH₂·HCl

Anilinium chloride   "Aniline hydrochloride"

Keep in mind that these compounds are really ionic—they are salts—even though the name and formula seem to indicate a loose association of molecules. The properties of the compounds (solubility, for example) are those characteristic of salts.

When amines are converted to salts, they become water-soluble.

Physiologically active amines are often converted to salts and thereby rendered water-soluble. For instance, procaine is soluble only to the extent of 0.5 g in 100 g of water. The chloride salt is soluble to the remarkable degree of 100 g in 100 g of water.

$$H_2N-\underset{}{\bigcirc}-\overset{O}{\underset{\|}{C}}-OCH_2CH_2-\underset{CH_2CH_3}{\overset{}{N}}-CH_2CH_3$$

Procaine

$$H_2N-\underset{}{\bigcirc}-\overset{O}{\underset{\|}{C}}-OCH_2CH_2-\underset{CH_2CH_3}{\overset{H}{N^+}}-CH_2CH_3 \quad Cl^-$$

Procaine hydrochloride (Novocaine)

Procaine hydrochloride, perhaps better known by the trade name Novocaine, is widely used as a local anesthetic.

Many of the most important local anesthetics, like procaine, are compounds derived from *p*-aminobenzoic acid. Figure 12.4 shows a number of these compounds. Note that the only compound that is not a derivative of *p*-aminobenzoic acid, lidocaine, does share some structural features with the other compounds shown. Lidocaine is usually the "local" of choice nowadays. It is highly effective yet has a fairly low toxicity.

Formation of Substituted Amides

Simple amides result from reaction of carboxylic acids and amines.

Carboxylic acids react with ammonia, when heated, to produce simple amides.

$$R-\overset{O}{\underset{\|}{C}}-OH + NH_3 \xrightarrow{heat} R-\overset{O}{\underset{\|}{C}}-NH_2 + H_2O$$

Substituted amides can be regarded as the product of the dehydration reaction of carboxylic acids and amines.

$$R-\overset{O}{\underset{\|}{C}}-OH + H-\overset{H}{\underset{|}{N}}-R' \longrightarrow R-\overset{O}{\underset{\|}{C}}-\overset{H}{\underset{|}{N}}-R' + H_2O$$

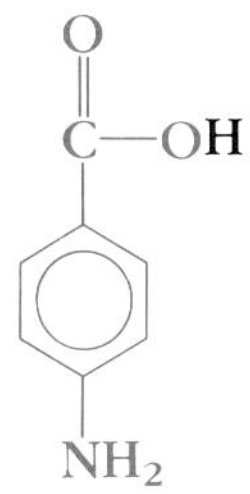

12.5 Chemical Properties of Amines

FIGURE 12.4 Some local anesthetics.

p-Aminobenzoic acid

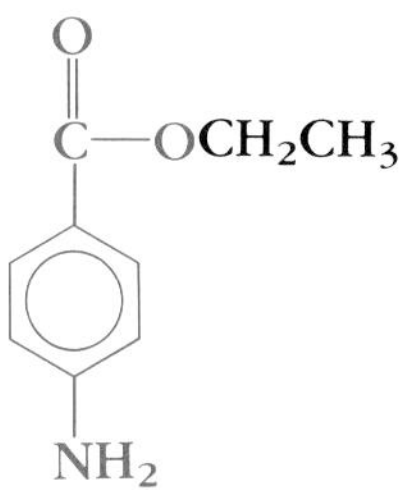

Ethyl *p*-aminobenzoate (Benzocaine)

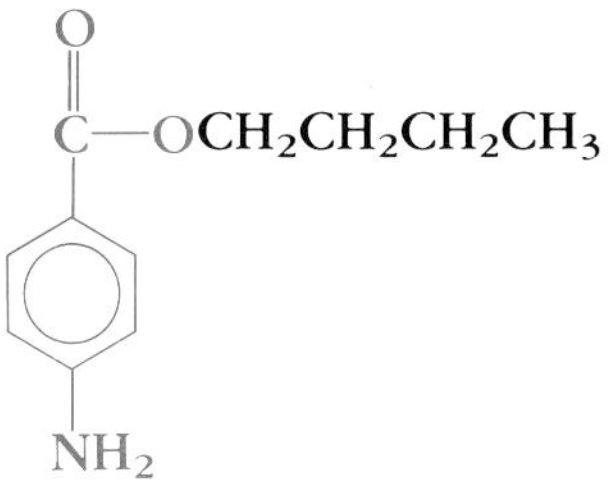

Butyl *p*-aminobenzoate (Butesin)

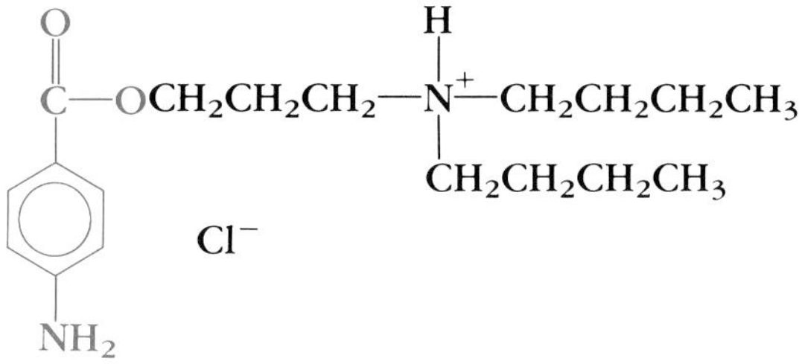

Butacaine

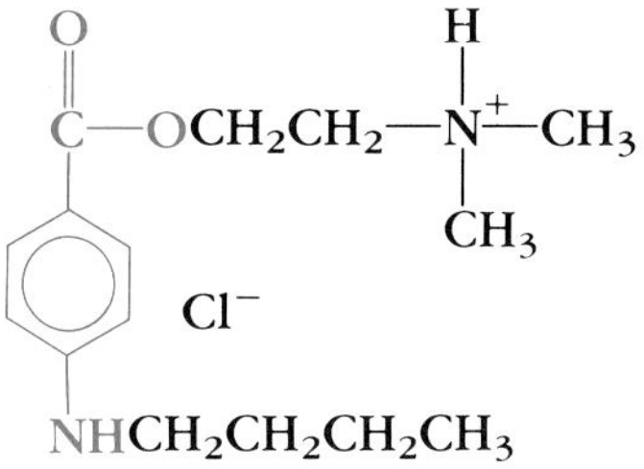

Tetracaine

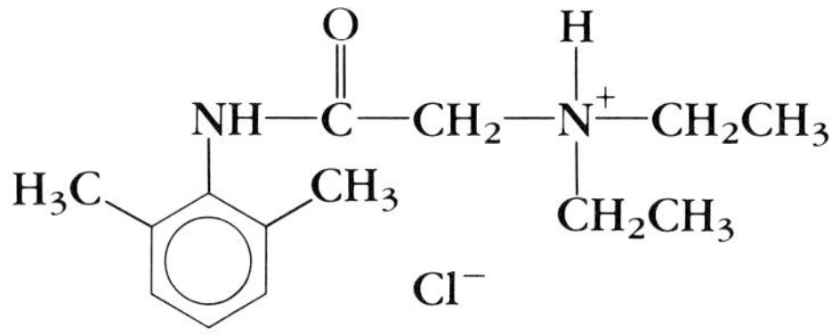

Lidocaine (Xylocaine)

In chemistry laboratories, this reaction is actually carried out using special reactive derivatives of the carboxylic acid instead of the acid itself. In living organisms, the formation of the amide bond is under the control of enzymes. Proteins are substituted amides formed through the reaction of amino acids such as alanine (Section 12.2). An amino acid molecule has both an amino group and a carboxyl group. When two molecules react, the amino group of one molecule reacts with the carboxyl group of the second.

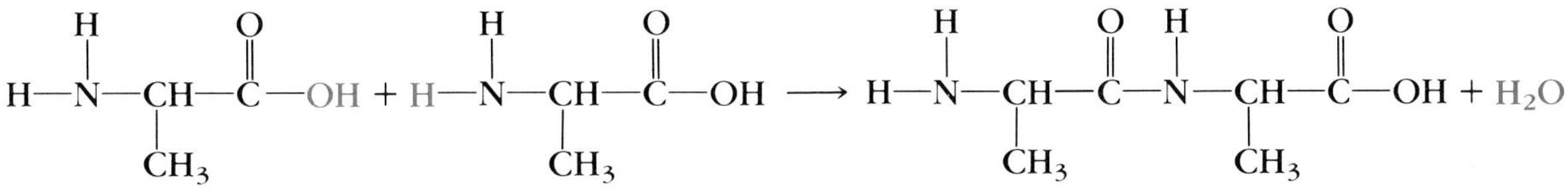

Notice that the product still has a free amino group at one end and a free carboxyl group at the other. That means that this compound can go on to react again. As we shall see when we discuss the chemistry of proteins, that is exactly what happens.

12.6 Physiological Properties of Selected Amines and Derivatives

Amines play an important role in the state of mental health.

Amines and derivatives of amines play major roles in the control of mood and the state of mental health in general. One theory of the biochemical basis for mental illness holds that the balance of two naturally occurring amines determines the state of one's mental health. One of these amines, norepinephrine, is a powerful stimulant. The other, serotonin, is a depressant.

$$\text{Norepinephrine} \qquad \text{Serotonin}$$

According to the theory, when something happens in the biochemistry of the body to cause an excess of serotonin in the brain, a person is depressed. The greater the excess, the greater the depression. If, on the other hand, norepinephrine is formed in excess, the person is in an elated, perhaps hyperactive state. In large excess, norepinephrine could cause a *manic* state (Figure 12.5).

The body produces another compound, closely related in structure to norepinephrine. Norepinephrine is a primary amine. Its *N*-methyl derivative is called epinephrine or, more commonly, **adrenaline**.

$$\text{Epinephrine (adrenaline)}$$

A tiny amount of adrenaline causes a great increase in blood pressure. When we are frightened or under stress, the flow of adrenaline prepares our bodies for fight or flight. Because culturally imposed inhibitions

12.6 Physiological Properties of Selected Amines and Derivatives

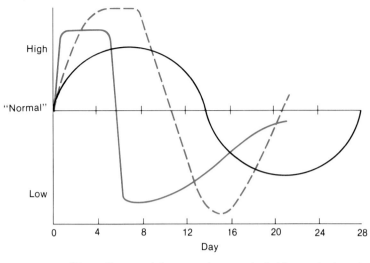

FIGURE 12.5
Effects of drugs on the mental state of an individual. The diagram is illustrative only and should not be interpreted literally.

prevent fighting or fleeing in most modern situations, the adrenaline-induced supercharge is not used. This sort of frustration has been implicated in some forms of mental illness.

Drugs are used medically to regulate abnormal highs and lows such as those that occur in manic-depressive psychoses. Lithium carbonate (Li_2CO_3) has been used in leveling out the dangerous manic stage. Psychic energizers such as imipramine can be used to combat deep depression.

Imipramine

For better or worse, there is now an enormous chemical arsenal available for combating various mental problems. For centuries, the people of India used the snakeroot plant, *Rauwolfia serpentina*, to treat a variety of ailments including maniacal forms of mental illness. In

1952, Emil Schlittler, of Switzerland, isolated from the plant an active alkaloid, which he named reserpine and which has the following impressive structure.

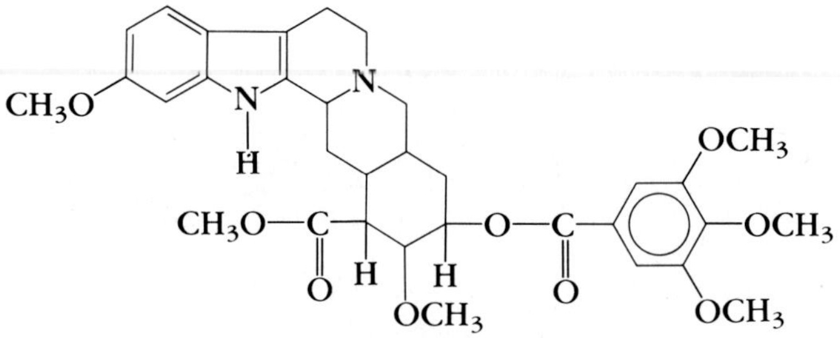

Reserpine

Reserpine was found to be so effective as a sedative that by 1953 it had replaced electroshock therapy for 90% of psychotic patients.

At about the same time, chlorpromazine, which had been used in France as an antihistamine, was tried on psychotic patients in the United States as a tranquilizer.

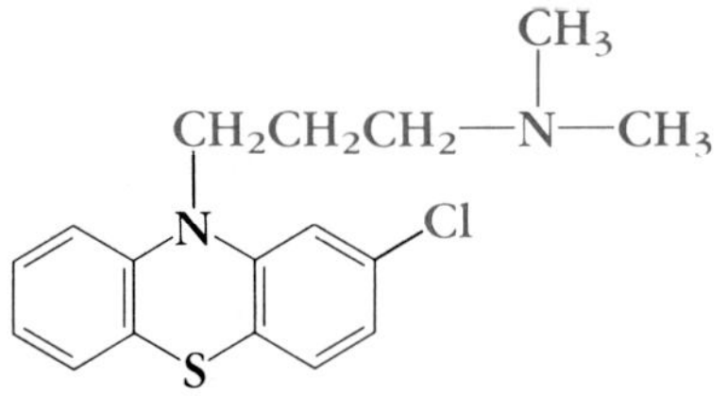

Chlorpromazine
(Thorazine)

It was found to be extremely effective against the symptoms of most forms of schizophrenia.

Many compounds related to chlorpromazine have been synthesized. Several of these have been found to have interesting physiological properties. Promazine itself is a tranquilizer but is not as potent as chlorpromazine.

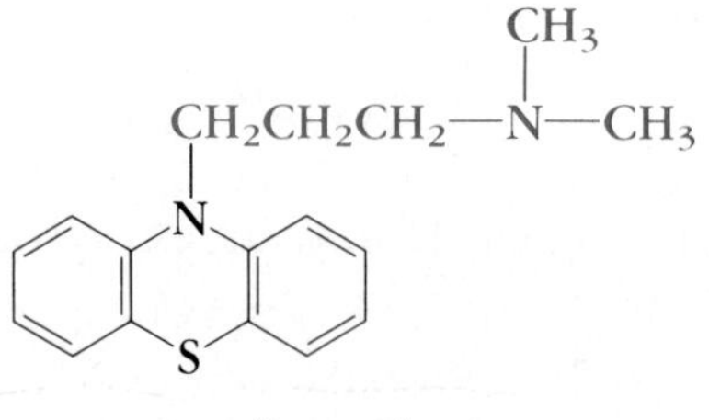

Promazine

12.6 Physiological Properties of Selected Amines and Derivatives

Thioridazine is a potent tranquilizer, reputed to be without some of the undesirable side effects of chlorpromazine.

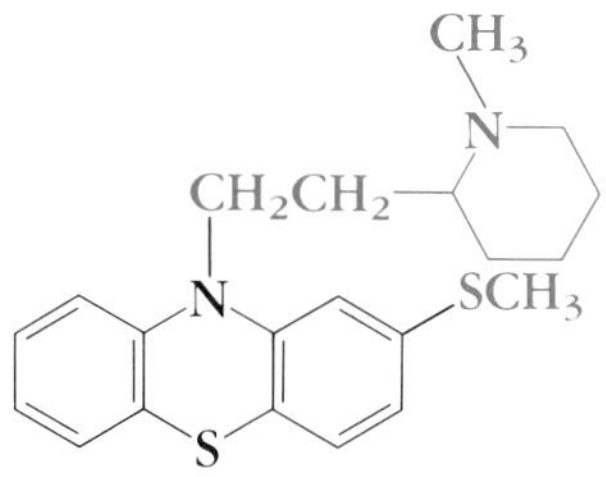

Thioridazine
(Mellaril)

As we noted earlier in this section, imipramine, which differs from promazine only in the replacement of a sulfur atom by a —CH_2CH_2— group, is not a tranquilizer at all, but an antidepressant. This indicates that slight structural changes sometimes result in profound changes in properties. We have a long way to go in understanding why drugs act as they do.

Some of the chemicals developed for control of mood have been increasingly subject to abuse by the people using them (Table 12.3). Among the more widely known stimulants ("uppers") are a variety of synthetic amines called the **amphetamines** (Figure 12.6). All are related to 2-phenylethylamine.

Amphetamines—synthetic amines that act as stimulants

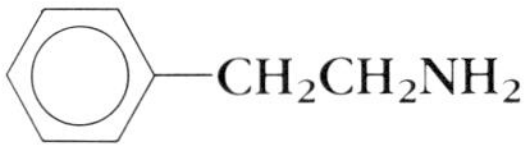

2-Phenylethylamine

Amphetamine has been extensively used for weight reduction.

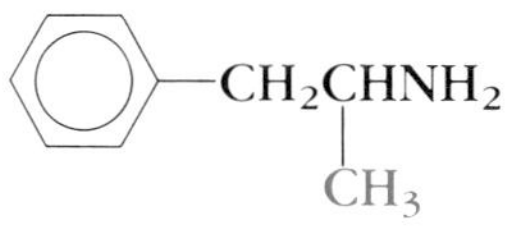

Amphetamine
(Benzedrine)

It has also been used to treat mild depression and narcolepsy, a rare form of sleeping sickness. Amphetamine induces excitability, restlessness, tremors, insomnia, dilated pupils, increased pulse rate and blood pressure, hallucinations, and psychoses. It is no longer recommended for weight reduction because any weight loss achieved is generally only temporary. The greatest problem, however, has been the diversion of vast quantities of amphetamines into the illegal drug market. Amphet-

TABLE 12.3 Toxicities of a Variety of Drugs[a]

Drug	LD_{50} (mg/kg body wt)[b]	Method of Administration	Experimental Animals
Local anesthetics			
Lidocaine	400	Subcutaneous	Mice
Procaine	50	Intravenous	Rats
Cocaine	17.5	Intravenous	Rats
Barbiturates			
Barbital	600	Oral	Mice
Pentobarbital	130	Intraperitoneal	Mice
Secobarbital	125 (MLD)	Oral	Rats
Phenobarbital	660	Oral	Rats
Amobarbital	575	Oral	Rabbits
Thiopental	120	Intraperitoneal	Rats
Narcotics			
Morphine	500	Subcutaneous	Mice
Codeine	120 (MLD)	Oral	Guinea pigs
Heroin	150	Subcutaneous	Rabbits
Meperidine	170	Oral	Rats
Stimulants			
Caffeine	200	Oral	Rats
Nicotine	55	Oral	Rats
	1.0 (MLD)	Intravenous	Rats
Amphetamine	5 (MLD)	Oral	Rats
	85	Oral	Rabbits
Methamphetamine	70	Intraperitoneal	Mice
Epinephrine	50	Oral	Mice
Mescaline	500	Intraperitoneal	Mice

[a] Comparisons of toxicities in different animals—and extrapolation to humans—are at best crude approximations. Method of administration can also have a profound effect on the observed toxicity.
[b] LD_{50} = lethal dose for 50% of the population of test animals; MLD = minimum lethal dose.

FIGURE 12.6
Amphetamine tablets. [Courtesy of the U.S. Bureau of Narcotics and Dangerous Drugs, Washington, DC.]

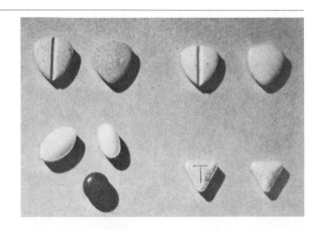

12.6 Physiological Properties of Selected Amines and Derivatives

amines are inexpensive; armed forces personnel, truck drivers, and college students have been among the heavy users.

Methamphetamine has a more pronounced psychological effect;

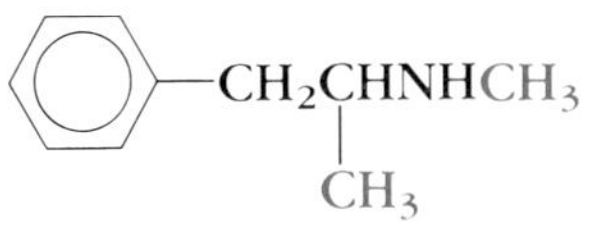

$$\text{C}_6\text{H}_5-\text{CH}_2\text{CHNHCH}_3 \quad | \quad \text{CH}_3$$

Methamphetamine
(Methedrine)

it is the "speed" that abusers inject into their veins. Such injections, at least initially, are said to give the individual a euphoric "rush." Shooting methamphetamine is dangerous, however, because it is quite toxic.

Amphetamines are similar in structure to the natural stimulants epinephrine and norepinephrine and may act by mimicking these brain amines.

Amphetamines are "uppers"; the **barbiturates** are "downers." The barbiturates are a family of related compounds that display a wide variety of properties. They produce mild sedation, deep sleep, and even death. The medicinal value of barbiturates was discovered at the turn of the century. A derivative called barbital was found to be useful in euthanizing pets.

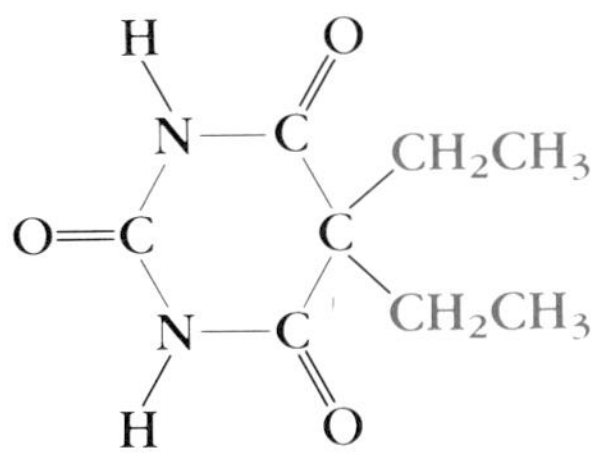

Barbital

Several thousand barbiturates have been synthesized through the years, but only a few have found widespread use in medicine. Pentobarbital (Nembutal) is employed as a short-acting hypnotic drug.

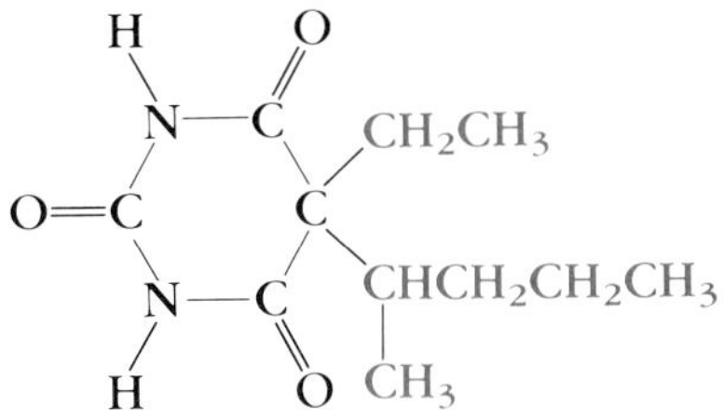

Pentobarbital

Before the discovery of modern tranquilizers, pentobarbital was widely used as a calmative against anxiety and other psychic disorders. Another short-acting barbiturate is secobarbital (Seconal).

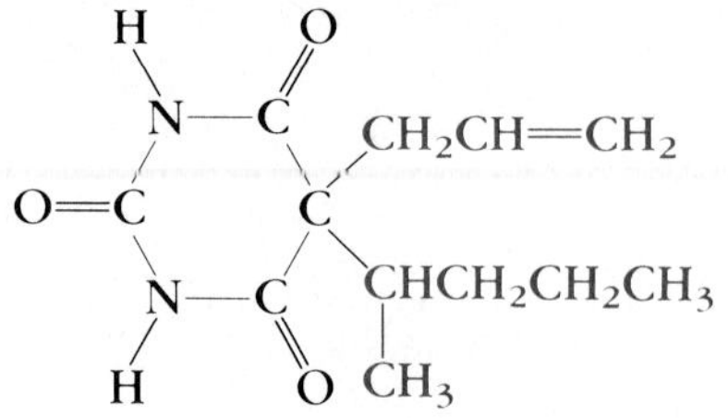

Secobarbital
(Seconal)

It is used as a hypnotic and sedative.

Phenobarbital (Luminal) is a long-acting drug used as an anticonvulsant for epileptics and people with brain damage. The action of amobarbital (Amytal) is intermediate in duration relative to secobarbital and phenobarbital.

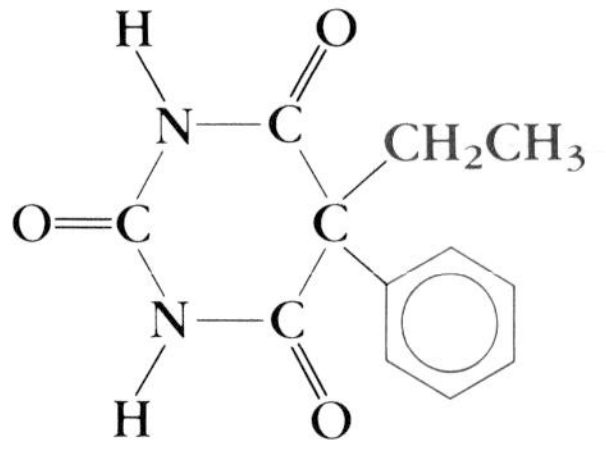

Phenobarbital
(Luminal)

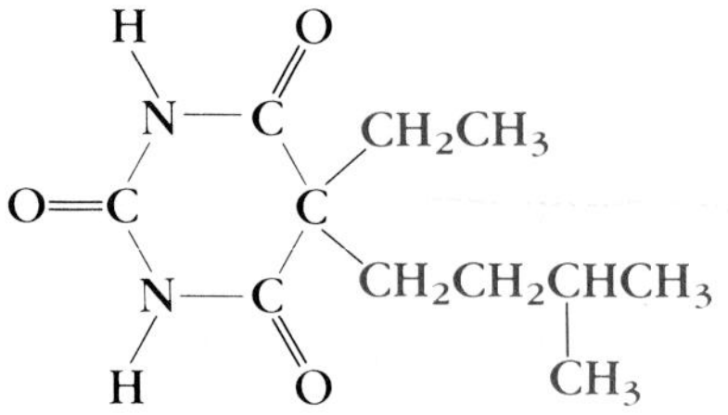

Amobarbital
(Amytal)

Thiopental (Pentothal), a compound that differs from pentobarbital only in the replacement of an oxygen atom by a sulfur atom, is widely used in anesthesia.

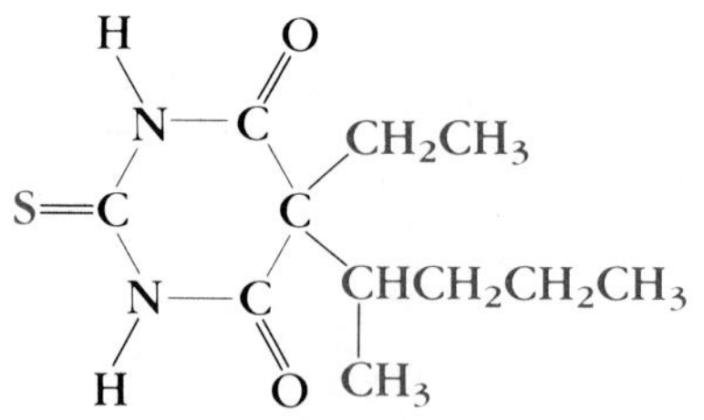

Thiopental
(Pentothal)

Thiopental has been investigated as a possible "truth" drug. Although it seems to aid psychiatric patients in recalling traumatic experiences and enables uncommunicative individuals to talk more freely, it does not prevent someone from withholding the truth or even from lying. No true "truth" drug exists.

In small doses (a few milligrams), the barbiturates act as sedatives. In larger doses (100 mg), the barbiturates induce sleep. At one time they were widely used as sleeping pills. The lethal dose is in the vicinity of 1500 mg (1.5 g). Barbiturates are often the drug of choice for suicide. News reports list the cause of death as "an overdose of sleeping pills." The potential for accidental overdoses is great. Groggy individuals may be unable to remember whether they took their sleeping pills and take more. Also, sleeping pills and alcohol exhibit a **synergistic effect**.* Both are depressants, and the combination is much more potent than the simple sum of their separate effects.

The barbiturates are strongly addictive. Habitual use leads to the development of a tolerance to the drug. One requires ever-larger doses to get the same degree of effect. Side effects are similar to those of alcohol. Abuse leads to hangovers, drowsiness, dizziness, and headaches. Withdrawal symptoms are often severe and may be accompanied by convulsions and delirium. In fact, some medical authorities now say that withdrawal from barbiturates is more likely to cause death than withdrawal from heroin.

The barbiturates are actually cyclic amides (not amines). The barbiturate ring resembles that of pyrimidine. Recent evidence indicates that the barbiturates may act by substituting for the pyrimidine bases in nucleic acids (Chapter 17).

Another class of widely used antianxiety drugs is the benzodiazepines, compounds that feature seven-membered heterocyclic rings. Of these, perhaps the best known are diazepam (Valium) and chlorodiazepoxide (Librium). For many years, Valium was the most prescribed drug in the United States. A related drug, flurazepam (Dalmane), is widely used to treat insomnia. It has replaced the barbiturates as the sleeping pill of choice.

Flurazepam (Dalmane)

Diazepam (Valium)

Chlordiazepoxide (Librium)

*Synergism is not limited to alcohol–barbiturate interactions. Fully one-half of the 100 most frequently prescribed drugs interact with alcohol, including antihistamines, tranquilizers, and medicines for hypertension. These alcohol–drug interactions lead to 2500 deaths per year in the United States, most of them not intentional.

The benzodiazepine derivatives and the carbamates were formerly called "minor tranquilizers." They are still widely used for the treatment of anxiety. Some studies have shown them to be remarkably effective. Others have found the drugs to be little better than a placebo. (A placebo is a pill that looks and tastes like the drug tablet but contains no active ingredient. The actual effect apparently depends upon the expectation of the patient). To the extent that these antianxiety agents really work, they do so simply by making people feel better by making them feel dull and insensitive. They do not solve any of the underlying problems that cause anxiety.

Like most other mind-altering drugs, Valium acts by fitting specific receptors. Presumably our bodies produce Valium-like compounds that fit these receptors. To date, no such compound has been found. Rather, scientists have found compounds, called β-carbolines (β=beta), which act on the brain's anxiety receptors to produce terror. There is yet so much to learn about the chemistry of the brain.

Anyway, what price tranquility? After 20 years of use, it was finally found that Valium is addictive. People trying to go off it after prolonged use go into painful withdrawal.

CHAPTER SUMMARY

Amines, nitrogen-containing organic bases, are classified according to their structure as shown below. The last structure is a substituted ammonium ion rather than an amine.

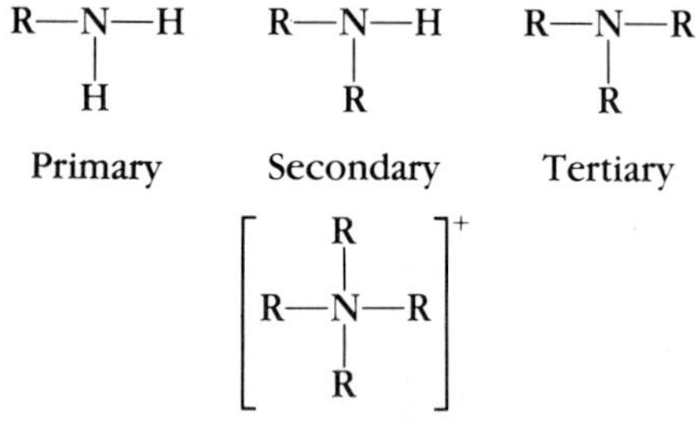

```
    R—N—H       R—N—H       R—N—R
      |           |           |
      H           R           R
   Primary     Secondary    Tertiary

              ⎡    R    ⎤ +
              ⎢    |    ⎥
              ⎢ R—N—R   ⎥
              ⎢    |    ⎥
              ⎣    R    ⎦
                Quaternary
                   ion
```

The names of aliphatic amines consist of the names of the alkyl groups attached to the nitrogen followed by the suffix -amine. The simplest aromatic amine is named aniline. Other aromatic amines are named as aniline derivatives or have special common names. The position of an alkyl group attached to the nitrogen atom in an aromatic amine is specified by the letter N.

Heterocyclic compounds are those in which an atom other than carbon is incorporated in a ring. Compounds in which all ring atoms are carbon are described as carbocyclic. Some examples of heterocyclic amines are shown in Figure 12.2.

Amines that are isolated from plants are called alkaloids.

The molecules of primary or secondary amines can form hydrogen bonds with one another, but these interactions are somewhat weaker than those found in alcohols. These amines have boiling points lower than those of the corresponding alcohols. Tertiary amines, which do not possess a hydrogen atom suitable for hydrogen bonding, boil at about the same temperature as comparable ethers. The smaller amines of all classes show appreciable solubility in water and also exhibit strong, objectionable odors.

The molecules of all classes of amines have an unshared pair of electrons, which gives the amines their characteristic basicity. Amines react with acids to form ammonium salts. The cation formed from the amine is named by changing the ending -amine to

-*ammonium*. The name of the salt, as usual, consists of the name of the cation followed by the name of the anion. The cation formed from aniline is the anilinium ion.

Amines may also react with carboxylic acids to form substituted amides.

$$R-\overset{O}{\underset{\|}{C}}-OH + R'-NH_2 \xrightarrow{heat} R-\overset{O}{\underset{\|}{C}}-NHR' + H_2O$$

A number of local anesthetics, including benzocaine and procaine, are based on *p*-aminobenzoic acid. Two amines, norepinephrine and serotonin, are naturally occurring, mood-controlling chemicals. Many other nitrogen-containing compounds, both naturally occurring and synthetic, are marketed as drugs. Included among these are the amphetamines ("uppers"), barbiturates ("downers"), potent tranquilizers like chlorpromazine, and minor tranquilizers like meprobamate. Some of these are strongly addictive, and many exhibit dangerous interactions with other drugs, including alcohol.

CHAPTER QUIZ

1. Which is a primary amine?
   a. NH_3 b. RNH_2 c. R_2NH d. R_3N
2. R_2NH and $R-CH(OH)R$ are
   a. both primary
   b. both secondary
   c. primary and secondary, respectively
   d. secondary and primary, respectively
3. How many different structures are possible for the amine C_2H_7N?
   a. 1 b. 2 c. 3 d. 4
4. In nomenclature the prefix "alpha" (α) refers to
   a. first listed atoms
   b. carbon next to functional group
   c. shape of the letter α
   d. first member of a series
5. The compound pyrrole is
   a. carbocyclic
   b. aliphatic
   c. linear
   d. heterocyclic

6. Which amines are not capable of intermolecular hydrogen bonding?
   a. primary b. secondary
   c. tertiary d. none
7. R_3N reacts with an acid to form
   a. an alcohol b. a salt
   c. an amine d. water
8. Carboxylic acids react with ammonia to produce
   a. amides b. amines
   c. esters d. alcohols
9. Which is a barbiturate?
   a. cocaine b. thiopental
   c. heroin d. nicotine
10. Codeine is a
    a. stimulant b. narcotic
    c. barbiturate d. local anesthetic

PROBLEMS

12.1 Structure and Classification of Amines

1. Write structural formulas for the eight isomeric amines that have the molecular formula $C_4H_{11}N$. Give each a common name and classify it as a primary, secondary, or tertiary amine.
2. Identify each of the following compounds as an amine, alcohol, phenol, or ether. In the case of an amine or alcohol, classify the compound as primary, secondary, or tertiary.
   a. $CH_3CH_2CH_2OH$
   b. $CH_3CH_2CH_2NH_2$
   c. $CH_3\overset{OH}{\underset{|}{C}}HCH_3$
   d. $CH_3\overset{NH_2}{\underset{|}{C}}HCH_3$
   e.
   f.
   g. $CH_3CH_2NHCH_2CH_3$
   h. $CH_3CH_2OCH_2CH_3$
   i. $CH_3-\underset{\underset{CH_3}{|}}{N}-CH_3$
   j. $CH_3-\overset{OH}{\underset{\underset{CH_3}{|}}{\overset{|}{C}}}-CH_3$
   k.
   l.

12.2 Simple Amines

3. Write structural formulas for the following.
   a. dimethylamine
   b. diethylmethylamine
   c. cyclohexylamine

4. Write structural formulas for the following.
   a. 3-aminopentane
   b. 1,6-diaminohexane

5. Write the structural formula of 2-aminoethanol.

6. Write structural formulas for the following.
   a. aniline
   b. *m*-bromoaniline
   c. *N,N*-dimethylaniline

7. Name the following compounds.
   a. $CH_3CH_2CH_2NH_2$
   b. $CH_3CH_2NHCH_3$
   c. $CH_3CH_2-N-CH_2CH_3$
      $\quad\quad\quad\quad\;\; |$
      $\quad\quad\quad\quad\;\; CH_2CH_3$

8. Name the following compounds.
   a. $O_2N-\bigcirc-NH_2$
   b. $\bigcirc-NHCH_2CH_3$

12.3 Heterocyclic Amines

9. Distinguish between a carbocyclic compound and a heterocyclic compound.

10. Write structural formulas for the following.
    a. pyridine
    b. purine
    c. pyrimidine

11. What is an alkaloid? Name several alkaloids.

12.4 Physical Properties of Amines

12. Which compound has the higher boiling point: butylamine or pentane? Why?

13. Which compound has the higher boiling point: butylamine or butyl alcohol? Why?

14. Which compound has the higher boiling point: trimethylamine or propylamine? Why?

15. Which compound is more soluble in water: $CH_3CH_2CH_3$ or $CH_3CH_2NH_2$? Why?

16. Which of the following compounds is more soluble in water? Why?

    $CH_3CH_2CH_2NH_2$ or
    $\quad\quad CH_3CH_2CH_2CH_2CH_2CH_2CH_2NH_2$

17. Which of the following compounds is more soluble in water? Why?

    $\quad NH_2 \quad\;\; CH_3 \quad\; CH_3$
    $\quad\;\; | \quad\quad\quad | \quad\quad\;\; |$
    $CH_2CH_2CHCH_2CHCH_3$
    or
    $\quad NH_2 \quad\;\; NH_2 \quad\; NH_2$
    $\quad\;\; | \quad\quad\quad | \quad\quad\;\; |$
    $CH_2CH_2CHCH_2CHCH_3$

12.5 Chemical Properties of Amines

18. Write structural formulas for the following.
    a. anilinium bromide
    b. "aniline hydrochloride"
    c. tetramethylammonium chloride

19. Name the following compound.

    $[CH_3CH_2NH_2CH_2CH_3]^+ \; Br^-$

20. Amine X is insoluble in water, yet it dissolves readily in aqueous hydrochloric acid. Explain.

21. Draw the structural formula of the salt formed.

    $CH_3NH_2 + HBr \rightarrow$

22. Draw the structural formula of the salt formed.

    $\bigcirc-NHCH_3 + HNO_3 \longrightarrow$

23. Draw the structural formula of the salt formed.

    $CH_3-N-CH_3 + H_2SO_4 \longrightarrow$
    $\quad\quad\;\; |$
    $\quad\quad\; CH_3$

24. Draw the structural formula of the salt formed.

    $\bigcirc\!\!\!\!{}_{N}^{H} + HCl \longrightarrow$

Problems

25. Draw the amide, if any, that is derived from hexanoic acid and butylamine.

26. Draw the amide, if any, that is derived from

    CH$_3$CH$_2$C(=O)—OH   and   CH$_3$—NH—CH$_3$

27. Draw the amide, if any, that is derived from benzoic acid and aniline.

28. Draw the amide, if any, that is derived from

    CH$_3$C(=O)—OH   and   CH$_3$—N(CH$_3$)—CH$_3$

29. Draw the carboxylic acid and amine from which the following amide was formed.

    CH$_3$CH$_2$N(CH$_3$)—C(=O)CH$_2$CH$_3$

30. Draw the carboxylic acid and amine from which the following amide was formed.

    C$_6$H$_5$—C(=O)—N(CH$_3$)—CH$_3$

31. Draw the carboxylic acid and amine from which the following amide was formed.

    C$_6$H$_5$—NHC(=O)CH$_2$CH$_3$

32. Draw the carboxylic acid and amine from which the following amide was formed.

    (piperidine)N—C(=O)—CH$_3$

33. A carboxyl group and an amino group combined to form the amide functional group in the following compound. Draw the two starting materials for the reaction.

    ? + ? ⟶ H$_2$N—CH$_2$C(=O)—NHCH$_2$C(=O)—OH

12.6 Physiological Properties of Selected Amines and Derivatives

34. What do we mean when we say the amphetamines are "uppers"? Why are barbiturates called "downers"?

35. What is the basic structure common to all barbiturate molecules? How is the basic structure modified to change the properties of individual barbiturate drugs?

36. What is synergism?

37. Which two naturally occurring amines are presently considered to play major roles in the biochemistry of mental health? What are their proposed roles?

38. Examine the structure of the reserpine molecule (p. 400). Identify the following.
    a. five ether functional groups
    b. two amine functional groups
    c. two ester functional groups

39. Acetbutolol has been proposed as a drug for the treatment of heart disease (angina and arrhythmias) and hypertension. The compound has five functional groups. Name the five families of organic compounds to which acetbutolol could be assigned.

    CH$_3$CH$_2$CH$_2$C(=O)—NH—C$_6$H$_3$(C(=O)CH$_3$)—O—CH$_2$CHCH$_2$NHCHCH$_3$
    with OH and CH$_3$ substituents

    Acetbutolol

40. Labetalol has been proposed as a drug for the treatment of angina and hypertension. Circle the four functional groups in the molecule, and name the families of organic compounds that incorporate these functional groups.

    C$_6$H$_5$—CH$_2$CH$_2$CHNHCH$_2$CH—C$_6$H$_3$(C(=O)NH$_2$)(OH)
    with CH$_3$ and OH substituents

    Labetalol

41. Refer to Table 12.3. Which is more toxic: procaine or cocaine? Can you use the data in Table 12.3 to compare the toxicities of lidocaine and cocaine?

42. If the minimum lethal dose (MLD) of amphetamine is 5 mg per kilogram of body weight, what would be the MLD for a 70-kg person? Can toxicity studies on animals always be extrapolated to human beings?

13

CARBOHYDRATES: STRUCTURE AND METABOLISM

DID YOU KNOW THAT...

- carbohydrates are made by plants from carbon dioxide and water?
- the term *carbohydrate* is actually a misnomer?
- living organisms can use only right-handed carbohydrates?
- deoxyribose is ribose without an oxygen atom?
- a diabetic person cannot process sugar normally?
- lactose is found in milk?
- termites can live on cellulose?
- digestion of carbohydrates occurs by hydrolysis?
- glycolysis means "splitting sugar"?

Almost everyone knows what carbohydrates are: they are what you eat or do not eat, depending on whose diet you follow. In fact, as any dietitian or nutritionist will tell you, carbohydrates must be included in any well-balanced diet. They are the body's primary source of energy. Energy is stored in the complex molecular structure of the carbohydrates at the time these compounds are synthesized by green plants from carbon dioxide and water. The entire biosphere, and that includes human beings, is dependent on this endothermic reaction called photosynthesis. When we metabolize the products of photosynthesis, the complex compounds rearrange back into simple molecules, and in the process they release their stored energy for our use. Subtle variations in the molecular structures of carbohydrates have profound effects on the utility of the compounds. We shall soon see how the difference in the orientation of a bond can mean the difference between nutrition and starvation.

13.1 Carbohydrates: Definitions and Classifications

Carbohydrate = *carbo* (carbon) + *hydr* (hydrogen) + *ate* (oxygen)

Carbohydrates are compounds of carbon, hydrogen, and oxygen. They include starches, sweet-tasting compounds called sugars, and structural materials such as cellulose (Figure 13.1). The term **carbohydrate** had its origin in a misinterpretation of the molecular formulas for many of these substances. For example, the formula for several important sugar isomers is $C_6H_{12}O_6$, but the formula can also be presented as that of a "carbon hydrate," $(C \cdot H_2O)_6$. These compounds are not hydrates of carbon, however; they are alcohols. All contain the hydroxyl (—OH) functional group, and most contain several such groups. Most also contain a real or latent carbonyl (C=O) group. By **latent carbonyl group**, we mean a functional group such as a hemiacetal or an acetal (Section 10.14) that can be more or less readily converted to a carbonyl group.

FIGURE 13.1
Wood, lettuce, potatoes, and cotton are all substances that are rich in carbohydrates. Sugar is a pure carbohydrate. [Photo © The Terry Wild Studio.]

13.3 Mirror-Image Isomerism

Carbohydrates are often called **saccharides** (from the Latin *saccharum*, "sugar"). William Proust, the English physician who first recognized the three general classes of foodstuffs that we now call carbohydrates, fats, and proteins, suggested that they be called the saccharine, the oily, and the albuminous. Simple carbohydrates, those that cannot be hydrolyzed to smaller units, are called **monosaccharides**. Carbohydrates that can be hydrolyzed to two monosaccharide units are called **disaccharides**, and carbohydrates that can be hydrolyzed to many monosaccharide units are called **polysaccharides**.

Saccharide—sugar

13.2 Monosaccharides: Further Classifications

Monosaccharides can also be classified by the number of carbon atoms per molecule. The *-ose* ending indicates a carbohydrate; the prefixes are ones we have encountered previously.

Number of Carbon Atoms	Class
3	Triose
4	Tetrose
5	Pentose
6	Hexose
7	Heptose

Still another system permits classification of monosaccharides on the basis of the type of carbonyl group present. Sugars with an aldehyde group are called **aldoses**, and those with a ketone group are called **ketoses**. The two systems are often combined. For example, an aldopentose is a monosaccharide with five carbon atoms and an aldehyde function. Similarly, a ketohexose has six carbon atoms and a ketone function.

13.3 Mirror-Image Isomerism

The simplest sugars are the **trioses**. The aldotriose glyceraldehyde and the ketotriose dihydroxyacetone are important intermediates in carbohydrate metabolism.

$$\begin{array}{cc}
\text{H} & \\
| & \\
\text{C}=\text{O} & \text{CH}_2\text{OH} \\
| & | \\
\text{CHOH} & \text{C}=\text{O} \\
| & | \\
\text{CH}_2\text{OH} & \text{CH}_2\text{OH} \\
\text{Glyceraldehyde} & \text{Dihydroxyacetone}
\end{array}$$

FIGURE 13.2
Mirror-image isomers cannot be superimposed. No matter how you turn the one isomer, it will never exactly match the other.

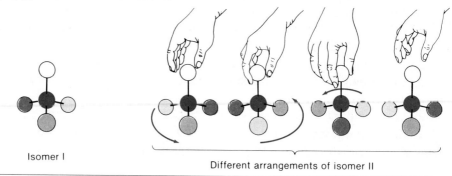

Glyceraldehyde exhibits a type of isomerism we have not discussed before. Glyceraldehyde exists in two forms, described as mirror images of one another.

$$
\begin{array}{c c c}
\text{I} & \text{Mirror plane} & \text{II}
\end{array}
$$

These drawings should be viewed as three-dimensional, with the wedge-shaped bond coming out of the page toward you and the dotted bond going back into the page away from you. Although it might seem that the two structures represent the same compound, they do not. Structure I cannot be made to coincide exactly with structure II. If you could grab one of the structures by its carbonyl (C=O) group and spin the remaining three legs of the tetrahedron around, you could *not* make the three remaining groups of the one model line up exactly with the corresponding three groups in the other model. The two structures represent two real and different compounds that are mirror-image isomers (Figure 13.2).

A chiral carbon atom has four different groups attached to it.

The glyceraldehydes contain a **chiral carbon atom**, that is, a carbon that is attached to four *different* groups. Compounds that incorporate this structural feature can exist as mirror-image isomers and are said to exhibit **chirality** or "handedness." (Your hands are mirror-image isomers of one another.)

Note that the following structures,

which are also mirror images of each other, are not mirror-image *isomers*. They represent the same molecule. This compound does *not* contain a chiral carbon atom because there are two identical groups (the two hydrogen atoms) attached to the central carbon atom.

As usual, chemists simplify the drawings of mirror-image isomers to save time. Instead of attempting to represent accurately the three-dimensional structure, they settle for two-dimensional drawings of isomers such as glyceraldehyde.

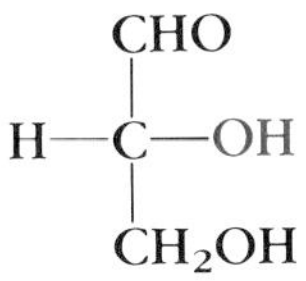

$$\begin{array}{cc} \text{CHO} & \text{CHO} \\ \text{H}-\text{C}-\text{OH} & \text{HO}-\text{C}-\text{H} \\ \text{CH}_2\text{OH} & \text{CH}_2\text{OH} \\ \text{III} & \text{IV} \end{array}$$

The two isomers are also distinguished by prefixes attached to their names. The two-dimensional formula drawn with the hydroxyl (OH) group to the right (III) is called D-glyceraldehyde (D for *dextro*, from the Latin *dexter*, "right"). When the hydroxyl group is drawn to the left (IV), the compound represented is L-glyceraldehyde (L for *levo*, from the Latin *laevus*, "left").

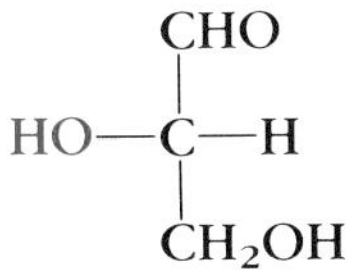

The natural sugars that we shall consider are all members of the D family, that is, they are related to D-glyceraldehyde. Because we will deal only with D sugars, we will not always explicitly specify the "handedness" of these compounds. Nonetheless, this form of isomerism is of major significance. For some as yet unknown reason, life on this planet developed in such a way that only D carbohydrates were biosynthesized and metabolized. The chemical factories we call our bodies are simply unable to process L sugars.

13.4 SOME COMMON MONOSACCHARIDES

We shall consider only a few of the simple carbohydrates, and these are presented in Figure 13.3. Each of these structures has more than one chiral carbon atom (marked by *). The arrangement of the bottom-most

FIGURE 13.3 Some common monosaccharides.

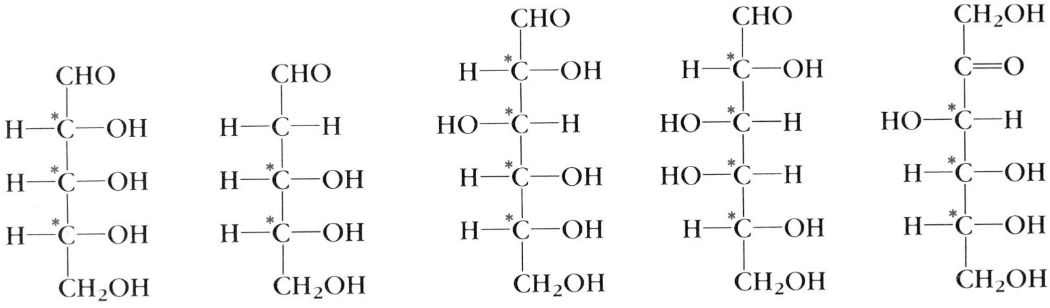

$$\begin{array}{ccccc}
\text{CHO} & \text{CHO} & \text{CHO} & \text{CHO} & \text{CH}_2\text{OH} \\
\text{H}-\overset{*}{\text{C}}-\text{OH} & \text{H}-\overset{*}{\text{C}}-\text{H} & \text{H}-\overset{*}{\text{C}}-\text{OH} & \text{H}-\overset{*}{\text{C}}-\text{OH} & \text{C}=\text{O} \\
\text{H}-\overset{*}{\text{C}}-\text{OH} & \text{H}-\overset{*}{\text{C}}-\text{OH} & \text{HO}-\overset{*}{\text{C}}-\text{H} & \text{HO}-\overset{*}{\text{C}}-\text{H} & \text{HO}-\overset{*}{\text{C}}-\text{H} \\
\text{H}-\overset{*}{\text{C}}-\text{OH} & \text{H}-\overset{*}{\text{C}}-\text{OH} & \text{H}-\overset{*}{\text{C}}-\text{OH} & \text{HO}-\overset{*}{\text{C}}-\text{H} & \text{H}-\overset{*}{\text{C}}-\text{OH} \\
\text{CH}_2\text{OH} & \text{CH}_2\text{OH} & \text{H}-\overset{*}{\text{C}}-\text{OH} & \text{H}-\overset{*}{\text{C}}-\text{OH} & \text{H}-\overset{*}{\text{C}}-\text{OH} \\
& & \text{CH}_2\text{OH} & \text{CH}_2\text{OH} & \text{CH}_2\text{OH} \\
\text{D-Ribose} & \text{D-2-Deoxyribose} & \text{D-Glucose} & \text{D-Galactose} & \text{D-Fructose}
\end{array}$$

chiral carbon atom determines whether the sugar is D or L. All these naturally occurring sugars are classified as D because the hydroxyl group attached to this bottom chiral carbon atom is pointing to the right just as it is in D-glyceraldehyde. D-Ribose and D-deoxyribose are **aldopentoses**; D-glucose and D-galactose are **aldohexoses**, and D-fructose is a **ketohexose**.

Ribose and the closely related compound deoxyribose are incorporated in a variety of complex molecules, including the nucleic acids, the stuff of which genes are made. As the name suggests, deoxyribose is ribose minus one oxygen atom; the hydroxyl group at the second carbon atom of ribose has been replaced by a hydrogen atom in deoxyribose.

Glucose is undoubtedly the most important hexose. This sugar is sometimes called *dextrose* or *blood sugar* and makes up about 0.065–0.110% of our blood. An approximately 5% solution of glucose in water is frequently used as an intravenous source of nourishment or quick energy. It has been estimated that half of the carbon atoms in the biosphere are tied up in glucose. Unfortunately for the hungry people of the world, most of it is stored in the form of a polysaccharide, cellulose (Section 13.8), which has little or no food value for humans.

Galactose is obtained, along with glucose, from the hydrolysis of milk sugar. It is also found in brain and nerve tissue as part of very complex molecules called cerebrosides and gangliosides. The only difference between galactose and glucose is the position of the hydroxyl group on the fourth carbon atom of these hexoses.

Fructose is the only ketose that we shall examine closely. It can be found with other sugars in honey and fruit juices; it is the only sugar found in the semen of men and bulls; and it can be obtained, along with glucose, from the hydrolysis of common table sugar. Fructose is sometimes called *levulose*. From the third through the sixth carbon atoms, its structure is identical to that of glucose.

13.5 More on the Structure of Monosaccharides

We have shown the monosaccharides as hydroxyaldehydes and ketones. In Chapter 10, we examined a reaction of aldehydes (or ketones) with alcohols. Usually, to bring about the reaction, one must allow an aldehyde molecule to come into contact with an alcohol molecule. With the sugars, however, such contact is inevitable because both functional groups are on the same molecule.

The reaction of an aldehyde with an alcohol produces a hemiacetal.

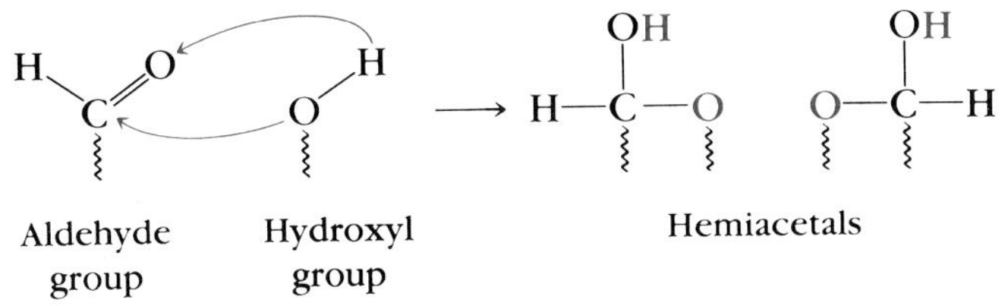

13.5 More on the Structure of Monosaccharides

$$\text{Aldehyde group} + \text{Hydroxyl group} \longrightarrow \text{Hemiacetals}$$

We show two products because the carbon atom is chiral; four different groups are now attached to the former carbonyl carbon. To make this point clearer, let us show the reaction for glucose.

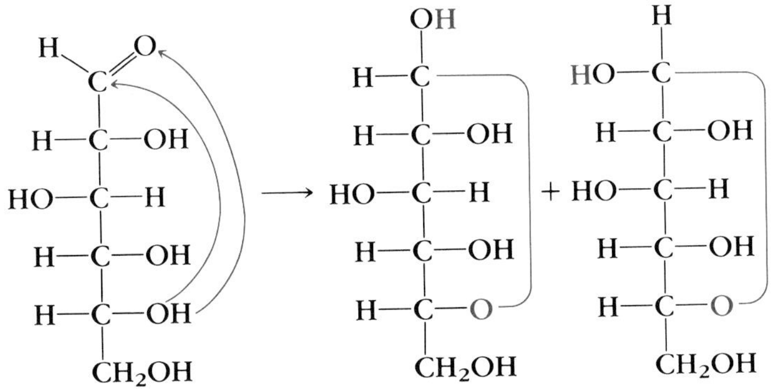

When the hydroxyl group at carbon number 5 reacts with the carbonyl group, the resulting —H and —OH groups at C-1 (the carbon in the number 1 position) can orient in two different ways.

"Now wait a minute," you say. "You have ignored a hydroxyl group right next to the carbonyl in order to play around with one located far down the chain of carbon atoms, and you had to write a long, ridiculous-looking bond to do it!" That is a valid point. In Section 10.14 we considered a similar compound, but there we arranged the molecule so that we took into account approximately correct bond angles. When we did that, the formula for the molecule did not look so awkward. Figure 13.4 shows a model of the aldehyde form of glucose; it is a floppy molecule and can fold back on itself. The center structure in Figure 13.5 is drawn to resemble the model. Notice how the hydroxyl group on C-5 is actually quite near the carbonyl carbon atom. When hemiacetal formation takes place, the originally doubly bonded oxygen atom may be pushed up or down, producing the two hemiacetal forms. The structure on the left, with the hydroxyl group on C-1 projected downward, is called the alpha (α) form. That on the right, with this hydroxyl group pointed upward, is the beta (β) form. The pure beta or the pure alpha form can be obtained as crystals. Once these are placed in solution, however, the forms interconvert. This interconversion is referred to as **mutarotation** (from the Latin *mutare*, "to change"). At equilibrium, the mixture is about 36% alpha and 64% beta. Less than

FIGURE 13.4
Models of the aldehyde form of glucose. In the folded one the oxygen atom on the fifth carbon is near the carbonyl carbon.

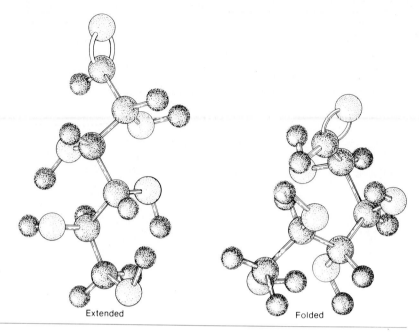

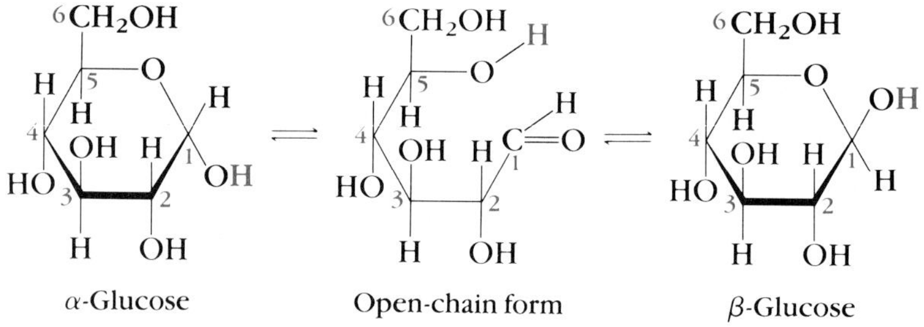

FIGURE 13.5
In aqueous solution, glucose exists as an equilibrium mixture of these three forms. α-Glucose accounts for about 36%; the open-chain form, for about 0.02%; and β-glucose, for about 64%.

FIGURE 13.6
Cyclic forms of common monosaccharides. Compare with Figure 13.3. The wavy bonds indicate that the hemiacetal or hemiketal hydroxyl group can lie above or below the plane of the ring. Note that the five-carbon sugars ribose and deoxyribose and the six-carbon ketose, fructose, preferentially form five-membered rings.

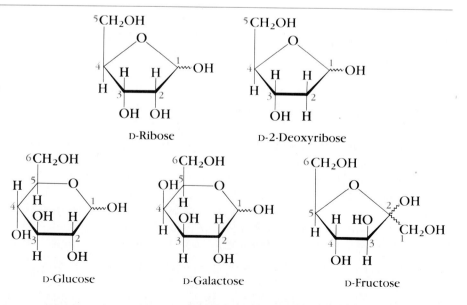

0.02% of the aldehyde form is present, but this is enough to give most of the characteristic reactions of aldehydes. Because the forms interconvert, all the molecules eventually can react through the aldehyde form.

Figure 13.6 presents a selection of monosaccharides in their most common cyclic forms.

The difference between alpha and beta forms may seem trivial, but such differences are often crucial in biochemical reactions, as we shall soon see.

13.6 Some Reactions of Monosaccharides

The formation of hemiacetals (discussed in Section 13.5) and of hemiketals illustrates the value of the concept of a functional group. Sugars, like other compounds containing hydroxyl and carbonyl groups, exhibit chemical properties characteristic of these functional groups.

Aldehydes are among the most easily oxidized functional groups in organic chemistry. Therefore, you should not be surprised to learn that the aldoses react with the weak oxidizing agents found in Tollens', Fehling's, and Benedict's reagents. What should surprise you is the fact that ketoses such as fructose also react with these reagents. The simple ketones we discussed in Chapter 10 are not readily oxidized. Why should the ketones in Chapter 13 be different? The answer is that the sugars are not simple ketones but hydroxy ketones, and this structural feature permits sugars to isomerize under certain conditions.

Many aldehydes and ketones (including simple ones) can undergo a rearrangement referred to as **tautomerization**.

$$-\underset{H}{\overset{}{C}}-\overset{O}{\underset{}{\overset{\|}{C}}}- \quad \underset{}{\overset{base}{\rightleftharpoons}} \quad -C=\underset{}{\overset{OH}{C}}-$$

The product is an *enol (both an alkene and an alcohol)*. The enol and the aldehyde or ketone are isomers or, more specifically, **tautomers**. They differ only in the position of a hydrogen atom and a double bond. Only aldehydes and ketones that have a hydrogen atom located on the carbon atom adjacent to the carbonyl group can undergo this isomerization. The adjacent carbon atom is termed the *alpha** carbon,

Tautomers: isomers that differ only in the position of a double bond and a hydrogen

---

* It is unfortunate that chemists chose to use the Greek alphabet for describing various ring forms of sugars and for designating positions along the chain of a carbonyl-containing compound. Thus, *alpha* is used to specify one of the two hemiacetal forms of cyclic sugar molecules and also the position adjacent to a carbonyl group. There is no connection between these two uses of the same term.

and the attached hydrogen atom is called an *alpha* hydrogen. For most aldehydes and ketones, the equilibrium strongly favors the carbonyl-containing isomer.

What does all this have to do with the oxidation of ketoses? If ketoses are treated with base (or basic reagents), they undergo tautomerization.

$$\underset{\underset{H}{|}}{\overset{\overset{HO}{|}}{-C}}-\overset{\overset{O}{\|}}{C}- \rightleftharpoons -\overset{\overset{HO}{|}}{C}=\overset{\overset{OH}{|}}{C}-$$

But ketoses not only have alpha hydrogen atoms, they also have alpha hydroxyl groups. Once the ketose has isomerized, the two hydroxyl groups are essentially equivalent. The enol can revert back to the ketone form by reforming the carbonyl group at its original position. However, the enol could also reform the carbonyl group at what was originally the hydroxyl position.

$$-\overset{\overset{HO}{|}}{C}=\overset{\overset{OH}{|}}{C}- \rightleftharpoons -\overset{\overset{O}{\|}}{C}-\underset{\underset{H}{|}}{\overset{\overset{OH}{|}}{C}}-$$

Thus, the complete equilibrium system looks like this:

$$\underset{\underset{H}{|}}{\overset{\overset{HO}{|}}{-C}}-\overset{\overset{O}{\|}}{C}- \rightleftharpoons -\overset{\overset{HO}{|}}{C}=\overset{\overset{OH}{|}}{C}- \rightleftharpoons -\overset{\overset{O}{\|}}{C}-\underset{\underset{H}{|}}{\overset{\overset{OH}{|}}{C}}-$$

Tollens', Fehling's, and Benedict's reagents are all basic solutions. If fructose is treated with one of these solutions, this equilibrium is established.

$$\begin{array}{c} H \\ | \\ H-C-OH \\ | \\ C=O \\ | \\ HO-C-H \\ | \\ H-C-OH \\ | \\ H-C-OH \\ | \\ CH_2OH \end{array} \rightleftharpoons \begin{array}{c} H \\ | \\ C-OH \\ \| \\ C-OH \\ | \\ HO-C-H \\ | \\ H-C-OH \\ | \\ H-C-OH \\ | \\ CH_2OH \end{array} \rightleftharpoons \begin{array}{c} H \\ | \\ C=O \\ | \\ CH-OH \\ | \\ HO-C-H \\ | \\ H-C-OH \\ | \\ H-C-OH \\ | \\ CH_2OH \end{array}$$

13.6 Some Reactions of Monosaccharides

That is why fructose (a ketose) is as readily oxidized as glucose (an aldose). As soon as the ketose comes into contact with a basic solution, some aldose is formed and reacts with the weak oxidizing agent. As the aldose isomer is removed from the equilibrium, more of it forms (Le Châtelier's principle). Eventually all of the sample could be oxidized. All monosaccharides are called **reducing sugars** because all are oxidized by weak oxidizing agents. (The sugars are called *reducing* sugars because they *reduce* the oxidizing agent.)

Reducing sugars are oxidized by weak oxidizing agents.

The equilibrium we have shown is oversimplified. Instead of involving the hydroxyl group at C-1, we could have involved C-3 in the equilibrium, which would ultimately result in the formation of a carbonyl group at C-3. That, in turn, could lead to isomerization at C-4. Once these sugars are placed in basic solution, enolization can ultimately convert any of the hydroxyl groups in the molecule to a carbonyl group. In carbohydrate metabolism, sugars undergo analogous isomerizations, permitting different isomers to form a common intermediate that can then enter a particular reaction pathway.

Diabetic persons fail to metabolize sugar normally, and in uncontrolled diabetes sugar appears in the urine (Section 13.11). A relatively simple screening test for diabetes detects the presence of this sugar. A urine sample is treated with Benedict's reagent, which contains copper(II) ion. If a red-orange precipitate of copper(I) oxide forms in the clear blue reagent solution, the presence of sugar is assumed, and further testing is indicated. Because a knowledge of glucose levels is so important to the treatment of diabetes, much effort has been put into the development of easy-to-use test methods. An elegant application of chemistry permits diabetic individuals to monitor their own condition. A plastic strip is impregnated with all the necessary chemicals. The strip is dipped in a urine sample, and the color that develops on the strip is compared to color standards to find the sugar concentration (Figure 13.7).

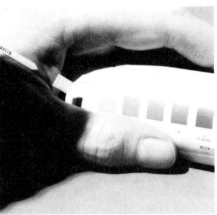

FIGURE 13.7
A simple test for sugar in urine. The reagent strip is dipped into the urine sample and then compared to a color chart that indicates sugar content. [Diastix is a registered trademark of Ames Division, Miles Laboratories, Inc., Elkhart, Ind. 46515.]

The third reaction of monosaccharides to be considered here has important consequences for the structure of disaccharides and polysaccharides. Hemiacetals react with alcohols to form acetals (Section 10.14). The cyclic hemiacetal forms of monosaccharides can also form acetals.

$$\text{An } \alpha\text{-aldohexose} + \text{ROH} \xrightarrow{\text{HCl}} \text{An } \alpha\text{-glycoside (an acetal)} + H_2O$$

The alcohol with which the hemiacetal reacts may be as simple as methanol or as complicated as another sugar molecule (Section 13.7). While hemiacetals are unstable and open to form free aldehyde groups, acetals are quite stable. Thus **glycosides**, as the acetals formed from sugars are called, do not open on their own. A sugar containing a hemiacetal group will reduce Benedict's and Fehling's reagents. If the carbonyl group of the sugar is tied up as an acetal, then it will not reduce these oxidizing agents.

Hemiacetal   Hemiacetal   Acetal

Remember: In a hemiacetal, one carbon atom is attached by single bonds to two different oxygen atoms. One of these oxygen atoms is part of a hydroxyl group. In acetals, one carbon atom is also attached by single bonds to different oxygen atoms, but neither of these oxygen atoms is part of a hydroxyl group.

13.7 Some Important Disaccharides

Disaccharides are acetals formed by the reaction of two monosaccharides. We shall look at three such compounds—maltose, lactose (milk sugar), and sucrose (table sugar).

Maltose is composed of two glucose units joined by an acetal linkage. The two sugar rings are connected through the first carbon atom of one ring and the fourth carbon atom of the other.

13.7 Some Important Disaccharides

Maltose (α form) — Alpha acetal linkage; Hemiacetal (α form)

Maltose (β form) — Alpha acetal linkage; Hemiacetal (β form)

The acetal linkage is fixed in the alpha form because acetals are not free to open and close. The right-hand ring can assume either the alpha or the beta form, since the first carbon atom in this ring is part of a hemiacetal functional group. Maltose is a reducing sugar because of this unstable hemiacetal group. Remember that the acetal link in maltose *must* be alpha. If it were beta, the compound would not be maltose.

In lactose, a disaccharide that is isolated from milk, a galactose unit is joined to a glucose unit through a beta acetal link extending from the first carbon atom of the galactose ring to the fourth carbon atom of the glucose ring.

Lactose (α form) — Galactose unit, Glucose unit; Beta acetal linkage; Hemiacetal (α form)

Lactose (β form) — Galactose unit, Glucose unit; Beta acetal linkage; Hemiacetal (β form)

Note that lactose is also a reducing sugar because the first carbon atom of the glucose unit is in the hemiacetal form. Like maltose and the monosaccharides, lactose exists in alpha and beta forms. The acetal link connecting the two rings of lactose must, of course, remain beta.

Sucrose is common table sugar, the carbohydrate most people call "sugar." It is obtained principally from sugarcane and sugar beets. It is also the first nonreducing sugar we have encountered. Sucrose is composed of a glucose unit and a fructose unit. Both monosaccharides use their latent carbonyl groups in making the connection between the two rings.

Only one isomer of the disaccharide sucrose exists. Both rings are locked in the arrangement shown. The compound has no hemiacetal groups, and sucrose will not reduce Benedict's or Fehling's reagent.

13.8 Polysaccharides

Starch is a polymer of glucose.

Amylose is a single-chain polymer of glucose.

Amylopectin is a branched-chain polymer of glucose.

Cellulose is also a polymer of glucose.

Plants make glucose by photosynthesis and store it in the form of **starch**, a polymer of glucose. Plant seeds, especially the cereal grains, and tubers, such as potatoes, are particularly rich in starch. These are the energy reserves for the plant and are sources of energy for animals.

Starch can be separated into two fractions, one called **amylose** and the other, **amylopectin**. Both forms contain long chains of glucose units with C-1 of each unit joined to C-4 of the next by an alpha linkage (the same arrangement one finds in maltose). In amylopectin, branches are formed at various distances along this chain. The branches are formed by links between the first and *sixth* carbon atoms of a pair of rings. Partial structures of both amylose and amylopectin are shown in Figure 13.8.

Animals store carbohydrate in a form closely related to amylopectin in structure. In animals, this branched, alpha-linked polyglucose molecule is referred to as **glycogen**.

Like starch and glycogen, **cellulose** is a polymer of glucose. It differs, however, in that the glucose units are joined by beta acetal linkages (Figure 13.9). This apparently minor difference has tremendous significance. Although most animals can digest and metabolize starch, only a limited number can obtain food value from cellulose. Cellulose is primarily the structural material of plants. Grazing animals and termites may live on cellulose-rich grasses and woods, but human beings are unable to extract nutritional value from the cellulose. Thus, millions of people around the world are starving in the midst of an abundance of glucose because the glucose is tied up as cellulose.

13.8 Polysaccharides

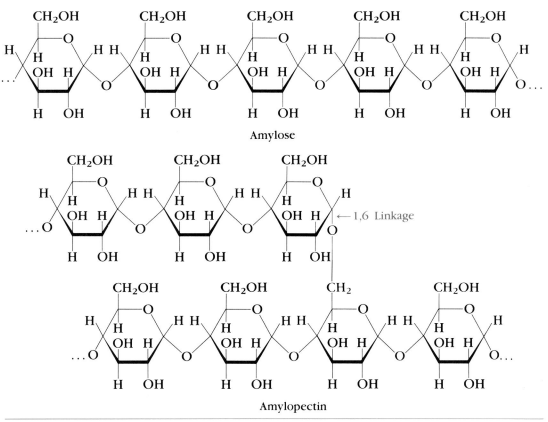

FIGURE 13.8
Structures of small segments of amylose and amylopectin molecules.

FIGURE 13.9
Structure of a small segment of a cellulose molecule.

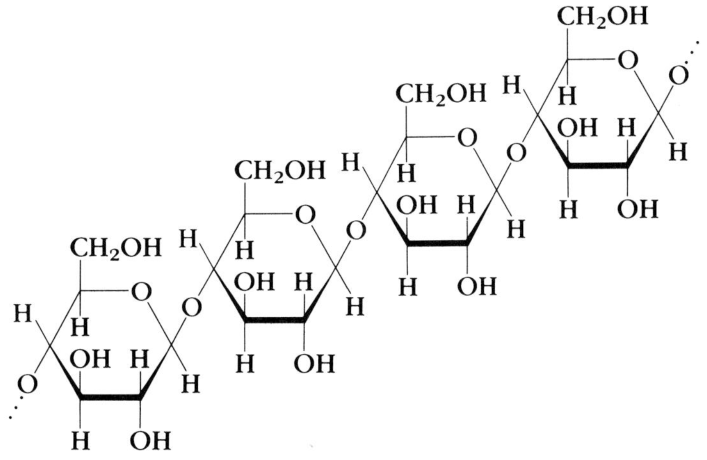

13.9 Digestion of Carbohydrates

Digestion is the process in which food is prepared for incorporation into the body. The process takes place in a long tunnel called the digestive tract that runs through the body. The structures that make up the digestive tract include the mouth, pharynx (throat), esophagus, stomach, small intestine, large intestine, rectum, and anus (Figure 13.10). Beginning in the mouth, but with most of the work being done in the small intestine, food is broken down into units small enough to pass through the walls of the small intestine and into the circulatory system. All the material that does not enter the body during the digestive process continues on through the large intestine and out of the body as waste.

In chemical terms, digestion is hydrolysis. Carbohydrates are hydrolyzed to monosaccharide units. The reaction for sucrose is

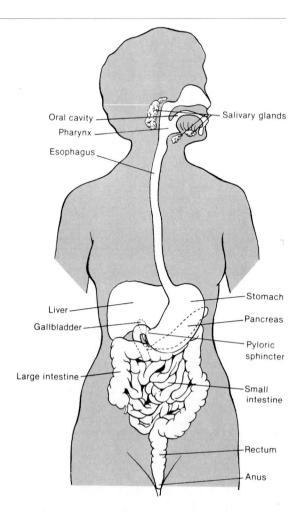

FIGURE 13.10
The human digestive system, showing the organs somewhat displaced for clearer viewing. The small intestine is much longer than indicated here.

13.10 Metabolism of Carbohydrates

[Sucrose + H—OH →(enzyme) Glucose + Fructose structural diagram]

Maltose and lactose are similarly hydrolyzed, and in polysaccharides such as starch, many acetal links must be opened. Enzymes are the biochemical catalysts designed to facilitate these reactions. Each hydrolysis reaction is governed by a specific enzyme—there is one for sucrose, one for maltose, one for lactose, and so on. Human beings do not possess an enzyme that will work on cellulose, and that is why cellulose emerges from the digestive tract relatively unchanged.

Once the monosaccharides make their way into the circulatory system, they are carried to various parts of the body and metabolized.

13.10 Metabolism of Carbohydrates

Figure 13.11 gives an overview of the metabolic processing of the monosaccharides. The simple carbohydrates are first converted to phosphoric acid esters in the liver. After this modification, they can interconvert among themselves. This interconversion is important because certain structures serve as entrance points into key metabolic pathways. The monosaccharides can be converted to glucose 1-phosphate and then to glycogen, the polymeric storage form of glucose in the body.

[Glucose 1-phosphate structural diagram with OPO_3^{2-} group]

Glucose 1-phosphate

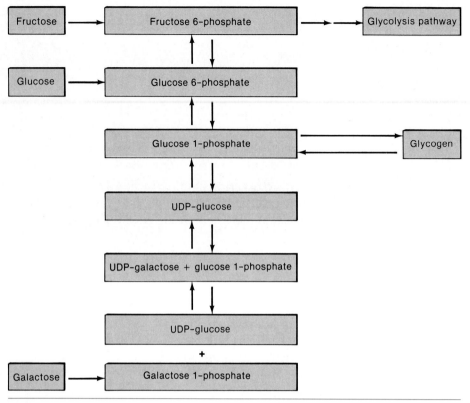

FIGURE 13.11
An overview of carbohydrate metabolism. UDP is uridine diphosphate, a complex molecule that serves as a carrier of sugar molecules.

Conversion to fructose 6-phosphate permits the monosaccharides to enter the glycolysis pathway.

$$^{2-}O_3POCH_2 \quad O \quad CH_2OH$$

Fructose 6-phosphate

In glycolysis a sugar is split.

The word **glycolysis** means "splitting of sugar." In this complex series of reactions, glucose is converted to two lactic acid molecules and releases some of its stored energy.

$$C_6H_{12}O_6 \longrightarrow 2\ C_3H_6O_3 + \text{energy}$$
$$\text{Glucose} \qquad \text{Lactic acid}$$

Figure 13.12 shows the intermediates involved in this conversion and emphasizes the complexity of the process. The reactions involved are familiar: oxidation–reduction (e.g., steps 6 and 11), keto–enol

13.10 Metabolism of Carbohydrates

tautomerism (e.g., steps 5 and 10), ester formation and hydrolysis (e.g., steps 1 and 2). The glycolysis pathway releases energy from glucose in the absence of oxygen (sometimes called *anaerobic* glycolysis, or glycolysis without air).

The energy released by the glucose molecule is stored in other molecules, represented in Figure 13.12 by the abbreviation ATP. We shall discuss ATP (and ADP) and the energy transformations that occur in metabolic processes in Chapter 16.

FIGURE 13.12
Intermediates in the anaerobic glycolysis pathway.

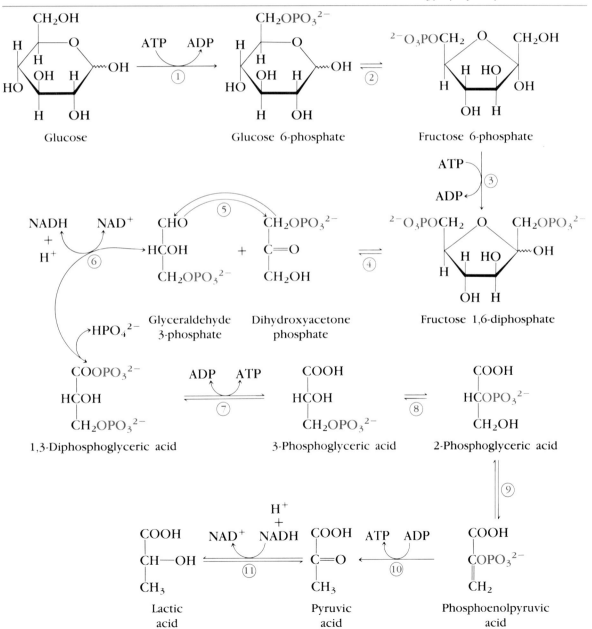

Steps 6 and 11 are oxidation–reductions. The NAD^+ and NADH shown in these steps serve as the oxidizing and reducing agents. We shall also consider these molecules in greater detail in Chapter 16.

Notice that in the last step of the glycolysis pathway (step 11), pyruvic acid is converted to lactic acid. Pyruvic acid can be channeled into a different metabolic pathway, one that does use oxygen to convert the pyruvic acid not to lactic acid but to carbon dioxide and water. In

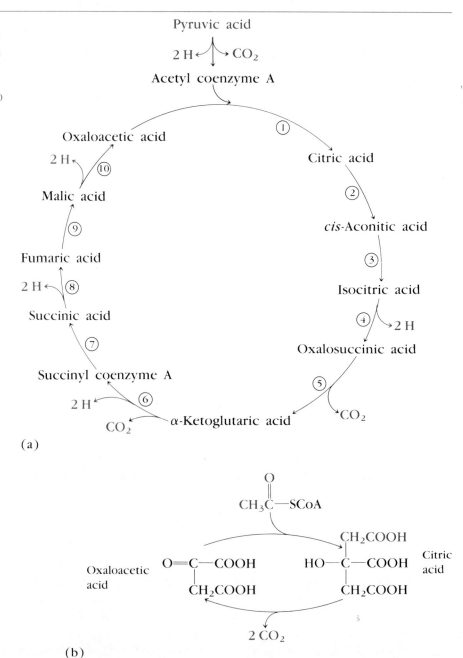

FIGURE 13.13
The Krebs cycle. (a) The conversion of pyruvic acid to carbon dioxide involves a large number of transformations. (b) This greatly condensed version of the cycle emphasizes the fate of the two carbon atoms in acetyl coenzyme A.

this *aerobic* oxidation, all the available energy is released from the original carbohydrate molecule. Even lactic acid that has been formed via glycolysis can be oxidized back to pyruvic acid and then carried through the aerobic process. The alternative pathway is called the **Krebs cycle**, and it *is* a cycle (Figure 13.13a).

In the Krebs cycle, citric acid is ultimately converted to oxaloacetic acid, which is then converted back to citric acid (Figure 13.13b). As pyruvic acid enters the Krebs cycle, one of its three carbon atoms is stripped off as carbon dioxide. The remaining two-carbon fragment is then tacked on to oxaloacetic acid, converting the four-carbon oxaloacetic acid to six-carbon citric acid. As the cycle progresses, two carbon atoms are removed in the form of carbon dioxide to re-form oxaloacetic acid. Overall, the original pyruvic acid molecule is converted to carbon dioxide:

$$\tfrac{5}{2} O_2 + \underset{\text{Pyruvic acid}}{C_3H_4O_3} \longrightarrow 3\,CO_2 + 2\,H_2O$$

The reaction as written above accurately describes the overall chemistry of aerobic oxidation. We would write this same equation for the complete combustion of pyruvic acid in a flame. The equation does not, however, convey the complexity of the process as it occurs in the body under the control of special enzymes and with an efficiency and a speed that chemists can only envy.

13.11 INSULIN AND DIABETES MELLITUS

A variety of hormones affect the blood sugar level, but the one familiar to most people is **insulin**. Insulin is synthesized in the pancreas, and its secretion into the bloodstream is stimulated by a rise in blood sugar levels. The hormone triggers an increase in the rate at which tissues and organs absorb the circulating glucose for storage, oxidation, or transformation. Insulin acts by increasing the transport of glucose across cell membranes. If glucose is to be stored, it is converted to glycogen in liver and muscles. [If those storage areas are full, the glucose can be converted to fat and stored in adipose tissue (fatty tissue).] Of course, the glucose can be oxidized for energy.

Insulin increases the transport of glucose across a cell membrane.

If insulin production is faulty, as it is in individuals who have **diabetes mellitus**, then control of blood sugar levels is lost. The result is an increase in the concentration of blood sugar. The kidneys react to this situation by dumping some of the sugar into the urine. The production of large quantities of sugar-containing urine is characteristic of persons with uncontrolled diabetes. To control the symptoms of the

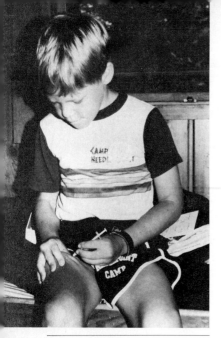

FIGURE 13.14
A diabetic administering insulin. [Courtesy of the American Diabetes Association, Minnesota Affiliate, Inc.]

disease, a diabetic person needs insulin (Figure 13.14). The meat-packing industry has long supplied insulin for this purpose. Pork and beef insulin, however, are not exactly the same as human insulin. They are certainly much better than no insulin at all, but the use of these substitutes can be accompanied by a number of serious side effects. What appears to be a breakthrough in this area has come from research in molecular biology (Section 17.5). Scientists can now create bacteria that will produce human insulin, such as Humulin, identical in every way to the insulin produced by the human pancreas (see Figure 17.17). Many of the side effects of pork and beef insulin can thereby be avoided. Some diabetic individuals have only mild cases of the disease and can produce their own insulin, though usually not enough. These individuals are treated with chemicals that simply stimulate the release of insulin from the pancreas. Some of these drugs have been shown to increase the risk of heart disease, and their use has become controversial. Insulin is a protein molecule (Chapter 15) and is destroyed by the digestive process if taken by mouth. It is usually injected subcutaneously. The synthetic drugs used to stimulate a person's own insulin production are not inactivated by the digestive process and can therefore be taken orally.

CHAPTER SUMMARY

Carbohydrates, one of the three major classes of foodstuffs, are polyhydroxy aldehydes or ketones and are sometimes referred to as saccharides. Those carbohydrates that cannot be hydrolyzed to simpler carbohydrates are called monosaccharides. On hydrolysis, disaccharides give two monosaccharide units and polysaccharides yield many monosaccharide units. The monosaccharides can be classified as aldoses (aldehydes) or ketoses (ketones) depending on the nature of their carbonyl group. They can also be classified according to their carbon content as follows: triose (3 C), tetrose (4 C), pentose (5 C), hexose (6 C), heptose (7 C), and so on.

Carbohydrates exhibit mirror-image isomerism because the compounds consist of chiral molecules. The chirality results from the presence of one or more carbon atoms attached to four *different* groups. The mirror-image isomers are classified into two families, D and L, depending on the orientation of the hydroxyl group attached to the highest-numbered chiral center. By convention, the structural formula for monosaccharides is oriented to place this carbon atom near the bottom of the molecule.

$$\begin{array}{cc}
\text{CHO} & \text{CHO} \\
| & | \\
-\text{C}- & -\text{C}- \\
| & | \\
-\text{C}- & -\text{C}- \\
| & | \\
\text{H}-\text{C}-\text{OH} & \text{HO}-\text{C}-\text{H} \\
| & | \\
\text{CH}_2\text{OH} & \text{CH}_2\text{OH} \\
\text{D sugar} & \text{L sugar}
\end{array}$$

Common monosaccharides are ribose, deoxyribose, glucose (blood sugar), galactose, and fructose (Figure 13.3). These compounds exist in open-chain, carbonyl-containing forms and in hemiacetal ring forms (Figure 13.6). The equilibrium among the open-chain form and two alternative ring forms (designated alpha and beta) strongly favors the ring forms. Interconversion among these forms is called mutarotation.

The carbonyl-containing form of saccharides can undergo a base-catalyzed isomerization referred to as keto–enol tautomerism. The result of this isomerization is an interconversion among carbohydrates with the same carbon content (for example, glucose and fructose). As a result of mutarotation and tautomerism, both monosaccharides and disaccharides that have a free carbonyl group or a latent carbonyl group (that is, the hemiacetal or hemiketal form) can be oxidized by Tollens', Fehling's, and Benedict's reagents and are therefore called reducing sugars. The hemiacetal or -ketal forms of carbohydrates can also react with alcohols to form acetals or ketals (referred to as *glycosides* in carbohydrate chemistry). Those carbohydrates that have all their carbonyl groups tied up as acetals or ketals are nonreducing sugars.

The important disaccharides are maltose (two 1,4-alpha-linked glucose units), lactose (milk sugar, 1,4-beta-linked galactose and glucose units), and sucrose (table sugar; glucose and fructose units both linked through their latent carbonyl groups).

The common polysaccharides are all polymers of glucose. Starch (a plant product) and glycogen (a related animal product) both use 1,4-alpha linkages to connect the glucose units to one another. Starch consists of two fractions, an unbranched form called amylose and a branched form called amylopectin. The extended 1,4-linked polymer chains of both amylopectin and glycogen exhibit occasional branching via a 1,6 linkage. Cellulose is a common polysaccharide whose structure consists of 1,4-beta-linked glucose units. Human beings can digest polysaccharides with alpha linkages but not those with beta linkages.

The digestion of carbohydrates begins in the mouth but is primarily accomplished in the small intestine. Digestion consists of the conversion of all carbohydrates to monosaccharides via hydrolysis catalyzed by appropriate enzymes.

The primary role of carbohydrates in the body is as a ready source of energy. After absorption into the blood from the intestine, the sugars may be converted to glycogen for storage or transported as glucose to cells in need of fuel. The two principal metabolic paths by which the energy stored in the glucose molecule is made available are anaerobic glycolysis and aerobic oxidation through the Krebs cycle. In anaerobic glycolysis, a glucose molecule is converted to two lactic acid molecules, and some of the released energy is trapped in adenosine triphosphate (ATP) molecules. The Krebs cycle is fed by acetyl coenzyme A and yields two molecules of carbon dioxide for every acetyl group fed into the cycle.

In diabetes mellitus, control over the movement of glucose into and out of cells is lost because of a deficiency of the hormone insulin. Some diabetic persons control the symptoms of the disease by injections of pork or beef insulin or of human insulin produced by specially developed bacteria. In some mild cases of the disease, oral drugs can be used to stimulate the production of the individual's own insulin.

CHAPTER QUIZ

1. Which is true for aldoses?
   a. They always occur with ketoses.
   b. They are sugars with an aldehyde group.
   c. They contain ketone groups.
   d. They are cellulose sugars.

2. A chiral carbon atom
   a. has a C=C double bond
   b. is a biochemical isotope
   c. is present in all sugars
   d. has four different groups attached to it

3. Glucose is a
   a. pentose
   b. hexose
   c. heptose
   d. tetrose

4. A hemiacetal results from the combination of an aldehyde with a(n) *mirror image*
   a. hydroxyl group
   b. acid group
   c. water group
   d. ether group

5. Mutarotation
   a. occurs in the solid state
   b. is the interconversion of alpha and beta forms
   c. occurs with all carbohydrates
   d. can be observed with the unaided eye

6. Tautomerism is
   a. the coexistence of aldehydes and ketones
   b. the optical activity of sugars
   c. a rearrangement
   d. the result of fermentation

7. Contrary to acetals, the hemiacetals have
   a. a double bond
   b. a hydroxide group
   c. an open-ring structure
   d. no reactivity
8. Maltose is the linkage of
   a. glucose and fructose
   b. two glucose units
   c. galactose and glucose
   d. fructose and galactose
9. Glycogen is
   a. cellulose
   b. animal starch
   c. amylose
   d. reduced sugar
10. Which statement is false?
    a. The Krebs cycle converts citric acid to oxaloacetic acid.
    b. Insulin aids in glucose transport across the cell membrane.
    c. Sucrose is a reducing sugar.
    d. Lactose is a disaccharide.

PROBLEMS

13.1 Carbohydrates: Definitions and Classifications

1. Define:
   a. disaccharide
   b. polysaccharide
   c. latent carbonyl group

13.2 Monosaccharides: Further Classification

2. Define:
   a. hexose
   b. aldose
3. Draw the structure of a ketotetrose.
4. Draw the structure of an aldoheptose.

13.3 Mirror-Image Isomerism

5. Define:
   a. triose
   b. chiral center
6. Draw formulas for D-glyceraldehyde and L-glyceraldehyde.

13.4 Some Common Monosaccharides

7. Identify these sugars as aldose or ketose.
   a. D-glyceraldehyde
   b. D-ribose
   c. D-deoxyribose
   d. D-galactose
   e. D-glucose
   f. D-fructose
   g. L-fructose
8. Identify each of the following as a triose, tetrose, pentose, or hexose.
   a. L-glucose
   b. D-deoxyribose
   c. D-fructose
   d. L-glyceraldehyde
9. For the following sugars, specify whether each one is a D sugar or an L sugar.

   a.
   $$\begin{array}{c} \text{CHO} \\ | \\ \text{H—C—OH} \\ | \\ \text{H—C—OH} \\ | \\ \text{H—C—OH} \\ | \\ \text{CH}_2\text{OH} \end{array}$$

   b.
   $$\begin{array}{c} \text{CHO} \\ | \\ \text{HO—C—H} \\ | \\ \text{H—C—OH} \\ | \\ \text{HO—C—H} \\ | \\ \text{CH}_2\text{OH} \end{array}$$

   c.
   $$\begin{array}{c} \text{CHO} \\ | \\ \text{H—C—OH} \\ | \\ \text{HO—C—H} \\ | \\ \text{CH}_2\text{OH} \end{array}$$

   d.
   $$\begin{array}{c} \text{CH}_2\text{OH} \\ | \\ \text{H—C—OH} \\ | \\ \text{CHO} \end{array}$$

10. From memory, draw formulas for the open-chain forms of D-glucose, D-galactose, and D-fructose.
11. Draw the cyclic structure for α-D-glucose.
12. A monosaccharide called mannose differs from glucose only in the arrangement of the groups at the second carbon. Draw the open-chain structure of D-mannose. Draw the cyclic structure for β-D-mannose.

13.6 Some Reactions of Monosaccharides

13. Define:
    a. mutarotation
    b. glycoside
    c. reducing sugar
    d. tautomerism
14. The structure of a methyl glycoside of glucose is

a. Is carbon 1 in the alpha or the beta arrangement?
b. Is the compound a reducing sugar?
c. Will it give a positive test with Benedict's reagent?

15. Draw an enol tautomer for acetaldehyde.

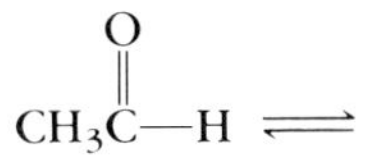

$$CH_3\overset{O}{\overset{\|}{C}}-H \rightleftharpoons$$

16. Draw two different enol tautomers for 2-butanone.

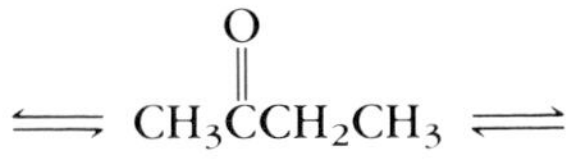

$$\rightleftharpoons CH_3\overset{O}{\overset{\|}{C}}CH_2CH_3 \rightleftharpoons$$

17. Draw an enol tautomer for D-glyceraldehyde.

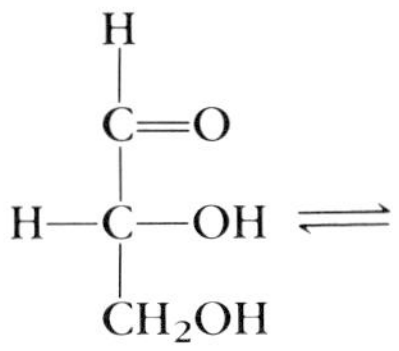

$$\begin{array}{c} H \\ | \\ C=O \\ | \\ H-C-OH \rightleftharpoons \\ | \\ CH_2OH \end{array}$$

13.7 Some Important Disaccharides

18. For each of these abbreviated sugar formulas, indicate whether the glycosidic link is alpha or beta.

a. 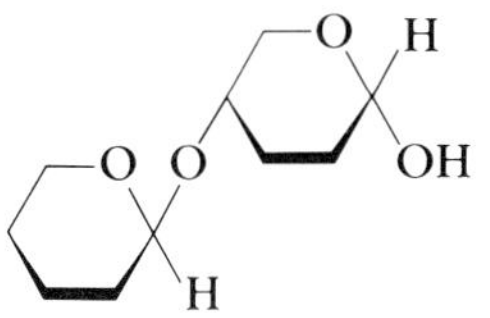
b. 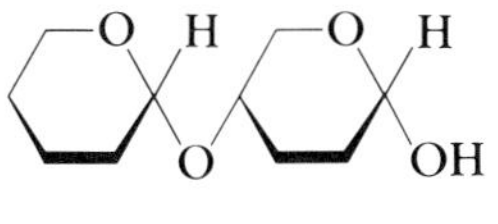
c. 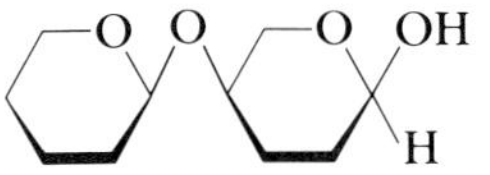
d.

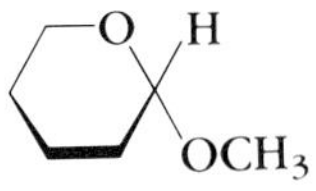

19. Which of the structures in Problem 18 contains a hemiacetal group? If present, is the hemiacetal group alpha or beta?

20. Which of the structures shown in Problem 18 is *not* a reducing sugar?

21. Indigo (the dye used to color blue jeans) occurs as the glucoside indican in *Indigofera tinctoria*. Is indican an alpha or beta glucoside?

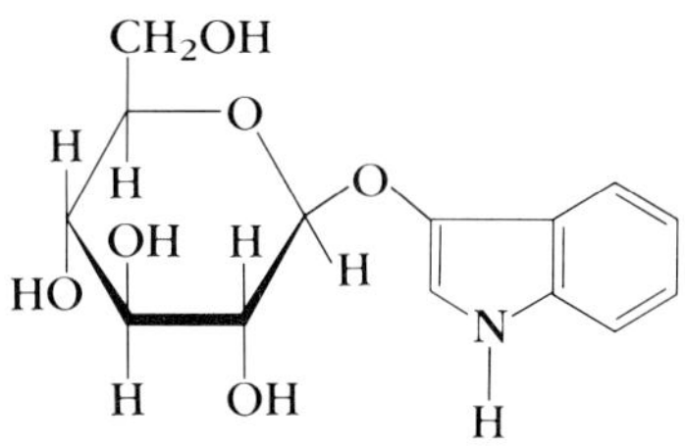

22. Identify these sugars by their proper names.
  a. blood sugar   **b.** milk sugar
  c. dextrose   **d.** levulose
  e. table sugar

23. Melibiose is a disaccharide that occurs in some plant juices. Its structure is

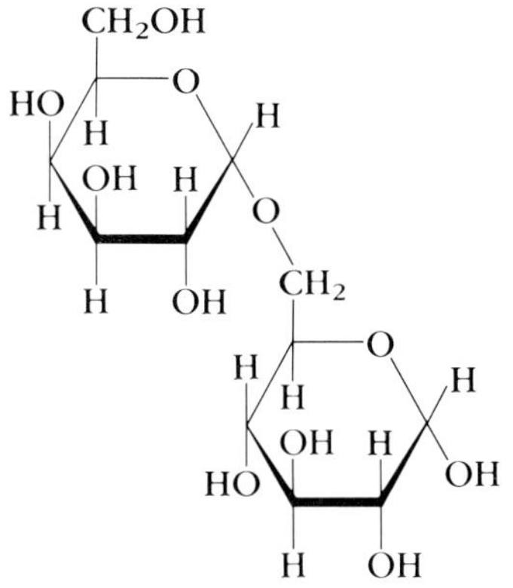

What monosaccharide units are incoporated in melibiose?

24. What type of linkage (alpha or beta) joins the two rings of melibiose (Problem 23)?

25. Is melibiose (Problem 23) a reducing sugar? If so, circle the hemiacetal function and indicate whether it is alpha or beta.

13.8 Polysaccharides

26. How do amylose and amylopectin differ? How are they similar?

27. How do amylose and cellulose differ? How are they similar?

28. How do amylopectin and glycogen differ? How are they similar?

13.9 Digestion of Carbohydrates

29. What monosaccharide is obtained from the hydrolysis of starch?
30. What monosaccharide is obtained from the hydrolysis of cellulose?
31. What monosaccharide is obtained from the hydrolysis of maltose?
32. What monosaccharides are obtained from the hydrolysis of lactose?
33. What monosaccharides are obtained from the hydrolysis of sucrose?
34. What is the general type of reaction in digestion?
35. What are the products of the digestion of monosaccharides? disaccharides? polysaccharides?
36. In what section of the digestive tract does most of the digestion of carbohydrates take place?
37. If a cracker that is rich in starch is chewed for a long time, it begins to develop a sweet, sugary taste. Why?
38. Why cannot human beings digest cellulose?

13.10 Metabolism of Carbohydrates

39. What is the storage form of carbohydrate in the body, and where is it found?
40. What is the ultimate product obtained from a glucose molecule that completes the glycolysis pathway?
41. What molecule is used to trap the energy released during anaerobic glycolysis of glucose?
42. Glycolysis is described as *anaerobic* and the Krebs cycle is described as *aerobic*. What does this mean?
43. When pyruvic acid is fed into the Krebs cycle, in what form do its three carbon atoms finally emerge?

13.11 Insulin and Diabetes Mellitus

44. What is the role of insulin in controlling blood sugar?
45. Why cannot someone who has diabetes take insulin by mouth?

14

LIPIDS: STRUCTURE AND METABOLISM

DID YOU KNOW THAT...

- fats are lipids?
- fats store energy with twice the efficiency of other foods?
- if energy were stored only in carbohydrates, we would weigh twice as much?
- our bodies have 10 trillion cells?
- fatty acids almost always contain an even number of carbon atoms?
- the hardening of arteries is due, in part, to lipo deposits?
- "hard" water contains alkaline earth ions?
- birth control pills are synthetic imitations of progesterone and estrone?
- prostaglandins might be useful in controlling high blood pressure or gastric ulcers?
- bile salts help dissolve lipids in water?
- during ketosis the odor of nail polish remover (acetone) may be detected on the breath?
- low-carbohydrate diets can lead to depression?

The food we eat is divided into three primary groups: carbohydrates (Chapter 13), proteins (Chapter 15), and **lipids**. The best-known lipids are the **fats**, and, since thin is in, the least desirable foods in the opinion of many people are those with high fat content. Gram for gram, fats pack more than twice the caloric content of carbohydrates or proteins. While that may be bad news for dieters, it is also evidence of the efficiency of nature's designs. The body has a limited capacity for storing carbohydrates. It can tuck away a bit of glycogen in the liver or in muscle tissue, but carbohydrates, primarily in the form of glucose, are meant to serve the body's immediate energy needs. If we intend to store energy reserves, then the more energy we can pack into a given space, the better off we are. The oxidation of fats supplies about 9 kcal/g, whereas the oxidation of carbohydrates supplies only 4 kcal/g. So we store fats, and our capacity for doing so is astounding. A recorded instance exists of a man weighing 486 kg (about 1000 lb). If all that "reserve energy" were stored as carbohydrates, he would have weighed well over a ton.

14.1 What Is a Lipid?

Of the three types of foodstuffs, two are classified by functional groups. Carbohydrates are polyhydroxy aldehydes or polyhydroxy ketones. Proteins, as we shall see, are polyamides. Lipids are not poly anythings. Many are esters or compounds that can form esters, but that is not what makes a lipid a lipid. A lipid is defined by its solubility. Water is the

A lipid is defined by its solubility.

FIGURE 14.1
Cooking oil, butter, birth control pills, and gallstones are rich in compounds that are classified as lipids.

major solvent in living systems, and most reactions of physiological importance take place in aqueous solutions. Thus, water insolubility is a noteworthy property if found in biologically important molecules. **Lipids** are those biomolecules that are more soluble in relatively nonpolar organic solvents than they are in water (Figure 14.1). Organic solvents typically used to extract lipids from body tissues include diethyl ether and chloroform.

Lipids vary greatly in function. In addition to providing our energy stores, they are important components of brain and nerve tissue, serve as protective padding and insulation for vital organs, act to regulate some body functions, dissolve certain vitamins, and, perhaps most important, make up the major part of the membranes of each of the 10 trillion cells in our bodies.

14.2 Fatty Acids

Many lipids are esters, which are derivatives of alcohols and acids. Fats and many other lipids are esters of long-chain carboxylic acids, frequently called **fatty acids**. The fatty acids almost always contain an even number of carbon atoms arranged in an unbranched chain. Some, the *unsaturated* fatty acids, contain one or more double bonds. Free fatty acids are rare in nature, but fats and other lipids provide a reservoir from which the acids can be obtained. A list of some common fatty acids and an important source of each are given in Table 14.1.

Esters of long-chain carboxylic acids (fatty acids) almost always contain an even number of carbon atoms arranged in an unbranched chain.

TABLE 14.1 Some Fatty Acids in Natural Fats

Number of Carbon Atoms	Condensed Structure	Name	Source
4	$CH_3CH_2CH_2COOH$	Butyric acid	Butter
6	$CH_3(CH_2)_4COOH$	Caproic acid	Butter
8	$CH_3(CH_2)_6COOH$	Caprylic acid	Coconut oil
10	$CH_3(CH_2)_8COOH$	Capric acid	Coconut oil
12	$CH_3(CH_2)_{10}COOH$	Lauric acid	Palm kernel oil
14	$CH_3(CH_2)_{12}COOH$	Myristic acid	Oil of nutmeg
16	$CH_3(CH_2)_{14}COOH$	Palmitic acid	Palm oil
18	$CH_3(CH_2)_{16}COOH$	Stearic acid	Beef tallow
16	$CH_3(CH_2)_5CH{=}CH(CH_2)\text{-}COOH$	Palmitoleic acid	Butter
18	$CH_3(CH_2)\text{-}CH{=}CH(CH_2)\text{-}COOH$	Oleic acid	Olive oil
18	$CH_3(CH_2)_4(CH{=}CHCH_2)_2(CH_2)_6COOH$	Linoleic acid	Soybean oil
18	$CH_3CH_2(CH{=}CHCH_2)_3(CH_2)_6COOH$	Linolenic acid	Fish oils

14.3 Fats: The Triglycerides

Chemically, fats are esters of fatty acids and glycerol, a trihydroxy alcohol. Three acid units are combined with the three hydroxyl groups of the glycerol molecule. If the fatty acids are all the same, as in tristearin,

$$CH_2O\text{-}H \quad HO\text{-}CC_{17}H_{35}(\text{=O}) \\ CH\text{-}O\text{-}H \quad HO\text{-}CC_{17}H_{35}(\text{=O}) \\ CH_2O\text{-}H \quad HO\text{-}CC_{17}H_{35}(\text{=O}) \longrightarrow \text{Tristearin} + 3\,H_2O$$

Glycerol — Stearic acid → Tristearin

the fat is a **simple triglyceride**. Most naturally occurring fats contain different fatty acid units in the same molecule.

$$CH_2O\text{-}CC_{17}H_{33}(\text{=O}) \quad \text{Oleate group}$$
$$CH\text{-}O\text{-}CC_{17}H_{35}(\text{=O}) \quad \text{Stearate group}$$
$$CH_2O\text{-}CC_{15}H_{31}(\text{=O}) \quad \text{Palmitate group}$$

A mixed triglyceride

These fats are called **mixed triglycerides**.

Animal fats generally contain both saturated and unsaturated fatty acid units, but the former predominate. At room temperature, animal fats are usually solids. At the body temperature of warm-blooded creatures, though, these fats are apt to exist in the liquid state. Plants, on the other hand, are not warm-blooded. They keep their fats in the liquid state by incorporating a higher proportion of unsaturated acid units. A double bond in these molecules causes a kink in the long hydrocarbon chain. Compare these structural formulas for stearic acid (from animal fat) and oleic acid (from vegetable fat).

14.3 Fats: The Triglycerides

Stearic acid

Oleic acid

Each angle in these "zigzag" formulas represents one carbon atom in the fatty acid chain. The fat molecules that incorporate unsaturated chains do not stack neatly; hence, the forces between molecules are weaker. Weaker forces correspond to lower melting points. The highly unsaturated fats (called **polyunsaturated fats**) characteristically isolated from vegetable sources are liquids at room temperature. Liquid fats are called **oils**, as in peanut oil, olive oil, or corn oil. Table 14.2 compares the fatty acid content of selected fats and oils.

The degree of unsaturation of a fat or an oil is usually measured in terms of the **iodine number**. Remember that carbon–carbon double bonds react with halogens, for example, bromine.

The iodine number measures the degree of unsaturation of a fat or oil.

$$\text{>C=C<} + Br_2 \longrightarrow -\underset{Br}{\overset{|}{C}}-\underset{Br}{\overset{|}{C}}-$$

TABLE 14.2 Fatty Acid Distribution Ranges (as %) in Several Fats and Oils

Fat or Oil	Lauric	Myristic	Palmitic	Stearic	Oleic	Linoleic	Others
Animal fats							
Beef tallow	0.2	2–3	25–30	21–26	39–42	1–3	1.2–1.8[a]
Lard		1	25–30	12–16	41–51	3–8	4.2–8.2[a]
Butter	1–4	8–13	25–32	8–13	22–29	3	6.6–15.5[b]
Vegetable oils							
Palm oil		1–6	32–47	1–6	40–52	2–11	
Corn		0–2	8–10	1–4	30–50	34–56	1–4[a]
Soybean		0.3	7–11	2–5	22–34	50–60	3–14[c]
Cottonseed		0–3	17–23	1–3	23–44	34–55	0–1

[a] Mainly the unsaturated fatty acid with 16 carbon atoms per molecule (palmitoleic acid).
[b] Mainly the acids with 4, 6, 8, or 10 carbon atoms per molecule.
[c] Mainly linolenic acid.

TABLE 14.3 Typical Iodine Numbers of Some Fats and Oils

Fat or Oil	Iodine Number
Butter	25–40
Beef tallow	30–45
Lard	45–70
Olive oil	75–95
Peanut oil	85–100
Cottonseed oil	100–117
Corn oil	115–130
Fish oils	120–180
Soybean oil	125–140
Safflower oil	130–140
Sunflower oil	130–145
Linseed oil	170–205

Iodine is also a halogen. The iodine number is the number of grams of iodine that will be consumed in a reaction with 100 g of fat or oil. The higher the iodine number is, the higher the degree of unsaturation in the fat. Table 14.3 lists the iodine numbers of commercially important fats and oils.

In one form of arteriosclerosis, or hardening of the arteries, lipids are deposited on the walls of the arteries (Figure 14.2). The deposits harden, and the arterial wall becomes less elastic, while the opening through the vessel becomes narrower. Statistical evidence exists that links diets rich in saturated fats to a higher incidence of the disease. Additional studies have failed to prove that a diet rich in saturated fats results in arteriosclerosis. Nonetheless, many concerned people have heeded the advice to avoid animal fats in favor of polyunsaturated vegetable fats. Advertisers who encourage you to buy corn oil margarine (prepared from a relatively unsaturated vegetable oil) rather than butter (a relatively saturated animal fat) acknowledge this concern. Because consumers prefer to spread margarine rather than pour it on their toast, vegetable oils are partially hydrogenated. Hydrogenation removes double bonds:

$$\mathrm{\underset{}{\overset{}{>}}C{=}C\underset{}{\overset{}{<}} + H_2 \xrightarrow{\text{catalyst}} -\underset{H}{\overset{|}{C}}-\underset{H}{\overset{|}{C}}-}$$

As double bonds are removed, the melting point of the oil goes up until

FIGURE 14.2
Photomicrograph of a cross section of a "hardened" artery, showing deposits of plaque. [Courtesy of Biomedical Graphics, University of Minnesota Hospitals, Minneapolis.]

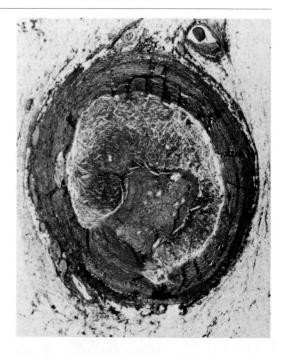

14.4 Soaps and the Saponification of Fats

TABLE 14.4 Cholesterol and Saturated Fat Levels in Some Foods

Food	Cholesterol (mg)	Saturated Fat (g)
Whole milk (8 oz)	33	5.1
Low-fat yogurt (4 oz)	2	0.1
Butter (1 tbsp)	31	7.1
Margarine (1 tbsp)	0	1.8
One egg	274	1.7
Lean pork (3 oz)	80	3.2
Shrimp (11 large)	96	0.2
Ground beef (3 oz)	88	7.6
Beef liver (3 oz)	372	2.5
Coconut oil (1 tbsp)	0	11.8
American cheese (1 oz)	27	5.6
White rice (1 cup)	0	0.1
Fruit	0	trace
Pound cake (1 slice)	68	5.9

the oil becomes a solid at room temperature. If too many double bonds are removed, the product is as saturated as an animal fat. Therefore, the hydrogenation is limited to the degree necessary to give the product the desired consistency.

Saturated fats are the greatest factor in raising blood cholesterol. The American Heart Association advises a cholesterol intake of no more than 300 mg/day. Table 14.4 shows the cholesterol and saturated fat levels of some typical foods.

14.4 SOAPS AND THE SAPONIFICATION OF FATS

Animal fats are available in large quantities as a by-product of the meat-packing industry. These fats provide the starting material for the synthesis of soap (Figure 14.3). As was noted in Chapter 11, the alkaline hydrolysis of fats is called *saponification*, literally, soap-making. Animal fat is cooked with lye (sodium hydroxide) to produce glycerol and the sodium salts of fatty acids, which are called **soaps**. The reaction of tristearin is illustrated in Figure 14.4. Potassium hydroxide is sometimes employed in place of sodium hydroxide. Potassium soaps are softer and produce a finer lather. They are used in shaving creams.

Dirt and grime usually adhere to skin, clothing, and other surfaces because they are combined with greases and oils—body oils, cooking fats, or similar substances—which act like a glue. Since oils are not miscible with water, washing with water alone does little good. Soap molecules have a "split personality." One end is ionic and dissolves in

FIGURE 14.3
Soap is made in large reaction vessels such as these, which are three stories deep. Such containers are stocked with 59 000 kg of fat and 34 000 kg of lye solution. The yield is about 68 000 kg of soap and 6400 kg of glycerol. [Courtesy of Lever Brothers Company.]

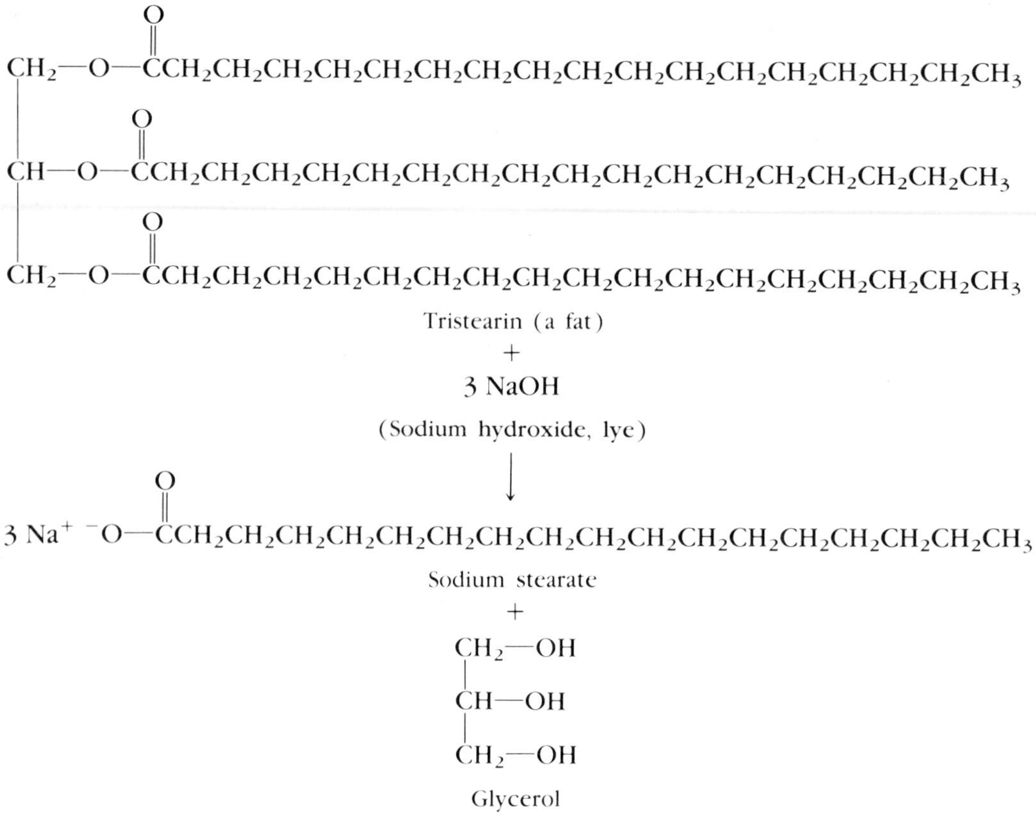

$$CH_2-O-\overset{\overset{O}{\|}}{C}CH_2CH_2CH_2CH_2CH_2CH_2CH_2CH_2CH_2CH_2CH_2CH_2CH_2CH_2CH_2CH_3$$

$$CH-O-\overset{\overset{O}{\|}}{C}CH_2CH_2CH_2CH_2CH_2CH_2CH_2CH_2CH_2CH_2CH_2CH_2CH_2CH_2CH_2CH_3$$

$$CH_2-O-\overset{\overset{O}{\|}}{C}CH_2CH_2CH_2CH_2CH_2CH_2CH_2CH_2CH_2CH_2CH_2CH_2CH_2CH_2CH_2CH_3$$

Tristearin (a fat)
+
3 NaOH
(Sodium hydroxide, lye)
↓

$$3\ Na^+\ ^-O-\overset{\overset{O}{\|}}{C}CH_2CH_2CH_2CH_2CH_2CH_2CH_2CH_2CH_2CH_2CH_2CH_2CH_2CH_2CH_2CH_3$$

Sodium stearate
+

$$CH_2-OH$$
$$CH-OH$$
$$CH_2-OH$$

Glycerol

FIGURE 14.4
Soap is produced by the reaction of animal fat with sodium hydroxide. Tristearin is a fat, and sodium stearate is a soap.

water. The other end is like a hydrocarbon and dissolves in the nonpolar oils (Figure 14.5). If we represent the ionic end of the soap molecule as a circle and the hydrocarbon end as a zigzag line, we can illustrate the cleansing action of soap schematically (Figure 14.6). The hydrocarbon "tails" stick into the oil. The ionic "heads" interact with the water. In this manner the oil is broken into tiny droplets and dispersed throughout the aqueous phase. The droplets do not coalesce because of the repulsion of the charged groups (the carboxyl anions) on their surfaces. With the oil no longer "gluing" it to the surface, the dirt can be easily removed.

Soap does not work well in hard water. **Hard water** is water that contains certain metal ions, particularly magnesium, calcium, and iron ions. Soap anions, that is, fatty acid anions, react with these metal cations to form greasy, insoluble curds.

$$2\ CH_3CH_2CH_2CH_2CH_2CH_2CH_2CH_2CH_2CH_2CH_2CH_2CH_2COO^-\ Na^+ + Ca^{2+} \longrightarrow$$
Soap (soluble)

$$(CH_3CH_2CH_2CH_2CH_2CH_2CH_2CH_2CH_2CH_2CH_2CH_2CH_2COO^-)_2\ Ca^{2+} + 2\ Na^+$$
Bathtub ring (insoluble)

14.4 Soaps and the Saponification of Fats

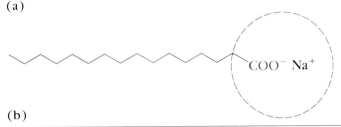

FIGURE 14.5
Sodium palmitate, a soap. (a) Structural formula. (b) A schematic representation.

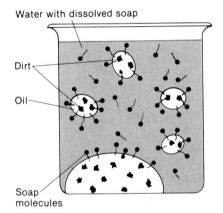

FIGURE 14.6
The action of soap in removing dirt.

These curds precipitate from solution as bathtub rings or "dulling soap films" and are no good at all as emulsifying agents.

Synthetic detergents imitate the split personality of soaps but do so by using molecules that do not precipitate in hard water. Several examples of synthetic detergents are shown in Figure 14.7.

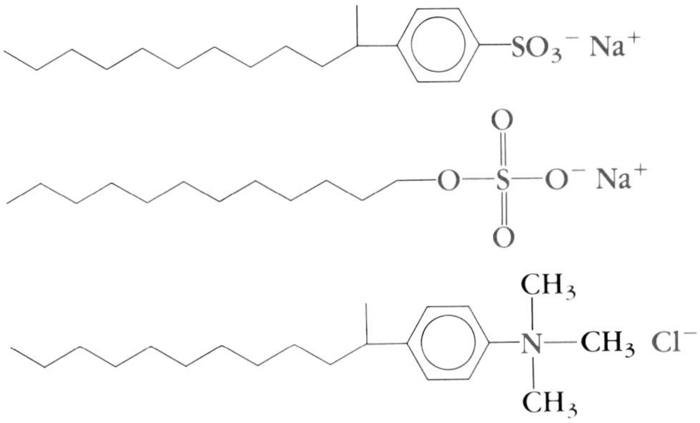

FIGURE 14.7
A selection of synthetic detergents. Notice the long hydrocarbon tail and the ionic functional group.

14.5 Phosphatides

Glycerol is an alcohol. It forms esters with acids. When the acids are carboxylic acids, the esters formed are fats. But glycerol can also form esters with inorganic acids, like phosphoric acid. The phosphatides are **phospholipids** (phosphorus-containing lipids) in which glycerol is esterified with two fatty acid groups and one phosphoric acid group. The structure of phosphoric acid

$$HO-\overset{\overset{O}{\|}}{\underset{\underset{OH}{|}}{P}}-OH$$

is such that it can react with more than one alcohol to form a double or triple ester. In the **phosphatides**, the phosphoric acid is esterified with one other alcohol molecule, usually an amino alcohol.

$$\begin{array}{l} CH_2O-C(=O)-(CH_2)_{14}CH_3 \\ CH-O-C(=O)-(CH_2)_{14}CH_3 \\ CH_2O-P(=O)(O^-)-O-CH_2CH_2-\overset{+}{N}R_3 \end{array}$$

Glycerol unit → CH—O—C ...
Fatty acid units
Phosphoric acid unit
Amino alcohol unit

Zwitterion—an intramolecular salt containing both a cationic and an anionic group

We have drawn the molecule as it exists, as an intramolecular salt with both cationic and anionic groups. Such molecules are called **zwitterions**. (We shall encounter zwitterions again in the next chapter, Sections 15.1 and 15.3, when we look at amino acids and proteins.) The two amino alcohols most commonly incorporated in phosphatides are ethanolamine and choline.

$$HOCH_2CH_2NH_2 \qquad HOCH_2CH_2\overset{\overset{CH_3}{|}}{\underset{\underset{CH_3}{|}}{N^+}}-CH_3$$

Ethanolamine   Choline
(2-aminoethanol)

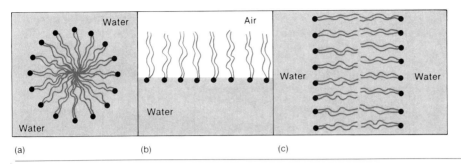

FIGURE 14.8
(a) Micelle, (b) monolayer, and (c) bilayer formed when polar lipids are added to the water.

Phosphatides containing ethanolamine (R = H in the phosphatide structure) are members of a class of lipids called **cephalins**. When choline is the amino alcohol (R = CH$_3$), the compounds are **lecithins**.

Both cephalins and lecithins are found as major components of cell membranes. These compounds have the same sort of split personality we saw in soap. They have a polar (zwitterion) end and a nonpolar (hydrocarbon) end. When "polar lipids" such as these are placed in water, they disperse, forming clusters of molecules called **micelles** (Figure 14.8a). The hydrocarbon "tails" of these lipids are directed inward, away from the polar aqueous phase; the hydrophilic ("water-loving") heads are directed outward into the water. Each micelle may contain thousands of polar lipid molecules. The illustration shows a cross section of a micelle, which is actually a three-dimensional, spherical structure. Other "satisfactory" arrangements are also possible. The polar lipids can form **monolayers** (layers one molecule thick) on the surface of water. The polar heads stick into the water, and the nonpolar tails stick up in the air (Figure 14.8b). Finally, **bilayers** can be formed in which the hydrophobic ("water-fearing") tails are sandwiched between the hydrophilic heads that face outward toward the water (Figure 14.8c). Bilayers resemble the membranes that enclose the components of living cells. The structure of a cell membrane is far more complex. For one thing, proteins are embedded within the membrane (Figure 14.9), and it appears that the proteins are involved in the selective transport of materials into and out of the cell.

Micelle—a cluster of polar lipids

Monolayer—a layer one molecule thick

Bilayer—a layer two molecules thick

FIGURE 14.9
A cell membrane model showing proteins in a lipid bilayer.

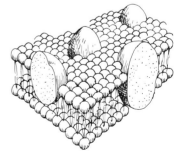

14.6 Nonsaponifiable Lipids: Steroids and Prostaglandins

All the lipids discussed so far are **saponifiable**; they react with aqueous bases to yield simpler components such as glycerol, fatty acids, amino alcohols, or sugars. An extraction of lipids from cellular material, however, also yields a small fraction that is not hydrolyzed into simpler components by alkali. The major portion of the **nonsaponifiable** lipid fraction obtained from animal cells consists of a class of compounds called **steroids**. These compounds include the bile salts, sex hormones as well as other hormones, and the most abundant steroid by far, *cholesterol*.

Cholesterol

Cholesterol is a common component of all animal tissues. The brain is about 10% cholesterol. Cholesterol is a major component of certain types of gallstones (Figure 14.10). It is also found in the lipid deposits in hardened arteries. Cholesterol serves as an intermediate in the biosynthesis of many other steroid molecules (Figure 14.11).

All steroids share cholesterol's system of four fused rings—three rings with six members and the fourth ring with five.

Cholic acid, in the form of a salt derivative, is found in bile, a digestive fluid produced in the liver. It serves an important role in the digestion of fats (Section 14.7). Corticosterone is a hormone produced in the cortex of the adrenal glands. Pregnenolone, progesterone, and estrone are all female sex hormones. Synthetic imitations of these

FIGURE 14.10
Gallstones are often composed largely of cholesterol. [Courtesy of Biomedical Graphics, University of Minnesota Hospitals, Minneapolis.]

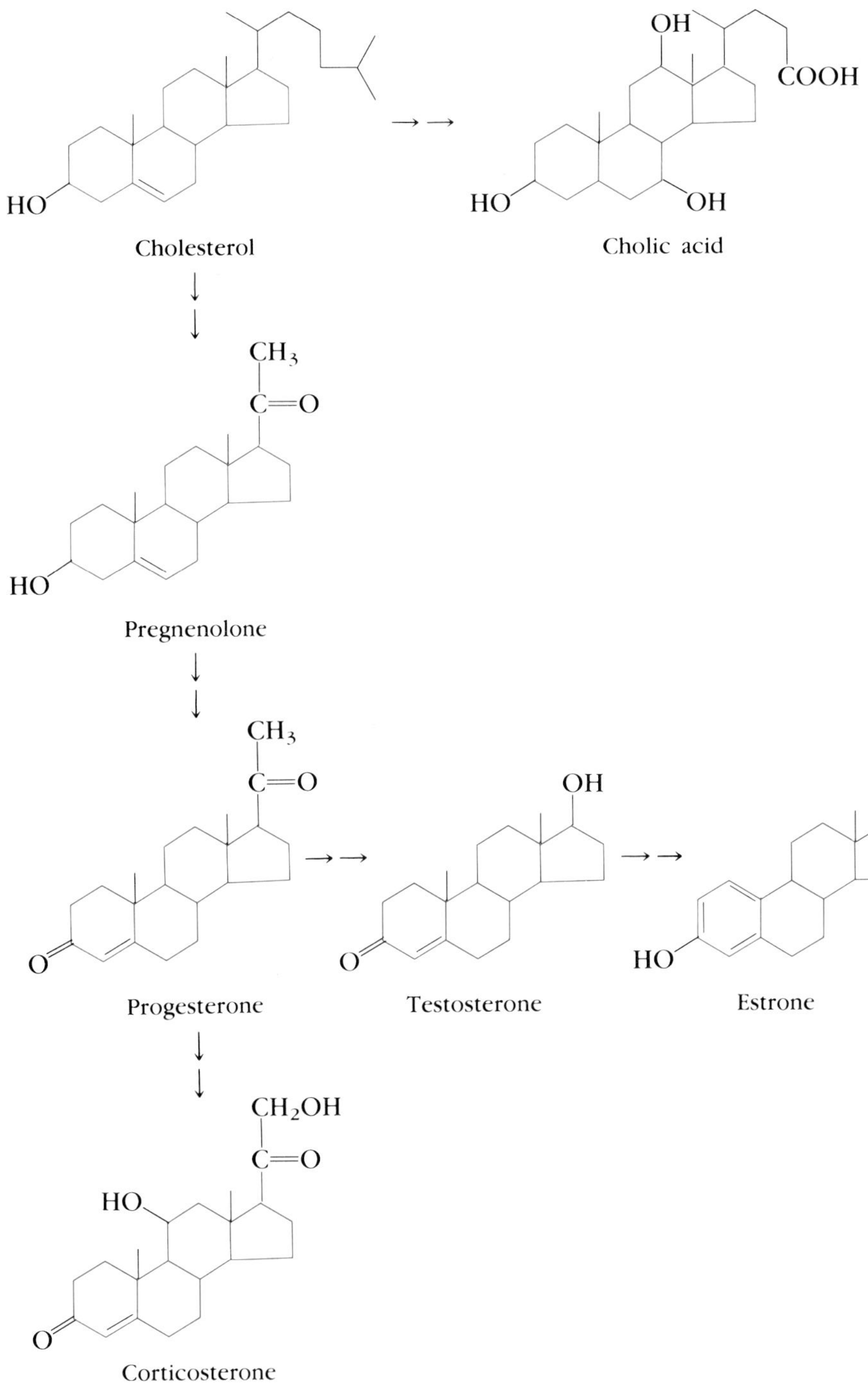

14.6 Nonsaponifiable Lipids: Steroids and Prostaglandins

FIGURE 14.11
The conversion of cholesterol to other biologically important steroids. Cholic acid is a bile acid. Progesterone is the pregnancy hormone. Corticosterone is an adrenal cortical steroid.

hormones are the active ingredients in birth control pills. Testosterone is a male sex hormone. The range of steroid functions in the body demonstrates that even a "minor" modification in chemical structure can result in a major change in the effect produced by the compound.

The **prostaglandins** are not steroids, but they do contain a ring and they are nonsaponifiable. Prostaglandin E_1 is a typical member of the class.

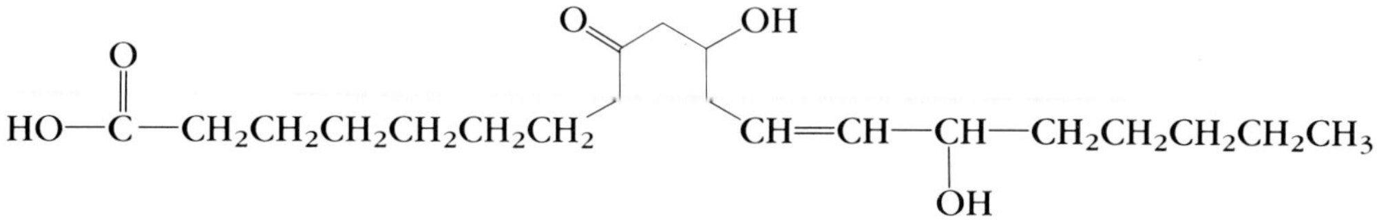

$$HO-\underset{\underset{O}{\|}}{C}-CH_2CH_2CH_2CH_2CH_2CH_2-\underset{\underset{O}{\|}}{C}-CH_2-CH(OH)-CH=CH-\underset{\underset{OH}{|}}{CH}-CH_2CH_2CH_2CH_2CH_3$$

Prostaglandin E_1

These compounds are widely distributed throughout the body but at extremely low concentrations. They are very potent biochemicals. In some ways, the prostaglandins act much as hormones do, regulating such things as smooth muscle activity, blood flow, and the secretion of various substances. These properties suggest that the prostaglandins or appropriate synthetic variations might prove useful in the treatment of high blood pressure or gastric ulcers or as an abortifacient (abortion inducer). It is not surprising, then, that many pharmaceutical companies have undertaken investigations of the prostaglandins. More than 6000 papers are published per year on these substances.

14.7 Digestion and Metabolism of Fats

Lipases—enzymes that split triglycerides

Like the carbohydrates, fats are digested primarily in the small intestine. As is also the case with carbohydrates, the hydrolysis of fats is catalyzed by special enzymes. In the presence of these enzymes, called **lipases**, triglycerides are split into fatty acids, glycerol, soaps (salts of fatty acids), and monoglycerides and diglycerides (Figure 14.12). Unlike carbohydrates, lipids are by definition insoluble in water. The digestive juices are aqueous solutions. To bring the lipids into contact with the enzymes dissolved in the digestive juices, it is necessary to emulsify the fats. This is done by the bile salts, one of which is sodium glycocholate.

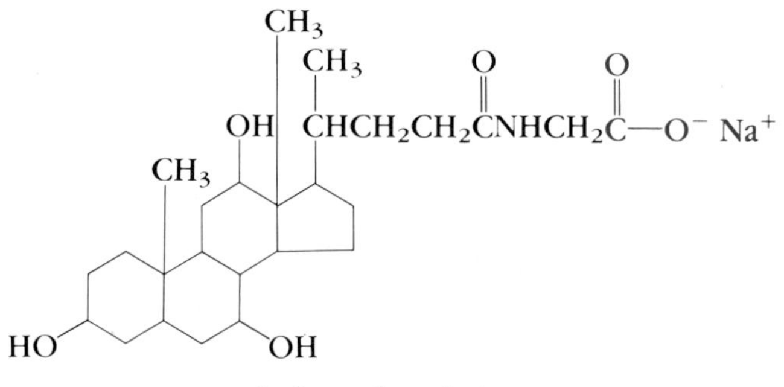

Sodium glycocholate

14.7 Digestion and Metabolism of Fats

The bile salts act in much the same way that soap does (Section 14.4). They break down large fat globules into smaller ones and keep the smaller globules suspended in the aqueous digestive medium. The greatly increased surface area of the fat particles and the opportunity afforded for more intimate contact with the lipase enzymes result in a much more rapid digestion of the fats.

As the hydrolysis products of fat digestion pass through the intestinal wall, they are reassembled as triglycerides. These reassembled fats are not taken directly into the bloodstream but are first carried through a secondary transportation network, the lymphatic system (see Figure 7.28, page 248), which ultimately does dump the fats into the blood circulatory system. Since blood is an aqueous solution, special arrangements must be made to keep the fats in circulation. The fats are tacked on to water-soluble proteins, and these complexes are efficiently transported by the blood to various storage locations.

Fats are stored throughout the body. Principally they are deposited in special kinds of connective tissue called **adipose tissue**. The storage places are called **fat depots**. Considerable fat is stored around vital organs such as the heart, kidneys, and spleen. There it serves as a protective cushion, helping to prevent injury to the organs. Fat is also stored under the skin, where it helps insulate against sudden temperature changes. The fat acts just like the insulation in the walls of a house,

Bile salts act a lot like soaps.

Fats are stored in adipose tissue throughout the body.

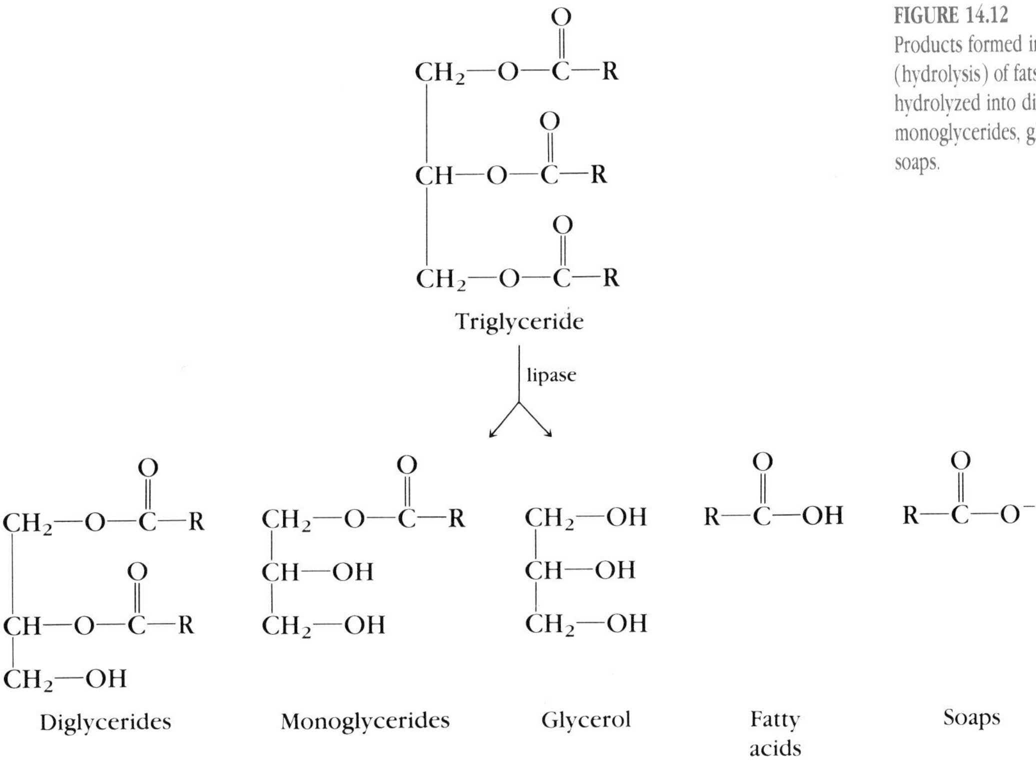

FIGURE 14.12
Products formed in the digestion (hydrolysis) of fats. Triglyceride (fat) is hydrolyzed into diglycerides, monoglycerides, glycerol, fatty acids, and soaps.

trapping body heat and preventing it from escaping to the outside. In some ways, the fatty tissue also acts like the furnace of a house. When the outside temperature drops, metabolic activity in the cells generates heat to compensate for that lost to the environment.

When the fat reserves are called on for energy, the fats are metabolized in a multistep process. The basic outline of the process is given in Figure 14.13. First, the fat molecules are once again hydrolyzed to glycerol and free fatty acids. Glycerol is taken into the glycolysis pathway (Section 13.10). The fatty acids are carried through a cyclic reaction sequence (Figure 14.14) in which the acids are chopped apart two carbon atoms at a time. It is the same fundamental chemistry we have encountered so often before—oxidation, hydration, and so on. The two-carbon fragment that is cut off with each turn of the *fatty acid cycle* appears in the form of acetyl coenzyme A (acetyl CoA). You remember the acetyl group,

$$CH_3\overset{\overset{O}{\|}}{C}-$$

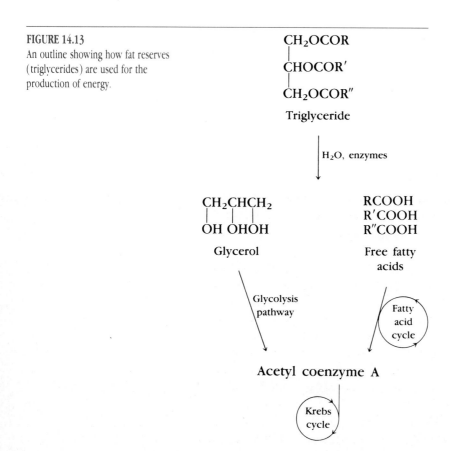

FIGURE 14.13
An outline showing how fat reserves (triglycerides) are used for the production of energy.

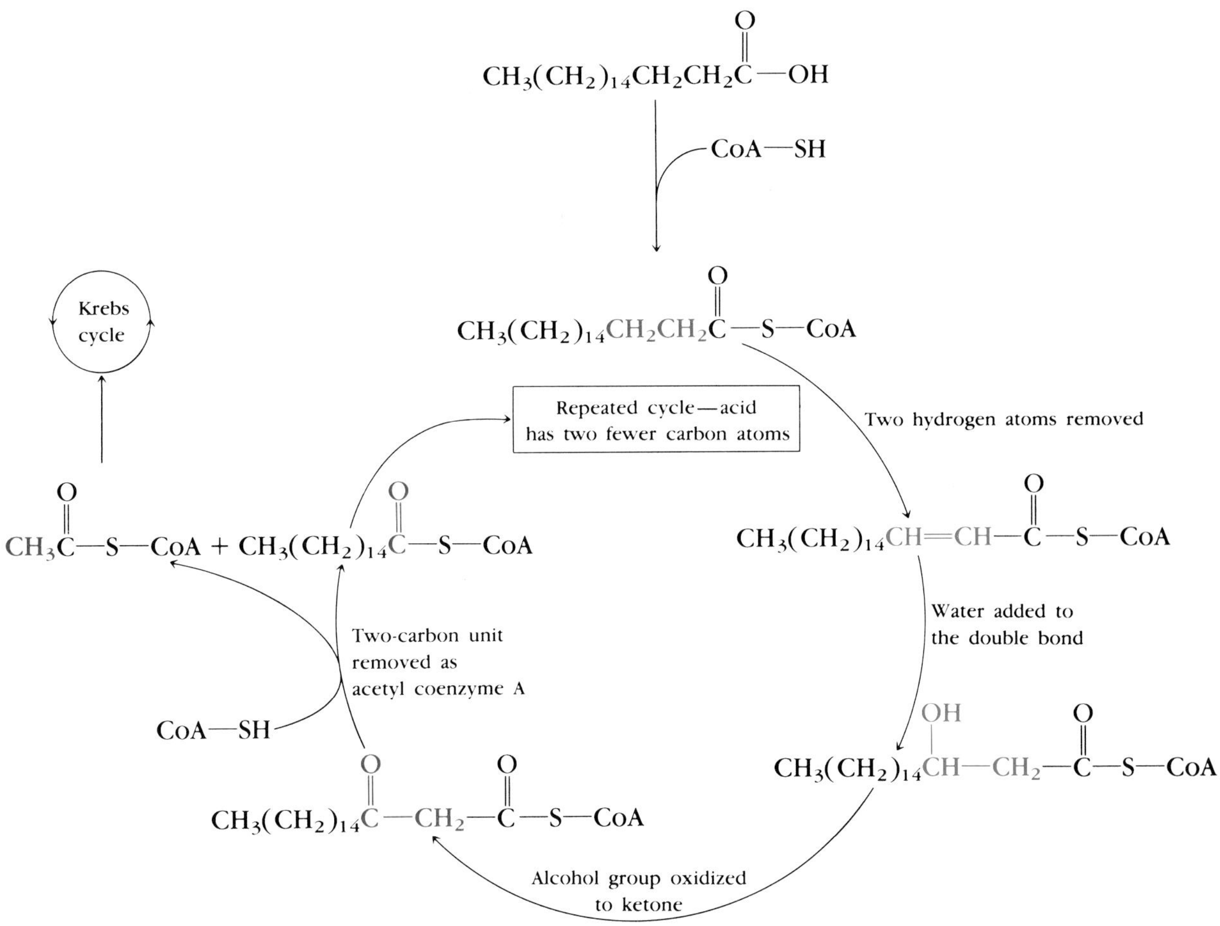

14.7 Digestion and Metabolism of Fats

FIGURE 14.14
The fatty acid cycle, showing one turn of stearic acid through the cycle.

Coenzyme A is a large molecule that contains a sulfhydryl group (—SH). The acetyl group attaches to coenzyme A through the sulfhydryl group, and in this form it enters the Krebs cycle (Section 13.10). The fatty acid cycle is perhaps more appropriately called the **fatty acid spiral** because the fatty acid derivative that passes through the cycle becomes smaller and smaller at each turn (Figure 14.15). On its last swing through the cycle, the acid yields two molecules of acetyl coenzyme A.

Not all the acetyl CoA formed by oxidation of fatty acids enters the Krebs cycle (Figure 14.16). Some can be used to synthesize new fatty acids in a process that is very nearly a reversal of the fatty acid cycle (which explains why most natural fatty acids contain an even number of carbon atoms). This also explains how carbohydrates can be converted to fats. Carbohydrates produce acetyl coenzyme A on their way to the Krebs cycle. This acetyl CoA can be diverted to fatty acid synthesis. Acetyl CoA is also used in the biosynthesis of steroids. It

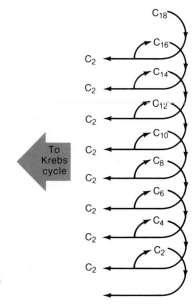

FIGURE 14.15
The fatty acid "cycle" is really a spiral, with two carbon atoms chopped off for each swing around the cycle.

is a remarkably versatile molecule. (Just in case you are curious, see Figure 14.17 for the structure of coenzyme A. Remember, all of that beautiful, complex molecular architecture was designed to carry a two-carbon acetyl unit from place to place.)

14.8 THE KETONE BODIES

Two related compounds, acetoacetic acid and β-hydroxybutyric acid, are intermediates in the fatty acid spiral. A third compound, acetone, is formed by the decarboxylation of acetoacetic acid. These three compounds are called the **ketone bodies** (even though one, β-hydroxybutyric acid, is *not* a ketone). The relationship among the three is shown in Figure 14.18.

The ketone bodies are normal components of the blood in concentrations of about 1 mg/100 mL. Much higher concentrations build up during starvation and in certain carbohydrate metabolism disorders

FIGURE 14.16
Acetyl coenzyme A plays a variety of roles in cellular chemistry.

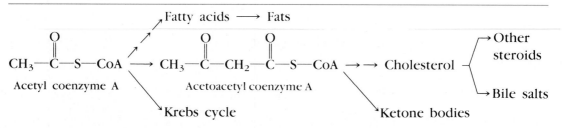

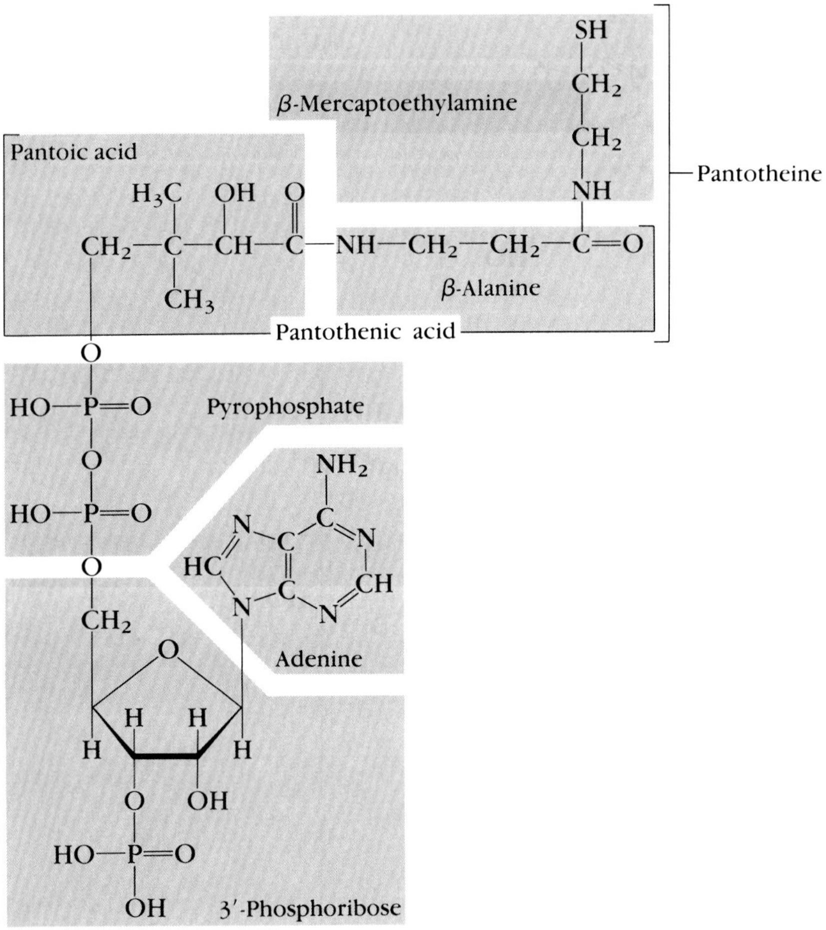

14.8 The Ketone Bodies

FIGURE 14.17
Coenzyme A.

such as diabetes mellitus. Elevated levels of the ketone bodies in the blood, called **ketosis**, may also result from diets that severely restrict carbohydrate intake. During ketosis, the odor of acetone may be detected on the breath. Uncontrolled ketosis leads to **acidosis**. Because two of the ketone bodies are carboxylic acids, high concentrations lead to a lowering of blood pH. Acidic blood cannot transport oxygen very well. In moderate to severe acidosis, "air hunger" sets in. Breathing becomes labored and very painful. The body also loses fluids as the kidneys, in an attempt to get rid of the acids, eliminate large quantities of water. The short oxygen supply and dehydration lead to depression. Even mild acidosis produces lethargy, loss of appetite, and a generally run-down feeling.

Ketosis—an elevated level of ketone bodies in the blood

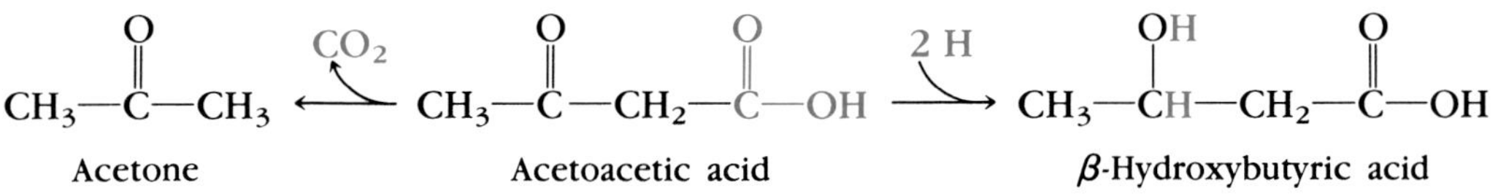

FIGURE 14.18
The three ketone bodies.

14.9 THE CHEMISTRY OF STARVATION

When the body is totally deprived of food, whether voluntarily or involuntarily, the condition is known as starvation. During total fasting, the glycogen reserves are rapidly depleted, and the body must call on its fat reserves. Ultimately, even the bone marrow (which is also a fat storage depot) is depleted, and it becomes red and jellylike rather than white and firm. In the early stages of a total fast, body proteins are metabolized at a relatively rapid rate. After several weeks, the rate of protein breakdown slows considerably as the brain adjusts to using fatty acid metabolites for some of its energy requirements. When no more fat reserves remain, the body must again draw heavily on structural protein for its energy requirements. The emaciated appearance of a starving individual is due to depleted muscle protein.

Only in the early stages of starvation do people have the energy to protest their condition by rioting. They soon become docile and placid as ketosis appears and acidosis rapidly follows. Oxygen deprivation leads to depression and lethargy. Even the ability to think clearly and make decisions is impaired. A starving person often dies quietly. Starva-

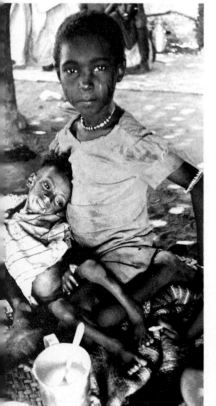

FIGURE 14.19
While much of the world suffers from starvation, many people in the industrialized nations show the effects of overindulgence in the "good life." [(a) © Reuters/Bettmann Newsphotos; (b) Nancy D'Antonio/Photo Researchers, Inc.]

tion is seldom the sole cause of death; weakened by starvation, a person succumbs to disease. Even "minor" diseases such as chickenpox and measles become life-threatening disorders. Barring disease, starvation alone will lead to death from circulatory failure as the heart muscle becomes too weak to pump blood.

One of the anomalies of our time is that, while much of the world suffers from hunger and malnutrition, the major dietary problem in industrialized nations is **obesity** (Figure 14.19). We eat too much. Whether the food is taken in as fat, carbohydrate, or protein, the excess is stored as fat. This fat shortens our life spans, gradually killing us by placing increased stress on the heart and circulatory system.

CHAPTER SUMMARY

Lipids are biomolecules that exhibit greater solubility in relatively nonpolar organic solvents than in water. They are used in the body as energy reserves, insulation, padding for vital organs, hormones, and cell membranes.

Triglycerides, or fats, are esters of fatty acids and glycerol. Fatty acids are long-chain carboxylic acids, the most common of which are listed in Table 14.1. A simple triglyceride incorporates a single kind of fatty acid, whereas a mixed triglyceride includes two or three different fatty acids.

Fatty acids may be described as saturated or unsaturated, depending on the absence or presence of double bonds. Vegetable fats (referred to as oils because they are liquids at room temperature) incorporate higher amounts of unsaturated fats and therefore exhibit lower melting points than animal fats. The level of unsaturation in a fat is reported as its iodine number, that is, the grams of iodine that react with 100 g of the triglyceride. The level of unsaturation of a fat or oil can be reduced by hydrogenation.

The basic hydrolysis of a fat is termed saponification. The salts of fatty acids obtained on saponification are called soaps. Soaps act as cleaning agents by dispersing and suspending oily dirt in water solutions. Soaps are deactivated by hard water because the cations (Ca^{2+}, Mg^{2+}, and Fe^{3+}) in hard water combine with the fatty acid anions to form water-insoluble salts. Synthetic detergents are designed not to precipitate in hard water.

Phosphatides are an example of phosphorus-containing lipids (phospholipids). Phosphatides are mixed esters incorporating glycerol, two fatty acid units, and an amino alcohol unit linked through a phosphate group. In cephalins, the amino alcohol is ethanolamine, and in lecithins the amino alcohol is choline. Phosphatides are zwitterions (internal salts); each molecule has an ionic end and a nonpolar end (a characteristic shared with soaps and detergents). Molecules with this structural feature can form micelles, monolayers, and bilayers. In all these arrangements, the ionic and nonpolar parts of the molecules are aligned with structurally related parts on neighboring molecules. Membranes in the body are more complex versions of bilayers.

Nonsaponifiable lipids are those that cannot be hydrolyzed by base; included in this class are steroids and prostaglandins. All steroids share a common fused four-ring structure. Included among the steroids are cholesterol, bile acids, and adrenal and sex hormones. The prostaglandins are potent biochemicals that exhibit hormone-like properties.

Fats are digested principally in the small intestines under the influence of enzymes called lipases. Bile salts emulsify fats to permit better contact between the fats and the digestive enzymes. Upon hydrolysis, fats yield glycerol, fatty acids and their salts, and monoglycerides and diglycerides. As these products pass through the intestinal membrane they are reassembled into triglycerides, which are carried to the bloodstream via the lymphatic system.

Fats are stored throughout the body in adipose tissue (fat depots). When called on to supply energy, fats are first split into glycerol and fatty acids. The glycerol feeds into the glycolysis pathway, and the fatty acids are degraded to acetyl coenzyme A via the fatty acid cycle. The acetyl CoA can be used to fuel the Krebs cycle, to form steroids, or to form new fatty acids.

Three compounds formed in association with fatty acid metabolism are referred to as ketone bodies. These include acetoacetic acid, β-hydroxybutyric acid, and acetone. Abnormally high levels of fat metabolism lead to a buildup of these compounds in the blood, a condition termed ketosis. Because two of the three compounds are carboxylic acids, high blood levels can lead to a drop in pH and a condition called acidosis.

CHAPTER QUIZ

1. A lipid is defined by its
   a. functional group
   b. action on the body
   c. chemical reactivity
   d. solubility

2. Fats are esters of fatty acids and
   a. aldehyde   b. glycerol
   c. ketone   d. alkane

3. The process of hydrogenation removes
   a. double bonds   b. hydrogen
   c. water   d. carbon links

4. Saponification involves the reaction of fats with
   a. hydroxide ions b. soap
   c. acid   d. water

5. The cleansing properties of soap are due to its
   a. saturated chain
   b. acid group
   c. solubility in both oil and water
   d. reactivity with a multitude of chemicals

6. One definition of "zwitterions" is
   a. synthetic soaps   b. calcium deposits
   c. intramolecular salts d. amino acids

7. Saponifiable lipids can produce all but which product?
   a. alkanes   b. glycerol
   c. fatty acids   d. sugars

8. Which of the following is a steroid?
   a. soap   b. cholesterol
   c. sugar   d. micelle

9. Hydrolysis of fats is catalyzed by
   a. alcohols   b. esters
   c. cholines   d. enzymes

10. Fat storage in the body
    a. occurs in many places
    b. is localized
    c. is always undesirable
    d. has no physical function

PROBLEMS

14.1 What Is a Lipid?

1. Define and illustrate a lipid.

14.2 Fatty Acids

2. Define and illustrate a fatty acid.

3. Which of these fatty acids are saturated and which are unsaturated?
   a. caproic acid   **b.** oleic acid
   c. stearic acid   **d.** palmitic acid
   e. linolenic acid **f.** caprylic acid

4. How many carbon atoms are there in a molecule of each of the compounds listed in Problem 3?

5. Arrange these fatty acids in order of increasing levels of unsaturation: linoleic acid, linolenic acid, oleic acid.

6. Write a structural formula for lauric acid.

7. Write a structural formula for oleic acid.

14.3 Fats: The Triglycerides

8. Define:
   a. simple triglyceride   **b.** mixed triglyceride
   c. fat   **d.** oil
   e. polyunsaturated fat   **f.** iodine number

9. Write a structural formula for glyceryl tristearate (tristearin).

10. In the determination of the iodine number of a fat, what functional group in the fat molecule reacts with the reagent?

11. Which would have the higher iodine number: tristearin or triolein?

12. Which would you expect to have the higher iodine number—corn oil or beef tallow? Explain your reasoning.
13. Which would you expect to have the higher iodine number—hard or liquid margarine? Explain your reasoning.
14. What fat would be formed by the complete hydrogenation of triolein? Of trilinolein?

14.4 Soaps and the Saponification of Fats

15. Define soap.
16. Write a structural formula for sodium oleate.
17. Write a structural formula for calcium myristate (see Table 11.2).
18. Why does soap lose its cleaning efficiency in hard water?
19. Synthetic detergents work well in hard water that deactivates soap. Why?
20. Draw the structural formulas of the products of this saponification reaction.

$$\begin{array}{l} CH_2-O-\overset{O}{\overset{\|}{C}}(CH_2)_6CH_3 \\ CH-O-\overset{O}{\overset{\|}{C}}(CH_2)_4CH_3 \xrightarrow{NaOH} \\ CH_2-O-\overset{O}{\overset{\|}{C}}(CH_2)_8CH_3 \end{array}$$

21. The structures of synthetic detergents are designed to give these compounds the same kind of cleaning action as soap. Which of the following compounds would be effective cleaning agents?

22. Write the equation for the saponification of glyceryl trilaurate.
23. How would the equation for Problem 22 differ if the compound being saponified were triolein?
24. What general structural features make soaps good cleaning agents?
25. Describe the process by which soap acts to remove oily dirt.
26. Classify the following lipids as saponifiable or nonsaponifiable.
    a. tristearin b. cholesterol
    c. a lecithin d. a prostaglandin

14.5 Phosphatides

27. Define:
    a. phospholipid b. phosphatide
    c. lecithin   d. cephalin
    e. micelle   f. monolayer
    g. bilayer
28. Which are phospholipids?
    a. tristearin   b. cholesterol
    c. phosphatides d. prostaglandin E_1
    e. cephalin   f. triolein
29. Which compounds are esters of fatty acids?
    a. cephalin   b. cholesterol
    c. vegetable oil d. prostaglandins
30. Both lecithins and cephalins are classed as phosphatides. How do lecithins differ from cephalins?
31. What general structural feature do phosphatides share with soaps and detergents?
32. Using a circle to represent the ionic group and zigzag lines to represent the two fatty acid units, draw the arrangement of phosphatide molecules in a micelle and in a bilayer.

a. $CH_3CHCH_2CHCH_2CHCH_2CHCH_2CHCH_2CHCH_2CHCH_2-O-\overset{O}{\underset{\|}{\overset{\|}{S}}}-O^- K^+$
   $\quad\ \ \, |\quad\ \ |\quad\ \ |\quad\ \ |\quad\ \ |\quad\ \ |\quad\ \ |\qquad\quad\ \, \|$
   $\quad\ \ CH_3\ CH_3\ CH_3\ CH_3\ CH_3\ CH_3\ CH_3\qquad\quad\ O$

b. $\underset{\underset{}{}}{\overset{OH}{|}}\ \underset{}{\overset{OH}{|}}\ \overset{OH}{|}\ \overset{OH}{|}\ \overset{OH}{|}\ \overset{OH}{|}\ \overset{OH}{|}\ \overset{OH}{|}\ \overset{O}{\overset{\|}{}}$
   $CH_2CH_2CHCH_2CHCH_2CHCH_2CHCH_2CHCH_2CHCH_2CHCH_2C-O^-\ Na^+$

c. $\underset{\underset{SO_3^-\ Na^+}{|}}{CH_2CH_2}\underset{\underset{}{}}{CHCH_2}\underset{\underset{SO_3^-\ Na^+}{|}}{CHCH_2CH_2}$
   $\qquad\qquad\ \ \, \underset{SO_3^-\ Na^+}{|}$

14.6 Nonsaponifiable Lipids: Steroids and Prostaglandins

33. Define:
    a. steroid
    b. prostaglandin
    c. saponifiable lipid
    d. nonsaponifiable lipid
34. Draw the basic steroid skeleton.
35. Which of the following compounds are classified as steroids?
    a. tristearin
    b. cholesterol
    c. testosterone
    d. prostaglandins
    e. cephalin
    f. cholic acid
36. Which is the most abundant steroid found in animals?
37. Name some of the steroid hormones.
38. What common steroid serves as a starting material for steroid hormone synthesis in the body?
39. List some possible future uses of prostaglandins as drugs.

14.7 Digestion and Metabolism of Fats

40. Define:
    a. adipose tissue
    b. fat depot
41. What are the enzymes that catalyze the hydrolysis of triglycerides called?
42. Salts of the bile acids are involved in the digestion of lipids. Relate the action of bile salts to that of soap.
43. If lipids are insoluble in water, how is it possible for them to be transported in the aqueous fluids of the body?
44. What purpose does the lymphatic transport system serve?
45. What is the end product of the fatty acid spiral?
46. Identify three metabolic pathways in which acetyl coenzyme A plays a role.
47. What happens to the glycerol formed from the hydrolysis of a fat?
48. Why do most naturally occurring fatty acids have an even number of carbon atoms?
49. Give three functions of adipose tissue in the body.

14.8 The Ketone Bodies

50. Define:
    a. ketone bodies
    b. ketosis
    c. acidosis
51. Why does acidosis result from ketosis?
52. Use equations to show how acetone and β-hydroxybutyric acid are formed from acetoacetic acid.

14.9 The Chemistry of Starvation

53. a. During a fast, what energy reserves are used first?
    b. What energy reserves supply the major part of the body's needs?
    c. What is the final source of energy called on by the body?
54. Body mass index is a person's weight in kilograms divided by the square of his or her height in meters. Calculate your BMI. (The ideal range is 20–25.)

15

Proteins: Structure and Metabolism

Did You Know That...

▶ proteins are found in all living organisms?

▶ natural amino acids have left-handed configurations?

▶ the adult human body can make all but eight of the amino acids needed for protein synthesis?

▶ plant proteins are usually deficient in some amino acids?

▶ some proteins have molecular weights in the millions?

▶ silk and wool are proteins?

▶ alcohol kills bacteria by disrupting their protein structure?

▶ a person who has ingested mercury should be given raw egg immediately?

▶ enzyme blood levels change as a result of heart attack, exercise, and anxiety?

▶ the brain has built-in sources of painkillers?

▶ acupuncture releases brain "opiates"?

Carbohydrates, lipids, and proteins—these are the three classes of foods. All are essential to life, but proteins are perhaps closest to the stuff of life itself. No living part of the human body—or of any other organism, for that matter—is completely without protein. Protein is present in the blood, in muscles, in the brain, and even in tooth enamel. The smallest cellular organisms, the bacteria, contain protein. Viruses, so small that they make bacteria look like giants, are nothing but large molecules of nucleoproteins. **Nucleoproteins** are combinations of proteins with nucleic acids, the other class of biomolecules that might rightly claim to be the stuff of life itself. Nucleic acids determine which living organism is constructed from the raw materials at hand. They do this by controlling protein synthesis. We shall examine nucleic acids and their role as carriers of the genetic message in Chapter 17. In the present chapter, we want to show you why proteins are the materials of choice when it comes to building protozoa, plants, and people (Figure 15.1).

Combinations of proteins with nucleic acids are called nucleoproteins.

15.1 Amino Acids: Structure and Physical Properties

To understand protein structure, one must first examine amino acids. Amino acids contain two functional groups, a carboxyl group (—COOH) and an amino group (—NH$_2$). In the amino acids that form proteins, the amino group is always bonded to the carbon atom next to the carboxyl group. This position next to the carboxyl group is called the *alpha* position, and the amino acids that we shall consider are therefore referred to as α-amino acids.

Amino acids always contain a carboxyl group (—COOH) and an amino group (—NH$_2$).

$$H_2N-\underset{|}{\overset{|\alpha}{C}}-\overset{O}{\underset{}{\overset{\|}{C}}}-OH$$

The above formula indicates the proper placement of these groups, but the structure is not really correct. Acids react with bases to form salts; the carboxyl group is acidic, and the amino group is basic. Therefore, these two functional groups interact, the acid transferring a proton to the base. The resulting product is a **zwitterion**, a compound like the phospholipids (Section 14.5) in which the anion and cation are part of the same molecule.

$$H_3\overset{+}{N}-\underset{\underset{H}{|}}{\overset{\overset{R}{|}}{C}}-\overset{O}{\underset{}{\overset{\|}{C}}}-O^-$$

FIGURE 15.1
Leather, silk, wool, and meat tenderizer are all rich in protein. Insulin is a protein hormone. [Photo © The Terry Wild Studio.]

While amino acids are sometimes written as nonionized structures, their physical properties suggest that they actually exist in the ionized form. They are crystalline solids at room temperature. Their water solubility varies but is generally greater than their solubility in nonpolar organic solvents such as chloroform and benzene. These properties are consistent with a zwitterion structure. For this reason, we shall ordinarily draw amino acids as zwitterions.

What makes one amino acid different from another is the identity of the R group, which is also attached at the alpha position. If this R group is anything but a hydrogen atom, the amino acid contains a chiral carbon atom and exists as a pair of mirror-image isomers. Nearly all naturally occurring amino acids have the L configuration. Note the similarity in the arrangements of L-glyceraldehyde and the amino acid L-serine.

$$\begin{array}{cc} \text{CHO} & \text{COO}^- \\ \text{HO}-\text{C}-\text{H} & \text{H}_3\overset{+}{\text{N}}-\text{C}-\text{H} \\ \text{CH}_2\text{OH} & \text{CH}_2\text{OH} \\ \text{L-Glyceraldehyde} & \text{L-Serine} \end{array}$$

The simplest α-amino acid and the only common one that does not exist as a pair of mirror-image isomers is glycine (R = H).

$$\text{H}_3\overset{+}{\text{N}}-\text{CH}_2-\overset{\overset{\text{O}}{\|}}{\text{C}}-\text{O}^-$$

The amino acids can be divided into four groups based on the structure of the side chain (the R group). Several amino acids have nonpolar hydrocarbon side chains; two examples are alanine (R = CH_3) and phenylalanine (R = $CH_2C_6H_5$). A second group has polar side chains that are not readily ionized. Included in this group are serine (R = CH_2OH) and cysteine (R = CH_2SH). The third and fourth groups have side chains that do form ions readily. One set, such as aspartic acid (R = CH_2COOH) and glutamic acid (R = CH_2CH_2COOH), has acidic side chains. Members of the final group have basic side chains. Lysine (R = $CH_2CH_2CH_2CH_2NH_2$) is a typical example. Structures of all the common α-amino acids are given in Table 15.1.

There are four groups of amino acids.
- nonpolar side chains
- polar side chains
- basic side chains
- acidic side chains

TABLE 15.1 Amino Acids That Occur Naturally in Proteins

Name	Abbreviation	Essential?	Structure
Nonpolar Side Chains			
Glycine	Gly	No	CH_2-COO^- $\|$ $^+NH_3$
Alanine	Ala	No	$CH_3-CH-COO^-$ $\|$ $^+NH_3$
Phenylalanine	Phe	Yes	$C_6H_5-CH_2-CH-COO^-$ $\|$ $^+NH_3$
Valine	Val	Yes	$CH_3-CH-CH-COO^-$ $\| \quad \|$ $CH_3 \ ^+NH_3$
Leucine	Leu	Yes	$CH_3CHCH_2-CH-COO^-$ $\| \quad\quad \|$ $CH_3 \quad\ ^+NH_3$
Isoleucine	Ile	Yes	$CH_3CH_2CH-CH-COO^-$ $\| \quad \|$ $CH_3 \ ^+NH_3$
Proline	Pro	No	$\begin{array}{c} CH_2-CH_2 \\ \| \quad\quad \| \\ CH_2 \quad CH-COO^- \\ \diagdown \ \diagup \\ ^+NH_2 \end{array}$
Methionine	Met	Yes	$CH_3-S-CH_2CH_2-CH-COO^-$ $\|$ $^+NH_3$
Polar, but Not Ionizable, Side Chains			
Serine	Ser	No	$HO-CH_2-CH-COO^-$ $\|$ $^+NH_3$
Threonine	Thr	Yes	$CH_3CH-CH-COO^-$ $\| \quad\ \|$ $OH \ ^+NH_3$
Asparagine	Asn	No	$H_2N-\overset{O}{\overset{\|}{C}}-CH_2-CH-COO^-$ $\|$ $^+NH_3$
Glutamine	Gln	No	$H_2N-\overset{O}{\overset{\|}{C}}-CH_2CH_2-CH-COO^-$ $\|$ $^+NH_3$
Cysteine	Cys	No	$HS-CH_2-CH-COO^-$ $\|$ $^+NH_3$
Cystine	Cys-S—S-Cys	No	$^-OOC-CH-CH_2-S-S-CH_2-CH-COO^-$ $\quad\quad\ \| \quad\quad\quad\quad\quad\quad\quad\quad\ \|$ $\quad\quad ^+NH_3 \quad\quad\quad\quad\quad\quad\quad\ ^+NH_3$
Hydroxyproline	Hyp	No	$\begin{array}{c} HO-CH-CH_2 \\ | \quad\quad | \\ CH_2 \quad CH-COO^- \\ \diagdown \ \diagup \\ ^+NH_2 \end{array} $

TABLE 15.1 (Continued)

Name	Abbreviation	Essential?	Structure
\multicolumn{4}{c}{Polar, but Not Ionizable, Side Chains (continued)}			
Tyrosine	Tyr	No	HO—⟨ ⟩—CH$_2$—CH(⁺NH$_3$)—COO⁻
3,5-Dibromotyrosine	—	No	HO—⟨Br,Br⟩—CH$_2$—CH(⁺NH$_3$)—COO⁻
3,5-Diiodotyrosine	—	No	HO—⟨I,I⟩—CH$_2$—CH(⁺NH$_3$)—COO⁻
Thyroxine	—	No	HO—⟨I,I⟩—O—⟨I,I⟩—CH$_2$—CH(⁺NH$_3$)—COO⁻
Tryptophan	Trp	Yes	(indole)—CH$_2$—CH(⁺NH$_3$)—COO⁻
\multicolumn{4}{c}{Basic Side Chains, Ionizable}			
Lysine	Lys	Yes	H$_3$⁺NCH$_2$CH$_2$CH$_2$CH$_2$—CH(NH$_2$)—COO⁻
Hydroxylysine	Hyl	No	H$_3$⁺NCH$_2$CH(OH)CH$_2$CH$_2$—CH(NH$_2$)—COO⁻
Arginine	Arg	a	H$_2$N—C(=⁺NH$_2$)—NHCH$_2$CH$_2$CH$_2$—CH(NH$_2$)—COO⁻
Histidine	His	b	(imidazole)—CH$_2$—CH(⁺NH$_3$)—COO⁻
\multicolumn{4}{c}{Acidic Side Chains, Ionizable}			
Aspartic acid	Asp	No	HOOC—CH$_2$—CH(⁺NH$_3$)—COO⁻
Glutamic acid	Glu	No	HOOC—CH$_2$CH$_2$—CH(⁺NH$_3$)—COO⁻

[a] Essential to growing children but not to adult human beings.
[b] Essential to rats but not to human beings.

15.2 THE ESSENTIAL AMINO ACIDS

The essential amino acids must be included in your diet.

The adult human body can synthesize all but eight of the amino acids needed for making proteins. Those eight (indicated in Table 15.1) are called **essential amino acids**. They must be included in the diet. (Linoleic acid and linolenic acid are similarly classified as **essential fatty acids** because they are required for health but cannot be synthesized by the body.) An **adequate protein** is one that supplies all the essential amino acids in the quantities needed for growth and repair of body tissues. Most proteins from animal sources incorporate all the essential amino acids in adequate amounts. Lean meat, milk, fish, eggs, and cheese contain adequate proteins. In fact, gelatin is about the only inadequate animal protein. On the other hand, one would have to eat a wide variety of plant materials to be sure of getting enough of all the essential amino acids. Most proteins from plant sources are deficient in one or more amino acids. For example, corn protein has inadequate amounts of lysine and tryptophan. People whose diet consists chiefly of corn may suffer from malnutrition even though the caloric content of the food is adequate. A protein-deficiency disease called kwashiorkor is common in parts of Africa where corn is the major food (Figure 15.2). Even a well-balanced all-vegetable diet is likely to be lacking in vitamin B_{12}, for this nutrient is not found in plants. Other nutrients scarce in all-plant diets include calcium, iron, riboflavin, and (for children not exposed to sunlight) vitamin D. These needs can be met by vitamin and mineral supplements or through a modified vegetarian diet that includes milk, eggs, cheese, and/or fish.

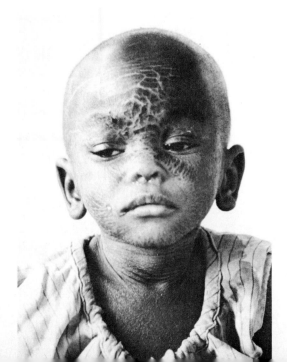

FIGURE 15.2
A lack of proteins and vitamins causes the deficiency disease known as kwashiorkor, which can be recognized by retarded growth, discoloration of the skin and hair, bloating, swollen belly, and mental apathy. [Courtesy of the World Health Organization, New York. Photo by Paul Almasy.]

15.3 AMINO ACIDS AS BUFFERS

Amino acids can act as either acids or bases. They (and the proteins) act as buffers in living organisms. In the presence of added acid, the carboxylate group of the zwitterion captures protons.

$$\overset{+}{H_3N}-\underset{R}{CH}-COO^- + H_3O^+ \longrightarrow \overset{+}{H_3N}-\underset{R}{CH}-COOH + H_2O$$

Amino acids can act as either acids or bases.

Note that the product is a positive ion. If base is added, protons are removed from the amino group of the zwitterion.

$$\overset{+}{H_3N}-\underset{R}{CH}-COO^- + OH^- \longrightarrow H_2N-\underset{R}{CH}-COO^- + H_2O$$

In this case, a negative ion is formed. In each instance, the amino acid acts to tie up added acid or base. Thus, the charge that an amino acid carries depends on the pH of the system.

At some intermediate pH value, an amino acid exists almost entirely as the zwitterion. That particular pH value is called the **isoelectric point**. At the isoelectric point, the positive and negative charges on an amino acid (or protein) just balance, and the molecule *as a whole* is electrically neutral. The isoelectric point does *not* necessarily coincide with pH 7 (the value we ordinarily associate with "neutrality"). Each amino acid has its own characteristic isoelectric point (Table 15.2). The simple amino acids with nonionizable side chains have isoelectric points ranging from pH 5.0 to pH 6.5. Basic amino acids (those in which the side chain incorporates a basic group) have isoelectric points at relatively high pH values. Acidic amino acids have isoelectric points at quite low pH values.

The isoelectric point is the pH value at which the amino acid exists as a zwitterion.

15.4 FORMATION OF PROTEINS: THE PEPTIDE BOND

Many reactions are possible for amino acids. All these compounds contain at least two organic functional groups, and many contain a third in the side chain. Nonetheless, one reaction is unquestionably the most important—the polymerization of amino acids to form proteins. Carboxylic acids react with amines to form amides (Section 12.5). The amino group of one amino acid molecule can react with the carboxyl group of another. A molecule of water is lost, and the two amino acids

TABLE 15.2 Isoelectric Points of Representative Amino Acids

Amino Acid	Class	pH
Aspartic acid	Acidic	2.77
Glutamic acid	Acidic	3.22
Cysteine	Simple	5.02
Methionine	Simple	5.75
Tryptophan	Simple	5.89
Glycine	Simple	5.97
Alanine	Simple	6.11
Lysine	Basic	9.74

are linked through an amide functional group called a **peptide bond**.

$$H_3\overset{+}{N}-\underset{R}{CH}-\overset{\overset{O}{\|}}{C}-O^- + H_3\overset{+}{N}-\underset{R}{CH}-\overset{\overset{O}{\|}}{C}-O^- \longrightarrow H_3\overset{+}{N}-\underset{R}{CH}-\overset{\overset{O}{\|}}{C}-NH-\underset{R}{CH}-\overset{\overset{O}{\|}}{C}-O^- + H_2O$$

Peptide bond

Peptide or protein—a polymer of amino acids

Notice that there is still a reactive amino group on the left and a carboxyl group on the right. Each of these can react further to join more amino acid units. This process can continue until thousands of units have joined to form a giant molecule—a polymer called a **peptide** or **protein**.

$$\cdots \underset{R}{CH}-\overset{\overset{O}{\|}}{C}-NH-\underset{R}{CH}-\overset{\overset{O}{\|}}{C}-NH-\underset{R}{CH}-\overset{\overset{O}{\|}}{C}-NH-\underset{R}{CH}-\overset{\overset{O}{\|}}{C}-NH-\underset{R}{CH}-\overset{\overset{O}{\|}}{C}-NH \cdots$$

When only two amino acids are joined, the product is called a **dipeptide**.

$$H_3\overset{+}{N}-CH_2-\overset{\overset{O}{\|}}{C}-NH-\underset{\underset{C_6H_5}{CH_2}}{CH}-\overset{\overset{O}{\|}}{C}-O^-$$

Glycylphenylalanine (a dipeptide)

When three amino acids are combined, the substance is a **tripeptide**.

$$H_3\overset{+}{N}-\underset{CH_2OH}{CH}-\overset{\overset{O}{\|}}{C}-NH-\underset{CH_3}{CH}-\overset{\overset{O}{\|}}{C}-NH-\underset{CH_2SH}{CH}-\overset{\overset{O}{\|}}{C}-O^-$$

Serylalanylcysteine (a tripeptide)

Polypeptide—more than 10 amino acids

Products with more than 10 amino acid units are often simply called **polypeptides**. When the molecular weight of a compound exceeds 10 000, the polymer is called a protein. Under these rules, insulin (molecular weight = 5700) is classified as a peptide, whereas hemoglobin (molecular weight = 64 500) is a protein. The distinction between peptides and proteins is an arbitrary one, and it is not always precisely applied.

15.5 The Sequence of Amino Acids in Proteins

TABLE 15.3 Isoelectric Points of Selected Proteins

Protein	Source	pH
Silk fibroin	Silk	2.2
Casein	Cow's milk	4.6
Gelatin	Pig's feet	4.8
Serum albumin	Human blood serum	4.9
Albumin	Egg white	4.9
Lactoglobin	Cow's milk	5.2
Insulin	Cow's pancreas	5.3
Hemoglobin	Human red blood cells	6.7
Ribonuclease	Pancreas	9.5
Lysozyme	Egg white	10.7

As with amino acids, every protein has a characteristic isoelectric point (Table 15.3). In proteins, however, since the alpha carboxyl and amino groups are tied up in the peptide bonds that hold the molecule together, it is the ionizable side chains of the amino acid units that establish the isoelectric points of proteins. At its isoelectric point the protein molecule as a whole is electrically neutral; it may contain many ionized groups, but the positively charged side chains are exactly balanced by negatively charged side chains. At a pH value other than the value at the isoelectric point, a protein carries a net positive or negative charge. Thus, the structure of a protein and its ability to carry out its physiological function are influenced by pH.

15.5 THE SEQUENCE OF AMINO ACIDS IN PROTEINS

For peptides and proteins to be physiologically active, it is not enough that they incorporate certain amounts of specific amino acids. The order or *sequence* in which the amino acids are connected is also of critical importance. Glycylalanine, for example, is not the same as alanylglycine.

$$H_3\overset{+}{N}-\underset{H}{\overset{H}{C}}H-\underset{}{\overset{O}{C}}-NH-\underset{CH_3}{\overset{CH_3}{C}}H-COO^- \qquad H_3\overset{+}{N}-\underset{CH_3}{\overset{CH_3}{C}}H-\underset{}{\overset{O}{C}}-NH-\underset{H}{\overset{H}{C}}H-COO^-$$

Glycylalanine Alanylglycine

Although the structural difference seems minor, the two substances do behave differently in the body.

Gly-Ala-Ser-Lys-Cys	Ala-Lys-Cys-Gly-Ser	Lys-Ala-Ser-Gly-Cys
Gly-Ala-Ser-Cys-Lys	Ala-Lys-Cys-Ser-Gly	Lys-Ala-Ser-Cys-Gly
Gly-Ala-Lys-Ser-Cys	Ala-Cys-Gly-Ser-Lys	Lys-Ala-Cys-Gly-Ser
Gly-Ala-Lys-Cys-Ser	Ala-Cys-Gly-Lys-Ser	Lys-Ala-Cys-Ser-Gly
Gly-Ala-Cys-Ser-Lys	Ala-Cys-Ser-Gly-Lys	Lys-Ser-Gly-Ala-Cys
Gly-Ala-Cys-Lys-Ser	Ala-Cys-Ser-Lys-Gly	Lys-Ser-Gly-Cys-Ala
Gly-Ser-Ala-Lys-Cys	Ala-Cys-Lys-Gly-Ser	Lys-Ser-Ala-Gly-Cys
Gly-Ser-Ala-Cys-Lys	Ala-Cys-Lys-Ser-Gly	Lys-Ser-Ala-Cys-Gly
Gly-Ser-Lys-Ala-Cys	Ser-Gly-Ala-Lys-Cys	Lys-Ser-Cys-Gly-Ala
Gly-Ser-Lys-Cys-Ala	Ser-Gly-Ala-Cys-Lys	Lys-Ser-Cys-Ala-Gly
Gly-Ser-Cys-Ala-Lys	Ser-Gly-Lys-Ala-Cys	Lys-Cys-Gly-Ala-Ser
Gly-Ser-Cys-Lys-Ala	Ser-Gly-Lys-Cys-Ala	Lys-Cys-Gly-Ser-Ala
Gly-Lys-Ala-Ser-Cys	Ser-Gly-Cys-Ala-Lys	Lys-Cys-Ala-Gly-Ser
Gly-Lys-Ala-Cys-Ser	Ser-Gly-Cys-Lys-Ala	Lys-Cys-Ala-Ser-Gly
Gly-Lys-Ser-Ala-Cys	Ser-Ala-Gly-Lys-Cys	Lys-Cys-Ser-Gly-Ala
Gly-Lys-Ser-Cys-Ala	Ser-Ala-Gly-Cys-Lys	Lys-Cys-Ser-Ala-Gly
Gly-Lys-Cys-Ala-Ser	Ser-Ala-Lys-Gly-Cys	Cys-Gly-Ala-Ser-Lys
Gly-Lys-Cys-Ser-Ala	Ser-Ala-Lys-Cys-Gly	Cys-Gly-Ala-Lys-Ser
Gly-Cys-Ala-Ser-Lys	Ser-Ala-Cys-Gly-Lys	Cys-Gly-Ser-Ala-Lys
Gly-Cys-Ala-Lys-Ser	Ser-Ala-Cys-Lys-Gly	Cys-Gly-Ser-Lys-Ala
Gly-Cys-Ser-Ala-Lys	Ser-Lys-Gly-Ala-Cys	Cys-Gly-Lys-Ala-Ser
Gly-Cys-Ser-Lys-Ala	Ser-Lys-Gly-Cys-Ala	Cys-Gly-Lys-Ser-Ala
Gly-Cys-Lys-Ala-Ser	Ser-Lys-Ala-Gly-Cys	Cys-Ala-Gly-Ser-Lys
Gly-Cys-Lys-Ser-Ala	Ser-Lys-Ala-Cys-Gly	Cys-Ala-Gly-Lys-Ser
Ala-Gly-Ser-Lys-Cys	Ser-Lys-Cys-Gly-Ala	Cys-Ala-Ser-Gly-Lys
Ala-Gly-Ser-Cys-Lys	Ser-Lys-Cys-Ala-Gly	Cys-Ala-Ser-Lys-Gly
Ala-Gly-Lys-Ser-Cys	Ser-Cys-Gly-Ala-Lys	Cys-Ala-Lys-Gly-Ser
Ala-Gly-Lys-Cys-Ser	Ser-Cys-Gly-Lys-Ala	Cys-Ala-Lys-Ser-Gly
Ala-Gly-Cys-Ser-Lys	Ser-Cys-Ala-Gly-Lys	Cys-Ser-Gly-Ala-Lys
Ala-Gly-Cys-Lys-Ser	Ser-Cys-Ala-Lys-Gly	Cys-Ser-Gly-Lys-Ala
Ala-Ser-Gly-Lys-Cys	Ser-Cys-Lys-Gly-Ala	Cys-Ser-Ala-Gly-Lys
Ala-Ser-Gly-Cys-Lys	Ser-Cys-Lys-Ala-Gly	Cys-Ser-Ala-Lys-Gly
Ala-Ser-Lys-Gly-Cys	Lys-Gly-Ala-Ser-Cys	Cys-Ser-Lys-Gly-Ala
Ala-Ser-Lys-Cys-Gly	Lys-Gly-Ala-Cys-Ser	Cys-Ser-Lys-Ala-Gly
Ala-Ser-Cys-Gly-Lys	Lys-Gly-Ser-Ala-Cys	Cys-Lys-Gly-Ala-Ser
Ala-Ser-Cys-Lys-Gly	Lys-Gly-Ser-Cys-Ala	Cys-Lys-Gly-Ser-Ala
Ala-Lys-Gly-Ser-Cys	Lys-Gly-Cys-Ala-Ser	Cys-Lys-Ala-Gly-Ser
Ala-Lys-Gly-Cys-Ser	Lys-Gly-Cys-Ser-Ala	Cys-Lys-Ala-Ser-Gly
Ala-Lys-Ser-Gly-Cys	Lys-Ala-Gly-Ser-Cys	Cys-Lys-Ser-Gly-Ala
Ala-Lys-Ser-Cys-Gly	Lys-Ala-Gly-Cys-Ser	Cys-Lys-Ser-Ala-Gly

FIGURE 15.3
These 120 pentapeptides incorporate the same set of five different amino acids; only the sequence varies. An astronomical number of variations is possible with more than 20 different amino acids and chain lengths extending to thousands of units.

15.5 The Sequence of Amino Acids in Proteins

When chemists describe peptides and proteins, they find it much simpler to indicate the amino acid sequence by using the abbreviations for the amino acids shown in Table 15.1. The sequence for glycylalanine is written Gly-Ala, and that for alanylglycine is Ala-Gly. This shorthand indicates that the peptide is arranged with the free amino group to the left and the free carboxyl group to the right.

As the length of the peptide chain increases, the possible sequential variations become almost infinite (Figure 15.3). This potential for many different arrangements is exactly what one needs in a material that is to be used to make such diverse things as hair and skin and eyeballs and toenails and 10 000 different enzymes. Just as we can make millions of different words with our 26-letter English alphabet, we can make millions of different proteins with 26 of the available amino acids. Just as one can write gibberish with the English alphabet, one can make nonfunctioning proteins by putting together amino acids in the wrong sequence. Yet while the correct sequence is ordinarily of utmost importance, it is not always absolutely required. A misspelled word may still convey its meaning, and a protein with a small percentage of "incorrect" amino acids may continue to function. It may not function as well, however. Sometimes a seemingly minor change can have a disastrous effect. Some people have hemoglobin with one incorrect amino acid unit in about 300. That "minor" error is responsible for *sickle cell anemia*, an inherited condition with frequently devastating effects (Figure 15.4).

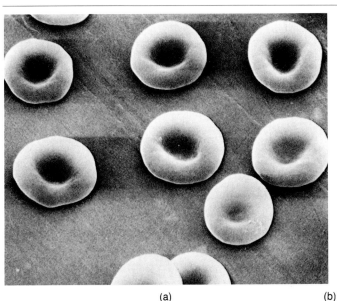

FIGURE 15.4
Changing just one amino acid in 300 in hemoglobin has a profound effect on the shape of the red blood cell that houses the hemoglobin. (a) Normal red blood cells. (b) A sickled cell. [Courtesy of Richard F. Baker, University of Southern California Medical School, Los Angeles.]

(a)　　　　(b)

15.6 More on the Structure of Proteins

The primary structure of a protein molecule is its amino acid sequence.

FIGURE 15.5
The sequence of amino acids in proteins varies from one species to another. The variation illustrated here is for a hormone involved in establishing protective coloration in animals.

Proteins are big polypeptides; some have formula weights ranging into the millions. Each kind of protein has its own characteristic composition, amino acid sequence, and three-dimensional shape. The structure of proteins is generally discussed at four organization levels. We have already considered one of these. The **primary structure** of protein molecule is its amino acid sequence. To specify the primary structure, one merely writes out the sequence of amino acids in the long-chain molecule (Figure 15.5). ("Merely" may not be the best choice of words here, when one considers the fact that some proteins contain as many as 10 000 amino acid units!) What "holds" the primary structure together are the peptide links between the amino acid units. The long protein chains are not simply arranged at random like tangled threads. To describe a protein's structure fully, it is necessary to discuss secondary, tertiary, and sometimes quaternary levels of organization for that protein. In a similar way, the cord on your telephone might be described

$H_3\overset{+}{N}$-Asp-Ser-Gly-Pro-Tyr-Lys-Met-Glu-His-Phe-Arg-Trp-Gly-Ser-Pro-Pro-Lys-Asp-$\overset{\overset{O}{\|}}{C}$—$O^-$
      1   2   3   4   5   6   7   8   9 10 11 12 13 14 15 16 17 18

Ox β-MSH

$H_3\overset{+}{N}$-Asp-Glu-Gly-Pro-Tyr-Lys-Met-Glu-His-Phe-Arg-Trp-Gly-Ser-Pro-Pro-Lys-Asp-$\overset{\overset{O}{\|}}{C}$—$O^-$
      1   2   3   4   5   6   7   8   9 10 11 12 13 14 15 16 17 18

Porcine β-MSH

$H_3\overset{+}{N}$-Asp-Glu-Gly-Pro-Tyr-Lys-Met-Glu-His-Phe-Arg-Trp-Gly-Ser-Pro-Arg-Lys-Asp-$\overset{\overset{O}{\|}}{C}$—$O^-$
      1   2   3   4   5   6   7   8   9 10 11 12 13 14 15 16 17 18

Horse β-MSH

$H_3\overset{+}{N}$-Asp-Glu-Gly-Pro-Tyr-Arg-Met-Glu-His-Phe-Arg-Trp-Gly-Ser-Pro-Pro-Lys-Asp-$\overset{\overset{O}{\|}}{C}$—$O^-$
      1   2   3   4   5   6   7   8   9 10 11 12 13 14 15 16 17 18

Monkey β-MSH

$H_3\overset{+}{N}$-Ala-Glu-Lys-Lys-Asp-Glu-Gly-Pro-Tyr-Arg-Met-Glu-His-Phe-Arg-Trp-Gly-Ser-Pro-Pro-Lys-Asp-$\overset{\overset{O}{\|}}{C}$—$O^-$
     1   2   3   4   5   6   7   8   9 10 11 12 13 14 15 16 17 18 19 20 21 22

Human β-MSH

15.6 More on the Structure of Proteins

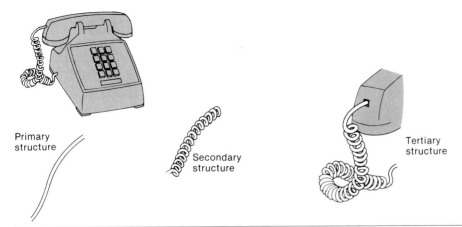

FIGURE 15.6
Three levels of structure of a telephone cord.

as a plastic-coated metal wire (primary structure) coiled into a helix (secondary structure), which folds into a looped arrangement (tertiary structure) (see Figure 15.6).

The **secondary structure** of a protein refers to the arrangement of chains about an axis. This organizational level is maintained by a pattern of hydrogen bonds involving the atoms of the peptide link.

Secondary structure—the arrangement of chains about an axis

One such arrangement is referred to as a *pleated sheet* and is found in silk.

$$
\begin{array}{c}
\diagup \qquad\qquad\qquad \diagdown \\
RHC \qquad\qquad\qquad CHR \\
C=O \cdots\cdots H-N \\
\cdots H-N \qquad\qquad C=O \cdots \\
CHR \qquad\qquad RHC \\
\cdots O=C \qquad\qquad N-H \cdots \\
N-H \cdots\cdots O=C \\
\diagup \qquad\qquad\qquad \diagdown
\end{array}
$$

The polypeptide chains exist in an extended zigzag arrangement (Figure 15.7). Hydrogen bonds between the carbonyl oxygen atom of one peptide bond and the N—H of a nearby peptide bond hold the adjacent chains together. This pattern of hydrogen bonds is quite regular. The properties of silk—strength and resistance to stretching— are a consequence of its structure. To break the fibers would involve rupturing thousands of hydrogen bonds or breaking covalent bonds. Since the chains are already fully extended, the fibers cannot easily be stretched.

The secondary structure of many proteins is not the pleated sheet but a helical arrangement. Wool is a good example of this type of

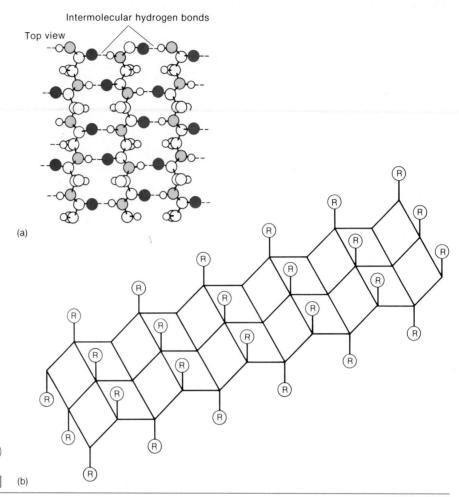

FIGURE 15.7
Pleated sheet conformation of polypeptide chains. (a) Ball-and-stick model. (b) Schematic drawing emphasizing the pleats. [From C. David Gutsche and Daniel J. Pasto, *Fundamentals of Organic Chemistry*, © 1975. Reprinted by permission of Prentice-Hall, Inc., Englewood Cliffs, NJ.]

arrangement. In wool (and hair and muscle), the amino acid units are arranged in a right-handed or **alpha helix** (Figure 15.8). Each turn of the helix requires 3.6 amino acid units. The N—H groups in one turn form hydrogen bonds to the carbonyl groups in the next (the dotted lines in the center model of Figure 15.8). In silk, the hydrogen bonds are between different peptide chains, that is, they are intermolecular hydrogen bonds. In wool, the hydrogen bonds are intramolecular; different parts of the same chain interact. (Remember: The covalent bond between the carbonyl carbon atom and the nitrogen atom

$$\overset{\overset{\displaystyle O}{\|}}{C}-N$$

is the peptide bond that maintains the primary structure. The hydrogen bond between the C=O in one amide group and the N—H in another is responsible for the secondary structure.)

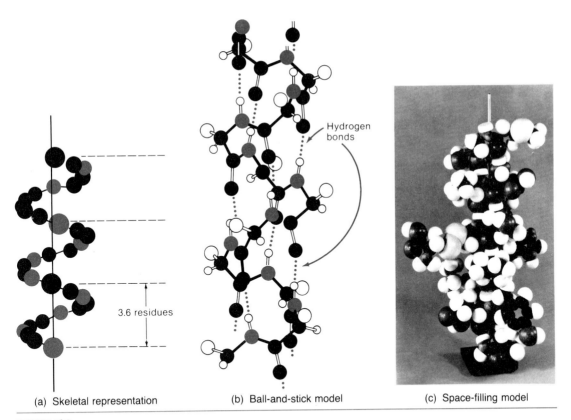

FIGURE 15.8
Three representations of the alpha-helical conformation of a polypeptide chain. (a) The skeletal representation best shows the helix. (b) Hydrogen bonding between turns of the helix is shown in the ball-and-stick model. (c) The space-filling model shows the actual shape of a short segment of the chain. [(a) and (b) from C. David Gutsche and Daniel J. Pasto, *Fundamentals of Organic Chemistry*, © 1975. Reprinted by permission of Prentice-Hall Inc., Englewood Cliffs, NJ. (c) courtesy of Science Related Material Inc., Janesville, WI.]

Unlike silk, wool can be stretched. (Think of how the coiled wire of the telephone cord can be stretched.) Such stretching results in an elongation of the hydrogen bonds joining the turns. When the stretching is discontinued, the helix returns to its original conformation, unless, of course, you have radically disrupted the structure—by treating the wool with hot water, for example (see Section 15.7).

The most abundant protein in our bodies is *collagen*, the tough protein of connective tissue. Collagen exists in neither an alpha helical form nor the pleated-sheet arrangement. Its structure is unique in that it consists of three protein chains intertwined in a **left-handed helix** (Figure 15.9). This structure is much more open than the tightly coiled alpha helix. Although there are no hydrogen bonds between the coils of an individual polypeptide chain, there are interchain hydrogen bonds holding the triple helix together. The triple helixes are collected in bundles, like the strands in a rope, and this arrangement gives collagen a fibrous structure well suited to its role as the protein of connective tissue.

FIGURE 15.9
The collagen molecule is a triple helix.

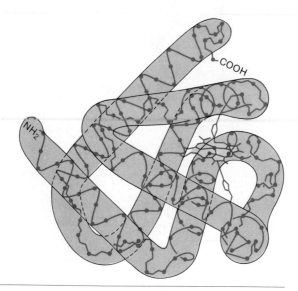

FIGURE 15.10

A skeletal representation of the tertiary structure of myoglobin. [Reprinted with permission from H. Neurath (Ed.), *The Proteins*, vol. 2. New York: Academic Press, 1964. Copyright © 1964 by Academic Press, Inc.]

Tertiary structure brings amino acids that are relatively far apart in the chain close together.

The **tertiary structure** of a protein relates to the spatial arrangements that bring amino acids that are relatively far apart in the protein chain into proximity. In describing tertiary structure, we frequently talk about how the molecule is *folded*. An example is the protein chain in myoglobin, a molecule that transports oxygen in muscles (Figure 15.10). Long sections of the myoglobin chain are coiled in the alpha-helix arrangement (secondary structure), and then the molecule is folded into a compact spherical shape (tertiary structure).

The tertiary structure is maintained by four different types of interactions (Figure 15.11). Hydrogen bonds again play a role; in this instance, however, the side chains of the amino acid units may also be involved (the hydroxyl group of serine is one example). Salt bridges may also form. A **salt bridge** is an ionic bond that forms when a carboxylic acid group in one side chain transfers its proton to an amino group in another side chain (see Figure 15.11). The amino acid units involved may be distant from one another in the primary sequence of the protein. The side chains come into contact only because of the folding or coiling of the protein chain. Such interactions also occur between

The tertiary structure can be maintained by hydrogen bonds, salt bridges, disulfide linkages, and/or hydrophobic interactions.

FIGURE 15.11

The tertiary structure of proteins is maintained by four different types of interactions.

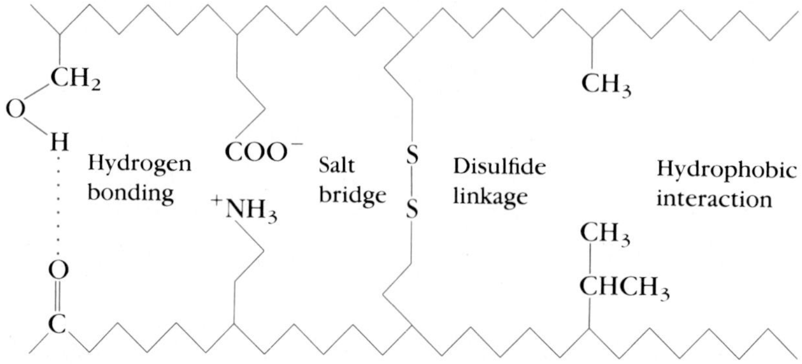

15.6 More on the Structure of Proteins

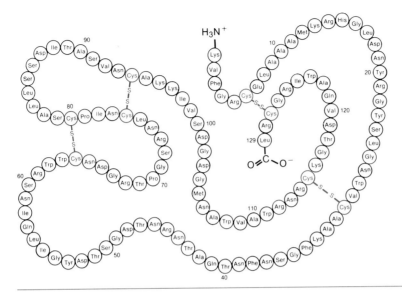

FIGURE 15.12

The enzyme lysozyme has four disulfide crosslinks. [From *Biochemistry*, 2nd ed., by Lubert Stryer, W. H. Freeman and Company. Copyright © 1981. After R. E. Canfield and A. K. Liu, *J. Biol. Chem.*, 240 : 2000 (1965); D. C. Philips, *Sci. Amer.*, 215 : 79 (1966).]

chains. A third interaction involves the formation of a covalent **disulfide linkage** (Figure 15.12). Disulfide linkages or bonds are formed when two *cysteine* units (whether on the same chain or two different chains) are oxidized to form a *cystine* unit.

$$\begin{array}{c} \text{SH} \\ | \\ \text{SH} \end{array} \quad \xrightleftharpoons[\text{reduction}]{\text{oxidation}} \quad \begin{array}{c} \text{S} \\ | \\ \text{S} \end{array}$$

Finally, the tertiary structure is maintained by hydrophobic interactions between nonpolar side chains. Proteins will fold in such a way as to bring nonpolar side chains inside the folded structure, in minimal contact with the aqueous solution surrounding the protein and in close contact with one another (Figure 15.13).

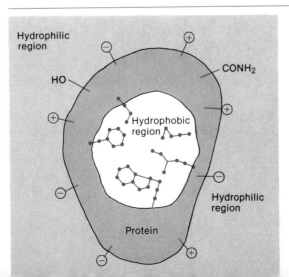

FIGURE 15.13

A folded protein chain might have a hydrophobic region on the inside and a hydrophilic region on the outside.

Quaternary structure—an aggregate of smaller units.

Some protein molecules consist of aggregates of smaller units. Hemoglobin is the most familiar example. A single hemoglobin molecule contains four polypeptide units. *Each unit* is roughly comparable to a myoglobin molecule. These subunits are arranged in a specific pattern (Figure 15.14). When we describe the **quaternary structure** of hemoglobin, we describe in detail the way in which the four units are packed together in the hemoglobin molecule.

FIGURE 15.14

The quaternary structure of hemoglobin. Shown to the right is hemoglobin with oxygen (oxyhemoglobin) and below is hemoglobin without oxygen (deoxyhemoglobin). The four protein subunits are designated α_1, α_2, β_1, and β_2. [Illustrations copyright by Irving Geis.]

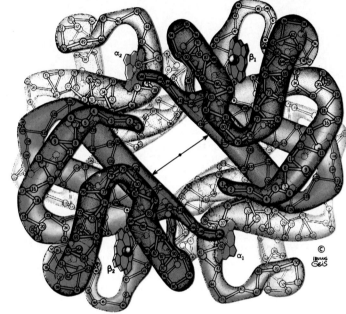

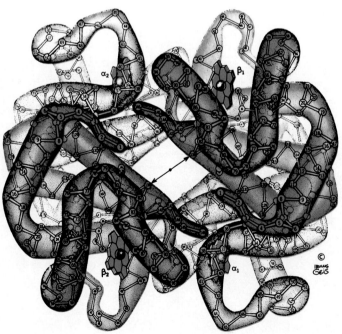

15.7 DENATURATION OF PROTEINS

In many ways, proteins are remarkable compounds. Their highly organized structures are truly masterworks of chemical architecture. But highly organized structures are also vulnerable, and this is true of many proteins. We shall define **denaturation** as the process in which a protein is rendered incapable of carrying out its assigned function. If the protein cannot do its job, we say that it has been *denatured*. (Sometimes denaturation is equated with the precipitation or coagulation of a protein. Our definition is a bit broader.) The process is sometimes reversible, but usually it is not. You have certainly observed the denaturation of egg protein. Egg white, which is clear when the egg is broken into the pan, turns opaque as the egg is frying. What you observe is the denaturation and coagulation of protein. No one has yet been able to reverse that process.

The primary structure of proteins is quite sturdy. In general, it takes fairly vigorous conditions for peptide bonds to be hydrolyzed (though chemists have devoted much effort to developing gentler methods, and enzymes manage to hydrolyze proteins with remarkable ease). At the secondary and tertiary levels (Figure 15.15), however, proteins are quite vulnerable to attack. Disulfide bonds are a bit resistant, but the critical hydrogen bonds can be disrupted in many ways.

How can you denature a protein? Heating it is one way. It works with eggs and also with bacteria, who, like you, need active protein to live. Autoclaving of surgical instruments sterilizes them through the action of heat on bacterial protein. Other forms of energy can be used, too, including ultraviolet light, which is occasionally used for sterilization.

Chemicals can also be used. Ethyl alcohol disrupts the hydrogen bonds in proteins and is used to sterilize the skin (by killing the bacteria present) when injections are to be given. A 70% aqueous solution is used, rather than the more commonly available 95% solution, because the higher concentration is so effective at coagulating the protein that it forms a surface "crust," which prevents the interior of the bacteria from being attacked (Figure 15.16). Acids and bases disrupt salt bridges and

Denaturation—the process in which a protein is rendered incapable of carrying out its function

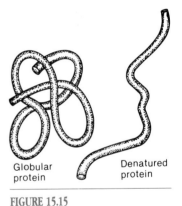

FIGURE 15.15
Denaturation of a protein. The globular protein (Section 15.8) is folded into the tertiary conformation necessary for its functioning. The denatured protein can assume various random conformations. It is not active but may, under proper conditions, refold to the active conformation.

FIGURE 15.16
Effect of alcohol on bacteria. Dark areas represent coagulated protein. (a) Bacteria before application of alcohol. (b) After application of 100% alcohol. (c) After application of 70% alcohol, which is more effective than 100% alcohol. [Reprinted with permission from L. Earle Arnow, *Introduction to Physiological and Pathological Chemistry*. 9th ed. St. Louis: C. V. Mosby Co., 1976.]

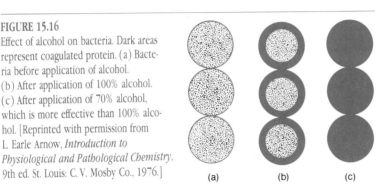

hydrogen bonds to cause precipitation of proteins. Tannic acid is sometimes applied to severe burns. The precipitated protein forms a crust over the wounds and thus slows the loss of body fluids.

Heavy metals (lead and mercury, for example) denature proteins and bring about their coagulation, probably by interacting with sulfhydryl group (—SH) or ionizable side chains. A person who has ingested mercury should be fed raw egg immediately. The mercury will then react with egg protein in the stomach rather than with other, more essential proteins in cells. This treatment requires that the person be induced to vomit, or that the stomach contents be pumped out, to prevent the ultimate digestion of the egg protein and the consequent release in the body of mercury ions. Quite clearly, the technique works only for acute poisonings and not for the far more common chronic mercury or lead poisoning.

Other ways of denaturing proteins (not discussed here) do exist. One point should be clear, however: the very complexity that makes proteins so versatile also makes them vulnerable.

15.8 Classification of Proteins

Of the many ways of classifying proteins, two that are related to structural characteristics will be considered here.

First, proteins can be described as either simple or conjugated. A **simple protein** is all protein; that is, it contains nothing but poly(amino acid) chains. A **conjugated protein** incorporates along with the poly(amino acid) structure some other organic or inorganic component, called a **prosthetic group**. Conjugated proteins may contain carbohydrates, lipids, metal ions, or, as we noted early in this chapter, nucleic acid. Hemoglobin is a conjugated protein. Its prosthetic group is *heme*.

The iron ion in the heme unit combines with oxygen. In this form oxygen is transported by hemoglobin through the circulatory system.

Proteins are also classified as fibrous or globular. **Fibrous proteins** are tough, water-insoluble structural materials. They include the **keratins**, the proteins of skin, hair, and nails; and **collagen**, the protein of connective tissue. The **globular proteins** are folded into roughly spherical conformations. These proteins are soluble in aqueous media, and they are usually more mobile and more easily denatured than fibrous protein. Hemoglobin and myoglobin are globular proteins. So, too, are the **serum albumins**, which help maintain fluid balance in the body and are also used to transport fatty acids in the blood. The **globulins** are another important group of globular proteins.

Gamma globulins are part of the body's defense mechanism. These proteins, which are also referred to as **antibodies**, are specifically tailored to attack invading foreign proteins or other large molecules foreign to the body. When viruses or disease-causing bacteria enter our bodies, specific antibodies are synthesized in an attempt to deactivate the invader. This process is referred to as the **immune response**. Unfortunately, when a kidney or a heart is transplanted to replace one weakened by disease, the host body does not recognize the organ as beneficial. It prepares antibodies to attack this foreign tissue just as if it were a deadly virus. To prevent rejection of the heart or kidney, powerful drugs are used to suppress the immune response and save the transplanted organ. At the same time, the body's ability to fight infection is also suppressed. The search is on for drugs that will selectively suppress the rejection of the organ but not the entire immune response.

15.9 ENZYMES

Enzymes are a highly specialized class of proteins. They control and direct the myriad chemical reactions that occur in living cells. Enzymes are biological catalysts produced by the cell. They have enormous catalytic power and are ordinarily highly specific, catalyzing only one reaction or a closely related group of reactions. Nearly all known enzymes are proteins. (Some forms of RNA can act as enzymes.)

Enzymes act by lowering the activation energy (Section 6.4) for a particular biochemical reaction. They do this by changing the reaction path. In the new route, the enzyme reacts with a compound called the **substrate** and bonds to it. This newly formed complex then separates

Enzymes are highly specialized.

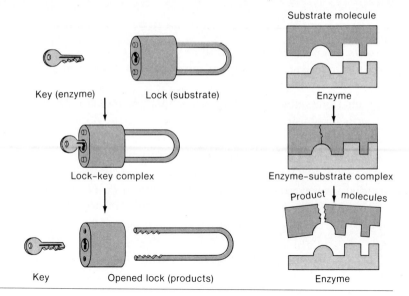

FIGURE 15.17
The lock-and-key model for enzyme action.

into products and the regenerated enzymes. Schematically, this may be written

$$\text{Enzyme} + \text{substrate} \rightleftharpoons \text{enzyme-substrate complex} \rightleftharpoons \text{enzyme} + \text{product}$$

Note that the steps are reversible. The enzyme catalyzes both the forward and reverse reactions.

Enzyme action is often explained by the analogy of a lock and key. The substrate must fit a portion of the enzyme, called the **active site**, quite precisely, just as a key must fit a tumbler lock in order to open it (see Figure 15.17). Not only must the enzyme and substrate fit precisely, but they are probably held together by electrical attraction. This requires that certain charged groups on the enzyme complement certain charged or partially charged groups on the substrate. Formation of new bonds between the enzyme and the substrate probably weakens bonds within the substrate. These weakened bonds can then be more easily broken to form products.

The active site on an enzyme is usually rather small, and only that small part comes into direct contact with the substrate. Incorporated in the active site are side chains of the enzyme, particularly those containing amino, carboxyl, hydroxyl, or sulfhydryl groups. The amino acid units involved need not be next to one another in the primary sequence of the enzyme. They must, however, be brought close together by the folding of the protein chain.

Portions of the enzyme molecule other than the active site may be involved in the catalytic process. Some interaction at a position remote from the active site can change the conformation of the enzyme and thus change its effectiveness as a catalyst. In this way it is possible to stop the catalytic action. Figure 15.18 offers a model for the inhibition of enzyme catalysis. An inhibitor molecule attaches to the enzyme

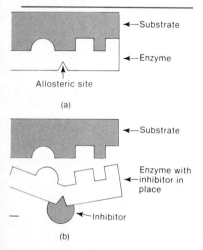

FIGURE 15.18
A model for the inhibition of an enzyme. (a) The enzyme and its substrate fit like lock and key. (b) With the inhibitor bound to the enzyme, the active site of the enzyme is distorted, and the enzyme will not bind its substrate.

at a position remote from the active site where the substrate is bound. The enzyme changes shape as it accommodates the inhibitor, and the substrate is no longer able to bind to the enzyme. This is one of the mechanisms by which cells "turn off" enzymes when their work is done.

Some enzymes consist entirely of protein chains. In others, another chemical component is necessary for the proper functioning of the enzyme. Such a component is called a **cofactor**. The cofactor may be a metal ion such as zinc (Zn^{2+}), manganese (Mn^{2+}), magnesium (Mg^{2+}), iron(II) (Fe^{2+}) or copper(II) (Cu^{2+}). If the cofactor is organic, it is called a **coenzyme**. By definition, coenzymes are nonprotein. The pure protein part of an enzyme is called the **apoenzyme**. Both the coenzyme and the apoenzyme must be present for enzymatic activity to take place.

$$\text{Coenzyme} + \text{apoenzyme} \longrightarrow \text{enzyme}$$
$$\text{Nonprotein} \qquad \text{Protein} \qquad\quad (\text{active})$$
$$(\text{inactive}) \qquad (\text{inactive})$$

Many coenzymes are vitamins or are derived from vitamin molecules.

Clinical analysis for enzymes in the body fluids or tissues is a common diagnostic technique in medicine. For example, blood levels of an enzyme involved in muscle metabolism rise in some forms of severe heart disease when the heart muscle has been damaged. On the other hand, many forms of strenuous (and healthful) physical activity also result in elevated blood levels of the enzyme. (Indeed, it is even possible for the level to increase in a person who hates needles and has tensed up while waiting for the blood sample to be taken!) Despite these variables, analysis for specific enzymes is an increasingly valuable diagnostic tool.

15.10 Digestion and Metabolism of Proteins

Like the carbohydrates and lipids, proteins are digested primarily in the small intestine. Here a wide variety of enzymes hydrolyze proteins to their constituent amino acids.

$$\cdots NH-\underset{\underset{R}{|}}{CH}-\underset{\underset{}{\|}}{\overset{O}{C}}-NH-\underset{\underset{R'}{|}}{CH}-\underset{\underset{}{\|}}{\overset{O}{C}}-NH-\underset{\underset{R''}{|}}{CH}-\underset{\underset{}{\|}}{\overset{O}{C}}\cdots \xrightarrow{H_2O, \text{ enzymes}}$$

$$\overset{+}{H_3N}-\underset{\underset{R}{|}}{CH}-\underset{\underset{}{\|}}{\overset{O}{C}}-O^- + \overset{+}{H_3N}-\underset{\underset{R'}{|}}{CH}-\underset{\underset{}{\|}}{\overset{O}{C}}-O^- + \overset{+}{H_3N}-\underset{\underset{R''}{|}}{CH}-\underset{\underset{}{\|}}{\overset{O}{C}}-O^- + \cdots$$

The product amino acids pass through the intestinal wall and into the bloodstream, where they join a circulating pool of amino acids. Un-

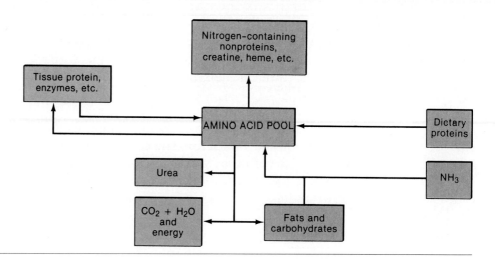

FIGURE 15.19
The amino acid pool and protein metabolism.

like carbohydrates and fats, the amino acids are not removed to storage areas. Membership in the circulating pool is temporary, however. New members are constantly added to the pool, and old ones are regularly withdrawn. The body can supplement the pool with amino acids synthesized from scratch or from the breakdown of tissue protein. It drains the pool to obtain raw material for new protein synthesis or for the synthesis of other nitrogen-containing compounds such as heme. It can also use the amino acids as a source of energy. Each day, some of the amino acids are broken down, and the nitrogen is eliminated from the body as urea. This represents a net drain of material from the pool. Nitrogen coming into the body as dietary protein compensates for this drain (Figure 15.19). It is interesting to note that hibernating bears burn mainly fat, thus minimizing the buildup of toxic urea produced when protein is metabolized. The urea that is produced is broken down, and the nitrogen is recycled to protein synthesis.

If amino acids are taken from the pool for something other than protein synthesis, then the elimination of the amino group is frequently the first step in the metabolic pathway. The amino group is replaced with a carbonyl group through a neatly matched set of reactions (Figure 15.20). In a reaction called **transamination**, the amino group of a particular amino acid is transferred to α-ketoglutaric acid.

$$\begin{array}{c} \text{COOH} \\ | \\ \text{CH}-\text{NH}_2 \\ | \\ \text{R} \end{array} + \begin{array}{c} \text{COOH} \\ | \\ \text{C}=\text{O} \\ | \\ \text{CH}_2 \\ | \\ \text{CH}_2 \\ | \\ \text{COOH} \end{array} \longrightarrow \begin{array}{c} \text{COOH} \\ | \\ \text{C}=\text{O} \\ | \\ \text{R} \end{array} + \begin{array}{c} \text{COOH} \\ | \\ \text{CH}-\text{NH}_2 \\ | \\ \text{CH}_2 \\ | \\ \text{CH}_2 \\ | \\ \text{COOH} \end{array}$$

An amino acid α-Ketoglutaric acid   An α-keto acid   Glutamic acid

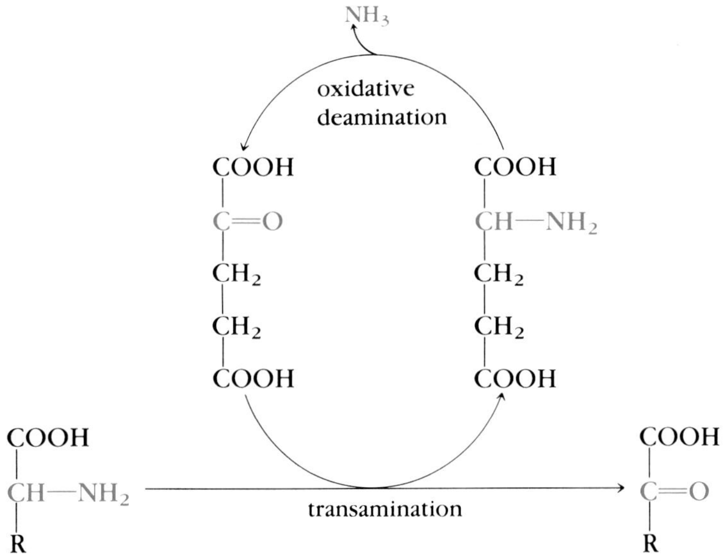

15.10 Digestion and Metabolism of Proteins

FIGURE 15.20 α-Ketoglutaric acid and glutamic acid as intermediates in the deamination of amino acids.

The product, glutamic acid, is reconverted to α-ketoglutaric acid by a reaction called **oxidative deamination**.

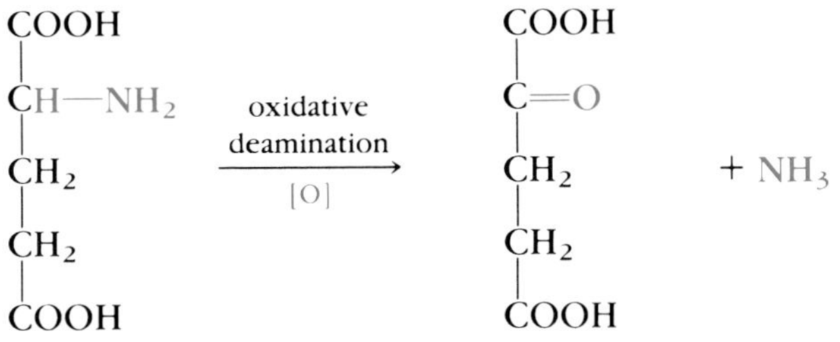

Glutamic acid → α-Ketoglutaric acid + NH_3 (oxidative deamination, [O])

Since the α-ketoglutaric acid is returned to its original form, the overall set of reactions has the effect of converting the amino group of the original amino acid to ammonia (actually, the ammonium ion). The body does not eliminate nitrogen from proteins as ammonia. As already mentioned, it converts the nitrogen, by means of another complex metabolic cycle, into the compound *urea*.

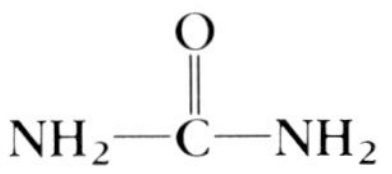

$$NH_2-\underset{\underset{O}{\|}}{C}-NH_2$$
Urea

The urea is transported to the kidneys and excreted in urine.

The keto acids produced through the transamination reaction make their way to the Krebs cycle. Each amino acid has a unique path to the Krebs cycle, with the transamination reaction operating at key junc-

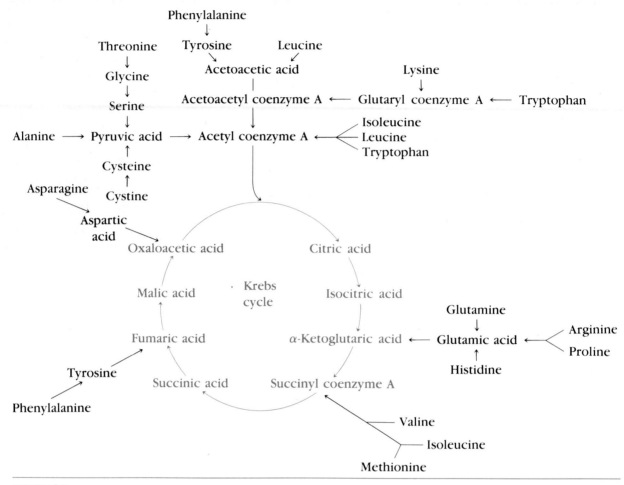

FIGURE 15.21
Metabolism of amino acids. All paths lead eventually to the Krebs cycle.

tures to remove nitrogen. Figure 15.21 hints at the complexity of the overall process by which amino acids contribute fuel to give us energy.

If carbohydrates are lacking in the diet (through voluntary restriction or starvation) or if glucose is not getting into cells (as in diabetes), the body responds by converting proteins to carbohydrates. Since amino acids can also be converted to fatty acids (through acetyl coenzyme A), proteins can make you fat. The amino acid pool can supply you with almost anything—proteins, fats, carbohydrates, energy, and even a tan (the coloring matter in your skin starts out as phenylalanine, an amino acid).

At several places in this chapter and in the two chapters preceding it, we have mentioned how metabolic pathways are connected. Figure 15.22 summarizes the interconnections of carbohydrate, lipid, and protein metabolism. This illustration does not have to be memorized, of course; we have included it to give you an opportunity to admire the elegance of nature's designs. Next time you eat a chocolate-coated peanut and caramel candy bar, stop and think of the machinery you are putting in motion.

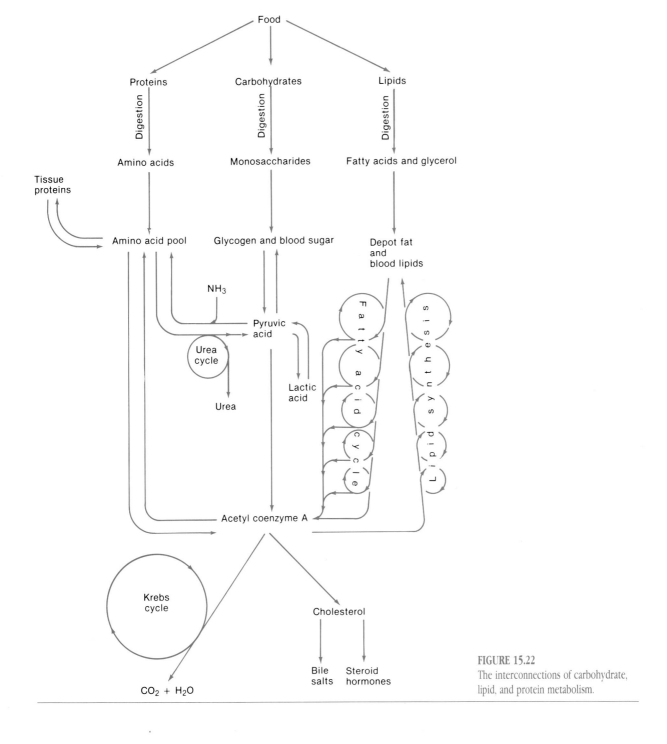

FIGURE 15.22
The interconnections of carbohydrate, lipid, and protein metabolism.

15.11 ENDORPHINS AND ENKEPHALINS

In 1973, two researchers at the Johns Hopkins University School of Medicine demonstrated that the brain contains specific receptor sites for morphine; that is, the structure of the brain incorporates special sites to which morphine can bind. That raised a question: Why should the human brain have receptors for a plant-derived drug like morphine? There seemed to be no good reason, so several investigators started a search for morphinelike substances that were produced within the body. Several such substances were soon found and determined to be small peptides. Those peptides containing 30 amino acid units were termed **endorphins** (for *endogenous morphines*, that is, "morphines" that are native to the brain). Smaller peptides containing five amino acid units were also isolated; these were called **enkephalins**. There are two enkephalins in the body, and they differ only in the amino acid at the end of the chain. Leu-enkephalin has the sequence Tyr-Gly-Gly-Phe-Leu and Met-enkephalin is Tyr-Gly-Gly-Phe-Met.

Enkephalins have been synthesized and shown to be potent pain relievers. Their use in medicine is quite limited, however, because after being injected they are rapidly broken down by the enzymes that hydrolyze proteins. It is hoped that modified forms that are more resistant to hydrolysis can be used as morphine substitutes for the relief of pain. Unfortunately, both the natural and modified enkephalins, like morphine, seem to be addictive.

Enkephalins and endorphins appear to be released as a response to pain deep in the body. Evidence exists that acupuncture anesthetizes by stimulating the release of brain "opiates." The long needles stimulate deep sensory nerves that cause the release of these peptides, which then block the pain signals.

The release of endorphins and enkephalins has also been used to explain other phenomena once thought to be largely psychological. A soldier wounded in battle feels no pain until the skirmish is over. His body has secreted its own painkiller. Research has shown that the level of endorphins in the blood of athletes increases after various levels of activity. Deep sensory nerves stimulated by vigorous exercise trigger the release of endorphins to block pain messages. Exercise such as distance running, which is extended over a long period of time, subjects the athlete to many of the same symptoms experienced by opiate users. Runners get a euphoric high during or after a hard run. They also seem to have withdrawal symptoms; they feel bad when they do not get their vigorous exercise. Apparently it is possible to get hooked on exercise.

Enkephalins are potent pain relievers.

CHAPTER SUMMARY

α-Amino acids are compounds that incorporate an amino and a carboxyl functional group attached to the same carbon atom. Amino acids exist as zwitterions, with the proton of the carboxyl group transferred to the amino group. This gives amino acids many of the properties of a salt. All amino acids except glycine contain a chiral carbon atom and can therefore exist as mirror-image isomers. The common amino acids belong to the L (levo) family.

Table 15.1 lists the names and structures of the common amino acids, which fall into four general groups based on the nature of the side chain attached to the alpha carbon atom. The four groups consist of those compounds with nonpolar side chains, with polar but nonionizable side chains, with acidic side chains, and with basic side chain. For convenience, amino acids are frequently represented by three-letter abbreviations of their names.

Essential amino acids are those required in the diet for proper protein synthesis in the body. An adequate protein is a nutrient protein that supplies all the essential amino acids in sufficient quantities. Most animal protein is adequate, but many plant proteins lack one or more of the essential amino acids.

Because they contain both an acidic and a basic group, amino acids act as buffers by picking up or releasing protons. The pH at which the amino acid (or protein) is electrically neutral is called the isoelectric point.

Peptides and proteins are polymers of amino acids in which the amino acid units are connected to one another by peptide (amide) bonds. Poly(amino acids) under 10 000 in molecular weight are classified as peptides, and those with weights above 10 000 are proteins. The number of amino acid units incorporated in smaller peptides may be designated by the prefixes di-, tri-, tetra-, and so on.

The structure of proteins can be described on four levels. The primary structure indicates the sequence of amino acids in the polymer chain. This structure is maintained by peptide bonds.

The secondary structure describes the arrangement of the chains about an axis. The alpha helix (for example, wool) and the pleated-sheet (silk) arrangements are among the most common. Collagen, the most abundant protein in our bodies, exhibits a unique helical structure in which individual chains are wound into relatively open, left-handed helixes. These chains are then coiled in groups of three to form a triple helix. The principal force maintaining secondary structure is hydrogen bonding between the N—H group of one peptide bond and the C=O of another.

Tertiary structure describes the way in which the protein chain is folded to bring relatively distant groups along the chain into proximity. This level of structure is maintained by salt bridges, disulfide bonds, hydrogen bonding between side chains, and hydrophobic interactions (Figure 15.12).

Quaternary structure describes the spatial relationship between protein subunits that are not covalently bonded to one another.

The structural complexity of proteins makes them subject to denaturation, a disruption of structure that destroys the biological activity of the protein. Denaturing agents include heat and other forms of energy, hydrogen-bonding agents such as ethanol, acids or bases, and heavy metal ions. Hydrogen bonds are most readily disrupted by denaturing agents, and peptide bonds are most resistant.

Proteins are classified in several ways. A conjugated protein consists of a poly(amino acid) part and some other organic or inorganic component called a prosthetic group (such as heme). A simple protein has no prosthetic group. Fibrous proteins, like the keratins and collagen, are water-insoluble and relatively tough. Globular proteins, such as hemoglobin or albumin, are roughly spherical and are soluble in aqueous solutions.

Antibodies are proteins manufactured by the body as a defense against large invading molecules that are foreign to the body (such as viruses). This immune response is a disadvantage in organ transplants because the transferred organ is also treated as unwanted foreign tissue.

Enzymes are proteins that serve as catalysts for all biochemical reactions. The material whose reaction the enzyme catalyzes is called the substrate. According to the lock-and-key theory of enzyme activity, the substrate fits precisely a portion of the enzyme called the active site where the actual catalytic action of the enzyme takes place. Catalytic activity may be regulated by alterations in the conformation (shape) of the enzyme caused by some interaction at a point distant from the active site. Some enzymes are simple proteins, but others require a cofactor such as a metal ion or a nonprotein organic molecule (called a coenzyme). The pure protein part of an enzyme is called the apoenzyme.

As with the other classes of foodstuffs, proteins undergo digestion primarily in the small intestine. In the digestive process, proteins are converted to amino acids, which are transferred to the blood. The dissolved amino acids form a circulating pool of raw material for protein synthesis or the synthesis of other nitrogen-containing compounds or for energy production.

Metabolism of protein as a source of energy requires the removal of the nitrogen. This is commonly accomplished through a matched set of reactions in which the amino group is first transferred to α-ketoglutaric acid to form glutamic acid (transamination) and is then removed from glutamic acid in a reaction called oxidative deamination. Ultimately the nitrogen atom is eliminated from the body in the form of urea. The keto acid formed from the original amino acid then travels one of several routes that lead to the Krebs cycle.

Endorphins and enkephalins are peptides found in the brain. They exhibit the physiological properties of opiates and are particularly effective in the relief of pain.

CHAPTER QUIZ

1. Using the analogy of language (which is composed of letters, words, sentences, etc.), amino acids would be
   a. letters
   b. words
   c. sentences
   d. paragraphs
2. Which is not a property of amino acids?
   a. They are crystalline solids.
   b. They are water-soluble.
   c. They are nonionic.
   d. Nearly all have the L configuration.
3. Alanine CH_3—CH—COO$^-$ has what kind of side chain?
    $|$
    $^+NH_3$
   a. nonpolar
   b. polar
   c. acidic
   d. basic
4. Which statement about amino acids is not true?
   a. They are unreactive.
   b. They are both acid and base.
   c. They form zwitterions.
   d. They can act as either acids or bases.

5. At the isoelectric point, amino acids
   a. have a pH of 7
   b. exist as the zwitterion
   c. are in the acid form
   d. all have the same pH
6. In protein chains, amino acids are stitched together via
   a. hydrogen bonding
   b. peptide bonding
   c. sulfide bridges
   d. secondary structure
7. The isoelectric point of casein is nearest which value?
   a. 2   b. 5   c. 8   d. 11
8. The secondary structure of proteins is held together by
   a. ionic bonds
   b. covalent bonds
   c. peptide bonds
   d. hydrogen bonds
9. The tertiary structure of proteins is held together by all but which of these bonds?
   a. metallic bonds
   b. hydrogen bonds
   c. salt bridge
   d. disulfide linkage
10. Which is not true about enzymes?
    a. They have high catalytic power.
    b. They are highly specific.
    c. They are proteins.
    d. They raise the activation energy of reactions.

PROBLEMS

15.1 Amino Acids: Structure and Physical Properties

1. What is the general structure for an α-amino acid?
2. Draw the side chains of the following amino acids.
   a. alanine   **b.** phenylalanine
   c. serine   **d.** aspartic acid
   e. cysteine   **f.** lysine
3. Write structural formulas for the following.
   a. glycine   **b.** alanine
   c. phenylalanine
4. Write a structural formula for an amino acid with an acidic side chain, and give the name of the compound.
5. Write a structural formula for an amino acid with a basic side chain, and give the name for the compound.

6. Write structural formulas for the following.
   a. cysteine
   b. cystine
7. To which family of mirror-image isomers do almost all naturally occurring amino acids belong?

15.2 The Essential Amino Acids

8. In general, what are the potential problems associated with a strict vegetarian diet?

15.3 Amino Acids as Buffers

9. Write a structural formula for the anion formed when glycine reacts with a base.
10. Write a structural formula for the cation formed when glycine reacts with an acid.

15.4 Formation of Proteins: The Peptide Bond

11. Write structural formulas for the following.
    a. glycylalanine
    b. alanylglycine
12. Write the structural formula for phenylalanyl-glycylalanine.
13. Define, describe, or illustrate each of the following.
    a. peptide bond
    b. essential amino acid
    c. isoelectric point
14. What is the difference between a polypeptide and a protein?
15. Amino acid units in a protein are connected by peptide bonds. What is another name for the functional group linking the units of proteins?

15.5 The Sequence of Amino Acids in Proteins

16. Translate the following abbreviated form into the structural formula for the peptide: Ser-Ala-Gly.
17. Write the abbreviated version of the structural formula shown below.

15.6 More on the Structure of Proteins

18. Describe the structure of silk, and explain how its properties reflect its structure. What name is given to the secondary structure of silk?
19. Describe the structure of wool, and relate this to wool's elasticity. What name is given to the secondary structure of wool protein?
20. Describe the structure of collagen.
21. Name the four kinds of interactions that maintain the tertiary structure of protein.
22. The following sets of amino acids are involved in maintaining the tertiary structure of a peptide. In each case, identify the type of interaction (see Problem 21) involved.
    a. aspartic acid and lysine
    b. phenylalanine and alanine
    c. serine and lysine
    d. cystine

15.7 Denaturation of Proteins

23. Describe some ways of denaturing a protein.
24. What level or levels of structure is (are) ordinarily disrupted in denaturation?
25. Is denaturation of protein usually reversible?

15.8 Classification of Proteins

26. Classify these proteins as fibrous or globular.
    a. albumin   b. antibodies
    c. collagen   d. hemoglobin
    e. keratins   f. myoglobin
27. Which class of proteins shows greater solubility in aqueous solutions, fibrous or globular?
28. Which class of proteins is more easily denatured, fibrous or globular?
29. Classify hemoglobin as a simple or conjugated protein. Explain your reasoning.
30. What proteins are involved in the immune response?

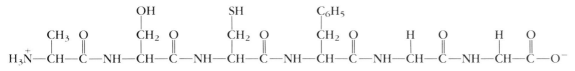

Structural formula for Problem 17

31. Why is the immune response a problem in organ transplantation?
32. Acquired immune deficiency syndrome (AIDS) is a highly lethal disease whose victims die of a variety of infections or of relatively rare forms of cancer. Relate the name of the disease to its effects.

15.9 Enzymes

33. Describe the lock-and-key model of enzyme action.
34. Alcohol dehydrogenase is an enzyme that catalyzes the conversion of ethanol to acetaldehyde. The active enzyme consists of a protein molecule and a zinc ion. Identify the following components of this reaction system.
    a. substrate
    b. cofactor
    c. apoenzyme
35. Can the zinc ion mentioned in Problem 34 be called a coenzyme? Explain your answer.
36. Succinate dehydrogenase is an enzyme that is active only in combination with a nonprotein organic molecule called flavin adenine dinucleotide (FAD). Is FAD a cofactor? Is it a coenzyme?
37. Enzymes are said to exhibit specificity. What does that mean?
38. Describe how an inhibitor may deactivate an enzyme. Can you propose another mechanism by which enzymes might be deactivated?

15.10 Digestion and Metabolism of Proteins

39. What are the products of protein digestion?
40. Draw the products of the hydrolysis of alanylglycine.
41. Aspartame, an artificial sweetener about 160 times as sweet as sucrose, is the methyl ester of a simple dipeptide. Draw the products of complete hydrolysis of aspartame.

$$H_3\overset{+}{N}-\underset{\underset{CH_2}{|}}{\underset{|}{CH}}-\overset{O}{\overset{\|}{C}}-NH-\underset{\underset{CH_2}{|}}{\underset{|}{CH}}-\overset{O}{\overset{\|}{C}}-O-CH_3$$
(with COO⁻ on first CH₂ and C₆H₅ on second CH₂)

Aspartame

42. A can of soda can be sweetened with about 0.25 g of aspartame. This amount of aspartame would release about 0.027 g of methanol on hydrolysis. The lethal dose of methanol for humans is about 25 g. How many cans of aspartame-sweetened soda would you have to drink to ingest a theoretically lethal dose of methanol?
43. The enzyme superoxide dismutase (SOD) is thought to have an antiaging effect in cells. Some "health food" stores sell SOD for oral ingestion with the idea that it will retard aging. Do you think it will work? Explain.
44. What is the amino acid pool? Are amino acids stored in the sense that glycogen and fats are stored? Explain.
45. Use simplified equations to show how α-ketoglutaric acid and glutamic acid serve as intermediates in the conversion of amino acid nitrogen to ammonia.
46. Which amino acid, upon transamination, would yield pyruvic acid?
47. Oxaloacetic acid is formed from aspartic acid via the transamination reaction.

$$\begin{array}{c} COOH \\ | \\ CH-NH_2 \\ | \\ CH_2 \\ | \\ COOH \end{array} \longrightarrow \text{oxaloacetic acid}$$

Draw the structural formula of oxaloacetic acid.

48. How does transamination serve to connect amino acid metabolism with carbohydrate and lipid metabolism?
49. If you take in more protein than your body requires to maintain normal activity, what happens to the excess?
50. In what form is the nitrogen of protein ultimately eliminated from the body?

15.11 Endorphins and Enkephalins

51. What triggered the search that led to the discovery of the endorphins and enkephalins?
52. Why are the endorphins and enkephalins compared to morphine?
53. What is the difference between Leu-enkephalin and Met-enkephalin

16 ENERGY AND LIFE

DID YOU KNOW THAT...

- only green plants can produce food?
- our cells can make high-energy molecules by using energy stored in foods?
- sunlight is the main source of energy on Earth?
- no natural or artificial system can function at 100% efficiency?
- chemical bonds store energy?
- "spontaneous" does not necessarily mean "immediate"?
- ATP is the universal energy currency?
- cyanide is lethal because it binds iron in the heme of cytochromes?
- mitochondria are supreme energy factories?
- fat molecules yield more energy than carbohydrates?
- muscle fatigue is the result of lactic acid buildup?
- CK levels in blood increase after a heart attack?

Life requires energy. Living cells are inherently unstable and avoid falling apart only because of a continual input of energy. Living organisms are restricted to using certain forms of energy. Supplying a plant with heat energy by holding it in a flame will do little to prolong its life. On the other hand, a green plant is uniquely able to tap the richest source of energy on Earth, sunlight. A green plant captures the radiant energy of the sun and converts it to chemical energy, which it stores in carbohydrate molecules. The plant can also convert those carbohydrate molecules to fat molecules and, with proper inorganic nutrients, to protein molecules.

A green plant is able to use energy directly from the sun.

Animals cannot use sunlight for energy.

Animals cannot directly use the energy of sunlight. They must eat plants, or other animals that eat plants, in order to get carbohydrates, fats, and proteins with their stored chemical energy. Once digested and transported to the cell, a food molecule can be used in either of two ways: as a building block to make new cell parts or repair old ones, or "burned" for energy.

In the last three chapters we have examined how cells metabolize different types of food molecules. Now let us focus on the life-sustaining energy transformations that accompany those metabolic reactions (Figure 16.1).

FIGURE 16.1
Energy transformations in biological systems. The energy of sunlight is captured in plants. Those plants then serve as fuel for animals. [Photo © The Terry Wild Studio.]

16.1 Life and the Laws of Thermodynamics

Thermodynamics is the study of the relationships between different forms of energy. Included within its scope are considerations of energy changes associated with chemical reactions. The laws of this quantitative science can be rigorously derived by using higher mathematics. That need not concern us here, for those aspects that are of interest in the chemistry of life can be rather simply stated and intuitively grasped. In fact, you first encountered thermodynamics in Chapter 6 when we discussed exothermic and endothermic reactions. At that time we concentrated on the transfer of *heat* energy. Our present discussion will not be so restricted.

The first law of thermodynamics states that energy is neither created nor destroyed. It can, however, be changed from one form to another. Life, for us, is a process that changes potential energy (the stored energy in our food) into the kinetic energy associated with breathing, the beating of our hearts, and general movement, whether it be walking, running, or twiddling our thumbs.

First law of thermodynamics: Energy is neither created nor destroyed.

The second law of thermodynamics states that when energy is converted from one form to another, some of it is "lost." This does not contradict the first law. Energy has not been destroyed; it has merely been lost in the sense that it is no longer useful. When you drive a car, for example, only about a fourth of the energy stored in the gasoline goes into moving the car down the road. The remainder is "lost" as heat. It may warm up the atmosphere around the car a bit, but it cannot be captured and put to work. We say that the car is about 25% efficient. The **efficiency** of a system, then, compares the amount of useful work accomplished to the amount of energy put into the system. Another statement of the second law is that no machine—or organism—can be 100% efficient. Neither technological improvement nor scientific progress will change this assessment. The efficiency of some processes can be increased, but it is impossible to achieve 100% efficiency.

Second law of thermodynamics: When energy is converted from one form to another, some of it is "lost."

The energy content of chemical compounds is changed when some chemical bonds are broken and new ones are formed. If, in this process, there is a net *release* of energy by the compounds, the reaction is said to be **exergonic**. If on the other hand, a net *input* of energy is required, the reaction is **endergonic**. To survive, all living organisms must carry out exergonic reactions that release sufficient energy to maintain life processes. Endergonic reactions must occur to counterbalance the energy-releasing processes. Reactions that store energy (such as the photosynthesis carried on by green plants) are endergonic. Because no process is 100% efficient, the endergonic reactions must store *more* energy than is required by organisms for survival. Some of the energy released by subsequent exergonic reactions will always be wasted.

Net release of energy—exergonic
Net input of energy—endergonic

16.2 Predicting Spontaneous Change: Free Energy

If you were holding a book and then released it, the book would drop to the floor spontaneously. If a book were lying on the floor, you would not stand patiently waiting for it to leap into your hands. You know that no matter how long you wait, the book will not jump from the floor back into your hands. The movement of book from hand to floor could be described as natural, and the movement of book from floor to hand as unnatural.

A chemical reaction can also be described as having a natural direction. Thermodynamicists have defined a term, *change in free energy* or ΔG, that indicates the natural direction of reactions. A negative value for ΔG (reported in kilocalories per mole) means that the reaction can occur naturally or spontaneously. Note that *spontaneously* does not necessarily mean *immediately*. A ΔG value does not indicate *when* the reaction will proceed but *whether* it can proceed naturally, that is, of its own accord. A positive value for ΔG means that the direction of the reaction is not natural. If the value for ΔG is zero, then the system being considered is at equilibrium.

16.3 Coupled Reactions

Many reactions essential to life are nonspontaneous.

Many reactions essential to life are nonspontaneous. How can these reactions be forced to proceed in directions that are not natural? How can you bring a book from the floor to your hand? Obviously, one way is to bend over and pick it up. Another way can take advantage of the natural process of a book falling to the floor. Suppose a book on the floor is connected by a taut rope to a larger book in your hand. All you have to do is let the rope slide over your hand as you let the book in your hand drop to the floor. What you have is a primitive pulley. As the larger book drops, the smaller book on the floor is lifted to your hand (Figure 16.2). The two processes have been *coupled*, and the spontaneous one drives the nonspontaneous one.

In a living cell, an endergonic reaction with a positive ΔG can be made to proceed if it is *coupled* with an exergonic reaction with a negative ΔG of larger absolute value. By *coupling* we mean that the two reactions share a common intermediate that transfers energy from one reaction to the other. (In the book analogy, the rope serves as the intermediate for transferring energy from one book to the other.)

Let us look at some specific chemistry to demonstrate this principle of coupled reactions. Consider the reaction of glucose and fructose to form sucrose.

Glucose + Fructose → Sucrose + H_2O

16.3 Coupled Reactions

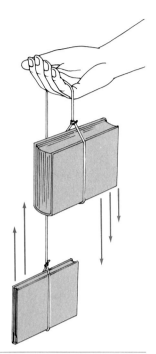

FIGURE 16.2
The coupling of a spontaneous and a nonspontaneous action.

This reaction has a positive ΔG; it is not spontaneous. The reaction must be coupled with another reaction that has a large negative ΔG. In some cells, the exergonic reaction is the hydrolysis of the compound adenosine triphosphate (ATP) to the compound adenosine diphosphate (ADP). We shall discuss the structure of these two compounds in the next section. For the moment, you need only know that the conversion of ATP to ADP is an exergonic reaction.

$$\text{Glucose} + \text{fructose} \longrightarrow \text{sucrose} + H_2O \quad \Delta G = +7.0 \text{ kcal/mole}$$
$$\text{ATP} + H_2O \longrightarrow \text{ADP} + HPO_4^{2-} \quad \Delta G = -7.3 \text{ kcal/mole}$$

The sum of the ΔGs for these reactions is negative, so if the two reactions could be coupled, the formation of sucrose would be thermodynamically favored. The reactions are coupled through the common intermediate, glucose 1-phosphate. When ATP reacts with glucose to form this intermediate, some of the energy stored in the bonds of ATP is trapped in the bonds of the intermediate.

$$\text{Glucose} + \text{ATP} \longrightarrow \text{glucose 1-phosphate} + \text{ADP}$$

Then the intermediate reacts with fructose, and this time some energy is trapped in the sucrose molecule.

$$\text{Glucose 1-phosphate} + \text{fructose} \longrightarrow \text{sucrose} + HPO_4^{2-}$$

Because glucose 1-phosphate is a product of the first reaction and a reactant in the second, it is not included in the equation for the overall reaction.

$$\text{Glucose} + \text{fructose} + \text{ATP} \longrightarrow \text{sucrose} + \text{ADP} + \text{HPO}_4^{2-}$$

16.4 ATP: Universal Energy Currency

Many cellular reactions, otherwise unfavored thermodynamically, can proceed if they are coupled with a high-energy reaction such as the hydrolysis of ATP or ADP. The structures of ATP (adenosine triphosphate), ADP (adenosine diphosphate), and AMP (adenosine monophosphate) are presented in Figure 16.3. Compounds of this general type will be discussed in great detail in Chapter 17. For the moment we can focus on the phosphate linkages. ATP is hydrolyzed in two steps, each of which releases considerable energy. The standard free-energy changes $(\Delta G°)^*$ are

$$\text{ATP} + \text{H}_2\text{O} \longrightarrow \text{ADP} + \text{HPO}_4^{2-} \qquad \Delta G° = -7.3 \text{ kcal/mol}$$

$$\text{ADP} + \text{H}_2\text{O} \longrightarrow \text{AMP} + \text{HPO}_4^{2-} \qquad \Delta G° = -6.5 \text{ kcal/mol}$$

$$\text{AMP} + \text{H}_2\text{O} \longrightarrow \text{adenosine} + \text{HPO}_4^{2-} \qquad \Delta G° = -2.2 \text{ kcal/mol}$$

ATP and ADP are high-energy compounds—they release large amounts of energy upon hydrolysis.

The free-energy changes associated with the first two transformations (involving the cleavage of P—O—P linkages) are significantly greater than that for the third transformation (cleavage of a P—O—C linkage). Because ATP and ADP release large amounts of energy upon hydrolysis, they are called **high-energy compounds**. AMP, which yields considerably less energy upon hydrolysis, is not regarded as a high-energy substance. The cutoff for the "high-energy" designation is rather arbitrarily set at about -5 kcal/mol.

ATP is the most important high-energy compound in cells and is referred to as the *energy currency* of the cell. It acts as a general repository of energy, trapping energy released when the fuel molecules of the cell are metabolized and then releasing that energy to drive otherwise unfavored reactions.

The standard free energies of hydrolysis for a number of important organic phosphates are given in Table 16.1.

---

*The standard free-energy change is the change in free energy measured under standard conditions of temperature, pressure, and concentration. Standard conditions, as defined by most thermodynamicists, do not coincide with conditions inside cells, where the reactions in which we are interested take place. Attempts to take into account intracellular conditions (which are not constant throughout the body) might yield values for the three hydrolysis steps of -8.8, -8.6, and -3.0 kcal/mol, respectively. The important point to note is that the first two values are higher than the third.

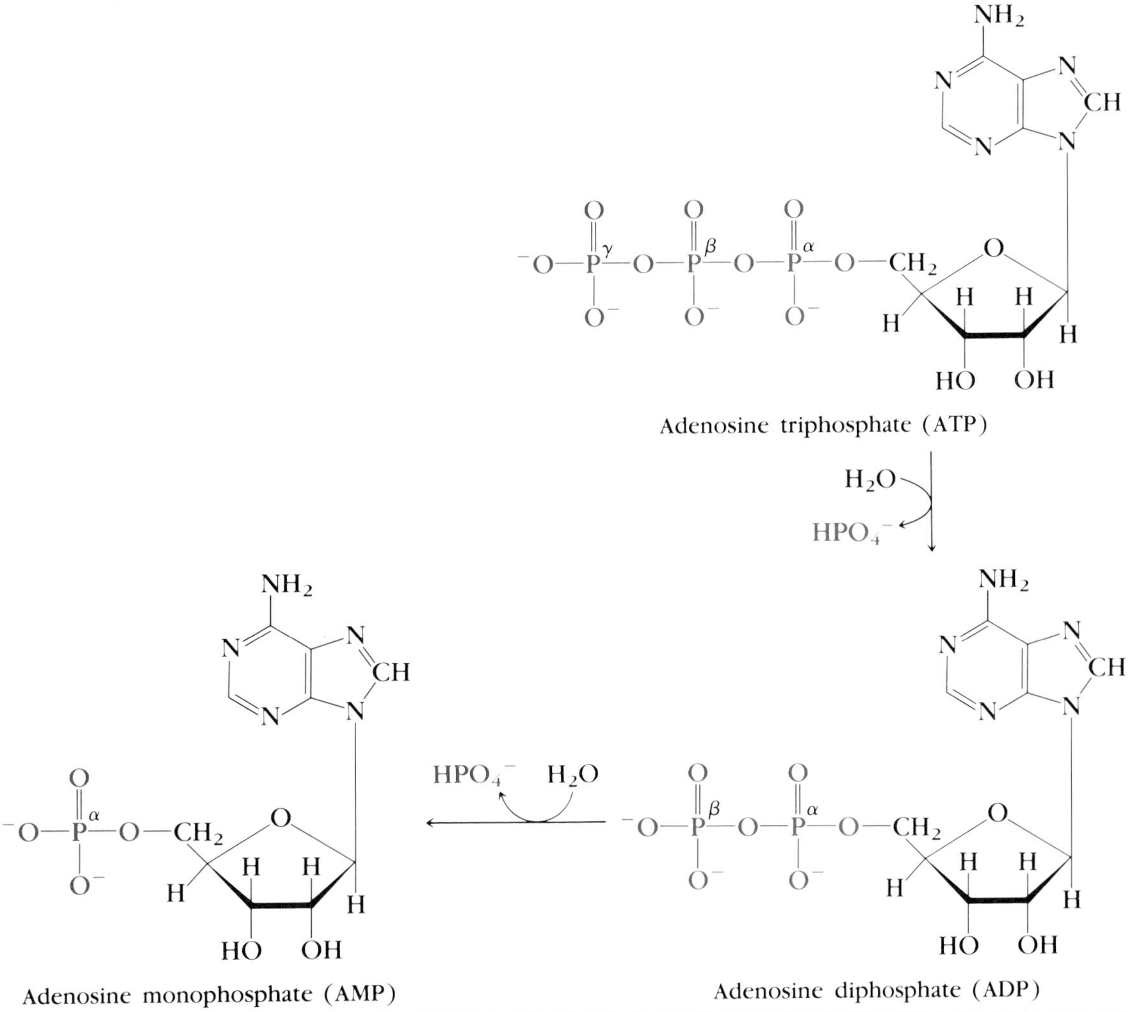

FIGURE 16.3
Stages in the hydrolysis of adenosine triphosphate.

TABLE 16.1 Free Energy of Hydrolysis of Some Phosphates

Compound	Hydrolysis Products	Approximate $\Delta G°$ (kcal/mol)
Phosphoenolpyruvate (PEP)	Ketopyruvic acid + HPO_4^{2-}	−12.8
1,3-Diphosphoglyceric acid	3-Phosphoglyceric acid + HPO_4^{2-}	−12.0
Creatine phosphate	Creatine + HPO_4^{2-}	−10.5
Acetyl phosphate	Acetic acid + HPO_4^{2-}	−10.0
ATP	ADP + HPO_4^{2-}	−7.3
ADP	AMP + HPO_4^{2-}	−6.5
Glucose 1-phosphate	Glucose + HPO_4^{2-}	−5.0
Fructose 6-phosphate	Fructose + HPO_4^{2-}	−3.8
Glucose 6-phosphate	Glucose + HPO_4^{2-}	−3.3
3-Phosphoglyceric acid	Glyceric acid + HPO_4^{2-}	−2.4
Glycerol 3-phosphate	Glycerol + HPO_4^{2-}	−2.2
AMP	Adenosine + HPO_4^{2-}	−2.2

16.5 Synthesis of ATP

ATP synthesis is coupled to a number of metabolic reactions.

The synthesis of ATP is coupled to a number of the metabolic reactions we considered in the preceding three chapters. Let us review a few of these.

1,3-Diphosphoglyceric acid is formed as an intermediate in the anaerobic glycolysis of carbohydrates. Note from Table 16.1 that this compound has a higher (more negative) $\Delta G°$ for hydrolysis than ATP; therefore, it can transfer a phosphate group to ADP to form ATP (Figure 16.4, step 7). The likelihood of the transfer is determined as follows: if the hydrolysis of ATP to ADP has a $\Delta G°$ of -7.3 kcal/mol, then the reverse reaction has a $\Delta G°$ of $+7.3$ kcal/mol.

$$ADP + HPO_4^{2-} \longrightarrow ATP + H_2O \qquad \Delta G° = +7.3 \text{ kcal/mol}$$

The hydrolysis of 1,3-diphosphoglyceric acid, according to Table 16.1, has a $\Delta G°$ of -12.0 kcal/mol.

$$\text{1,3-Diphosphoglyceric acid} + H_2O \longrightarrow \text{3-phosphoglyceric acid} + HPO_4^{2-}$$
$$\Delta G° = -12.0 \text{ kcal/mol}$$

If the two equations are added together to yield a net equation, the result is

$$\text{1,3-Diphosphoglyceric acid} + ADP \longrightarrow \text{3-phosphoglyceric acid} + ATP$$

The $\Delta G°$ of the coupled reactions is $-12.0 + 7.3 = -4.7$ kcal/mol. It is negative, and the coupled reactions are thermodynamically favored.

In step 10 of the glycolysis pathway (Figure 16.4) the conversion of phosphoenolpyruvic acid (PEP) to pyruvic acid is coupled with the formation of ATP from ADP.

$$PEP + ADP \longrightarrow \text{pyruvic acid} + ATP$$

According to Table 16.1, the hydrolysis of PEP has a $\Delta G°$ of -12.8 kcal/mol. Since this value is higher than that for the hydrolysis of ATP (-7.3 kcal/mol), the coupled reactions are thermodynamically favored; that is, PEP (higher negative $\Delta G°$) can transfer its phosphate group to form ATP (lower negative $\Delta G°$).

In contrast, the first step of the glycolysis pathway (Figure 16.4) indicates that the conversion of glucose to glucose 6-phosphate is coupled to the conversion of ATP to ADP. In this step, instead of forming ATP, the hydrolysis of ATP is used to drive the reaction.

$$\text{Glucose} + ATP \longrightarrow \text{glucose 6-phosphate} + ADP$$

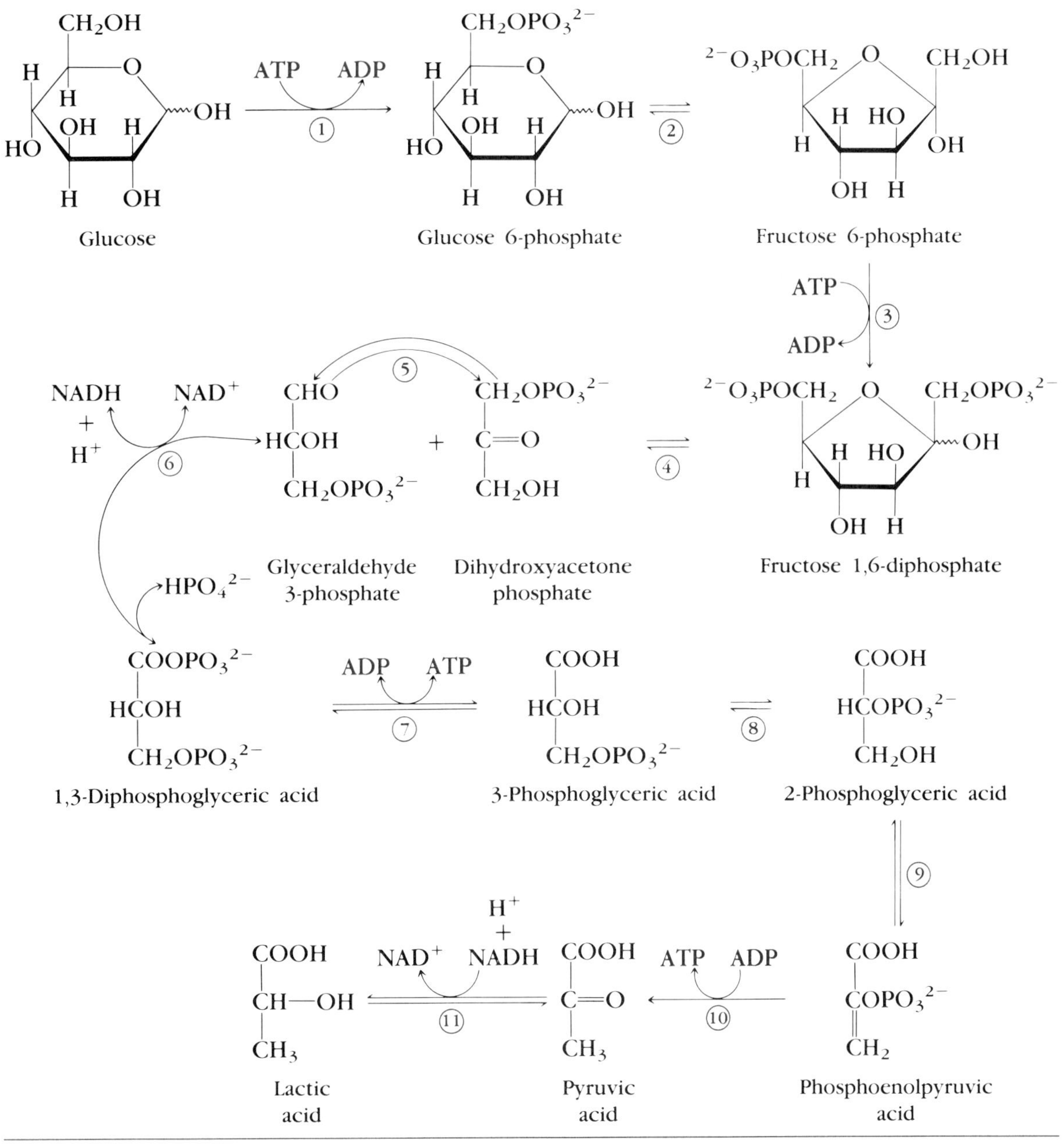

16.5 Synthesis of ATP

FIGURE 16.4
The anaerobic glycolysis pathway. ADP–ATP conversions occur at steps 1, 3, 7, and 10.

We could have predicted this from an examination of Table 16.1. The standard free energy of hydrolysis for glucose 6-phosphate is -3.3 kcal/mol. That is lower than the value for ATP. Therefore, we should expect the phosphate group to be transferred from ATP to form glucose 6-phosphate.

An analogous coupled reaction occurs in step 3 (Figure 16.4) of the glycolysis pathway. In glycolysis some ATP is used up and some is formed. In fact, the progression of one molecule of glucose through this metabolic pathway to form two molecules of lactic acid results in a net yield of two molecules of ATP. Some of the energy in the original glucose molecule is trapped in ATP molecules that can be used to drive other life-sustaining processes, ranging from the contraction of muscles to the transmission of nerve impulses to a whole spectrum of metabolic reactions. Two molecules of ATP for every molecule of glucose is a rather poor return on investment, but fortunately, ATP production associated directly with the conversion of glucose to lactic acid represents only a minor source of the body's energy currency. A much more productive route to ATP molecules is discussed next.

16.6 Oxidative Phosphorylation

At the end of Chapter 15 we presented a diagram (Figure 15.22) that illustrates the interrelationship of carbohydrate, protein, and fat metabolism. Tying everything together was acetyl coenzyme A, a common intermediate for all three groups and the fuel for the Krebs cycle (Figure 16.5). None of the steps of the Krebs cycle is shown coupled to

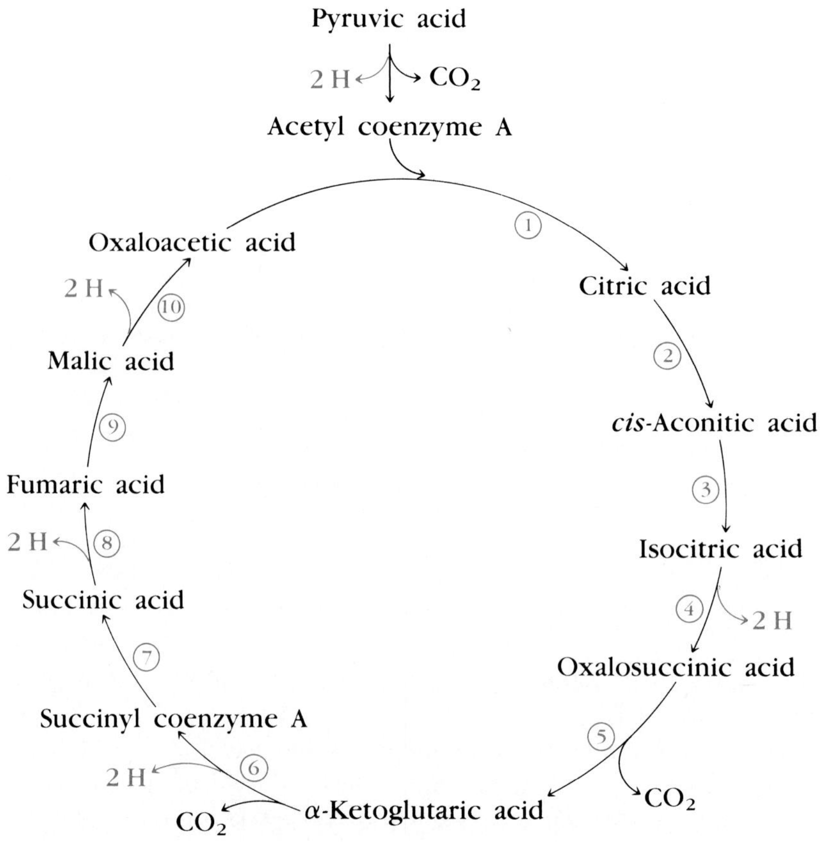

FIGURE 16.5
The Krebs cycle. At several places in the cycle, intermediates lose hydrogen and are oxidized. These intermediates serve as reducing agents for NAD^+ and FAD.

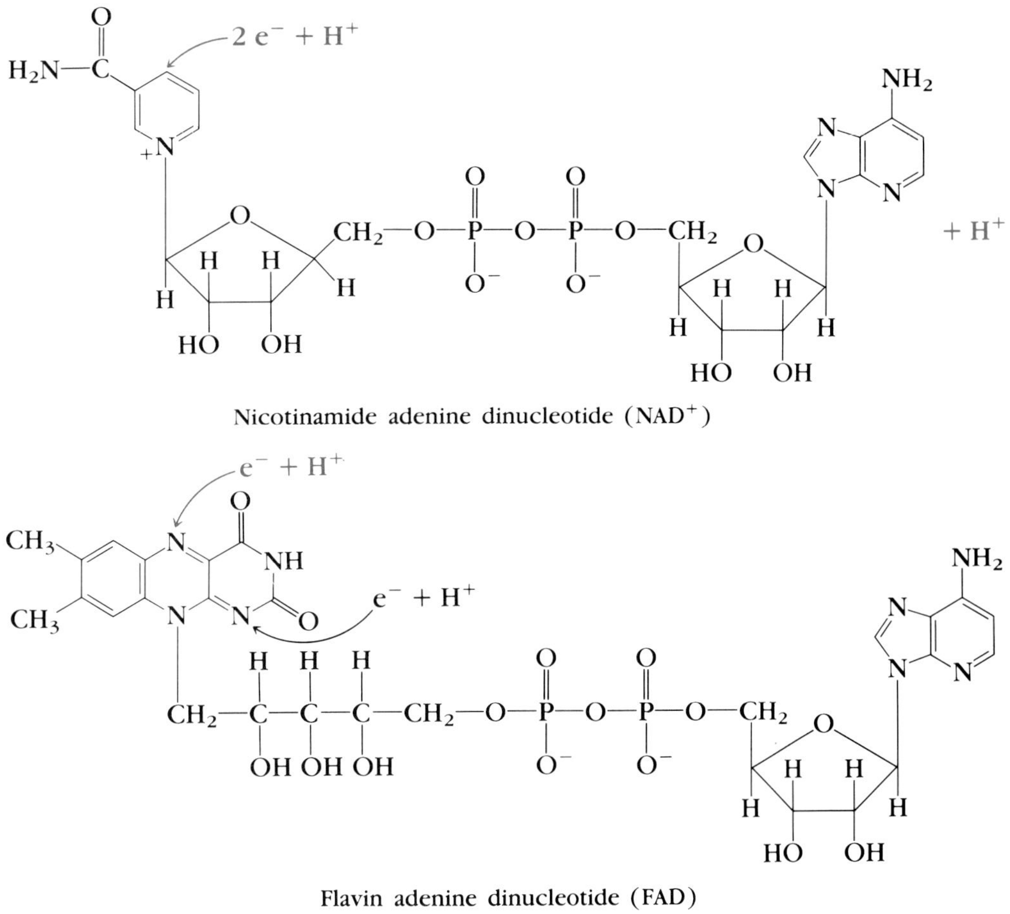

16.6 Oxidative Phosphorylation

Nicotinamide adenine dinucleotide (NAD$^+$)

Flavin adenine dinucleotide (FAD)

FIGURE 16.6
Oxidizing agents of the Krebs cycle.

the synthesis of ATP. In steps 4, 6, 8, and 10, the intermediates are undergoing oxidation, indicated in the diagram by the release of two hydrogen atoms at each of these steps. The diagram also indicates that as pyruvic acid feeds into the cycle it also undergoes oxidation to acetyl coenzyme A.

In Chapter 13 we termed the Krebs cycle *aerobic oxidation.* Aerobic means that oxygen is required, yet the diagram of the cycle in Figure 16.5 shows no oxygen. While the Krebs cycle rolls along, oxygen remains backstage but plays a vital role nonetheless. The hydrogen atoms we draw as by-products when a metabolite is oxidized must actually be transferred to an oxidizing agent. Oxygen is a wonderful oxidizing agent, but it does not pick up these hydrogen atoms directly. Instead, the hydrogen atoms are initially transferred to one of the two closely related molecules pictured in Figure 16.6, nicotinamide adenine dinucleotide and flavin adenine dinucleotide. By now such complex structures should not be intimidating; biochemists usually represent such structures by the initial letters of the names of the compounds. Thus, nicotinamide adenine dinucleotide is referred to, even in equations, as NAD^+, and flavin adenine dinucleotide is abbreviated as FAD. The important thing to remember is that these molecules serve as the oxidizing agents for the Krebs cycle by accepting the hydrogen atoms released by the Krebs cycle intermediates.

$$NAD^+ + 2\,H \longrightarrow NADH + H^+$$

$$FAD + 2\,H \longrightarrow FADH_2$$

The supplies of these oxidizing agents are limited. In order for the Krebs cycle to keep on turning, the reduced forms of these compounds, NADH and $FADH_2$, must be reoxidized. The ultimate oxidizing agent is—you guessed it—oxygen. The transfer of the hydrogen atoms from NADH and $FADH_2$ to oxygen (to form water) is coupled with the synthesis of ATP. The process is referred to as **oxidative phosphorylation**. The overall reaction for NADH can be written

The vast majority of ATP molecules are produced during oxidative phosphorylation.

$$NADH + H^+ + 3\,ADP + 3\,HPO_4^{2-} + \tfrac{1}{2}O_2 \longrightarrow NAD^+ + 3\,ATP + H_2O$$

As in most biochemical processes, the actual oxidation involves a whole series of reactions (Figure 16.7), which constitute what is called the **respiratory chain**. What travels down this chain are electrons. (Remember that oxidation and reduction can be regarded as the transfer of hydrogen atoms or electrons.) The **cytochromes** (Figure 16.7) are iron-containing, globular proteins, most of which contain one or more heme units. The iron may be in either the +2 or +3 oxidation state. The oxidized form (with Fe^{3+}) can accept an electron; the reduced form (Fe^{2+}) can donate one. Thus, electrons are transported from one cytochrome to another. Eventually, electrons are passed to molecules of oxygen (the reagent we replenish in our body with each breath), reducing the oxygen to water. The electrons originate ultimately in the metabolites undergoing oxidation in the Krebs cycle. For example, when isocitric acid is oxidized to oxalosuccinic acid by NAD^+, its electrons are passed via NADH through the respiratory chain to oxygen.

Hydrogen cyanide and its salts have long been recognized as extremely lethal poisons. The gas released in execution chambers is hydrogen cyanide, and a number of potent pesticides incorporate cyanide salts. The cyanide ion (CN^-), because of its small size, quickly makes its way into cells, where it binds to the iron in the heme of one of the cytochromes. This immediately shuts down the respiratory chain by inhibiting electron transfer, and cell respiration then ceases. The period from ingestion to death is only a few minutes, which is what makes this poison so lethal—little time is left to institute lifesaving measures.

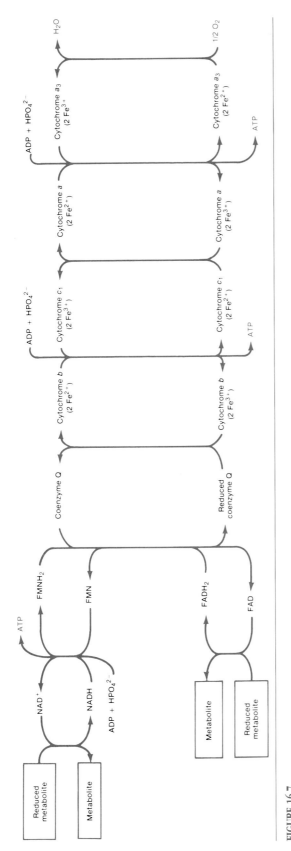

FIGURE 16.7
The electron transport system in oxidative phosphorylation.

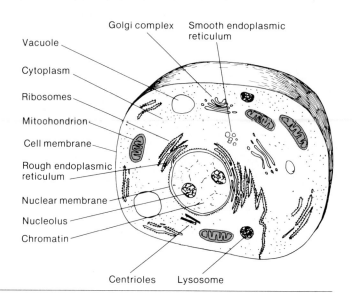

FIGURE 16.8
An idealized "typical" animal cell.

16.7 MITOCHONDRIA

In human beings, oxidative phosphorylation takes place in cell organelles called **mitochondria** (Figures 16.8 and 16.9). The number of mitochondria in a particular cell reflects the energy requirements of the cell, and the mitochondria can reproduce themselves if the energy requirements of the cell increase. Within the mitochondria are contained the enzymes of the respiratory chain and the enzymes required for the various metabolic pathways. Fuel molecules move into the mitochondria, where they are oxidized and release their energy to ATP. The ATP molecules, originally formed within, make their way out of the organelle and diffuse throughout the cell, where they can release their stored energy to drive necessary endergonic reactions.

The mitochondria are exquisitely designed energy factories. In recognition of the central role that this organelle plays in cellular energy production, the mitochondria are called the *powerhouses of the cell.*

FIGURE 16.9
(a) Electron micrograph of a mitochondrion. (b) Three-dimensional representation. [Photo courtesy of E.A. Munn.]

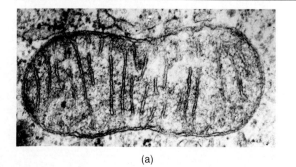

(a)

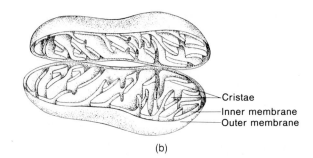

(b)

16.8 Efficiency of Cellular Energy Transformations

The overall oxidation of NADH by oxygen is accompanied by the release of a large amount of energy, and much of that energy is trapped in ATP molecules. Carbohydrates, fats, and proteins can ultimately fuel the Krebs cycle. Here is a comparison of the overall yield of ATP from the complete oxidation of an average fat molecule and of glucose. Remember that the vast majority of ATP molecules are produced during oxidative phosphorylation.

$$2\ C_{55}H_{102}O_6 + 155\ O_2 + 874\ ADP + 874\ HPO_4^{2-} \longrightarrow 110\ CO_2 + 874\ ATP + 976\ H_2O$$
   Fat

$$C_6H_{12}O_6 + O_2 + 36\ ADP + 36\ HPO_4^{2-} \longrightarrow 6\ CO_2 + 36\ ATP + 42\ H_2O$$
   Glucose

A single fat molecule produces more than ten times as much ATP as a single glucose molecule. Of course, the fat molecule contains 55 carbon atoms and the glucose molecule only 6. The fat molecule produces about 8 ATP molecules per carbon atom oxidized, and the glucose molecule yields 6 per carbon atom. The glucose gives a lower yield of ATP per carbon atom because the carbon in glucose is in a higher oxidation state. The oxidation number of carbon in fat is about -1.6. (See if you can calculate this number from the formula for the fat, $C_{55}H_{102}O_6$.) The oxidation number of carbon in glucose is 0. Both fats and carbohydrates are oxidized to carbon dioxide (CO_2), in which carbon has an oxidation number of $+4$. Thus, the carbon atoms in glucose go from 0 to $+4$, whereas the carbon atoms in the fat change from -1.6 to $+4$. More ATP is formed from the oxidation of a fat carbon atom because the fat carbon atom has a longer way to go.

How efficiently does ATP trap the energy released in these oxidation reactions? If standard free-energy values are used as the basis for the calculation, the efficiency of the process is about 38%. When actual intracellular conditions of temperature, pressure, and concentration are taken into account, efficiencies of 60% or better are calculated.

We know from the second law of thermodynamics that the process cannot be 100% efficient. What happens to the energy that is not trapped? It is released as heat to the surroundings, that is, to the cell, and this heat maintains body temperature. If we are exercising strenuously and our metabolism speeds up to provide the necessary energy for muscle contraction, then more heat is also produced. We begin to sweat to dissipate some of that heat. As the sweat evaporates, the excess heat is carried from the body by the departing water vapor.

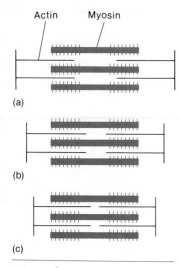

FIGURE 16.10
Diagram of actomyosin complex in muscle. (a) Extended muscle. (b) Resting muscle. (c) Partially contracted muscle.

Energy from fats is used for low- or moderate-intensity activity.

Fats and carbohydrates are used in prolonged heavy work.

The energy of carbohydrates is used for short bursts of high-intensity activity.

16.9 Muscle Power

The stimulation of muscle causes it to contract; that contraction is work and requires energy. The immediate source of energy for muscle contraction is ATP. The energy stored in this molecule powers the physical movement of muscle tissue. Two proteins, **actin** and **myosin**, play important roles in this process. Together actin and myosin form a loose complex called **actomyosin**, the contractile protein of which muscles are made (Figure 16.10). When ATP is added to isolated actomyosin, the protein fibers contract. It seems likely that the same process occurs in vivo, that is, in muscle in living animals. Not only does myosin serve as part of the structural complex in muscles, it also acts as an enzyme for the removal of a phosphate group from ATP. Thus, it is directly involved in liberating the energy required for the contraction.

In the resting person, muscle activity (including that of the heart muscle) accounts for only about 15–30% of the energy requirements of the body. Other activities, such as cell repair or the transmission of nerve impulses or even the maintenance of body temperature, account for the remaining energy needs. During intense physical activity, the energy requirements of muscle may be more than 200 times the resting level.

Fats are the major source of energy for sustained activity of low or moderate intensity. Prolonged, heavy work is fueled primarily by aerobic oxidation of both fats and carbohydrates. Anaerobic metabolism of carbohydrates supplies a significant proportion of the energy required for short bursts of high-intensity activity. The carbohydrate tapped for such all-out efforts is the glycogen stored in muscle tissue. To rapidly generate the ATP required for strenuous work, muscle glycogen is hydrolyzed, and the resulting glucose is converted to lactic acid via anaerobic glycolysis. The generation of energy by this pathway is self-limiting. As lactic acid builds up in muscle tissue, the pH drops and deactivates enzymes required for glycolysis. The muscle's response to stimuli becomes weaker, and in extreme cases there may be no response at all. In this state the muscle is described as fatigued.

When muscles use anaerobic pathways, they incur an **oxygen debt**. It is as if the body regards oxidation as the only proper source of energy for muscular activity and uses anaerobic metabolism as a temporary expedient. As soon as it can, the body oxidizes some of the resulting lactic acid back to pyruvic acid and ultimately to carbon dioxide and water. The energy released in the process is used to convert the rest of the lactic acid back to glycogen, which is re-stored for future use.

When is this oxygen debt repaid? Just as soon as the high level of muscular activity ceases. Sprinters, while running a 100-m dash, breathe in oxygen but still obtain only a fraction of the energy required for their intense muscular activity through aerobic processes. When the race is over, the sprinters continue to take in great gulps of air to repay the oxygen debt incurred during the race. We continue to breathe hard even after we stop vigorous activity because our body chemistry is still catching up and needs more of a critical reagent.

On the other hand, long-distance runners derive only a small percentage of their energy needs from glycolysis. Aerobic oxidation (the Krebs cycle coupled to oxidative phosphorylation) provides most of the required ATP. Some glycogenolysis (hydrolysis of glycogen) and glycolysis (anaerobic metabolism of glucose) do occur, and after long periods of this moderate muscle activity, lactic acid does build up and muscles do become fatigued. But since fats and some carbohydrate are supplying most of the energy through aerobic oxidation, the buildup of lactic acid takes much longer.

Muscle tissue seems to have been designed to provide for both short, intense bursts of activity and sustained, moderate levels of activity. Muscle fibers are divided into two categories, described as *fast twitch* and *slow twitch*. Table 16.2 lists some characteristics of these different types of muscle fibers. The Type I fibers described in Table 16.2 are called on during activity of light or moderate intensity. The respiratory capacity of these fibers is high, which means they can provide much energy via aerobic pathways, or, to put it another way, they are geared to oxidative phosphorylation. Notice that for Type I fibers myoglobin levels are also high. Myoglobin is the heme-containing protein in muscle that transports oxygen (as hemoglobin does in the blood). Aerobic oxidation requires oxygen, and this muscle tissue is geared to supply high levels of oxygen. The capacity of Type I muscle fibers for glycogenolysis is low. This tissue is not geared to anaerobic generation of energy and does not require the hydrolysis of glycogen. The number of mitochondria in Type I muscle tissue is high, as we would expect, because oxidative phosphorylation takes place in the mitochondria.

Two types of muscle fibers—fast twitch and slow twitch

TABLE 16.2 A Comparison of Types of Muscle Fiber

	Type I	Type IIB[a]
Category	Slow twitch	Fast twitch
Color	Red	White
Respiratory capacity	High	Low
Myoglobin level	High	Low
Catalytic activity of actomyosin	Low	High
Capacity for glycogenolysis	Low	High
Number of mitochondria	High	Low

[a] A Type IIA fiber exists that resembles Type I in some respects and Type IIB in others. We will discuss only the two types described in the table.

FIGURE 16.11
Metabolism fuels the effort. [Renee Lynn/Photo Researchers, Inc.]

The catalytic activity of the actomyosin complex is low. Remember that in addition to being the structural unit in muscle that actually undergoes contraction, myosin is responsible for catalyzing the hydrolysis of ATP to provide energy for the contraction. Low catalytic activity means that the energy is parceled out more slowly, which is not good if you want to lift 200 kg but is perfect for a 5-mile jog (Figure 16.11).

The characteristics of the Type IIB fibers described in Table 16.2 are just the opposite of those of Type I fibers. Low respiratory capacity, low myoglobin levels, and fewer mitochondria all argue against aerobic oxidation. A high capacity for glycogenolysis and the high catalytic activity of actomyosin allow this tissue to generate ATP rapidly via glycolysis and also to hydrolyze that ATP rapidly in intense muscle activity. Thus, this type of muscle tissue gives you the capacity to do short bursts of vigorous work. We say *bursts* because this type of muscle tissue fatigues relatively quickly. A period of recovery in which lactic acid is cleared from the muscle is required between brief periods of activity.

The fields of sports medicine and exercise physiology have done much to increase our understanding of muscle action. Endurance exercise training (for example, jogging for long distances) increases the size and number of mitochondria in muscle. An increase in the level of enzymes is required for the transport and oxidation of fatty acids, for the Krebs cycle, and for oxidative phosphorylation. The increase in mitochondrial enzymes is much greater for Type I fibers (used in prolonged, moderate-intensity activity) than for Type IIB fibers (brief, intense activity). Endurance training also increases myoglobin levels in skeletal muscles, providing for faster oxygen transport. These changes

FIGURE 16.12
The butterfly stroke, a test of the respiratory capacity of muscle. [David Lissy/The Picture Cube.]

FIGURE 16.13
Strength exercises build muscle mass but do not increase respiratory capacity of muscles. [Photo © Betty Lane/Photo Researchers, Inc.]

can be observed shortly after training begins, that is, within a week or two. Muscle changes resulting from endurance training do not necessarily include a significant increase in the size of the muscle, in contrast to the effect of strength exercises such as weight lifting. Weight lifting (presumably fueled primarily by anaerobic glycolsis) does not result in the mitochondrial changes we have just described. The mitochondria of heart muscle, which is working constantly anyway, also undergo no change during endurance training (Figures 16.12 and 16.13).

When we say that energy for muscle contraction is supplied by oxidative (aerobic) or glycolytic (anaerobic) processes (or both), we mean that the ATP used up by muscular activity is replaced through these pathways. In addition, muscles contain a relatively large concentration of creatine phosphate. As ATP is used up, creatine phosphate reacts with ADP to form more ATP (compare the standard free energies of hydrolysis for creatine phosphate and ATP, Table 16.1). Thus, creatine phosphate serves as a ready reserve for the quick restoration of ATP levels.

$$HN=C\begin{matrix}NH-PO_3^{2-}\\ \\NCH_2COOH\\ |\\CH_3\end{matrix} + ADP \longrightarrow HN=C\begin{matrix}NH_2\\ \\NCH_2COOH\\ |\\CH_3\end{matrix} + ATP$$

Creatine phosphate ⟶ Creatine

A standard clinical test performed on individuals suspected of having had a myocardial infarction involves the measurement of CK levels in the blood. CK is **creatine kinase**, the enzyme that catalyzes the phosphate transfer shown above. The heart is a muscle, and any unusual activity of this muscle (as in a "heart attack") would be accompanied by elevated CK levels. Since other muscular activity would also increase the CK level in the blood, the results of this test must always be interpreted most carefully.

We hope we have made clear the central role ATP plays in many life-sustaining metabolic reactions. Our cells can make ATP, using the energy stored in foods. But only green plants can make the food. They also replenish the oxygen we breathe. Have you thanked a green plant lately?

CHAPTER SUMMARY

Thermodynamics is the study of the relationships among different forms of energy. The first law of thermodynamics states that energy can be neither created nor destroyed. The second law states that the conversion of energy from one form to another cannot be accomplished with 100% efficiency.

A chemical reaction that results in a net release of energy is an exergonic reaction. Metabolic reactions that release energy to maintain life processes are exergonic. A reaction that requires a net input of energy is endergonic; photosynsthesis is an endergonic reaction.

The thermodynamic function ΔG (change in free energy) is defined in such a way that a natural (that is, spontaneous) reaction has a negative value for ΔG. A reaction with a positive ΔG is not spontaneous, that is, will not occur of its own accord.

In living systems, required endergonic reactions with $+\Delta G$s must be coupled with exergonic reactions having $-\Delta G$s in order to proceed. The coupling is done through a common intermediate. If the sum of the two ΔGs is negative, then the coupled reactions are favored.

The hydrolysis of ATP serves as the exergonic half of many coupled reactions in living systems. Because the ΔG for the hydrolysis of ATP is more negative than -5 kcal/mol, this compound (like others with similar values) is referred to as a high-energy compound. When high-energy food molecules are metabolized, some of the energy stored in these molecules is trapped by the coupled synthesis of ATP. Examples of this process can be found in the anaerobic glycolysis pathway.

The major synthesis of ATP occurs in oxidative phosphorylation. This metabolic pathway is coupled to the Krebs cycle through oxidizing agents like NAD^+ and FAD. As intermediates in the Krebs cycle are oxidized, NAD^+ and FAD are reduced. The reduced forms of the reagents are then reoxidized by oxygen through the respiratory chain associated with oxidative phosphorylation. The chain consists of a series of iron-containing proteins that carry electrons from NADH and $FADH_2$ to oxygen. Oxidative phosphorylation is inhibited by cyanide ion.

It is not possible to trap 100% of the energy released in the metabolism of food molecules in the bonds of ATP. The lost energy maintains the temperature of the body.

Mitochondria are cell organelles in which biological fuels are metabolized and oxidative phosphorylation takes place.

The immediate source of energy for muscle contraction is ATP. The hydrolysis of ATP is catalyzed by the protein complex actomyosin, which also functions as the structural material of muscles. Different types of muscle tissues appear to function under different conditions. Type I muscle fibers are geared to the use of energy provided by aerobic oxidation and are called on for sustained, low-to-moderate activity. Type IIB fibers make major use of glycolysis for energy and operate during bursts of vigorous activity. The accumulation of the end product of glycolysis, lactic acid, results in muscle fatigue. By using glycolysis, muscles incur an oxygen debt, which is repaid when the lactic acid is cleared from the muscle by aerobic metabolism.

Cells also contain high levels of the high-energy compound creatine phosphate. If ATP is used up, creatine phosphate can transfer its phosphate group to ADP and thus re-form the ATP that is required for further muscle contraction.

CHAPTER QUIZ

1. The second law of thermodynamics states that
   a. energy changes are never 100% efficient
   b. energy is neither created nor lost
   c. all energy comes from the sun
   d. all chemical reactions are driven by energy

2. Coupled reactions
   a. are parallel reactions
   b. are related only in name
   c. involve an exergonic process and an endergonic process
   d. form a product from two different reactants

3. Which molecule's hydrolysis produces more energy?
   a. ATP   b. ADP   c. AMP   d. HPO_4^{2-}

4. Which ΔG values would result in a thermodynamically favorable process?
   a. $-12 + 7$ b. $+12 - 7$
   c. $+12 + 7$ d. $-12 + 17$

5. Which is the most productive route to ATP molecules?
   a. anaerobic glycolysis
   b. aerobic glycolysis

    c. oxidative phosphorylation
    d. oxidation–reduction
6. Muscle contractile proteins are
    a. mitochondria
    b. actomyosin
    c. myoglobin
    d. creatine kinase
7. Fat molecules produce most ATP per carbon atom because
    a. they are longer chains
    b. the carbon oxidation number is lower for fats
    c. they are the most oxidized molecules
    d. the oxidation product is not carbon dioxide
8. Short bursts of activity are fueled by
    a. fats
    b. AMP
    c. aerobic oxidation
    d. anaerobic metabolism
9. Which property is characteristic of fast-twitch muscles?
    a. red color
    b. high catalytic activity of actomyosin
    c. high number of mitochondria
    d. high myoglobin level
10. Myoglobin is the oxygen-transporting protein in
    a. blood   b. cells
    c. intestine   d. muscle

PROBLEMS

16.1 Life and the Laws of Thermodynamics

1. State the first law of thermodynamics.
2. Energy is stored in a flashlight battery. Eventually the battery runs down. Has the energy been destroyed? Explain your reasoning.
3. State the second law of thermodynamics.
4. Which is more efficient: a vehicle that travels 10 km on 1 L of gasoline or a similar vehicle that travels 30 km on 5 L of gasoline?
5. Lightning can cause the following reaction in air.

$$N_2 + O_2 + \text{lightning} \rightarrow 2\,NO$$

In a firefly, the compound luciferin is converted to dehydroluciferin with the production of light.

$$\text{Luciferin} \rightarrow \text{dehydroluciferin} + \text{light}$$

Which reaction is exergonic, and which is endergonic?

16.2 Predicting Spontaneous Change: Free Energy

6. What is the sign of ΔG for a spontaneous reaction? When a reaction reaches equilibrium, what is the sign of ΔG?
7. What is meant by $\Delta G°$ (*standard* free-energy change)?
8. The free-energy change for a reaction is negative. Is the rate of the reaction fast or slow? Explain.

16.3 Coupled Reactions

9. What is a coupled reaction?
10. In a coupled reaction, which absolute value should be greater, the free-energy change associated with the exergonic reaction or the free-energy change associated with the endergonic reaction? Why?

16.4 ATP: Universal Energy Currency

11. Using Table 16.1, calculate the $\Delta G°$ for these reactions:
    a. PEP + acetic acid $\rightarrow$
        ketopyruvic acid + acetyl phosphate
    b. AMP + acetyl phosphate $\rightarrow$
        ADP + acetic acid
    c. creatine phosphate + ADP $\rightarrow$
        ATP + creatine
    d. ATP + glycerol $\rightarrow$
        glycerol 3-phosphate + ADP
12. What is the structural difference among ATP, ADP, and AMP?
13. Why is ATP referred to as the energy currency of the cell?
14. Referring to Table 16.1, indicate which of these compounds are classified as high-energy phosphates.
    a. ATP
    b. AMP
    c. PEP
    d. creatine phosphate
    e. glucose 1-phosphate
    f. glucose 6-phosphate

16.5 Synthesis of ATP

15. Is there a net consumption or synthesis of ATP in anaerobic glycolysis?

16.6 Oxidative Phosphorylation

16. Is there a net consumption or synthesis of ATP in oxidative phosphorylation?
17. What are the oxidizing agents used in the Krebs cycle?
18. What is the electron acceptor at the end of the respiratory chain? To what product is this compound reduced?
19. What is the function of the cytochromes in the respiratory chain?
20. Summarize the movement of electrons from the Krebs cycle through the oxidative phosphorylation chain by ordering the following compounds involved in the transport. Start with the compound in which the electrons originate, and end with the final compound to which the electrons are transferred: O_2, FAD, pyruvic acid, cytochromes.
21. Why is the cyanide ion so toxic?
22. What is the function of the mitochondria?

16.8 Efficiency of Cellular Energy Transformations

23. Which provides the greater number of ATP molecules: glycolysis of a glucose molecule or aerobic oxidation of a glucose molecule?
24. A fat yields more ATP molecules per carbon oxidized than a carbohydrate does. Why?

16.9 Muscle Power

25. Is there a net consumption or synthesis of ATP during muscle contraction?
26. What are the two functions of the actomyosin–protein complex?
27. Which energy reserves (that is, carbohydrates or fats) are tapped in intense bursts of vigorous activity?
28. Which energy reserves (carbohydrates or fats) are mobilized to fuel prolonged low levels of activity?
29. Which type of metabolism (aerobic or anaerobic) is primarily responsible for providing energy for intense bursts of vigorous activity?
30. Which type of metabolism (aerobic or anaerobic) is primarily responsible for providing energy for prolonged low levels of activity?
31. What is meant by *oxygen debt*?
32. Identify Type I and Type IIB muscle fibers as
    a. fast twitch or slow twitch
    b. suited to aerobic oxidation or to anaerobic glycolysis
33. Explain why high levels of myoglobin and mitochondria are appropriate for muscle tissue that is geared to aerobic oxidation.
34. Why does the high catalytic activity of actomyosin in Type IIB fibers suggest that these are the muscle fibers engaged in brief, intense physical activity?
35. Why can the muscle tissue that utilizes anaerobic glycolysis for its primary source of energy be called on only for *brief* periods of intense activity?
36. Which type of muscle fiber is most affected by endurance training exercises? What changes occur in the muscle tissue?
37. Birds use large, well-developed breast muscles for flying. Pheasants can fly 80 km/hr, but only for short distances. Great blue herons can fly only about 35 km/hr but can cruise great distances. What kinds of fibers would each have in its breast muscles?
38. What role does creatine phosphate play in supplying energy for muscle contraction?
39. Why could elevated levels of creatine kinase (CK) in the blood be taken as a possible indication of a heart attack?
40. A person collapses while shoveling snow, and a blood analysis reveals elevated levels of creatine kinase. Why would it be unwise to conclude that the individual had had a heart attack on this evidence alone?

17 NUCLEIC ACIDS

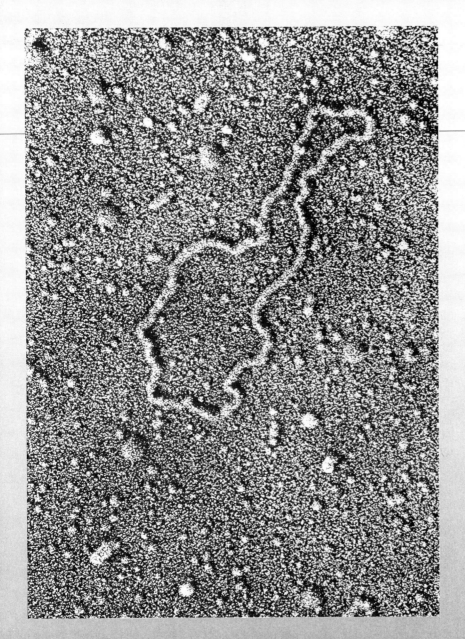

DID YOU KNOW THAT...

▶ biology is more complex than mathematics, and mathematics is more abstract than biology?

▶ nucleic acids store the pattern of life?

▶ nucleic acids are found in every living cell?

▶ nucleic acids use only four different bases as building blocks?

▶ nucleic acids may have molecular weights reaching into the billions?

▶ the base sequence in nucleic acids constitutes "the secret of life"?

▶ a single cell carries all hereditary information for the offspring?

▶ the double helix is a fragile structure held together only by hydrogen bonding?

▶ synthetic insulin and growth hormone have been tested since the early 1980s?

In contrast to what many people believe, the complexity of the sciences increases as one proceeds from physics to chemistry to biology. Because the language of physics is mathematics, most people regard physics as the most difficult of the sciences. Yet physical phenomena can be described with mathematical precision because the relationships involved are comparatively simple. We can write an equation that accurately describes the behavior of gases or of subatomic particles. A functioning cell defies such analysis.

Nonetheless, the cell is slowly yielding its secrets. One of these secrets, perhaps the most important one, is the method by which the cell stores and transmits information on how to reproduce itself. Nucleic acids are the molecules that store the patterns of life. Through nucleic acids these patterns are passed from one generation to the next. Nucleic acids also control the synthesis of proteins, including the enzymes that mediate those biochemical reactions that make an organism what it is.

As we have noted so often, with understanding comes control. The biochemists and molecular biologists who are unraveling these mechanisms are also learning how to manipulate the structure of living matter. The repair of defective genes, the design of precise molecular medicines, and control—for better or worse—of our heredity may lie in the future. While the twentieth century may be remembered as the nuclear age, the twenty-first century could well become the age of molecular biology.

17.1 THE STRUCTURE OF NUCLEIC ACIDS

Nucleic acids are found in every living cell. They are complex molecules, as might be expected from the role they play as the information and control centers of the cell. There are actually two kinds of nucleic acids (both polymers): **ribonucleic acid (RNA)**, found in all parts of the cell, and **deoxyribonucleic acid (DNA)**, found primarily in the cell nucleus. To understand how the nucleic acids function, we must become familiar with their structure. As we found for carbohydrates and proteins, the easiest way to approach the structure of polymers is to examine first the structure of the monomer units that combine to form the polymer. For nucleic acids, these building blocks are called **nucleotides**.

> RNA: ribonucleic acid
> DNA: deoxyribonucleic acid

> Nucleotides consist of three parts: a sugar, a heterocyclic amine base (these two parts together are called a nucleoside), and a phosphate group.

Nucleotides themselves consist of three parts. The first of these parts is a sugar, either ribose or deoxyribose.

Ribose   Deoxyribose

17.1 The Structure of Nucleic Acids

These sugars differ only in the functional group located at carbon 2. Ribose incorporates a hydroxyl group at this position; deoxyribose lacks this hydroxyl. Ribose is found in ribonucleic acid (RNA), and deoxyribose is found in deoxyribonucleic acid (DNA).

The sugars are substantially modified before they are incorporated into nucleic acids. First, the hemiacetal OH group (located at carbon 1) is replaced by a heterocyclic amine base (Section 12.3). The five organic bases most commonly found in combination with the sugar unit are

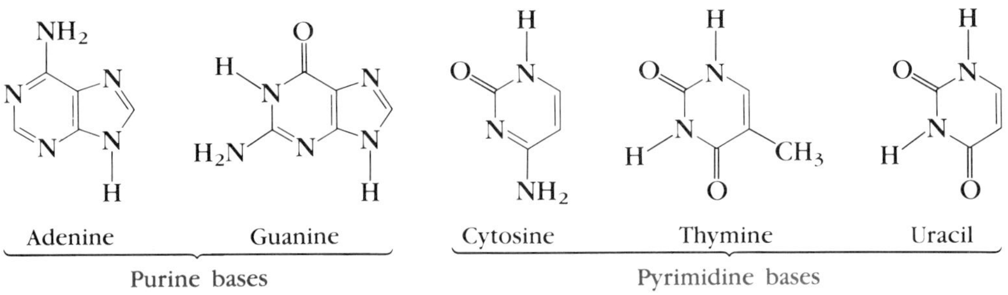

Adenine	Guanine	Cytosine	Thymine	Uracil
Purine bases		Pyrimidine bases		

The bases with two fused rings, adenine and guanine, are classsified as purines. Cytosine, thymine, and uracil are pyrimidines.

Figure 17.1 shows the eight sugar–base combinations, called **nucleosides**, that are ordinarily encountered. The bases adenine, guanine, and cytosine may be bound to either ribose or deoxyribose. Uracil is found only in combination with ribose, however, and thymine is found only with deoxyribose. Thus, uracil is found only in RNA, and thymine, only in DNA.

Nucleosides are converted to the structures called **nucleotides** (Figure 17.2) by formation of a phosphate ester at the C-5 hydroxyl group of the sugar ring. The conversion of the nucleoside cytidine to the nucleotide cytidine monophosphate is shown below.

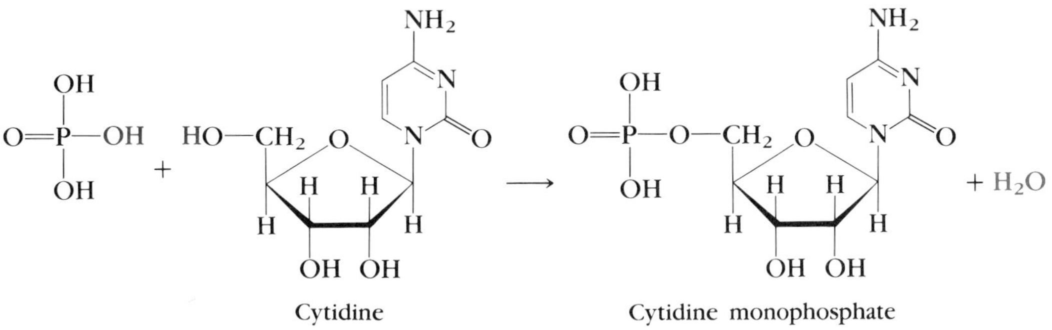

Cytidine + phosphate → Cytidine monophosphate + H$_2$O

How are the nucleotides combined to form a nucleic acid chain? When we considered the structure of phospholipids in Chapter 14, we

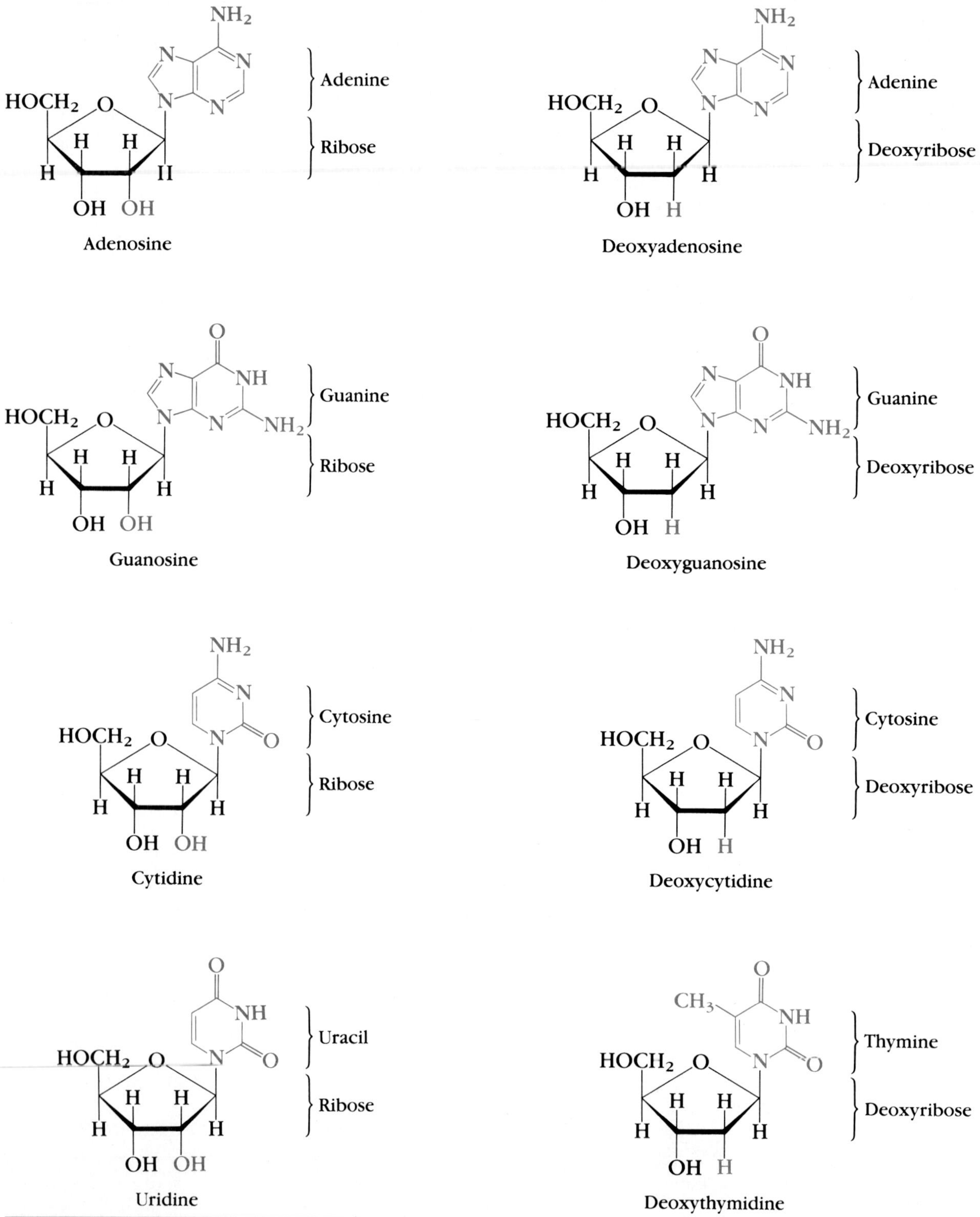

FIGURE 17.1
Eight common nucleosides.

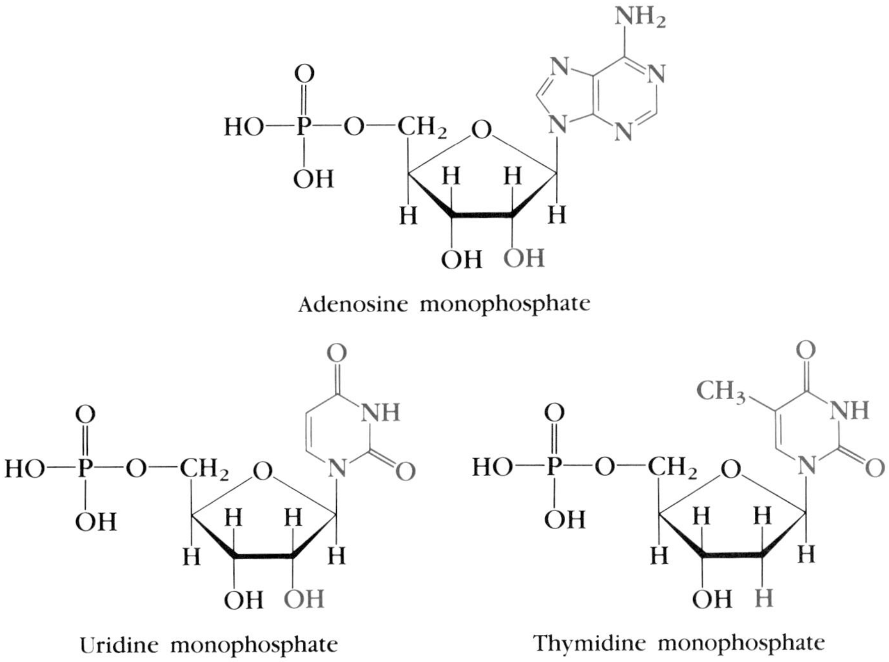

17.1 The Structure of Nucleic Acids

Adenosine monophosphate

Uridine monophosphate

Thymidine monophosphate

FIGURE 17.2 Structures of three representative nucleotides.

noted that phosphoric acid can form multiple esters by reacting with more than one alcohol functional group. Phosphoric acid's ability to form double esters permits the polymerization of nucleotides. The phosphate unit on one nucleotide forms an ester link with the C-3 hydroxyl group on the sugar ring of a second nucleotide.

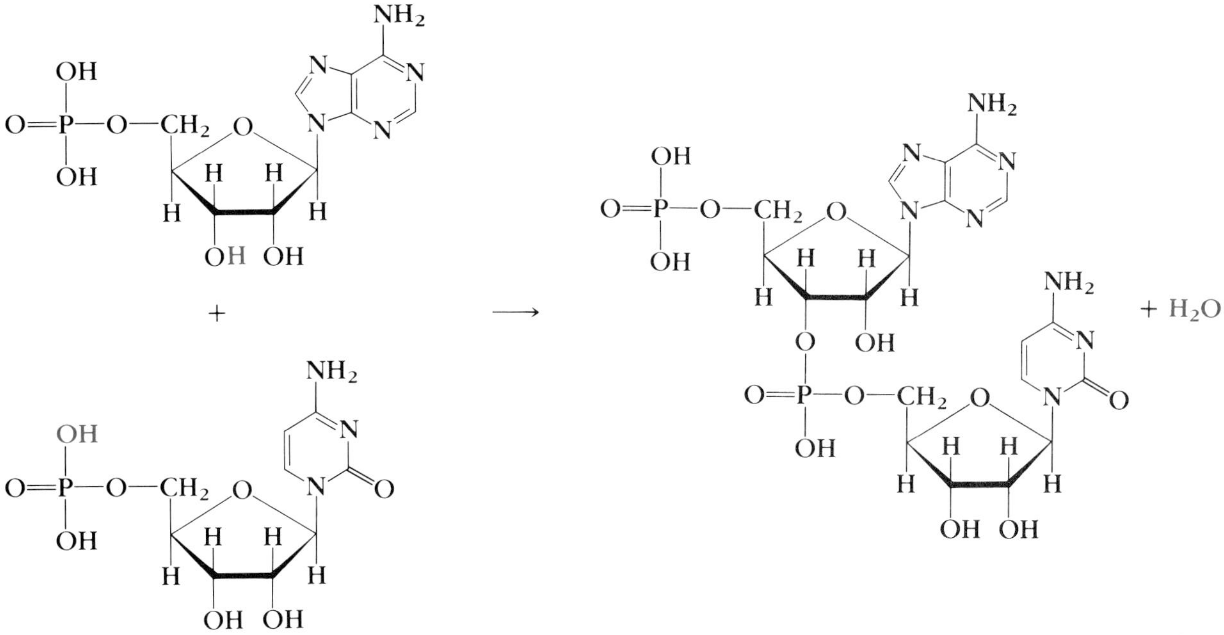

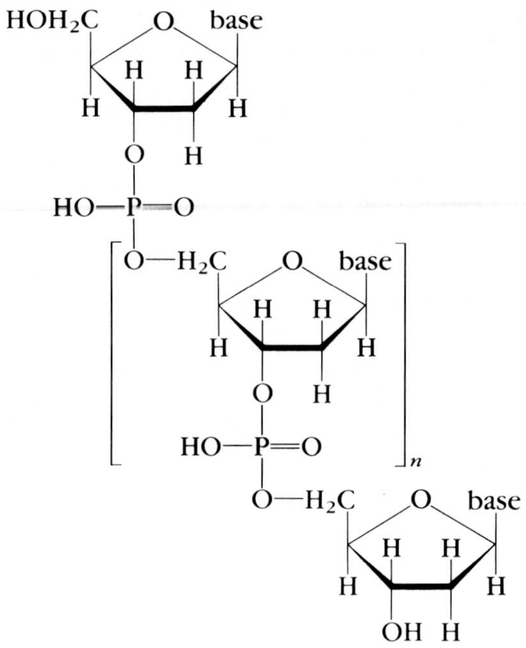

FIGURE 17.3
The polymeric backbone of a nucleic acid, shown here for deoxyribonucleic acid.

The product goes on to react with additional nucleotide units to build a long polymer chain (Figure 17.3). The backbone of the polymer consists of alternating phosphate and sugar units. The heterocyclic bases are branched off this backbone.

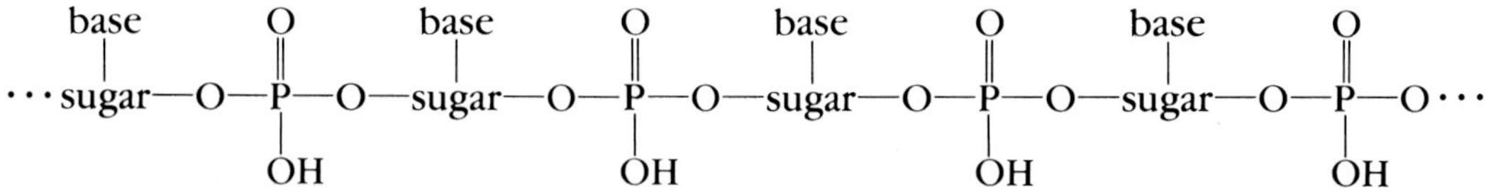

As we have seen, the sugar in DNA is 2-deoxyribose, and the one in RNA is ribose. The bases in DNA are adenine, guanine, cytosine, and thymine. Those in RNA are adenine, guanine, cytosine and uracil (Table 17.1). The phosphate group in the polymer chain still possesses one

TABLE 17.1 Components of DNA and RNA

Component	DNA	RNA
Purine bases	Adenine	Adenine
	Guanine	Guanine
Pyrimidine bases	Cytosine	Cytosine
	Thymine	Uracil
Pentose sugar	2-Deoxyribose	Ribose
Inorganic acid	Phosphoric acid	Phosphoric acid

17.2 The Secondary Structure of Nucleic Acids: Base Pairing

ionizable hydrogen atom (Figure 17.3). This is what makes these compounds nucleic *acids*. In solutions, however, or when combined with basic proteins in complexes called nucleoproteins, the acid may be ionized.

Nucleic acids resemble proteins in one respect: to specify completely the primary structure of a nucleic acid, one must specify the *sequence* of bases. Unlike the proteins, which incorporate more than 20 different amino acids, a nucleic acid uses only four different bases. The molecular weights of nucleic acids are often much greater than those of proteins, however, ranging into the billions for mammalian DNA. Along these great chains, the four bases may be arranged in essentially infinite variations. That is a crucial feature of these molecules because it is the base sequence that is used to store the multitude of information needed to build living organisms. Before we examine that aspect of nucleic acid chemistry, let us consider one more important feature of nucleic acid structure.

17.2 THE SECONDARY STRUCTURE OF NUCLEIC ACIDS: BASE PAIRING

In experiments designed to probe the structure of DNA, it was determined that the molar amount of adenine (A) in DNA corresponded to the molar amount of thymine (T). Similarly, the molar amount of guanine (G) is essentially the same as that of cytosine (C). To maintain this balance, the bases in DNA must be paired, A to T and G to C. But how? At the midpoint of this century, it was quite clear that the answer to this question would bring with it a Nobel prize. Although many illustrious scientists worked on the problem, two scientists who were relatively unknown in the world of science announced in 1953 that they had worked out the structure of DNA. Using data that involved quite sophisticated chemistry, physics, and mathematics, and working with models not unlike a child's construction set, James D. Watson and Francis Crick determined that DNA must be composed of two helixes wound about one another. The phosphate and sugar backbone of the polymer chains form the outside of the structure, which is rather like a spiral staircase. The heterocyclic amines are paired on the inside—with guanine always opposite cytosine and adenine always opposite thymine. In our staircase analogy, these base pairs are the steps (Figures 17.4 and 17.5).

Why do the bases pair in this precise pattern, always A to T and T to A, always G to C and C to G? The answer is, because of hydrogen bonding and a truly elegant molecular design. Figure 17.6 shows the two sets of base pairs. You should notice two things. First, a pyrimidine is paired with a purine in each case, and the long dimensions of both

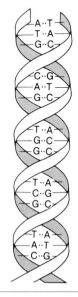

FIGURE 17.4
The DNA double helix as portrayed by Watson and Crick.

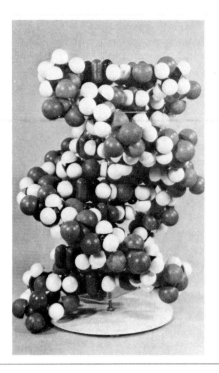
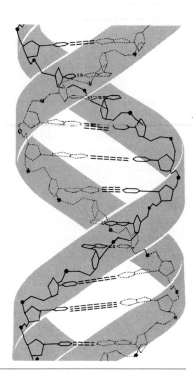

FIGURE 17.5
Two models of the DNA double helix. [Photograph courtesy of Science Related Materials, Inc., Janesville, WI.]

pairs are identical (1.085 nm). If two pyrimidines were paired, the combination would be narrower than the purine–pyrimidine combination (Figure 17.7). If two purines were paired, the combination would be wider than the purine–pyrimidine pair. Staircases are not made with stairs of random widths, and neither is DNA. In order for the two strands of the **double helix** to fit neatly, a pyrimidine must always be paired with a purine.

The second thing you should notice in Figure 17.6 is the hydrogen bonding between the bases in each pair. When guanine is paired with cytosine, three hydrogen bonds can be drawn between the bases. No

When guanine is paired with cytosine, three hydrogen bonds can form.

FIGURE 17.6
Base pairing of thymine to adenine and of cytosine to guanine.

17.2 The Secondary Structure of Nucleic Acids: Base Pairing

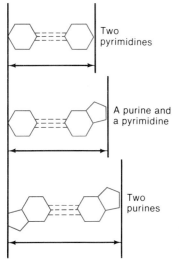

FIGURE 17.7
Difference in widths of possible base pairs.

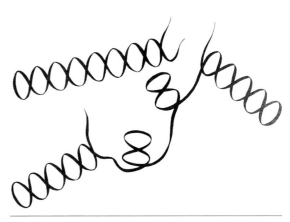

FIGURE 17.8
RNA occurs as single strands that can form double-helical portions by internal base pairing.

other pyrimidine–purine pairing will permit such extensive interaction. Indeed, in the combination shown in the figure, both pairs of bases fit like lock and key.

The molecules of RNA consist of single strands of the nucleic acid. Some internal (intramolecular) base pairing may occur in sections where the molecule folds back on itself. Portions of the molecule may exist in double-helical form (Figure 17.8)

Watson and Crick (Figure 17.9) received the Nobel prize in 1962 for discovering, as Crick put it, "the secret of life." The structure these two scientists proposed was accepted almost immediately by other scientists around the world because it answers so many crucial questions. It can explain how cells are able to divide and go on functioning, how genetic data are passed on to new generations, and even how proteins are built to required specifications. It all depends on the base pairing.

Watson and Crick discovered the structure of DNA in 1953 and received the Nobel prize in 1962.

FIGURE 17.9
James D. Watson and Francis Crick, who proposed the double helix model of DNA. [Courtesy of Harvard University Biological Laboratories, Cambridge, MA.]

17.3 DNA: Self-replication

Cats have kittens that grow up to be cats. Bears have cubs that grow up to be bears. How is it that each species reproduces after its own kind? How does a fertilized egg "know" that it should develop into a kangaroo and not a koala?

The physical basis of heredity has been known for a long time. Most higher organisms reproduce sexually. A sperm cell from the male unites with an egg cell from the female. The fertilized egg so formed must carry all the information needed to make the various cells, tissues, and organs necessary for the functioning of a new individual. For human beings, that single cell must carry the information for the making of legs, liver, lungs, heart, head, hair, and hands—in short, all the instruction ever needed for growth and maintenance of the individual. In addition, if the species is to survive, information must be set aside in germ cells—both sperms and eggs—for the production of new individuals.

Chromosomes contain the hereditary material.

The hereditary material is found in the nuclei of all cells, concentrated in elongated, threadlike bodies called chromosomes (Figure 17.10). The number of chromosomes varies with the species. Human body cells have 46 chromosomes. Germ cells carry only half the number of chromosomes of the body cells. Thus, in sexual reproduction the entire complement of chromosomes is achieved only when the egg and sperm combine; a new individual receives half its hereditary material from each parent.

Chromosomes are made of nucleoproteins. The nucleic acid in chromosomes is DNA, and it is the DNA that is the primary hereditary material (Figure 17.11). Arranged along the chromosomes are the basic units of heredity, the genes. Structurally, genes are sections of the DNA molecule (some viral genes contain only RNA). When cell division occurs, each chromosome produces an exact duplicate of itself.

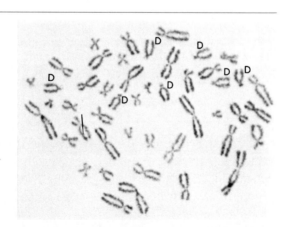

FIGURE 17.10
Human chromosomes in the metaphase stage, just before division. Note that there are 47 chromosomes instead of the normal 46. This cell came from a boy who had an extra chromosome 13. Since numbers 13, 14, and 15 are indistinguishable in this photograph, they are all labeled D, and there are 7 D chromosomes instead of the expected 6. [Courtesy of Eeva Therman, University of Wisconsin, Madison.]

17.3 DNA: Self-Replication

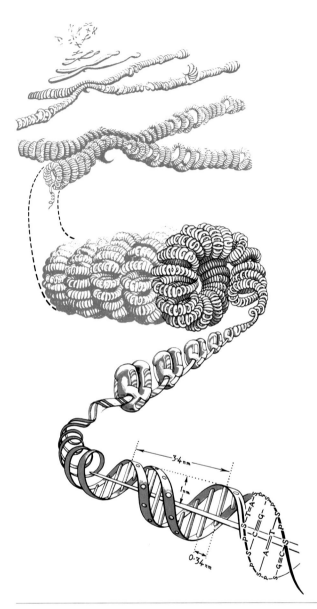

FIGURE 17.11
Chromosomes and DNA. Each chromosome is made up of two chromatids united at the centromere. A chromatid is a protein-coated strand of multicoiled DNA. [Adapted from Roger Warwick and Peter L. Williams, *Gray's Anatomy*, 36th British ed. London: Longmans Co. Ltd., 1980, 15, with permission of Churchill Livingstone.]

Transmission of genetic information therefore requires the **replication** (copying or duplication) of DNA molecules. The Watson–Crick double helix provides a ready model for this process. If the two chains of the double helix are pulled apart, and the hydrogen bonds holding the base pairs together are broken, then each chain can direct the synthesis of a new DNA chain. In the cellular fluid surrounding the DNA are all the necessary nucleotides (monomers). Synthesis begins with the base on a nucleotide pairing with its complementary base on the DNA strand (Figure 17.12). Keep in mind that adenine can pair only with thymine, and guanine only with cytosine. Each base unit in the

In order to pass on genetic information the DNA molecule must be replicated.

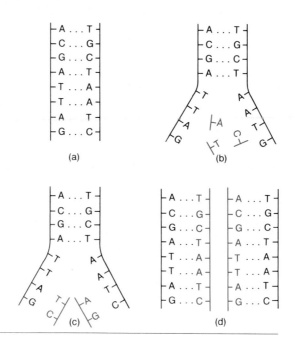

FIGURE 17.12
DNA replication. (a) The orginal double helix, flattened out here for clarity. (b) The helix beginning to split. Some free nucleotides from the cell are shown. (c) Nucleotides from the cell beginning to pair with bases on each original strand. (d) The two new double helixes, each identical to the original.

The sequence of bases along the DNA chain encodes the directions for building an organism.

separated strand can pick up only a unit identical to the one it paired with before. Each of the separating chains serves as a template, or pattern, for the formation of a new complementary chain.

As the nucleotides align, they react with one another to form the sugar–phosphate backbone of the new chain. In this way, each strand of the original DNA molecule forms a duplicate of its former partner. Whatever information was encoded in the original DNA double helix is now contained in each of the replicates. When the cell divides, each daughter cell gets one of the DNA molecules and all the information that was available to the parent cell. The entire process of replication is controlled by appropriate enzymes.

DNA can be compared to a book containing directions for putting together a model airplane or knitting a sweater. Knitting directions store information as words on paper. Letters of the alphabet are arranged in a certain way (e.g., "knit one, purl two"), and these words direct the knitter to carry out a particular operation with needles and yarn. If all the directions are correctly followed, the ball of yarn becomes a sweater.

How is information stored in DNA? The sequence of bases along the DNA chain encodes the directions for building an organism. Just as *saw* means one thing in English and *was* means another, the sequence of bases CGT means one thing, and GCT means something else. Although there are only four "letters"—the four bases—in the genetic code of DNA, their sequence along the long strands can vary so widely that essentially unlimited information storage is available. Each cell carries in its DNA all the information it needs to determine all the hereditary characteristics of even the most complex organism.

17.4 RNA: Protein Synthesis and the Genetic Code

Even if we accept the fact that DNA carries a message, how this message is read and acted on is still a problem. There are no little knitters swimming around in the cellular fluid with yarn in hand. There are, however, special RNA molecules that act as the cell's equivalent of knitters. In place of yarn and sweaters, the cell works with amino acids and proteins. Proteins serve as building material and also, most important, as enzymes (Chapter 15). Enzymes control all the metabolic reactions that occur in the complex chemical factory we call a living organism. The synthesis of the proper proteins is the key to genetic control.

In the first step of this synthesis, information in DNA is transferred to a special type of RNA molecule. The process is called **transcription**, and the special RNA molecule is called **messenger RNA (mRNA)**. (This is a little like xeroxing a page of instructions from a book of instructions.) One can envision a DNA double helix being partially unzipped. Then a limited portion of one of the DNA strands directs the synthesis of the single-stranded mRNA molecule (Figure 17.13). Once again the process involves a buildup of complementary nucleotides along the single DNA strand. The base sequence of DNA specifies the base sequence of mRNA. Thymine in DNA calls for adenine in mRNA, cytosine specifies guanine, guanine calls for cytosine, and adenine requires uracil. Remember that in RNA molecules, uracil is used in place of DNA's thymine. Notice the similarity in structure of these two bases (Section 17.1).

Messenger RNA (mRNA) transfers information from DNA to RNA.

DNA Base	Complementary RNA Base
Adenine	Uracil
Thymine	Adenine
Cytosine	Guanine
Guanine	Cytosine

The next step in protein synthesis involves the reading of the blueprint now encoded in the mRNA molecule and its **translation**

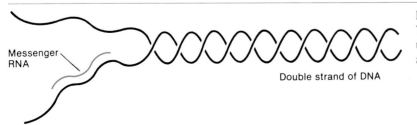

FIGURE 17.13
The DNA double helix is partly unwound, and messenger RNA is formed along a segment of one strand.

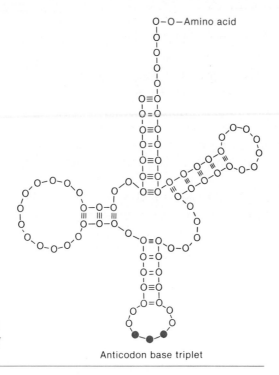

FIGURE 17.14
A given transfer RNA can carry only one kind of amino acid, which is specified by the base triplet in the anticodon.

Transfer RNA (tRNA) translates the specific base sequence of an mRNA molecule into the specific amino acid sequence.

into a protein structure. The mRNA travels from the nucleus to the cytoplasm of the cell, where the cellular parts called ribosomes are located. The ribosomes are themselves constructed of nucleoproteins. The mRNA becomes attached to the ribosomes, and it is here that the genetic code is deciphered.

In the cytoplasm is another kind of RNA, called **transfer RNA (tRNA)**. The tRNA is responsible for translating the specific base sequence of an mRNA molecule into the specific amino acid sequence of a protein molecule. A tRNA molecule has a looped structure (Figure 17.14). On one of these loops is a critical set of three bases (called the **triplet**). One of the trailing ends of the tRNA is attached to an amino acid. A given tRNA molecule can carry only one kind of amino acid. All tRNA molecules with the same base triplet carry the same kind of amino acid. For example, all tRNA molecules whose base triplet is CCU carry the amino acid glycine. A tRNA molecule with the base triplet GCG always has an arginine molecule attached. There are many more ways of sequencing three bases than there are kinds of amino acids. Thus, one kind of amino acid is usually associated with more than one triplet sequence. Cysteine is carried by tRNA molecules with either ACA or ACG triplets.

We are now ready to build proteins. Strung out along some ribosomes in the cytoplasm of a cell is an mRNA molecule. The sequence of

17.4 RNA: Protein Synthesis and the Genetic Code

bases along this molecule was determined by the DNA of a gene in the cell nucleus. Floating around in the cytoplasm surrounding the mRNA are tRNA molecules, each carrying its own amino acid. Let us suppose that a portion of the mRNA base sequence reads

···CGCGGAGGC···

The first three bases (CGC) could pair, through hydrogen bonding, with a nucleic acid that had a GCG sequence. There, floating by, is just such a nucleic acid, a tRNA molecule with the amino acid arginine attached. The triplet of bases of tRNA pairs up with the first three bases of the mRNA.

···CGCGGAGGC···
GCG
/
Arg

The next three bases of the mRNA, GGA, pair up with a CCU tRNA molecule, which carries the amino acid glycine.

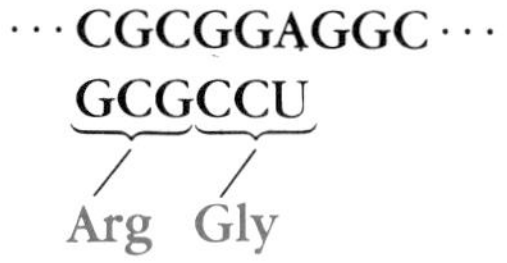

···CGCGGAGGC···
GCGCCU
/ /
Arg Gly

When the two tRNA molecules are appropriately lined up along the mRNA, a peptide bond forms between the two amino acids.

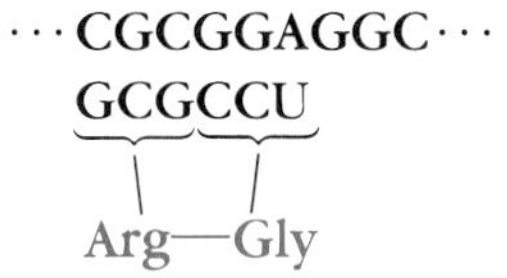

···CGCGGAGGC···
GCGCCU
\ /
Arg—Gly

The next three bases on the mRNA (GGC) pair with a CCG tRNA molecule

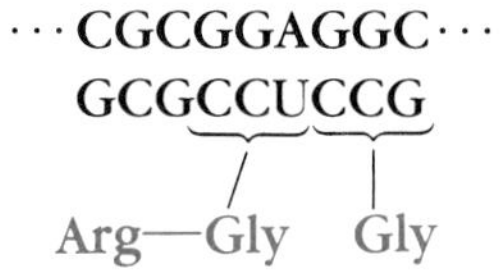

···CGCGGAGGC···
GCGCCUCCG
/ |
Arg—Gly Gly

The amino acid carried by the tRNA is then joined by a peptide bond to

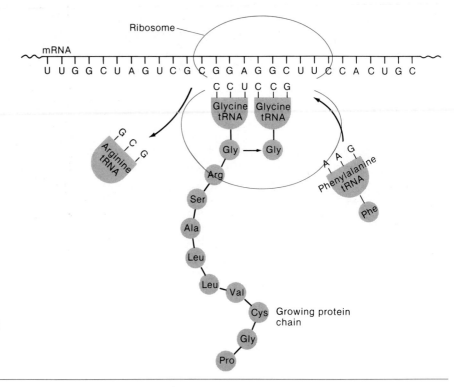

FIGURE 17.15
The synthesis of a protein molecule. [Adapted with permission from Abraham Mazur and Benjamin Harrow, *Textbook of Biochemistry*, 10th ed. Philadelphia: W. B. Saunders, 1971.]

the previously linked amino acids. The first tRNA molecule, having delivered its amino acid, returns to the cytoplasm.

$$\cdots \text{CGCGGAGGC} \cdots$$
$$\underbrace{\text{CCU}}\underbrace{\text{CCG}}$$
$$\text{Arg}—\text{Gly}—\text{Gly}$$

In this way the protein chain is gradually built up. The chain is released from the tRNA and mRNA as it is formed (Figure 17.15).

Each base triplet strung consecutively along the mRNA molecule is called a **codon**. The base triplets of the tRNA molecules are called **anticodons**. A codon on mRNA always pairs with its complementary tRNA anticodon. A complete dictionary of the genetic code has been compiled. Table 17.2 shows which amino acids are called for by all the possible mRNA codons. The amino acids serine and leucine are each specified by six different codons. Two amino acids, tryptophan and methionine, have only one codon each. Note that three codons, UAA, UAG, and UGA, are stop signals, calling for termination of the protein chain.

All proteins for muscle, hair, skin, and the all-important enzymes are formed by the mechanism outlined above. Next time you look in a

17.5 Genetic Engineering

TABLE 17.2 The Genetic Code for Protein Synthesis

	Second Base								
First Base	U		C		A		G		Third Base
U	UUU	Phe	UCU	Ser	UAU	Tyr	UGU	Cys	U
	UUC	Phe	UCC	Ser	UAC	Tyr	UGC	Cys	C
	UUA	Leu	UCA	Ser	UAA	Stop	UGA	Stop	A
	UUG	Leu	UCG	Ser	UAG	Stop	UGG	Trp	G
C	CUU	Leu	CCU	Pro	CAU	His	CGU	Arg	U
	CUC	Leu	CCC	Pro	CAC	His	CGC	Arg	C
	CUA	Leu	CCA	Pro	CAA	Gln	CGA	Arg	A
	CUG	Leu	CCG	Pro	CAG	Gln	CGG	Arg	G
A	AUU	Ile	ACU	Thr	AAU	Asn	AGU	Ser	U
	AUC	Ile	ACC	Thr	AAC	Asn	AGC	Ser	C
	AUA	Ile	ACA	Thr	AAA	Lys	AGA	Arg	A
	AUG	Met	ACG	Thr	AAG	Lys	AGG	Arg	G
G	GUU	Val	GCU	Ala	GAU	Asp	GGU	Gly	U
	GUC	Val	GCC	Ala	GAC	Asp	GGC	Gly	C
	GUA	Val	GCA	Ala	GAA	Glu	GGA	Gly	A
	GUG	Val	GCG	Ala	GAG	Glu	GGG	Gly	G

mirror, remember that what you see depends ultimately on something as fragile as a hydrogen bond.

17.5 Genetic Engineering

Over 3000 human diseases have a genetic component. Over the last decade or so, researchers have linked specific genes to specific diseases. Now the ability to use this information to diagnose and cure genetic diseases appears to be within our grasp. By determining the location of a gene on the DNA molecule, scientists have been able to identify and isolate genes with specific functions.

A gene is a rather elusive substance. There are approximately 10 000 genes on each human chromosome, and isolating the one gene that is defective (in the case of genetic disease) is understandably difficult. One detection approach is to treat the DNA with enzymes called **restriction endonucleases**. This method yields segments of genetic material called **restriction fragment length polymorphisms**, or RFLPs (pronounced "rif lips"). These segments are much easier to work with because they contain fewer genes. The pattern of RFLPs resulting from enzyme treatment is an inherited characteristic. RFLP patterns characteristic of certain families can be isolated. If the pattern of a relative matches that of a person with a genetic disease, that relative will probably develop the disease. Thus it is possible to identify and even predict the occurrence of a genetic disease.

A future hope of genetic engineering is that we will be able to introduce a functioning gene into a person's cells, thus correcting the action of a defective gene. The first attempts to do this were made in 1990.

Recombinant DNA

All living organisms (except some viruses) have DNA as their hereditary material. It should be possible, then, to place a gene from one organism into the genetic material of another. **Recombinant DNA** (rDNA) technology does just that. First, the gene is identified. Then it is isolated, by the RFLP process, and placed in a separate piece of DNA. The recombined DNA is then transferred into a bacterium or other suitable organism. The final step, called **cloning**, produces many copies of the modified bacterium, which can then produce relatively large amounts of the protein coded for by the gene.

By working backward from the amino acid sequence of the protein, scientists can work out the base sequence of the gene that codes for the protein. When isolated, the gene is spliced into a special kind of bacterial DNA called a **plasmid**. The recombined plasmid is then inserted into the host organism (Figure 17.16). Once inside the host, the plasmid replicates, making multiple exact copies (or clones) of

FIGURE 17.16
A schematic outline of recombinant DNA technology. [Adapted with permission from *Biotechnology*, ACS Department of Government Relations and Science Policy, December 1985. Copyright 1985, American Chemical Society.]

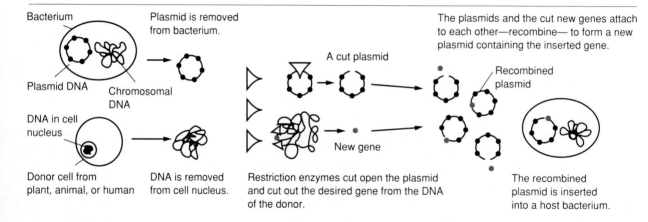

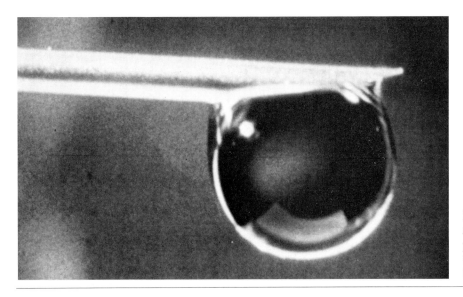

FIGURE 17.17
A globule of the first human insulin produced through recombinant DNA technology. [Courtesy of Eli Lilly and Company.]

itself. As the rDNA-engineered bacteria multiply, they become virtual factories for producing the desired protein.

Many valuable materials, difficult to obtain in any other way, are now made using rDNA technology. People with diabetes formerly had to use insulin from pigs or cattle. Now human insulin, a protein coded for by human DNA, is being produced by the cell machinery of bacteria (Figure 17.17). All newly diagnosed insulin-dependent diabetic persons in the United States are now treated with DNA-produced human insulin.

Human growth hormone, used to treat children who fail to grow properly, was formerly available only in tiny amounts obtained from cadavers. Now it is readily available through rDNA technology. This technology also yields interferon, a promising anticancer agent. Scientists have even designed bacteria that will "eat" the oil released in an oil spill.

Concern over the potential for disaster in this type of research has lessened somewhat in recent years. Initially, scientists worried about the possibility of producing a deadly "artificial" organism. What if a gene that causes cancer was spliced into the DNA of a bacterium that normally inhabits our intestine? We would have no natural immunity against such an organism. To protect against such a development, strict guidelines for recombinant DNA research have been instituted.

The new molecular genetics has already resulted in some impressive achievements. Its possibilities are mind-boggling—elimination of genetic defects, a cure for cancer, a race of geniuses, and who knows what else? Knowledge gives power; it does not necessarily give wisdom. Who will decide what sort of creature the human species should be? The greatest problem we shall likely face in our use of bioengineering is that of choosing who is to play God with the new "secret of life."

CHAPTER SUMMARY

The genetic material of the cell consists of complex molecules called nucleic acids. The compounds are polymers constructed from monomeric units that incorporate heterocyclic amine bases, sugars, and phosphate groups. These monomers, called nucleotides have the general structure

Z = H (DNA) or OH (RNA)

Nucleosides are nucleotides minus the phosphate group. Figure 17.1 shows the common nucleosides incorporated (as their phosphate esters) into nucleic acid polymers.

The two types of nucleic acids are deoxyribonucleic acid (DNA), in which the sugar unit is deoxyribose, and ribonucleic acid (RNA), which incorporates the sugar ribose. The purine bases adenine (A) and guanine (G) and the pyrimidine base cytosine (C) are found in both types of nucleic acids. The pyrimidine thymine (T) is found only in DNA, and the pyrimidine uracil (U) is found only in RNA. In the polymers, the phosphate group of one nucleotide unit is esterified with the C-3 hydroxyl group of the sugar. The backbone of the polymer, therefore, consists of alternating sugar and phosphate groups. The sequence of a nucleic acid is the order of the bases attached as substituents along this backbone.

A major difference between RNA and DNA is that RNA molecules consist of a single polymer strand, whereas DNA molecules consist of two polymer strands intertwined in a double helix. The two strands interact through hydrogen bonds formed between two bases located opposite one another on the two strands. Some base pairing is also observed in RNA molecules, but the bases are attached to the single RNA strand and are brought into proximity when the strand bends back on itself. In both types of nucleic acids, G and C form base pairs. In DNA, A and T form base pairs, and in RNA, A and U form base pairs. These are the only pairings possible.

In most species, the primary genetic material is DNA. The sequence of bases is the code that determines the structure of proteins required by the organism. This genetic message is conserved during the process of cell division because each new cell gets its own copy of each DNA molecule. When the cell divides, the DNA undergoes replication, a process in which each DNA strand acts as a template for the formation of a complementary strand. The accurate formation of the complementary strand depends on precise base pairing, A with T and G with C. Each of the daughter molecules of DNA formed on replication contains one of the parent DNA strands.

When protein synthesis is required, the genetic message in DNA is trnscribed to an RNA molecule. A segment of one strand of the DNA molecule (which constitutes a gene) acts as a template for the formation of a messenger RNA (mRNA) molecule. Once formed, the mRNA leaves the nucleus and attaches to cellular parts called ribosomes. There the mRNA directs the synthesis of proteins. Triplets of bases on mRNA are called codons and form hydrogen bonds with triplets of bases called anticodons on transfer RNA (tRNA) molecules. Transfer RNA molecules with a given anticodon carry a specific amino acid. Thus, a specific mRNA codon is associated with a specific amino acid on tRNA, and the sequence of bases along mRNA (and, ultimately, in DNA) determines the sequence of amino acids in a protein. This process is called translation.

Genetic engineers are using recombinant DNA technology to splice together genes that are not normally found together. In this way, it has been possible to develop bacteria that produce human proteins such as human growth hormone, interferon, and insulin.

CHAPTER QUIZ

1. Most DNA is found in the
   a. cell nucleus
   b. entire cell
   c. fluid outside the cell
   d. cell membrane

Problems

2. The building blocks of nucleic acids are
   a. proteins   b. DNA
   c. RNA   d. nucleotides
3. Nucleosides are combinations of
   a. sugars and phosphates
   b. bases and phosphates
   c. sugar and bases
   d. sugar, base, and proteins
4. Uracil can combine with
   a. thymine   b. guanine
   c. deoxyribose   d. ribose
5. The polymeric chain of nucleic acids consists of units of
   a. phosphate and sugar
   b. base and phosphate
   c. purine and sugar
   d. pyrimidine and sugar
6. Which base pairing does not occur in DNA?
   a. A–T   b. G–T   c. G–C   d. T–A
7. How many chromosomes are there in a normal human body cell?
   a. 2   b. 18   c. 46
   d. infinite combinations are possible.
8. Which base pairing occurs only in RNA?
   a. A–U   b. T–A   c. C–G   d. G–C
9. The process of information transmission from DNA to RNA is called
   a. replication   b. transcription
   c. translation   d. transfer
10. The special molecule that receives information from DNA is called
    a. mRNA   b. tRNA
    c. triplet   d. codon

PROBLEMS

17.1 The Structure of Nucleic Acid

1. Name the two kinds of nucleic acids.
2. Which of the two kinds of nucleic acids is concentrated in the nucleus of the cell?
3. a. What sugar is incorporated in the RNA polymer?
   b. What is the sugar unit in DNA?
4. Compare DNA and RNA with respect to the major bases present in each type of nucleic acid.
5. For each of the following, indicate whether the compound is a nucleoside, a nucleotide, or neither.

   a. 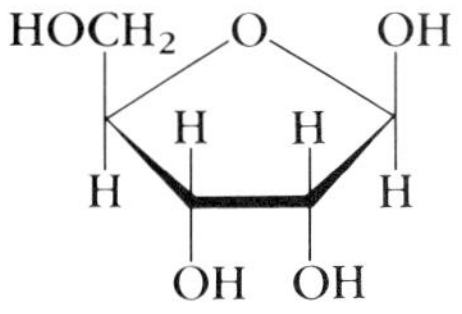 [structure: ribose with OH at anomeric position]

   b. 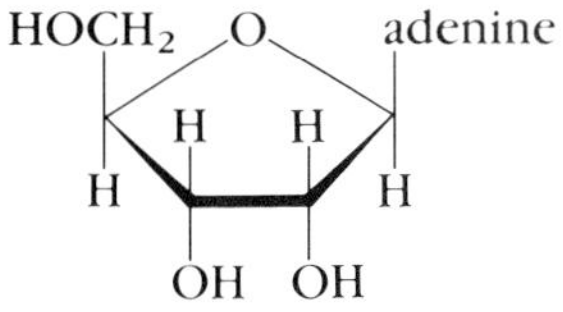 [structure: ribose with adenine]

   c. 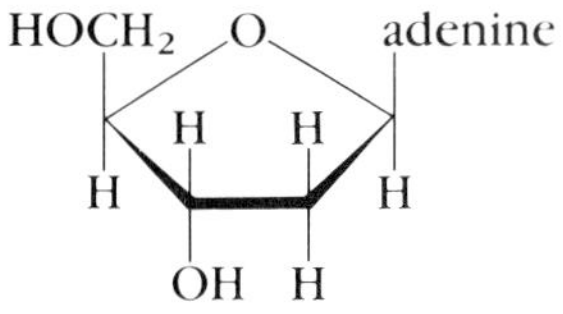 [structure: deoxyribose with adenine]

   d. 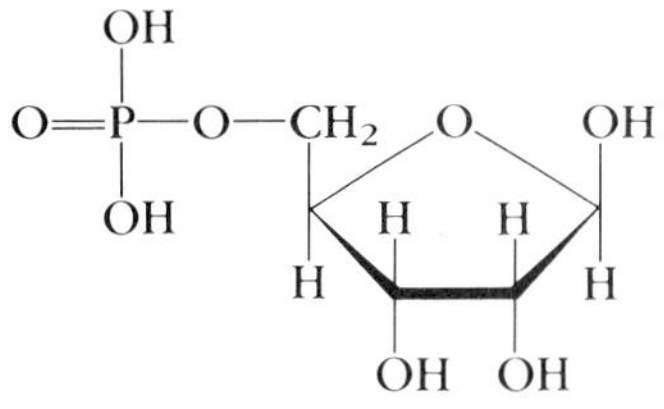 [structure: phosphate–ribose with OH]

   e. 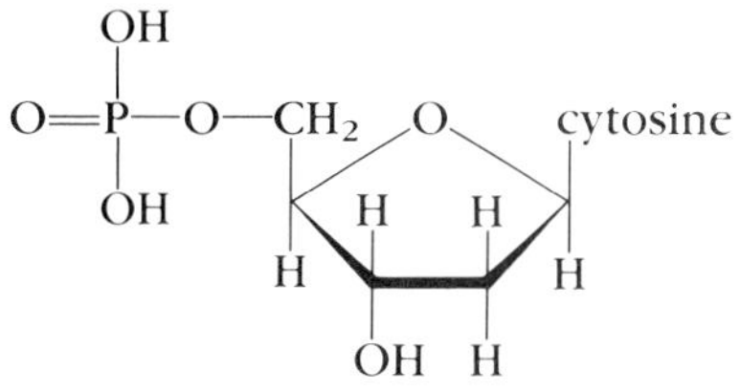 [structure: phosphate–deoxyribose with cytosine]

   f. 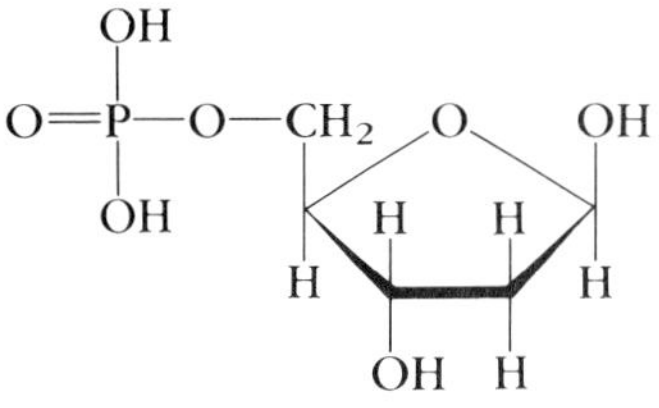 [structure: phosphate–deoxyribose with OH]

6. For each of the structures shown in Problem 5, indicate whether the sugar unit is ribose or deoxyribose.

7. Indicate whether each of the following nucleosides is incorporated in DNA or RNA or neither.

   a. HOCH₂―O―adenine (with OH, OH on ring)
   b. HOCH₂―O―adenine (with OH, H on ring)
   c. HOCH₂―O―cytosine (with OH, OH on ring)
   d. HOCH₂―O―thymine (with OH, H on ring)
   e. HOCH₂―O―uracil (with OH, H on ring)
   f. HOCH₂―O―guanine (with OH, OH on ring)

8. Which base is purine, and which is pyrimidine?

   a. (pyrimidine ring structure)   b. (purine ring structure)

9. For each compound in Problem 7, indicate whether the base is a purine or a pyrimidine.

10. Answer the questions for the molecule shown.

    (structure of thymidine with CH₃ on base, OH on 5' CH₂, OH and H on sugar)

    a. Is the compound a nucleic acid, a nucleotide, or a nucleoside?
    b. Is the base a pyrimidine or a purine?
    c. Would the compound be incorporated in DNA or RNA?

11. Answer questions a–c of Problem 10 for the following compound.

    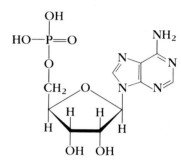

12. Using a schematic representation, show a length of nucleic acid polymer chain, indicating the positions of the sugar, phosphoric acid, and base units.

13. The primary structure of a protein is defined by the sequence of amino acids. What defines the primary structure of nucleic acid?

17.2 The Secondary Structure of Nucleic Acids: Base Pairing

14. In an RNA molecule, which base would pair with each base listed?
    a. adenine   b. guanine
    c. uracil   d. cytosine

15. How do DNA and RNA differ in secondary structure?

16. With the same sort of schematic representation used in Problem 12, show the overall design of the double helix.

17. Why is it structurally important in the DNA double helix that a purine base always pairs with a pyrimidine base?
18. What kind of intermolecular force is involved in base pairing?
19. In DNA, which base would be paired with the base listed?
    a. cytosine
    b. adenine
    c. guanine
    d. thymine

17.3 DNA: Self-replication
20. Describe the process of replication.
21. In replication, a parent DNA molecule produces two daughter molecules. What is the fate of each strand of the parent DNA double helix?

17.4 RNA: Protein Synthesis and the Genetic Code
22. We say that DNA controls protein synthesis, yet most DNA resides within the cell nucleus whereas protein synthesis occurs outside of the nucleus. How does DNA exercise its control?
23. Explain the role of messenger RNA in protein synthesis.
24. Explain the role of transfer RNA in protein synthesis.
25. Which nucleic acid or acids is (are) involved in the process referred to as transcription?
26. Which nucleic acid or acids is (are) involved in the process referred to as translation?
27. Which nucleic acid contains the codon?
28. Which nucleic acid contains the anticodon?
29. The base sequence along one strand of DNA is

    $\cdots$ ATTCG $\cdots$

    What would be the sequence of the complementary strand of DNA?
30. What sequence of bases would appear in the messenger RNA molecule copied from the original DNA strand shown in Problem 29?

31. If the sequence of bases along a messenger RNA strand is

    $\cdots$ UCCGAU $\cdots$

    what was the sequence along the DNA template?
32. What are the complementary triplets on tRNA for the following triplets on mRNA?
    a. UUU   b. CAU
    c. AGC   d. CCG
33. What are the complementary triplets on mRNA for the following triplets on tRNA molecules?
    a. UUG   b. GAA
    c. UCC   d. CAC
34. Using Table 17.2, identify the amino acids carried by the tRNA molecules in Problem 32.
35. Using Table 17.2, identify the amino acids carried by the tRNA molecules in Problem 33. Remember that Table 17.2 lists the mRNA codons.
36. Refer to Table 17.2. What amino acid sequence would result if the base sequence on mRNA is
    a. $\cdots$ UUACCUCGA $\cdots$
    b. $\cdots$ GCGUCAUAA $\cdots$
    c. $\cdots$ CCCCCCCC $\cdots$
37. If the DNA base sequence

    $\cdots$ TTACTCTCA $\cdots$

    acts as a template for mRNA formation, what amino acid sequence would eventually be produced from the mRNA?
38. What is the relationship between the cell parts called chromosomes, the units of heredity called genes, and the nucleic acid DNA?
39. Certain genes are implicated in cancer. Some of these *oncogenes* can be activated by single point mutations (changing of single base in DNA). Ordinarily the DNA triplet GGT codes for the amino acid proline. What amino acid is formed when the DNA triplet GGT is changed by mutation to TGT? to GTT? to CGT?

17.5 Genetic Engineering
40. What is the basic process in recombinant-DNA technology?
41. Discuss some applications of genetic engineering.

42. Chemical bonds can be broken by ultraviolet radiation, gamma rays, certain chemical substances, and so on. What are the biological implications of these facts? (Hint: What effect might the breaking of chemical bonds have on DNA?)

43. What are RFLPs? How do RFLP patterns indicate that a person probably will develop a genetic disease?

44. List the four steps in rDNA technology.

45. What is a plasmid?

18

Vitamins, Minerals, and Hormones

Did You Know That...

▶ vitamins, hormones, and most minerals are effective in trace amounts?
▶ minerals make up only 4% by weight of the human body?
▶ there are about 25 elements essential to life?
▶ anemia results in a general weakening of the body?
▶ calcium ions are necessary for the coagulation of blood?
▶ the function of some minerals is not understood?
▶ vitamin A deficiency causes blindness?
▶ a vitamin overdose can have adverse effects?
▶ removal of a portion of the pituitary gland results in "wasting away"?

By now you should be convinced that your health depends on the maintenance of a proper balance of chemicals in your body. In this last chapter, we want to introduce you to a diverse group of chemicals that share two things in common. They are required for good health and, with a few notable exceptions, they are effective in trace quantities.

From the time you were a small child, you have probably been told that you need vitamins and minerals for good health. That is true, and we would like to show you why. You are probably also aware that there are substances called hormones in your body. We have already considered some hormones in this text—insulin, for example. Now we want to give you a more general view of the role of hormones in the maintenance of health and body function.

Several minerals are used as structural materials or are distributed so widely throughout the body that they constitute a significant (though still small) fraction of the body's mass. Most minerals, however, and all the vitamins and hormones are notable for the minuteness of their effective amounts. But do not let the concentrations of these materials fool you. They are potent chemicals and are responsible for critical aspects of body fuction. We also need just the right amount. You cannot assume that if a small amount does a good job, then a large amount will do a better job. As is true for so much of biological chemistry, a proper balance is the key to proper function.

18.1 Minerals

Minerals are inorganic compounds that are vital to health.

As preceding chapters of this text have emphasized, many of the chemicals in living organisms are organic. But there are also inorganic chemicals that are vital to health, called **minerals**. Minerals are estimated to represent about 4% of the weight of a human body. Some of these, such as the chlorides (Cl^-), phosphates PO_4^{3-}), bicarbonates (HCO_3^-), and sulfates (SO_4^{2-}), occur in the blood and in other body fluids. Others, such as iron in hemoglobin and phosphorus in nucleic acids, are constituents of very complex organic compounds.

Minerals essential to one or more living organisms include the following elements.

Arsenic (As)	Iodine (I)	Selenium (Se)
Calcium (Ca)	Iron (Fe)	Silicon (Si)
Chlorine (Cl)	Magnesium (Mg)	Sodium (Na)
Chromium (Cr)	Manganese (Mn)	Sulfur (S)
Cobalt (Co)	Molybdenum (Mo)	Tin (Sn)
Copper (Cu)	Phosphorus (P)	Vanadium (V)
Fluorine (F)	Potassium (K)	Zinc (Zn)

18.1 Minerals

Together with the structural elements carbon, hydrogen, nitrogen, and oxygen, these make up the 25 chemical elements essential to life. Other elements are sometimes found in body fluids and tissues but are not known to be essential. These include aluminum (Al), lithium (Li), nickel (Ni), and boron (B). Eventually, it may be discovered that one or more of these are essential also.

Minerals serve many purposes in the body. For example, small amounts of iodine are necessary for the proper functioning of the thyroid gland. (When we discuss hormones in Section 18.3, you will see how the iodine is used.) A deficiency of iodine has dire effects, of which goiter is perhaps the best known (Figure 18.1). Iodine is available naturally in seafood. But, to guard against iodine deficiency, a small amount of sodium iodide (NaI) is often added to table salt (NaCl). The use of iodized salt has greatly reduced the incidence of goiter (Figure 18.2).

Iron ions are found in combination with a number of complex biomolecules. The most familiar is the oxygen-transporting compound hemoglobin, which incorporates iron as the iron(II) ion (Fe^{2+}). When too little iron is present, there is a shortage of oxygen supplied to body tissues, and anemia, which results in a general weakening of the body, results. The loss of blood that occurs with menstruation results in a small loss in the body's reserves of iron. Thus, for women between puberty and menopause, higher dietary intake of iron is necessary. Foods especially rich in iron compounds include red meat and liver.

Calcium and phosphorus are necessary for the proper development of bones and teeth. Growing children need about 1.5 g of each per day. Milk is a good source of these elements. Adults also require these elements, although that fact is less widely recognized. Calcium ions are also necessary for the coagulation of the blood (to stop bleeding) and

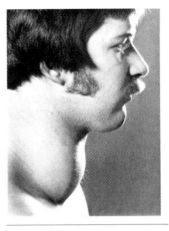

FIGURE 18.1
A man affected by goiter. The swollen thyroid gland in the neck results from a lack of the trace element iodine in the diet. [Photo by Joseph J. Mentrikoski, Department of Medical Photography, Geisinger Medical Center.]

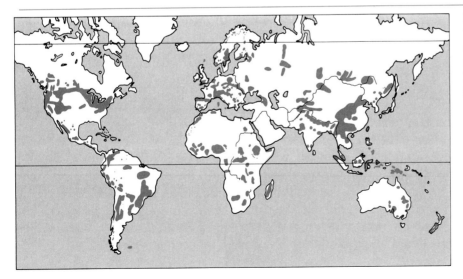

FIGURE 18.2
The areas of the world where goiter is endemic. The addition of sodium iodide or potassium iodide to table salt has largely eradicated goiter in the United States.

for maintenance of the rhythm of the heartbeat. Phosphorus plays many roles in the body. It is incorporated in phospholipids, in nucleic acids, and in carbohydrates that are to be metabolized for energy. Without phosphorus compounds, we could not get *any* quick energy from those quick-energy foods.

Sodium chloride in moderate amounts is essential to life. It plays an important role in the exchange of fluids between cells and blood plasma, for example. The presence of this salt increases water retention by the body and may result in swelling of tissue or high blood pressure (hypertension). An estimated 30 million people in the United States have hypertension. Most physicians agree that our diets generally contain too much salt, and they frequently advise patients who have hypertension to reduce their salt intake.

Copper, zinc, cobalt, manganese, molybdenum, and magnesium, along with iron and calcium, serve as cofactors for essential enzymes (Section 15.9). The functions of other minerals are only partially understood, which explains why bioinorganic chemistry is a flourishing area of research.

18.2 VITAMINS

FIGURE 18.3
Inflammation and abnormal pigmentation characterize pellagra, caused by niacin deficiency. [Courtesy of the World Health Organization, New York.]

Why are British sailor called "limeys"? Since early times, sailors have been plagued by a disease called *scurvy*. In 1747, British Navy Captain James Lind showed that the disease could be prevented by the inclusion of fresh fruit and vegetables in the diet. A convenient fresh fruit to carry on long voyages (there was no refrigeration in those days) was the lime. British ships put to sea with barrels of limes aboard, and the sailors ate a lime or two every day. Thus, they came to be known as "lime eaters" or simply "limeys."

In 1897, Christiaan Eijkman, a Dutch scientist, showed that polished rice lacked something found in the hulls of natural rice. Lack of that something caused the disease *beriberi*, which was quite a problem in the Dutch East Indies at the time.

A British scientist, F. G. Hopkins, fed a synthetic diet of carbohydrates, fats, proteins, and minerals to a group of rats. The rats were unable to sustain healthy growth. Again, something was missing.

In 1912, Casimir Funk, a Polish biochemist, coined the word *vitamine* (from the Latin word *vita*, meaning "life") for these missing factors. Funk thought that all the factors contained the amino group. In the United States the final *e* was dropped from the designation after it was found that not all the factors were amines. The generic term became **vitamin**. Eijkman and Hopkins shared the 1929 Nobel Prize in medicine and physiology for their important discoveries.

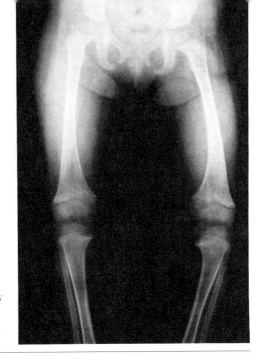

FIGURE 18.4
This X ray shows the typical distortion of leg bones in rickets, a disease caused by a deficiency of vitamin D. [Photo by Joseph J. Mentrikoski, Department of Medical Photography, Geisinger Medical Center.]

Vitamins are specific *organic* compounds that our bodies need for proper functioning. Absence or shortage of a vitamin results in a vitamin-deficiency disease (Figures 18.3 and 18.4). Our bodies cannot synthesize these compounds; therefore, vitamins must be included in the diet (Figure 18.5). Vitamins are required in much smaller amounts than are the basic foodstuffs (carbohydrates, fats, and proteins), and a well-balanced diet ordinarily provides all necessary vitamins in sufficient quantities for health. Some of the vitamins, with their structures, sources, and deficiency symptoms, are listed in Table 18.1.

As you can see from the table, the vitamins do not share a common chemical structure. They can, however, be divided into two broad categories: the **fat-soluble group**, including A, D, E, and K, and the **water-soluble group**, made up of the B complex and vitamin C. All the fat-soluble vitamins incorporate a high proportion of hydrocarbon

Vitamins are organic compounds that are vital for our bodies to function properly.

Vitamins are divided into two broad categories: fat-soluble vitamins and water-soluble vitamins.

FIGURE 18.5
The effect of ascorbic acid on bone formation. (a) Cross section of bone-forming cells from a guinea pig dying of scurvy. Note that there are no connective tissue fibers. (b) Cross section of bone-forming cells from another guinea pig with scurvy after 72 hours of treatment with ascorbic acid. Note that connective tissue fibers have appeared. [Courtesy of The Upjohn Company, Kalamazoo, MI.]

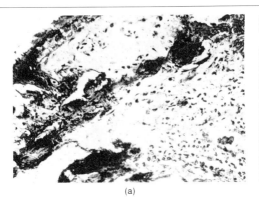
(a)

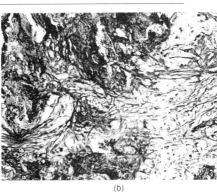
(b)

TABLE 18.1 The Vitamins

Structure and Name	Sources	Deficiency Symptoms
Fat-Soluble Vitamins		
Vitamin A (retinol)	Fish, liver, eggs, butter, cheese; also a vitamin precursor in carrots and other vegetables	Night blindness
Vitamin D_2 (calciferol)	Cod liver oil, irradiated ergosterol (milk supplement)	Rickets
Vitamin E (α-tocopherol)	Wheat germ oil, green vegetables, egg yolks, meat	Sterility, muscular dystrophy
Vitamin K_1 (phylloquinone)	Spinach, other green leafy vegetables	Hemorrhage

Water-Soluble Vitamins

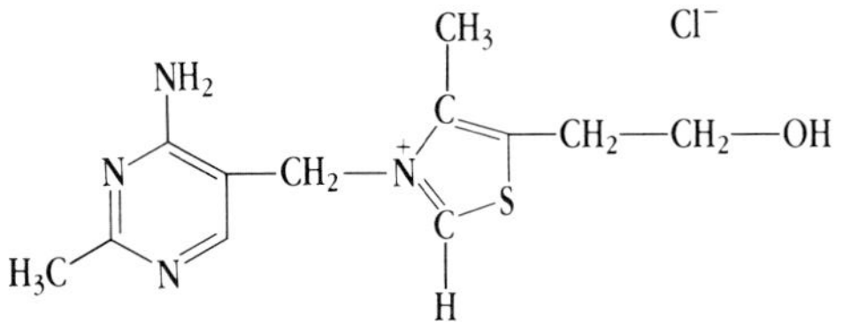

Vitamin B₁ (thiamine)

Germ of cereal grains, legumes, nuts, milk, and brewers' yeast

Beriberi—polyneuritis resulting in muscle paralysis, enlargement of heart, and ultimately heart failure

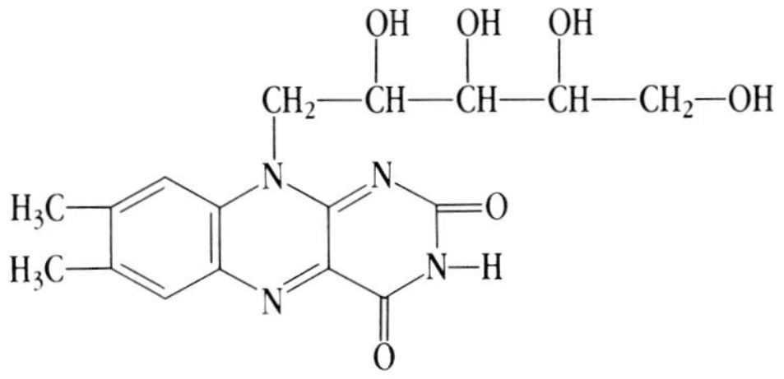

Vitamin B₂ (riboflavin)

Milk, red meat, liver, egg white, green vegetables, whole wheat flour (or fortified white flour), and fish

Dermatitis, glossitis (tongue inflammation)

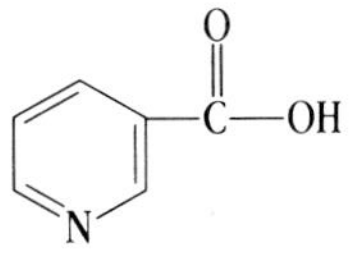

Niacin (Nicotinic acid, Nicotinamide)

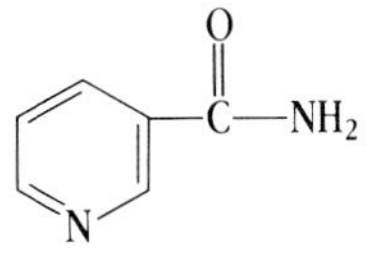

Red meat, liver, collards, turnip greens, yeast, and tomato juice

Pellagra—skin lesions, swollen and discolored tongue, loss of appetite, diarrhea, various mental disorders (Figure 18.3)

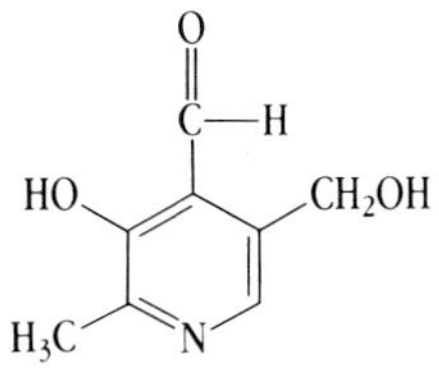

Vitamin B₆ (Pyridoxal, Pyridoxol, Pyridoxamine)

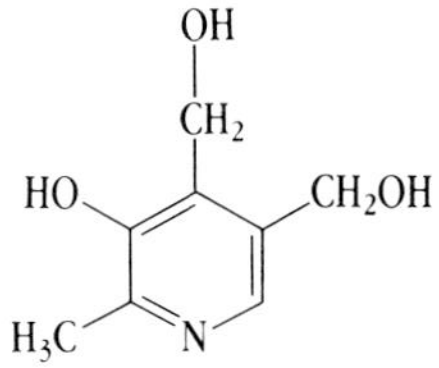

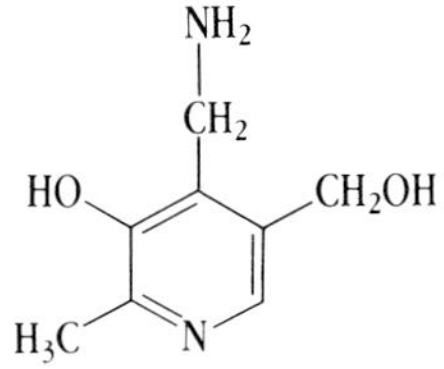

Eggs, liver, yeast, peas, beans, and milk

Dermatitis, apathy, irritability, and increased susceptibility to infections; convulsions in infants

(continued)

TABLE 18.1 (Continued)

Water-Soluble Vitamins (continued)

Structure and Name	Sources	Deficiency Symptoms
Pantothenic acid	Liver, eggs, yeast, and milk	(Possibly) emotional problems and gastrointestinal disturbances
Biotin	Beef liver, yeast, peanuts, chocolate, and eggs (although this vitamin cannot be synthesized by humans, it is a product of their intestinal bacteria)	Dermatitis
Vitamin C (ascorbic acid)	Citrus fruits, tomatoes, green peppers	Scurvy
Folic acid	Liver, kidney, mushrooms, yeast, and green leafy vegetables	Anemias (folic acid is used in the treatment of megaloblastic anemia, a condition characterized by giant red blood cells)

Pernicious anemia

Liver, meat, eggs, and fish
(not found in plants)

Vitamin B_{12} (cyanocobalamine)

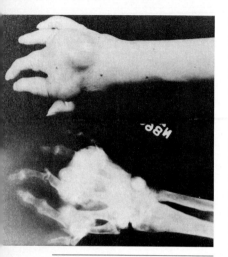

FIGURE 18.6
Too much vitamin D causes the deposit of bonelike material in joints. [Courtesy of Marc Moldawer M.D., The Methodist Hospital, Texas Medical Center, Houston.]

structural elements. One or two oxygen atoms are present, but the compounds as a whole are nonpolar. In contrast, a water-soluble vitamin contains a high proportion of the electronegative atoms oxygen and nitrogen, which can form hydrogen bonds to water, and therefore the molecule as a whole is soluble in water.

Fat-soluble vitamins dissolve in the fatty tissue of the body, and reserves of these vitamins can be stored here for future use. For example, while on an adequate diet, an adult can store several years' supply of vitamin A. If the diet becomes deficient in vitamin A, these reserves are mobilized, and the adult remains free of the deficiency disease for quite a while. On the other hand, a small child, who has not had the opportunity to build up a store of the vitamin, soon exhibits symptoms of the deficiency. Many children in developing countries are permanently blinded by vitamin A deficiency. Health workers in these countries often carry injectable solutions of vitamin A for emergency treatment of such cases. Because the fat-soluble vitamins are efficiently stored in the body, overdoses of these vitamins can result in adverse effects. Large excesses of vitamin A cause irritability, dry skin, and a feeling of pressure inside the head. Massive doses of the vitamin administered to pregnant rats result in malformed offspring. Vitamin D, like vitamin A, is fat-soluble. The effects of large overdoses are even more severe than with vitamin A. Too much vitamin D can cause pain in the bones, nausea, diarrhea, and weight loss (Figure 18.6). Bonelike material may be deposited in kidney tubules, in blood vessels, and in heart, stomach, and lung tissue. The amounts of both vitamin D and A in nonprescription capsules are regulated by the U.S. Food and Drug Administration.

The body has a very limited capacity to store water-soluble vitamins. It excretes anything over the amount that can be immediately used. Water-soluble vitamins must be taken in at frequent intervals, whereas a single dose of a fat-soluble vitamin can be used by the body over several weeks. A significant portion of the vitamin content of some foods can be lost when the food is cooked in water and then drained. The water-soluble vitamins go down the drain with the water.

In 1970, Linus Pauling (Figure 18.7) proposed a controversial treatment for the prevention of colds. Despite Pauling's two Nobel prizes (for chemistry in 1954 and for peace in 1962), his proposal was greeted with skepticism—even ridicule—by many in the scientific and medical communities. Pauling suggested that massive doses of vitamin C would prevent or greatly reduce the symptoms of the common cold. Daily doses of 250–15 000 mg were recommended, depending on the person and the circumstances. But studies have indicated that, with some exceptions, individuals normally excrete all but a small portion of such massive doses. Who is right, Pauling or his critics? The issue is still not settled. Pauling has theorized that human beings are actually a mutated form of mammals because, unlike most mammals, human

FIGURE 18.7
Linus Pauling, winner of two Nobel prizes.

beings cannot synthesize ascorbic acid (vitamin C). A 70-kg goat can produce 13 g of the compound daily, yet the recommended daily dietary allowance for human beings is 45 mg. This "abnormally" low level, according to Pauling, may be responsible for sudden infant death syndrome and for human vulnerability to viral diseases (including colds), cancer, heart and vascular diseases, and even drug addiction. Clinical tests of vitamin C therapy have produced conflcting results. In 1981, Pauling himself received a major grant from the National Cancer Institute to test his theories. It may be years before we know whether Pauling is right or not.

18.3 Hormones

Like vitamins, **hormones** are organic compounds. Unlike vitamins, hormones can be synthesized in the body. They are synthesized in the endocrine glands (Figure 18.8) and then discharged directly into the circulatory system. They serve as "chemical messengers." Hormones released in one part of the body signal profound physiological changes in other parts of the body. They cause reactions to speed up or slow down. In this way they control growth, metabolism, reproduction, and many other functions of body and mind.

Hormones are organic compounds that are synthesized in the body and used for regulation.

If we consider hormones as messengers, then the pituitary gland (hypophysis) must be viewed as the central dispatcher or control. Many of the pituitary hormones control the production of hormones by other endocrine glands. The removal of portions of the pituitary gland results in atrophy or a wasting away of other endocrine glands. Shut down the central control and you ultimately shut down much of the endocrine system. The pituitary itself responds to hormone signals from the hypothalamus. The hypothalamus acts as a go-between for the nervous system and the endocrine system and is triggered by nerve impulses that cause the gland to secrete hormones called **releasing factors**. The releasing factors make their way to the pituitary gland,

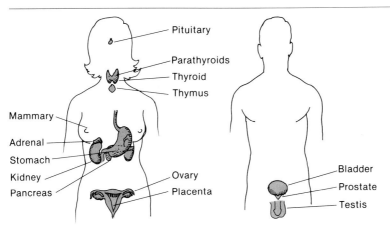

FIGURE 18.8

Approximate locations of endocrine glands in the human body.

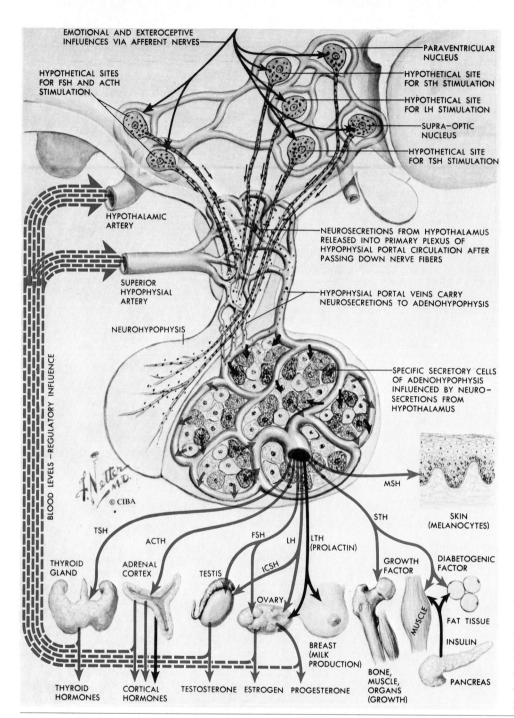

FIGURE 18.9
Schematic diagram of the interrelationship of the hypothalamus, the pituitary, and the organs on which they work. [Reproduced with permission from the CIBA Collection of Medical Illustrations by Frank Netter, M.D. Copyright © 1965 by the CIBA Pharmaceutical Company, Division of CIBA-Geigy Corporation. All rights reserved.]

where the production of pituitary hormones is stimulated. Hormonal regulation of bodily processes, then, may involve this sequence of actions: a nerve impulse signals the hypothalamus, which signals the pituitary, which signals some target endocrine gland, which signals some target tissue, which then responds in a specific way (Figure 18.9). Many important human hormones and their physiological effects are listed in Table 18.2.

18.3 Hormones

TABLE 18.2 Some Human Hormones and Their Physiological Effects

Name	Gland and Tissue	Chemical Nature	Effect
Various releasing and inhibitory factors	Hypothalamus	Peptide	Triggers or inhibits release of pituitary hormones
Human growth hormone (HGH)	Pituitary, anterior lobe	Protein	Controls the general body growth; controls bone growth
Thyroid-stimulating hormone (TSH)	Pituitary, anterior lobe	Protein	Stimulates growth of thyroid gland and production of thyroxine
Adrenal cortex-stimulating hormone (ACTH)	Pituitary, anterior lobe	Protein	Stimulates growth of the adrenal cortex and production of cortical hormones
Follicle-stimulating hormone (FSH)	Pituitary, anterior lobe	Protein	Stimulates growth of follicles in ovaries of females, sperm cells in testes of males
Luteinizing hormone (LH)	Pituitary, anterior lobe	Protein	Controls production and release of estrogens and progesterone from ovaries, testosterone from testes
Prolactin	Pituitary, anterior lobe	Protein	Maintains the production of estrogens and progesterone, stimulates the formation of milk
Vasopressin	Pituitary, posterior lobe	Protein	Stimulates contractions of smooth muscle; regulates water uptake by the kidneys
Oxytocin	Pituitary, posterior lobe	Protein	Stimulates contraction of the smooth muscle of the uterus; stimulates secretion of milk
Parathyroid	Parathyroid	Protein	Controls the metabolism of phosphorus and calcium
Thyroxine	Thyroid	Amino acid derivative	Increases rate of cellular metabolism
Insulin	Pancreas, beta cells	Protein	Increases cell use of glucose; increases glycogen storage
Glucagon	Pancreas, alpha cells	Protein	Stimulates conversion of liver glycogen to glucose
Cortisol	Adrenal gland, cortex	Steroid	Stimulates conversion of proteins to carbohydrates

(*continued*)

TABLE 18.2 (*Continued*)

Name	Gland and Tissue	Chemical Nature	Effect
Aldosterone	Adrenal gland, cortex	Steroid	Regulates salt metabolism; stimulates kidneys to retain Na^+ and excrete K^+
Epinephrine (adrenaline)	Adrenal gland, medulla	Amino acid derivative	Stimulates a variety of mechanisms to prepare the body for emergency action including the conversion of glycogen to glucose
Norepinephrine (noradrenaline)	Adrenal gland, medulla	Amino acid derivative	Stimulates sympathetic nervous system; constricts blood vessels, stimulates other glands
Estradiol	Ovary, follicle	Steroid	Stimulates female sex characteristics; regulates changes during menstrual cycle
Progesterone	Ovary, corpus luteum	Steroid	Regulates menstrual cycle; maintains pregnancy
Testosterone	Testis	Steroid	Stimulates and maintains male sex characteristics

Some hormones are amino acid derivatives. Thyroxine is an example.

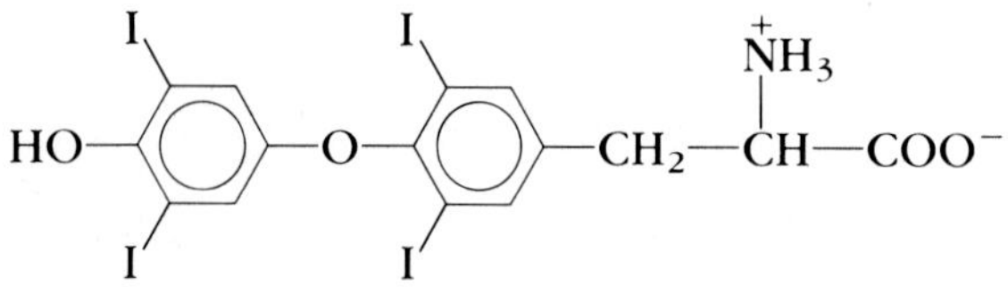

$$HO-\underset{I}{\overset{I}{\bigcirc}}-O-\underset{I}{\overset{I}{\bigcirc}}-CH_2-\underset{\underset{NH_3}{|}}{CH}-COO^-$$

Thyroxine (thyroid hormone)

Now you can see where the iodine required in our diets ends up (Section 18.1) and why the radioisotope iodine-131 is used for diagnosis and therapy involving the thyroid gland (Section 4.9). Other hormones, such as insulin, are peptides. Still others are steroids, some of which we discussed in Chapter 14.

To understand the operation of the endocrine system, let us look at an example of hormonal control, the menstrual cycle (Figure 18.10). To begin the cycle, the pituitary gland releases follicle-stimulating hormone (FSH), which triggers the production of the hormone estradiol by the ovary. FSH also triggers the maturing of an ovarian follicle.

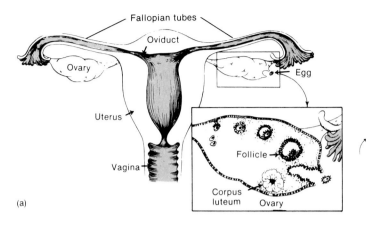
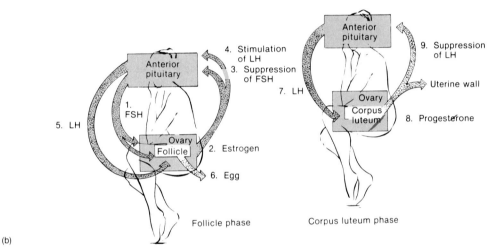
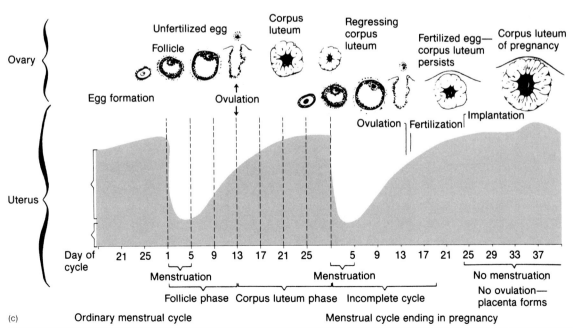

FIGURE 18.10
The menstrual cycle is controlled by hormones. The illustration opposite shows the normal menstrual cycle. Pregnancy—or the pseudo-pregnancy caused by the pill—prevents ovulation. (a) The female reproductive system. (b) Hormonal control of the female menstrual cycle. (c) Changes in the ovary and uterus. [Adapted with permission from Philip D. Sparks et al., *Student Study Guide for the Biological Sciences,* 3rd ed. Minneapolis: Burgess, 1973.]

The follicle is a structure within the ovary in which the ovum or egg develops. While the follicle matures, estradiol and its metabolites cause the uterus to begin to regenerate the lining lost at the end of the previous menstrual cycle. As the levels of estradiol build up in the blood, production of FSH by the pituitary gland is suppressed. The pituitary now begins to produce luteinizing hormone (LH). LH triggers ovulation—the release of the ovum from the mature follicle. The ovum begins a journey from the ovary, through the fallopian tube, to the uterus. The follicle that ruptured to release the ovum undergoes a change in structure and becomes the *corpus luteum*. The corpus luteum produces another hormone, progesterone, which suppresses production of LH and causes changes in the lining of the uterus. The uterine lining becomes richly supplied with blood and prepares in other ways for a possible pregnancy. If the ovum is not fertilized, the corpus luteum degenerates, the egg does not implant itself in the uterine lining, and hormone production by the ovary falls and triggers the onset of menstruation (the sloughing-off of the uterine lining). The low levels of both estrogens (estradiol and its metabolites) and progesterone signal the pituitary gland to start the cycle all over again.

If a sperm had managed to penetrate the ovum, then the fertilized egg would have been implanted in the uterine lining. The placenta develops to nurture the growing embryo and also produces hormones that stimulate the continued production of hormones by the corpus luteum. As long as hormone levels are maintained, menstruation does not take place, a new cycle does not begin, and no further ovulation occurs during the pregnancy.

FIGURE 18.11
Some synthetic steroids.

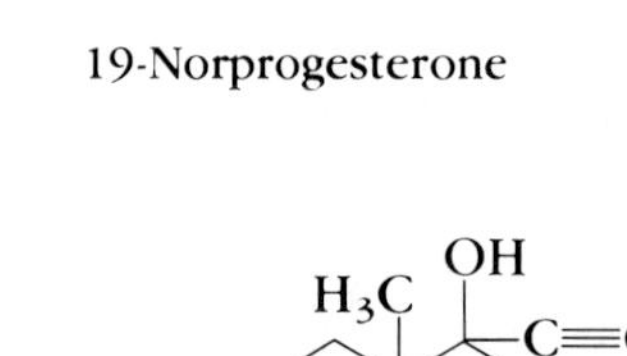

19-Norprogesterone

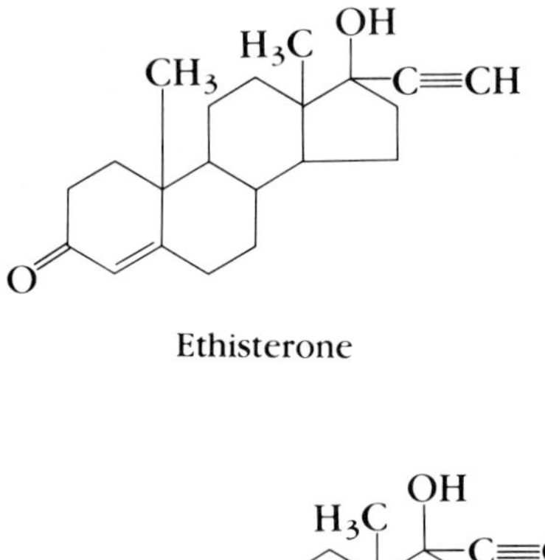

Ethisterone

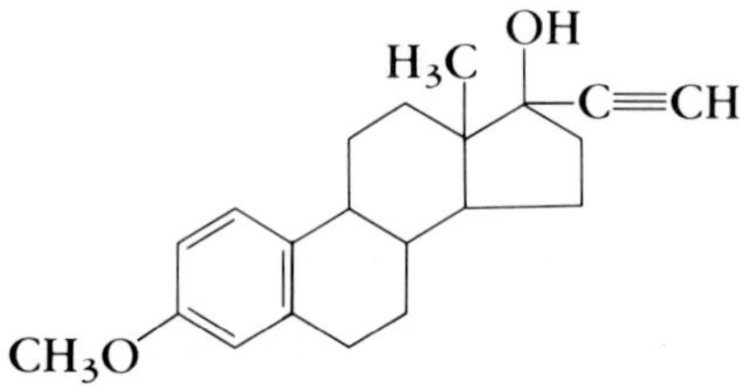

Mestranol

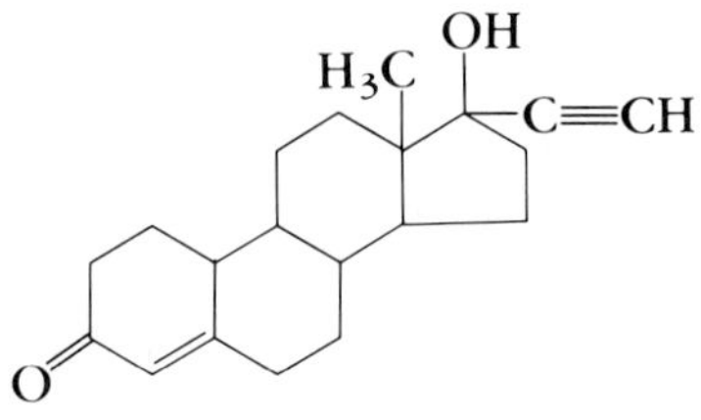

Norlutin

As details of the hormonal control of the female reproductive cycle were worked out, it became clear that voluntary control of the cycle is possible. Synthetic hormones, which could be taken orally and remain effective, were developed (Figure 18.11). Birth control pills work in a variety of ways. Some prevent ovulation, others prevent sperm and egg from meeting, and still others prevent implantation of the fertilized egg in the uterine lining. The medication may combine an estrogen with a progestin (a synthetic progesterone), or it may incorporate the progestin only. Strong evidence exists that the estrogens are responsible for a number of the undesirable side effects that appear in some women, including hypertension, acne, and abnormal bleeding. The pill may also cause blood-clotting in some women, but so does pregnancy. Such blood clots can block a blood vessel and cause death by stroke or coronary heart attack. The death rate associated with the use of the pill is about 3 in 100 000. This is only 10% of that associated with childbirth—about 30 in 100 000. For the general population, the pill is probably as safe as aspirin. Women who have blood with an abnormal tendency to clot should not take the pill, nor should those who experience any serious side effects. The so-called "minipill," containing only a small amount of progestin and no estrogen, was designed to avoid the problems associated with the combination pills.

18.4 Cyclic AMP: An Opportunity to See How Far We Have Come

Now you have an idea of what hormones do. The last topic we will take up in this text relates to how hormones do it. We want to consider this question primarily because the answer involves so many aspects of biochemistry.

First, let us look at the object of our consideration, **cyclic AMP**.

AMP stands for adenosine monophosphate, one of the nucleotides discussed in the last chapter. The molecule above is called cyclic AMP

TABLE 18.3 Hormonal Action on Cyclic AMP Levels

Tissue	Hormone	Principal Response
Bone	Parathyroid hormone	Calcium resorption
Muscle	Epinephrine	Glycogenolysis
Fat	Epinephrine	Lipolysis
	Adrenocorticotrophic hormone	Lipolysis
	Glucagon	Lipolysis
Brain	Norepinephrine	Discharge of Purkinje cells
Thyroid	Thyroid-stimulating hormone	Thyroxine secretion
Heart	Epinephrine	Increased contractility
Liver	Epinephrine	Glycogenolysis
Kidney	Parathyroid hormone	Phosphate excretion
	Vasopressin	Water reabsorption
Adrenal	Adrenocorticotrophic hormone	Hydrocortisone secretion
Ovary	Luteinizing hormone	Progesterone secretion

Modified from "Cyclic AMP," Ira Pastan. Copyright © 1972 by Scientific American, Inc. All rights reserved.

(cAMP) because the phosphate group has formed an intramolecular double ester. The hydroxyl groups at C-3 and C-5 of the sugar ring have both reacted to form ester links with the phosphate groups, thus creating an additional ring in this nucleotide.

In many cases, when a hormone is released, the target tissue shows a sudden increase in the level of AMP. Table 18.3 shows the range of hormones and tissues in which this effect has been noted. It has also been noted that many hormones never make their way into the cells of the target tissue but are stopped by the membranes surrouding the cells. Still the hormone triggers profound changes in the chemistry of the target cells. How?

Recall the structure of cell membranes (Section 14.5). They consist of phospholipid bilayers in which proteins are embedded. In the target cells, one of the protein particles has a receptor site for the appropriate hormone; that is, a portion of the protein is able to bind the hormone when the hormone arrives at the cell. The binding site is located on the outside of the cell membrane. When the hormone is bound to the protein, the conformation of the protein changes. (Recall the action of an inhibitor on an enzyme, Section 15.9.) As the shape of the binding protein changes, adjacent proteins, also embedded in the bilayer, change their shape also. One of these adjacent proteins is an enzyme responsible for the formation of AMP. A change in its conformation activates the enzyme, and AMP is produced on the inside of the cell (Figure 18.12).

18.4 Cyclic AMP: An Opportunity to See How Far We Have Come

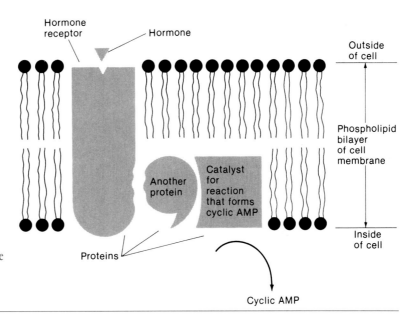

FIGURE 18.12
The binding of a hormone to a receptor site on the outside of a cell results in the activation of an enzyme that controls synthesis of cyclic AMP on the inside of the cell.

What does cyclic AMP do? In the cytoplasm of the cell, AMP locates a protein to which it can bind. The binding of the AMP changes the conformation of this protein, which adjusts by splitting off a subunit of itself (Figure 18.13). This subunit, once released, is an active enzyme ready to catalyze the reactions that produce the effects we observe. Via this mechanism, the hormone epinephrine ultimately causes the polysaccharide glycogen to convert to the monosaccharide glucose.

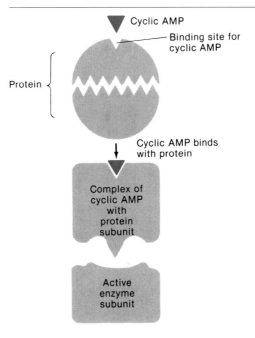

FIGURE 18.13
When cyclic AMP binds to a protein within a cell, the protein dissociates into subunits, some with catalytic activity.

There. We have mentioned hormones and nucleotides and enzymes and carbohydrates just to explain why you still have enough energy to pick up this book even though you have not eaten for several hours. Think of the complex chemistry involved in making sure that your blood sugar levels stay fairly constant. In a healty individual, all this happens routinely, in perfect balance, and with no conscious direction. That is just one of the myriad interrelated biochemical processes that life requires. That it all works, and works so well, is nothing short of miraculous.

CHAPTER SUMMARY

Vitamins and minerals are dietary ingredients (organic and inorganic, respectively) that are required in much smaller quantities than the major classes of foodstuffs. Hormones are organic compounds that are produced in small amounts by the body's endocrine glands.

Minerals essential for human health include iodine, which is required for the formation of thyroid hormone and the prevention of goiter; iron, which is incorporated in a variety of complex biomolecules, including hemoglobin; calcium and phosphorus, which are incorporated in the structural material of bone and teeth; and ions like Na^+ and Cl^-, among others, that are dissolved in body fluids. A number of ions (such as copper, zinc, and manganese) act as enzyme cofactors.

Vitamins are specific organic compounds that are required for the health of an organism but cannot be synthesized by the organism. Human vitamins are divided into two general categories: fat-soluble vitamins (A, D, E, and K) and water-soluble vitamins (the B complex group and vitamin C). Fat-soluble vitamins incorporate a high percentage of nonpolar groups, and water-soluble vitamins are characteristically polar molecules incorporating several oxygen or nitrogen atoms. Because water-soluble vitamins are soluble in the fluids of the body, overdoses are typically quickly excreted from the body. On the other hand, fat-soluble vitamins can build up in the fatty tissues of the body, and thus toxic overdoses are a possibility. Because excesses are rapidly excreted, water-soluble vitamins must be replenished frequently. Stored fat-soluble vitamins can provide for the body's needs over longer periods of time.

Table 18.1 lists the vitamins of both classes along with their deficiency symptoms.

Hormones are organic compounds secreted by the endocrine glands as chemical messengers. Hormones are transported via the blood to target organs, where they trigger specific effects. Overall control of the endocrine system is centered in the hypothalamus, which secretes releasing factors in response to nerve impulses. The releasing factors stimulate the production of various hormones by the pituitary gland, and these hormones in turn control the production of hormones by other endocrine glands. Important human hormones and their effects are listed in Table 18.2. The female reproductive cycle offers an example of hormonal control.

Cyclic AMP is a cyclic ester of adenosine monophosphate that is involved in the transmission of hormonal messages. The binding of a hormone to a receptor located on the outside of a cell membrane can result in the release of AMP inside the cell. The AMP then activates an enzyme responsible for catalyzing the reactions that are associated with the particular hormone.

CHAPTER QUIZ

1. Which is not known to be an essential mineral?
   a. As   b. Ni   c. Mo   d. V
2. Beriberi is the result of a deficiency of which vitamin?
   a. A   b. B   c. C   d. D
3. Vitamins are classified by
   a. structure   b. functional group
   c. solubility   d. reactivity

4. Which is true for water-soluble vitamins?
   a. They have long body storage lives.
   b. They must be taken frequently.
   c. They are made in the body.
   d. They are inorganic.
5. Egg yolk is a good source of which vitamin?
   a. A   b. B   c. D   d. E
6. How are hormones different from vitamins? They are
   a. organic   b. water-soluble
   c. synthesized by the body
   d. complex structures
7. Many hormones are
   a. acids   b. bases
   c. peptides   d. metal complexes
8. The effect of the hormone cortisol is to
   a. regulate salt metabolism
   b. stimulate conversion of proteins to carbohydrates
   c. increase the rate of cellular metabolism
   d. control the metabolism of P and Ca
9. Which is not a synthetic hormone?
   a. progesterone   b. mestranol
   c. norlutin   d. ethisterone
10. Which is not true about cyclic AMP?
    a. It contains phosphate.
    b. It is an organic structure.
    c. It changes conformation of proteins.
    d. It regulates smooth muscles.

PROBLEMS

18.1 Minerals

1. Three classes of compounds were discussed in this chapter: minerals, vitamins, and hormones. By definition, which of these is classified as organic, and which as inorganic?
2. Which of the three classes of compounds discussed in this chapter are required in the diet?
3. Indicate a biological function for each of the following.
   a. iodine   b. iron   c. calcium
4. Phosphorus is found throughout the body in a variety of biologically important compounds. Name some of the phosphorus-containing compounds discussed in Chapter 13–18.

5. Which of the following minerals would you expect to find in relatively large amounts in the human body? Explain your reasoning.
   a. Ca   b. Cl   c. Co   d. Mo
   e. Na   f. P   g. Zn

18.2 Vitamins

6. Compare and contrast vitamins, minerals, and hormones with respect to effective amounts.
7. Match the compound with its designation as a vitamin.

Compound	Designation
a. Ascorbic acid	1. Vitamin A
b. Calciferol	2. Vitamin B_{12}
c. Cyanocobalamine	3. Vitamin C
d. Retinol	4. Vitamin D
e. Tocopherol	5. Vitamin E

8. Which of the following are B vitamins?
   a. folic acid   b. riboflavin
   c. insulin   d. thiamine
   e. niacin
9. Identify the vitamin deficiency associated with these diseases.
   a. scurvy
   b. rickets
   c. night blindness
10. In each case, identify the deficiency disease associated with a diet lacking in the indicated vitamin.
    a. vitamin B_1 (thiamine)
    b. niacin
    c. vitamin B_{12} (cyanocobalamine)
11. What is the structural difference between water-soluble and fat-soluble vitamins?
12. Identify the following vitamins as water-soluble or fat-soluble.
    a. vitamin A   b. vitamin B_6
    c. vitamin B_{12} d. vitamin C
    e. vitamin K
13. Identify the following vitamins as water-soluble or fat-soluble.
    a. calciferol   b. niacin
    c. riboflavin   d. tocopherol

14. Structural formulas for two vitamins are shown below. Identify each of them as water-soluble or fat-soluble.

$$HOCH_2C\underset{CH_3}{\overset{CH_3}{|}}{-}\underset{}{\overset{OH}{\underset{|}{CH}}}{-}\underset{}{\overset{O}{\underset{||}{C}}}{-}NHCH_2CH_2COOH$$

Formula a

$$\text{(ring structure)}\ CH=CHC(CH_3)=CHCH=CHC(CH_3)=CHCH_2OH$$

Formula b

15. Which is more likely to be dangerous—an excess of a water-soluble vitamin or an excess of a fat-soluble vitamin? Why?
16. Could a "one-a-month" vitamin pill satisfy all human requirements? Explain your answer.
17. On what basis did some scientists object to the vitamin C therapy for colds proposed by Linus Pauling? Can you offer any supporting arguments for Pauling's position?

18.3 Hormones

18. Match the hormone to the gland that produces it.

Hormone	Gland
a. Thyroxine	1. Corpus luteum
b. FSH	2. Pancreas
c. Progesterone	3. Thyroid
d. Insulin	4. Pituitary
e. Epinephrine	5. Adrenal

19. What gland produces releasing factors?
20. What gland is the target of releasing factors?
21. Describe the general sequence of events by which a nerve impulse is translated by the endocrine system into a changed physiological state.

22. Match the hormone with the effect.

Hormone	Effect
a. FSH	1. Stimulates male sex characteristics
b. Prolactin	2. Stimulates female sex characteristics
c. Thyroxine	3. Regulates cell uptake of glucose
d. Testosterone	4. Stimulates milk production
e. Estradiol	5. Increases rate of cellular metabolism
f. Insulin	6. Stimulates growth of follicle in ovary

23. Answer the following questions about the menstrual cycle.
    a. What are the roles of the pituitary hormones FSH and LH?
    b. What is the relationship of the ovarian follicle and the corpus luteum?
    c. Which ovarian hormone affects the lining of the uterus in the early stages of the cycle? Which produces the changes in this lining in the later stages of the cycle?
    d. What triggers the start of a new cycle?
24. What is the general structural classification of the compounds incorporated in birth control pills?
25. Which of the two categories of synthetic hormones, estrogens or progestins, is associated with undesirable side effects? Name some of these side effects.

18.4 Cyclic AMP: An Opportunity to See How Far We Have Come

26. Draw the structure of cyclic AMP.
27. When a hormone causes an increase in AMP levels within the cells, where is the binding site for the hormone?
28. Explain how the binding of a hormone at its receptor site can release AMP within a target cell.
29. Outline a mechanism by which AMP may increase the rates of cellular reactions.

APPENDIX A THE INTERNATIONAL SYSTEM OF MEASUREMENT

Metric measurement was discussed in some detail in Chapter 1. Further discussion and additional tables are provided here.

The standard unit of length in the International System of measurement is the **meter**. This distance was once meant to be 0.000 000 1 of the Earth's quadrant, that is, of the distance from the North Pole to the equator measured along a meridian. The quadrant was difficult to measure accurately. Consequently, for many years the meter was defined as the distance between two etched lines on a metal bar (made of a platinum-iridium alloy) kept in the International Bureau of Weights and Measures at Sèvres, France. Today, the meter is defined even more precisely as being 1 650 763.73 times the wavelength of the orange-red line in the spectrum of krypton-86.

The primary unit of mass is the **kilogram** (1 kg = 1000 g). It is based on a standard platinum-iridium bar kept at the International Bureau of Weights and Measures. The **gram** is a more convenient unit for many chemical operations.

The derived SI unit of volume is the cubic meter. The unit more frequently employed in chemistry, however, is the **liter** (1 L = 0.001 m^3). All other SI units of length, mass, and volume are derived from these basic units.

APPENDIX A   THE INTERNATIONAL SYSTEM OF MEASUREMENT

TABLE A.1   Some SI Prefixes and Their Relationship to the Basic Units

Prefix	Abbreviation	Connotation
Pico-	p	$0.000\,000\,000\,001 \times$ (or $10^{-12} \times$)
Nano-	n	$0.000\,000\,001 \times$ (or $10^{-9} \times$)
Micro-	u	$0.000\,001 \times$ (or $10^{-6} \times$)
Milli-	m	$0.001 \times$ (or $10^{-3} \times$)
Centi-	c	$0.01 \times$ (or $10^{-2} \times$)
Deci-	d	$0.1 \times$ (or $10^{-1} \times$)
Deka-	da	$10 \times$ (or $10^{1} \times$)
Hecto-	h	$100 \times$ (or $10^{2} \times$)
Kilo-	k	$1\,000 \times$ (or $10^{3} \times$)
Mega-	M	$1\,000\,000 \times$ (or $10^{6} \times$)
Giga-	G	$1\,000\,000\,000 \times$ (or $10^{9} \times$)
Tera-	T	$1\,000\,000\,000\,000 \times$ (or $10^{12} \times$)

TABLE A.2   Some Metric Units of Length

1 kilometer (km) = 1000 meters (m)
1 meter (m) = 100 centimeters (cm)
1 centimeter (cm) = 10 millimeters (mm)
1 millimeter (mm) = 1000 micrometers (um)

TABLE A.3   Some Metric Units of Mass

1 kilogram (kg) = 1000 grams (g)
1 gram (g) = 1000 milligrams (mg)
1 milligram (mg) = 1000 micrograms (ug)

TABLE A.4   Some Metric Units of Volume

1 liter (L) = 1000 milliliters (mL)
1 milliliter (mL) = 1000 microliters (uL)
1 milliliter (mL) = 1 cubic centimeter (cm^3)

Appendix A   The International System of Measurement

TABLE A.5 Some Common Metric Conversions

Length

1 mile (mi) = 1.61 kilometers (km)	1 km = 0.621 mi
1 yard (yd) = 0.914 meter (m)	1 m = 1.094 yd
1 inch (in.) = 2.54 centimeters (cm)	1 cm = 0.394 in.

Mass

1 pound (lb) = 454 grams (g)	1 g = 0.0022 lb
1 ounce (oz) = 28.4 grams (g)	1 g = 0.035 oz
1 pound (lb) = 0.454 kilogram (kg)	1 kg = 2.20 lb
1 grain (gr) = 0.0648 gram (g)	1 g = 15.43 gr
1 carat (car) = 200 milligrams (mg)	1 mg = 0.005 car

Volume

1 U.S. quart (qt) = 0.946 liter (L)	1 L = 1.057 qt
1 U.S. pint (pt) = 0.473 liter (L)	1 L = 2.114 pt
1 fluid ounce (fl oz) = 29.6 milliliters (mL)	1 mL = 0.0338 fl oz
1 gallon (gal) = 3.78 liters (L)	1 L = 0.264 gal

TABLE A.6 Some Conversion Units for Pressure

1 millimeter of mercury (mm Hg) = 1 torr
1 atmosphere (atm) = 760 millimeters of mercury (mm Hg)
= 760 torr
1 atmosphere (atm) = 29.9 inches of mercury (in. Hg)
= 14.7 pounds per square inch (psi)
= 101 kPa

TABLE A.7 Some Conversion Units for Energy

1 calorie (cal) = 4.184 joules (J)
1 British thermal unit (Btu) = 1053 joules (J)
= 252 calories (cal)
1 food "Calorie" = 1 kilocalorie (kcal)
= 1000 calories (cal)
= 4184 joules (J)

TABLE A.8 Some Temperature Equivalents[a]

Phenomenon	Fahrenheit (°F)	Celsius (°C)
Absolute zero	−459.69	−273.16
Nitrogen boils/liquefies	−320.4	−195.8
Carbon dioxides solidifies/sublimes	−109.3	−78.5
Bitter cold night, northern Minnesota	−40	−40
Cold night, Indiana	0	−18
Water freezes/ice melts	32	0
Pleasant room temperature	72	22
Body temperature	98.6	37.0
Very hot day	100	38
Water boils/steam condenses	212	100
Temperature for baking biscuits	450	232

[a] Keep these equations in mind:

$$°F = \frac{9}{5}(°C) + 32 \qquad °C = (°F - 32)\frac{5}{9}$$

APPENDIX B SIGNIFICANT FIGURES

Unlike counting, measurement is never exact. You can *count* exactly 10 people in a room. If you asked each of those people to *measure* the length of the room to the nearest 0.01 m, however, the values they determine are likely to differ slightly. Table B.1 presents such a set of measurements.

Note that all 10 students agree on the first three digits of the measurement; differences occur in the fourth digit. Which values are correct? Actually, all are accurate within the accepted range of uncertainty for this physical measurement. The accuracy of measurement depends on the type of measuring instrument and the skill and care of the person making the measurement. Measured values are usually recorded with the last digit regarded as uncertain. The data in Table B.1 allow us to state that the length of the room is between 14.1 m and 14.2 m, but we are not sure of the fourth digit. The measurements in the table have four *significant figures,* which means that the first three are known with confidence and the fourth conveys an approximate value. **Significant figures** include all digits known with certainty plus one uncertain digit.

In any properly reported measurement, all nonzero digits are significant. The zero presents problems, however, because it can be used in two ways: to position the decimal point or to indicate a measured value. For zeros, follow these rules.

TABLE B.1 A Set of Measurements of the Length of a Room

Student	Length (m)	Student	Length (m)
1	14.14	6	14.14
2	14.15	7	14.17
3	14.17	8	14.17
4	14.14	9	14.16
5	14.16	10	14.17

1. A zero between two other digits is always significant.
   Examples: The number 1107 contains four significant figures.
   The number 50.002 contains five significant figures.
2. Zeros to the left of *all* nonzero digits are not significant.
   Examples: The number 0.000**163** has three significant figures.
   The number 0.0**6801** has four significant figures.
3. Zeros that are *both* to the right of the decimal point *and* to the right of nonzero digits are significant.
   Examples: The number 0.**2000** has four significant figures.
   The number 0.0**50120** has five significant figures.
   The number **802.760** has six significant figures.
4. Zeros in numbers such as 40 000 (that is, zeros to the right of *all* nonzero digits in a number that is written without a decimal point) may or may not be significant. Without more information, we simply do not know whether 40 000 was measured to the nearest unit or ten or hundred or thousand or ten thousand. To avoid this confusion, scientists use exponential notation (Appendix C) for writing numbers. In exponential notation, 40 000 would be recorded as 4×10^4 or 4.0×10^4 or 4.0000×10^4 to indicate one, two, and five significant figures, respectively.

Addition and Subtraction

In addition and subtraction, the result should contain no more digits to the right of the decimal point than the quantity that has the least number of or fewest digits to the right of the decimal point. Align the quantities to be added or subtracted on the decimal point, then perform the operation, assuming that blank spaces are zeros. Determine the correct number of digits after the decimal point in the answer, and round off to this number. In rounding off, you should increase the last significant figure by one if the following digit is five through nine.

EXAMPLE B.1

Add the following numbers: 49.146, 72.13, 5.9432.
  Align the numbers on the decimal point and carry out the addition.

$$\begin{array}{r} 49.146 \\ 72.13 \\ 5.9432 \\ \hline 127.2192 \end{array}$$

Appendix B Significant Figures

The quantity with the fewest digits after the decimal point is 72.13. The answer should have only two digits after the decimal point. Since the third digit after the decimal point is 9, the second digit after the decimal point should be rounded up to 2. Correct answer: 127.22.

EXAMPLE B.2

Add the following numbers: 744, 2.6, 14.812.

$$\begin{array}{r} 744 \\ 2.6 \\ 14.812 \\ \hline 761.412 \end{array} \longrightarrow 761$$

The first quantity has no digits to the right of the decimal point (which is understood to be at the right of the digits). The answer must therefore be rounded so that it, too, contains no digits to the right of the decimal point.

EXAMPLE B.3

Subtract 9.143 from 71.12496.

$$\begin{array}{r} 71.124\,96 \\ -9.143 \\ \hline 61.981\,96 \end{array} \longrightarrow 61.982$$

Since the second quantity has only three digits to the right of the decimal point, so must the answer.

Multiplication and Division

In multiplication and division, our answers can have no more significant figures than the factor that has the least number of significant figures. In these operations, the *position* of the decimal point makes no difference.

EXAMPLE B.4

Multiply 10.4 by 3.1416.

$$10.4 \times 3.1416 = 32.67264 \longrightarrow 32.7$$

The answer has only three significant figures because the first term has only three.

EXAMPLE B.5

Divide 5973 by 3.0.

$$\frac{5.973}{3.0} = 1.991 \longrightarrow 2.0$$

The answer has only two significant figures because the divisor has only two.

Exact Values

Some quantities are not measured but defined. A kilometer is defined as 1000 meters: 1 km = 1000 m. Similarly, 1 foot can be defined as 12 inches: 1 ft = 12 in. The "1 km" should not regarded as containing one significant figure; nor should "12 in." be considered to have two significant figures. In fact, these values can be considered to have an infinite number of significant figures (1.000 000 000 000 000 0 . . .) or, more correctly, to be *exact*. Such defined values are frequently used as conversion factors in problems (Chapter 1). When you are determining the number of significant figures for the answer to a problem, ignore such exact values. Use only the measured quantities in the problem to determine the number of significant figures in the answer.

PROBLEMS

Perform the indicated operations and give answers with the proper number of significant figures.

1. a. 48.2 m + 3.82 m + 48.4394 m
    b. 151 g + 2.39 g + 0.0124 g
    c. 15.436 mL + 9.1 mL + 105 mL
2. a. 100.53 cm − 46.1 cm
    b. 451 g − 15.46 g
    c. 19.71 L − 10.4 L
3. a. 73 m × 1.340 m × 0.41 m
    b. 0.137 cm × 1.43 cm
    c. 3.146 cm × 5.4 cm
4. a. $\dfrac{5.179 \text{ g}}{4.6 \text{ mL}}$ b. $\dfrac{4561 \text{ g}}{3.1 \text{ mol}}$ c. $\dfrac{40.00 \text{ g}}{3.2 \text{ mL}}$
5. $\dfrac{1.426 \text{ mL} \times 373 \text{ K}}{204 \text{ K}}$

APPENDIX C EXPONENTIAL NOTATION

Scientists often use numbers that are so large—or so small—that they boggle the mind. For example, light travels at 30 000 000 000 cm/s. There are 602 300 000 000 000 000 000 000 carbon atoms in 12 g of carbon. On the small side, the diameter of an atom is about 0.000 000 000 1 m. The diameter of an atomic nucleus is about 0.000 000 000 000 001 m.

It is obviously difficult to keep track of the zeros in such quantities. Scientists find it convenient to express such numbers as *powers of ten*. Tables C.1 and C.2 contain partial lists of such numbers.

The speed of light is usually expressed as 3×10^{10} (i.e., $3 \times 10 \times 10 \times 10 \times 10 \times 10 \times 10 \times 10 \times 10 \times 10 \times 10$) cm/s. The mass of an atom of cesium (Cs) is expressed as 2.21×10^{-22} g, that is, as

$$2.21 \times \frac{1}{10\,000\,000\,000\,000\,000\,000\,000} \text{ g}$$

Numbers such as 10^6 are called exponential numbers, where 10 is the *base* and 6 is the *exponent*. Numbers in the form 6.02×10^{23} are said to be written in *scientific notation*.

TABLE C.1 Positive Powers of Ten

$10^0 = 1$
$10^1 = 10$
$10^2 = 10 \times 10 = 100$
$10^3 = 10 \times 10 \times 10 = 1000$
$10^4 = 10 \times 10 \times 10 \times 10 = 10\,000$
$10^5 = 10 \times 10 \times 10 \times 10 \times 10 = 100\,000$
$10^6 = 10 \times 10 \times 10 \times 10 \times 10 \times 10 = 1\,000\,000$
⋮
$10^{23} = 100\,000\,000\,000\,000\,000\,000\,000$

TABLE C.2 Negative Powers of Ten

$10^{-1} = 1/10 = 0.1$
$10^{-2} = 1/100 = 0.01$
$10^{-3} = 1/1\,000 = 0.001$
$10^{-4} = 1/10\,000 = 0.000\,1$
$10^{-5} = 1/100\,000 = 0.000\,01$
$10^{-6} = 1/1\,000\,000 = 0.000\,001$
⋮
$10^{-13} = 1/10\,000\,000\,000\,000 = 0.000\,000\,000\,000\,1$

APPENDIX C EXPONENTIAL NOTATION

Exponential numbers are often used in calculations. The most common operations are multiplication and division. Two rules must be followed.

1. To *multiply* exponentials, *add* the exponents.
2. To *divide* exponentials, *subtract* the exponents.

These rules can be stated algebraically as

$$(x^a)(x^b) = x^{a+b}$$

$$\frac{x^a}{x^b} = x^{a-b}$$

Some examples follow.

$$(10^6)(10^4) = 10^{6+4} = 10^{10}$$
$$(10^6)(10^{-4}) = 10^{6+(-4)} = 10^{6-4} = 10^2$$
$$(10^{-5})(10^2) = 10^{(-5)+2} = 10^{-5+2} = 10^{-3}$$
$$(10^{-7})(10^{-3}) = 10^{(-7)+(-3)} = 10^{-7-3} = 10^{-10}$$
$$\frac{10^{14}}{10^6} = 10^{14-6} = 10^8$$
$$\frac{10^6}{10^{23}} = 10^{6-23} = 10^{-17}$$
$$\frac{10^{-10}}{10^{-6}} = 10^{(-10)-(-6)} = 10^{-10+6} = 10^{-4}$$
$$\frac{10^3}{10^{-2}} = 10^{3-(-2)} = 10^{3+2} = 10^5$$
$$\frac{10^{-8}}{10^4} = 10^{(-8)-4} = 10^{-12}$$
$$\frac{10^7}{10^7} = 10^{7-7} = 10^0 = 1$$

Problems involving both a coefficient and an exponential are solved by multiplying (or dividing) coefficients and exponentials separately.

Appendix C Exponential Notation   571

> **EXAMPLE C.1**
>
> To what is the following expression equivalent?
>
> $$(1.2 \times 10^5)(2.0 \times 10^9)$$
>
> First, mulitply the coefficients.
>
> $$1.2 \times 2.0 = 2.4$$
>
> Then multiply the exponentials.
>
> $$10^5 \times 10^9 = 10^{5+9} = 10^{14}$$
>
> The complete answer is
>
> $$2.4 \times 10^{14}$$

> **EXAMPLE C.2**
>
> To what is the following expression equivalent?
>
> $$\frac{8.0 \times 10^{11}}{1.6 \times 10^4}$$
>
> First, divide the coefficients
>
> $$\frac{8.0}{1.6} = 5.0$$
>
> Then divide the exponentials.
>
> $$\frac{10^{11}}{10^4} = 10^{11-4} = 10^7$$
>
> The answer is
>
> $$5.0 \times 10^7$$

EXAMPLE C.3

Give an equivalent for the following expression.

$$\frac{1.2 \times 10^{14}}{4.0 \times 10^6}$$

Before carrying out the division, it is convenient to rewrite the dividend (the numerator) so that the coefficient is larger than that of the divisor (the denominator).

$$1.2 \times 10^{14} = 12 \times 10^{13}$$

Note that the coefficient was made larger by a factor of 10, and the exponential was made smaller by a factor of 10. The quantity as a whole is unchanged. Now divide.

$$\frac{12 \times 10^{13}}{4.0 \times 10^6} = 3.0 \times 10^7$$

EXAMPLE C.4

Give an equivalent for the following expression.

$$\frac{(3 \times 10^7)(8 \times 10^{-3})}{(6 \times 10^2)(2 \times 10^{-1})}$$

In problems such as this, you can carry out the multiplications specified in the numerator and in the denominator separately and then divide the resulting numbers.

$$(3 \times 10^7)(8 \times 10^{-3}) = 24 \times 10^4$$
$$(6 \times 10^2)(2 \times 10^{-1}) = 12 \times 10^1$$
$$\frac{24 \times 10^4}{12 \times 10^1} = 2 \times 10^3$$

The multiplications and divisions in problems like this can be carried out in any convenient order.

Appendix C Exponential Notation

Only one other mathematical function involving exponentials is of importance to us. What happens when you raise an exponential to a power? You just multiply the exponent by the power. To illustrate:

$$(10^3)^3 = 10^9$$
$$(10^{-2})^4 = 10^{-8}$$
$$(10^{-5})^{-3} = 10^{15}$$

If the exponential is combined with a coefficient, the two parts of the number are dealt with separately, as in the following example.

$$(2 \times 10^3)^2 = 2^2 \times (10^3)^2 = 4 \times 10^6$$

For a further discussion of—and more practice with—exponential numbers, see one of the following references.

▶ Goldish, Dorothy M., *Basic Mathematics for Beginning Chemistry*, 4th ed. New York: Macmillan, 1990.
▶ Loebel, Arnold B., *Chemical Problem Solving by Dimensional Analysis*, 3rd ed. Boston: Houghton Mifflin, 1987.

PROBLEMS

1. Express each of the following in scientific notation.
   a. 0.00001
   b. 10 000 000
   c. 0.0034
   d. 0.000 0107
   e. 4 500 000 000
   f. 406 000
   g. 0.02
   h. 124×10^3

2. Carry out the following operations. Express the numbers in scientific notation.
   a. $(4.5 \times 10^{13})(1.9 \times 10^{-5})$
   b. $(6.2 \times 10^{-5})(4.1 \times 10^{-12})$
   c. $(2.1 \times 10^{-6})^2$
   d. $\dfrac{4.6 \times 10^{-12}}{2.1 \times 10^3}$
   e. $\dfrac{9.3 \times 10^9}{3.7 \times 10^{-7}}$
   f. $\dfrac{2.1 \times 10^5}{9.8 \times 10^7}$
   g. $\dfrac{4.3 \times 10^{-7}}{7.6 \times 10^{22}}$

Glossary

absolute scale A temperature scale in which the zero point is the coldest temperature possible, or absolute zero.

adsorbed hydrogen A condensed form of hydrogen collected in large volumes on the surface of certain metals such as platinum, palladium, and nickel.

acetal Product of hemiacetal–alcohol reaction; a 1,1-diether.

acid A proton donor.

acid anhydride A substance that reacts with water to produce an acid; a nonmetal oxide.

acid rain Rain having a pH less than 5.6.

acidosis Condition that results when the pH of the blood falls below 7.35 and oxygen transport is hindered.

actin A major protein in muscle, active in relaxation and contraction of the muscle.

activation energy Minimum energy necessary to initiate a reaction.

active site Portion of the enzyme into which substrate fits.

actomyosin The contractile protein of which muscles are made; contains actin and myosin.

addition polymerization A process in which monomers add to one another in such a way that the polymeric product contains all the atoms of the starting monomers.

adequate protein A protein that supplies all the essential amino acids in the quantities needed for the growth and repair of body tissues.

adipose tissue Connective tissue where fat is stored.

adrenaline Epinephrine, a secondary amine causing increase in blood pressure.

aerobic process A process that occurs in the presence of oxygen.

aerosol Particles of 1 micrometer diameter, or less, dispersed in air.

aflatoxins Toxic, carcinogenic compounds produced by molds growing on stored peanuts and grains.

agonist A molecule that fits and activates a specific receptor.

alchemy A mystical chemistry that flourished in Europe during the Middle Ages (A.D. 500 to 1500).

alcohols Organic compounds containing the —OH group.

aldehyde A compound that has a hydrogen attached to a carbonyl group.

aldohexose A six-carbon aldehyde monosaccharide.

aldopentose A five-carbon aldehyde monosaccharide.

aldose Sugar with an aldehyde group.

alkaloid A nitrogen-containing organic compound obtained from plants.

alkalosis A physiological condition in which the pH of the blood rises above 7.45.

alkali metal An element in Group IA of the periodic table.

alkaline earth metal An element in Group IIA of the periodic table.

alkane A hydrocarbon with only single bonds; a saturated hydrocarbon.

alkene A hydrocarbon containing one or more double bonds.

alkoxy RO— group.

alkyne A hydrocarbon containing one or more triple bonds.

alkyl A saturated hydrocarbon.

allotropes Different forms of the same element in the same state.

alloy A Mixture of two or more elements, at least one of which is a metal; an alloy has metallic properties.

alpha ray A radioactive beam consisting of helium nuclei.

amide An organic compound having the functional group CON in which the carbon is double-bonded to the oxygen and single-bonded to the nitrogen.

amine A compound that contains the elements carbon, hydrogen, and nitrogen; can be viewed as derived from ammonia by replacement of one, two, or three of the hydrogens by alkyl groups.

amino acid An organic compound that contains both an amino group and a carboxylic acid group; amino acids combine to produce proteins.

amino group An NH_2 unit.

amphetamines Synthetic amines used as stimulants.

amylase An enzyme that catalyzes the hydrolysis of starches.

amylopectin A starch with branched chains of glucose units.

amylose A starch with the glucose units joined in a continuous chain.

anabolic steroid A drug that aids in the building (anabolism) of body proteins and thus of muscle tissue.

anaerobic process A process that occurs in the absence of oxygen.

analgesic A pain reliever.

androgen A male sex hormone.

anhydrous Without water.

aniline A benzene ring with a primary amine attached.

anion A negatively charged ion.

anode The electrode at which oxidation occurs.

antagonist A drug that blocks the action of an agonist by blocking the receptor(s).

antibiotic A soluble substance produced by a mold or bacterium that inhibits growth of other microorganisms.

anticoagulant A substance that inhibits the clotting of blood.

anticodon The sequence of three adjacent nucleotides in a tRNA molecule that is complement to a codon on mRNA.

antidiuretic A water-conserving substance.

antihistamine A substance that relieves the symptoms of allergies: sneezing, itchy eyes, and runny nose.

anti-inflammatory Pertaining to substance that inhibits inflammation.

antimetabolite A compound that inhibits the synthesis of nucleic acids.

antipyretic A fever-reducing substance.

antiseptic A compound applied to living tissue to kill or prevent the growth of microorganisms.

apoenzyme Protein part of an enzyme.

aqueous Water-containing.

aromatic compound Any organic compound that contains a benzene ring.

arteriosclerosis Hardening of the arteries.

artificial transmutation A process by which one element is changed into another by an artificial means.

atmosphere The gaseous envelope surrounding the Earth (or other planet). A pressure equal to 760 torr.

atom Smallest characteristic particle of an element.

atomic mass unit The unit of relative atomic weights.

atomic number The number of protons in the nucleus of an atom of an element.

atomic theory A model that offers a logical explanation for the law of multiple proportions and the law of constant composition by stating that all elements are composed of atoms, all atoms of a given element are identical, but the atoms of one element differ from the atoms of any other element; that atoms of different elements can combine to give compounds and a chemical reaction involves a change not in the atoms themselves but in the way atoms are combined to form compounds.

Avogadro's number 6.02×10^{23}.

background radiation Ever-present radiation from cosmic rays and from natural radioisotopes in air, water, soil, and rocks.

barbiturate A depressant anticonvulsant drug.

barometer An instrument to measure atmospheric pressure.

base A proton acceptor.

basic anhydride A substance that reacts with water to produce a basic solution; a metal oxide.

basic research The search for knowledge for its own sake.

battery A series of electrochemical cells.

bends A disease resulting from dissolved nitrogen in the blood.

beriberi The disease caused by a deficiency of thiamine.

beta ray A beam of electrons emitted from atomic nuclei.

binary compound A compound containing two elements.

binding energy Energy derived from the conversion of mass to energy when neutrons and protons are put together to form nuclei.

bleach A compound used to remove unwanted color from fabrics, hair, or other materials.

blood sugar Glucose, a simple sugar circulated in the bloodstream.

bond Force that holds atoms together to form a molecule or crystal.

breeder reactor A nuclear reactor that is designed to convert nonfissile uranium-238 to fissile plutonium.

broad-spectrum antibiotic An antibiotic effective against a wide variety of microorganisms.

broad-spectrum insecticide An insecticide that kills many kinds of insects.

bromination Addition or substitution of bromine.

buffer A compound that consumes either acid or base to keep the pH of a solution essentially constant.

calorie The amount of heat required to raise the temperature of 1 g of water 1 °C.

Calorie 1000 calories (or 1 kcal); used to measure the energy content of foods.

cancer A malignant tumor; a tumor that grows and invades other tissues.

carbocyclic Rings containing only carbon atoms.

carbohydrate A compound consisting of carbon, hydrogen, and oxygen; a starch or sugar.

carbonyl functional group A carbon atom doubly bonded to an oxygen atom.

carboxyl group —COOH, the functional group of the organic acids.

carboxylic acid An organic compound that contains the —COOH functional group.

catalyst A substance that increases the rate of a chemical reaction without itself being used up.

catalytic converter An automotive device containing a catalyst for oxidizing carbon monoxide and hydrocarbons to carbon dioxide.

catalytic reforming A process that converts low-octane alkanes to high-octane aromatic compounds.

catenation A process in which carbon atoms join together to form long chains of hundreds or even thousands of atoms.

cathode The electrode at which reduction occurs.

cathode ray A stream of high-speed electrons.

cation A positively charged ion.

cephalin A phosphatide lipid.

chain reaction A self-sustaining change in which one or more products of one event cause one or more new events.

chemical bond The force of attraction that holds atoms together in compounds.

chemical equation A before-and-after description in which chemical formulas and coefficients represent a chemical reaction.

chemical properties Descriptions of the ways in which a substance reacts with other substances to change its composition.

chemistry The study of matter and the changes it undergoes.

chemotherapy The use of chemicals to control or cure diseases.

chiral carbon A carbon atom that is attached to four different groups.

chlorination Addition or substitution of chlorine.

chlorofluorocarbon A carbon compound that contains chlorine and fluorine.

chlorophyll A green plant pigment that absorbs energy from sunlight for use in the synthesis of carbohydrates from carbon dioxide and water.

chromosomes Threadlike bodies in cells that contain the hereditary materials.

cloning Reproducing in identical form.

codon A sequence of three adjacent nucleotides in mRNA that codes for one amino acid.

coenzyme Organic cofactor.

cofactor Enzyme activator.

collagen Proteins of connective tissue.

colligative properties Properties that depend on the number of particles in solution.

colloid Solute particles intermediate in size between solution solutes and suspended matter.

combustion Oxidation via exothermic reaction; burning.

compound A pure substance made up of two or more elements combined in fixed proportions.

condensation The reverse of vaporization; the change from the gaseous state to the liquid state.

condensation polymerization A process in which a portion of the monomer molecule is split out as the polymer is formed and not incorporated in the final polymer.

condensed structural formula An organic chemical formula that shows the atoms of hydrogen right next to the carbons to which they are attached.

configuration The arrangement of an atom's electrons in space.

continuous spectrum A spectrum in which there is a continuous variation from one color to another.

corrode Eat away by chemical action.

cosmic rays Radiation that comes from outer space and consists of high-energy atomic nuclei.

covalent bond A bond formed by a shared pair of electrons.

creatine kinase (CK) An enzyme that catalyzes phosphate transfer.

crenation Shriveling of a cell.

critical mass The mass of an isotope above which a self-sustaining chain reaction can occur.

curie A measure of the rate of disintegration of a radioactive material.

crystal A solid with plane surfaces at definite angles.

cyclic Ring-containing.

cytochrome Iron-containing globular protein.

dehydration Removal of water.

denature Alter properties by adding other chemicals or by subjecting to heat or radiation.

density The amount of mass (or weight) per unit volume.

deoxyribonucleic acid (DNA) The type of nucleic acid found primarily in the nuclei of cells.

depressant A drug that slows down both physical and mental activity.

destructive distillation The process of decomposition by heat with volatile substances distilled off.

deuterium An isotope of hydrogen with a proton and a neutron in the nucleus (mass of 2 amu).

dextro isomer A right-handed isomer.

diabetes A disease characterized by an abnormal increase in blood sugar.

dialysis Separation of small molecules and ions from large ones.

dioxins Chlorinated cyclic compounds once found as contaminants in herbicides.

dipeptide A compound composed of two bonded amino acids.

dipole A molecule that has a positive end and a negative end.

dipole interactions The attractive intermolecular forces that exist among polar covalent substances.

diprotic acid An acid that can donate two protons per molecule.

diffusion The movement of one substance into another by molecular motion.

dispersion forces The momentary, usually weak, attractive forces between molecules.

distillation The boiling off of the volatile compounds such as alcohol or water, leaving behind the solids and high-boiling compounds.

disulfide unit A covalent linkage through two sulfur atoms.

diuretic A substance that increases the body's output of urine.

double bond The sharing of two pairs of electrons between two atoms.

drug abuse Use of a drug for its intoxicating effect or for a purpose other than the intended use.

ductile Able to be drawn into wire.

efficiency Ratio of the amount of useful work to expended energy input.

elastomer A polymer with rubberlike properties.

electrolyte A compound that, when melted or taken into solution, conducts an electric current.

electron The unit of negative charge.

electron configuration The arrangement of electrons about a nucleus.

electron-dot symbol A representation in which the outer electrons of an atom are indicated by dots.

electronegativity The ability of an atom to attract electron density toward itself when joined to another atom by a chemical bond.

element A fundamental substance in which all atoms have the same number of protons.

emollient An oil or grease used as a skin softener.

emulsion A suspension of submicroscopic (colloidal) particles of fat or oil in water.

endergonic reaction Reaction in which the products have a higher energy content than the reactants.

endpoint The point at which an indicator dye changes color during a titration.

endorphin A naturally occurring peptide that bonds to the same receptor site as an opiate drug.

endothermic reaction A chemical reaction to which energy must be supplied as heat.

energy The capacity for doing work.

energy level Region of electron density around the nucleus.

enkephalin Compound composed of a peptide chain of five amino acid units; morphinelike substance produced by the body.

entropy A measure of the randomness of a system.

enzyme A biological catalyst.

equilibrium The condition under which the rates of the forward and reverse reactions are equal.

equivalent Grams of solute that will produce a mole of ionic charge in solution.

equivalent point Titration point where the number of base equivalents equals that of acid equivalents.

essential amino acid One of eight amino acids not produced in the body that must be included in the human diet.

ester A compound derived from carboxylic acids and alcohols. The —OH of the acid is replaced by an —OR group.

estrogen A female sex hormone.

ether A compound having two alkyl groups attached to an oxygen atom.

excited state The state in which an atom is supplied energy and an electron is moved from a lower to a higher energy level.

exergonic reaction Reaction the products of which have lower energy content than the reactants.

exothermic reaction A chemical reaction that releases heat.

experiment (v.) Try out a new idea of activity. (n.) An investigation in which variables are controlled.

facts Pieces of information that are verified by repeated testing and remain the same.

families of elements Vertical columns of the periodic table; also called groups.

fast-twitch fibers Muscle fibers that are stronger and larger and most suited for anaerobic activity.

fat A compound formed by the reaction of glycerol with three fatty acid units; a triglyceride.

fat-soluble vitamins Nonpolar vitamins with a high proportion of hydrocarbon structural elements; dissolve in the fatty tissue of the body and are stored for future use.

fatty acid A carboxylic acid that contains 4–20 or more carbon atoms in a chain.

fermentation The process by which yeast produces alcohol.

first law of thermodynamics Energy is neither created nor destroyed.

fission The splitting of a nucleus into two large fragments.

fixed nitrogen Nitrogen combined with another element.

fluorescence A phenomenon in which after exposure to sunlight, certain chemicals continue to glow even when taken into a dark room.

force Energy required to overcome inertia.

formalin 40% formaldehyde solution.

formula mass The sum of the atomic masses of all the atoms represented in the chemical formula.

free radical A reactive neutral chemical species that contains an unpaired electron.

freezing The reverse of melting; the change from liquid to solid state.

freon A carbon compound containing fluorine as well as chlorine; Du Pont trade name for chlorofluorocarbon.

fructose A simple sugar found in fruits and honey or made by isomerization of glucose.

fruit sugar Fructose.

fuels Substances that burn readily with the release of significant amounts of energy.

fuel cell A device in which chemical reactions are used to produce electricity directly from fuels and oxygen.

functional group A group of atoms that confers characteristic properties on a family of organic compounds.

fundamental particles Basic units from which more complicated structures can be fashioned; protons, neutrons, and electrons.

fusion Combination of nuclei to form a larger one.

gamma rays Rays similar to X-rays that are emitted from radioactive substances; have higher energy and are more penetrating than X-rays.

gas A state of matter in which the substance maintains neither shape nor volume.

general anesthetic A depressant that acts on the brain to produce unconsciousness as well as insensitivity to pain.

globulin A globular protein.

glucose The simple sugar that is circulated in the bloodstream; also called dextrose or blood sugar.

glycogen An animal starch composed of branched chains of glucose units

glycol Dihydroxy hydrocarbon.

glycolysis Splitting of sugar.

greenhouse effect The retention of the sun's heat energy by the Earth as a result of excess carbon dioxide or other substances in the atmosphere that causes an increase in the Earth's atmospheric and surface temperature.

ground state The state of an atom in which all electrons are in the lowest possible energy level.

group A vertical column of the periodic table; a family of elements.

half-life The time during which half of a given radioactive sample disintegrates.

hallucinogen A drug that produces visions and sensations that are not part of reality.

halogen An element in Group VIIA of the periodic table.

hard water Water containing ions of calcium, magnesium, and iron.

hazardous waste A substance that, when improperly managed, can cause or contribute to death or illness or threatens human health or the environment.

heat A measure of a quantity of energy, how much energy a sample contains.

heat capacity The amount of heat needed to change the temperature of 1 g of a material by 1 °C.

heat of vaporization The amount of heat involved in the evaporation or condensation of 1 g of a material.

hemiacetal Product of aldehyde–alcohol reaction.

hemolysis The rupture of red blood cells.

herbicide A chemical used to kill weeds.

heterocyclic compound A cyclic compound in which one or more atoms in the ring are not carbon.

heterogeneous mixture A mixture with definite separate component phases.

high-fructose corn syrup A sweetener made by the hydrolysis of starch and the isomerization of a part of the resulting glucose to fructose.

homogeneous mixture A solution with its solute dissolved and mixed evenly throughout.

hormone A chemical messenger that is secreted into the blood by an endocrine gland.

humectant A moistening agent.

hydration Addition of water to an alkene.

hydrocarbon An organic compound that contains only carbon and hydrogen.

hydrogen bond The dipole interaction between a hydrogen atom bonded to F, O, or N in a donor molecule and an F, O, or N atom in a receptor molecule.

hydrolysis The reaction of a substance with water; literally, a splitting by water.

hydronium ion (H_3O^+) A water molecule to which a hydrogen ion (H^+) has been added; the characteristic ion of an aqueous acid.

hydrosphere The oceans, seas, rivers, and lakes of the Earth.

hydroxide ion (OH^-) Responsible for the properties of base in water.

hydroxyl group The —OH group.

hyperacidity An excess of acid in the stomach.

hypertonic solution Solution of relatively high osmolarity.

hypotonic solution Solution of relatively low osmolarity.

immiscible Cannot be mixed; pertaining to liquids that are insoluble in each other.

immune response Synthesis of antibodies in response to invading bacteria.

indicator A dye the color of which depends on the acidity of the solution.

inorganic chemistry The study of the compounds of all elements other than carbon.

insulin A hormone affecting blood sugar level.

interionic forces The electrostatic forces between ions.

iodine number The number of grams of iodine that will be consumed by 100 g of fat or oil; an indication of the degree of unsaturation.

ion A charged atom or group of atoms.

ionic bond The chemical bond that results when electrons are transferred from one atom to another; the electrostatic interaction of oppositely charged ions.

ionization energy The energy needed to remove an electron from an atom.

ionizing radiation Radiation that causes the formation of ions from neutral particles.

isoelectric point The pH value at which an amino acid exists as a zwitterion.

isomerization The conversion of a compound from one isomeric form to another.

isomers Compounds that have the same molecular formula but different structural formulas and properties.

isotonic solutions Solutions of the same osmotic pressure.

isotopes Atoms that have the same number of protons but different numbers of neutrons.

joule The SI unit of heat.

ketone Compound with two alkyl groups attached to a carbonyl group.

keratin The tough fibrous protein that comprises most of the outermost layer of the epidermis and is the main component of hair and nails.

ketal Product of ketone–alcohol reaction.

ketohexose A six-carbon monosaccharide with a ketone group.

ketone bodies Acetoacetic acid, β-hydroxybutyric acid, and acetone.

ketose Sugar with a ketone group.

ketosis Elevated levels of ketone bodies in the blood.

kilocalorie Measurement of energy content; equal to 1000 calories.

kilogram The SI unit of mass; a quantity slightly greater than 2 lb.

kinetic energy The energy of motion.

kinetic molecular theory A model that uses the motion of molecules to explain the behavior of the three states of matter.

LD_{50} The dosage that would be lethal to 50% of the population of test animals.

lactose The sugar found in milk; composed of two simpler sugars, glucose and galactose.

lactose intolerance The inability of some individuals to break down the sugar lactose; caused by the absence of a necessary enzyme.

law of conservation of energy The amount of energy within the universe is constant; energy cannot be created or destroyed, only transformed.

law of conservation of mass Matter is neither created nor destroyed during a chemical change.

law of constant composition A compound always contains elements in certain definite proportions. Also called *law of definite proportions*.

law of multiple proportions Elements may combine in more than one proportion to form more than one compound, for example, CO and CO_2.

lecithin Cephalin that contains choline.

levo isomer A left-handed isomer.

limiting reagent The reactant that is used up first in a reaction, after which the reaction ceases no matter how much remains of other reactants.

line spectrum The pattern of colored lines emitted by each element.

lipases Enzymes that catalyze the hydrolysis of fats.

lipids Biomolecules that are more soluble in organic solvents than in water.

liquid A state of matter in which the substance assumes the shape of its container, flows readily, and maintains a fairly constant volume.

liter A measurement of volume equal to a cubic decimeter.

local anesthetic A substance that renders a part of the body insensitive to pain while leaving the patient conscious.

macromolecule A molecule with a very high molecular mass; a polymer.

marijuana A preparation made from the leaves, flowers, seeds, and small stems of the *Cannabis* plant.

mass A measure of the quantity of matter.

mass number The sum of the numbers of proton and neutrons in the nucleus of an atom of an element.

matter Stuff of which all materials are made; has mass and occupies space.

mechanism A series of individual steps in a chemical reaction that gives the net overall change.

melting The process in which a substance changes from the solid to the liquid state.

messenger RNA(mRNA) The type of RNA that contains the codons for a protein; mRNA travels from the nucleus of the cell to a ribosome.

metabolism Set of all the chemical reactions in the body that break down large molecules for energy and component parts and build large molecules from component parts.

metals The group of elements to the left of the heavy stepped line on the periodic table.

metastable Pertaining to an isotope that can release some energy and form a more stable isotope.

meter The SI unit of length, slightly longer than 3 feet.

methemoglobin Hemoglobin in which the iron ion has a 3+ charge.

micelle A cluster of molecules.

micronutrients Substances needed by the body only in tiny amounts.

mineral acids Acids derived from inorganic materials.

miscible Pertaining to liquids that are completely soluble in each other.

mitochondria Cell organelles.

mixture Matter with a variable composition.

molarity Moles of solute per liter of solution.

mole A quantity of a chemical substance that contains 6.02×10^{23} units of the substance.

molecule Discrete group of atoms held together by shared pairs of electrons.

monomer A molecule of relatively low molecular mass. Monomers are combined to make polymers.

monounsaturated fatty acid Fatty acid that contains one double bond per molecule.

mutarotation Interconversion of alpha and beta forms of sugars.

mutation Changes in the molecules of heredity in reproductive cells.

myosin The main protein in muscle; provides the energy for muscle contraction by hydrolyzing ATP.

narcotic A drug that produces both narcosis and relief of pain.

neuron Nerve cell.

neurotransmitter A chemical that carries an impulse across the synapse from one nerve cell to the next.

neutralization The reaction of an acid and a base to produce a salt and water.

neutron A fundamental particle with a mass of approximately 1 amu and no electric charge.

noble gases Generally unreactive gases that appear on the far right of the periodic table.

nonbonding pair Pair of electron not involved in a bond.

nonmetals The group of elements to the right of the heavy stepped line on the periodic table.

nonpolar covalent bond A covalent bond in which there is an equal sharing of electrons.

nuclear fusion Combination of two small nuclei to produce one larger nucleus.

nucleic acid The molecule of heredity, found in every living cell; built from nucleotides.

nucleoprotein A combination of a protein with a nucleic acid.

nucleoside Sugar–base combination.

nucleotide A combination of a heterocyclic amine, a pentose sugar, and phosphoric acid; the monomer unit of nucleic acid.

nucleus Concentrated, positively charged matter at the center of an atom; composed of protons and neutrons.

octet rule Atoms seek an arrangement that will surround them with eight electrons in the outermost energy level.

oil Substance formed from glycerol and three unsaturated fatty acids; liquid at room temperature.

oncotic pressure Colloid osmotic pressure.

orbital Space occupied by directional location of electrons.

organic chemistry The study of the compounds of carbon.

osmotic pressure The pressure needed to prevent the net flow of solvent from dilute to concentrated solutions.

oxidation An increase in the oxidation state of an element.

oxide A binary compound of oxygen and another element.

oxidized Increased in oxidation state.

oxidizing agent A substance that causes oxidation and is itself reduced.

oxygen debt An oxygen imbalance resulting from an anaerobic pathway.

paraffin An alkane; a saturated hydrocarbon.

partial pressure The pressure of an individual gas in a mixture of gases.

pathogenic Disease-causing.

peptide bond The amide linkage that bonds amino acids in chains of proteins, polypeptides, and peptides.

perfluorocarbon A compound in which all hydrogen atoms have been replaced by fluorine atoms.

periodic table A systematic arrangement of the elements in columns and rows; elements in a given column have similar properties.

periods The horizontal rows of the periodic table.

pH scale An exponential scale of acidity; below 7, acidic; 7, neutral; above 7, basic.

phenol A benzene ring with a hydroxyl group attached.

phenyl Benzene ring as substituent.

photoscan A permanent visual record showing the differential uptake of a radioisotope by various tissues.

photosynthesis The chemical process used by green plants to convert solar energy into chemical energy by reducing carbon dioxide.

physical properties Properties that describe qualities that can be demonstrated without changing the composition of the substance.

placebo A substance that looks and tastes like a real drug but has no active ingredients.

plasmid A circular piece of DNA that occurs outside the nucleus in bacteria.

plasmolysis The rupture of a cell.

polar covalent bond A covalent bond in which more than half of the bond's negative charge is concentrated around one of the two atoms.

polarity Charge separation occuring in a covalent bond where electrons are shared unequally.

polyatomic ion A charged particle containing two or more covalently bonded atoms.

polycyclic Pertaining to organic molecule containing several fused rings.

polymer A molecule with a large molecular mass; a chain formed of repeating smaller units.

polypeptide A polymer of amino acids; usually of lower molecular mass than protein.

polyunsaturated Containing two or more double bonds.

positron A positively charged particle with the mass of an electron.

potential energy Energy by virtue of position or composition.

product A substance produced by a chemical reaction and whose formula follows the arrow in a chemical equation.

progestin A compound that mimics the action of progesterone.

proof Twice the percentage of alcohol by volume.

property A quality or trait belonging to a particular form of matter.

prostaglandins Hormonelike compounds derived from arachidonic acid that are involved in increased blood pressure, the contraction of smooth muscle, and other physiological processes.

prosthetic group Organic or inorganic component introduced in conjugated protein.

protein Amino acid polymer.

proton The unit of positive charge in the nucleus of an atom.

proton donor An acid; a substance that gives up an H^+ (proton).

psychedelic drug Drug that induces colorful visions.

psychotomimetic drug Drug that induces psychosis.

psychotropic drug Drug that affects the mind.

pure substance Matter with a definite, or fixed, composition.

radioactivity Spontaneous emission of alpha, beta, and/or gamma rays by the disintegration of the nuclei of atoms.

radioisotope Radioactive isotope.

rate The speed of a reaction.

reactant A starting material or original substance in a chemical change; one whose formula precedes the arrow in a chemical equation.

recombinant DNA DNA in an organism that contains genetic material from another organism.

Recommended Daily Allowance (RDA) The recommended level of a nutrient necessary for a balanced diet.

redox reaction A reaction in which oxidation and reduction occur; an oxidation–reduction reaction.

reducing agent A substance that causes reduction and is itself oxidized.

reduction A decrease in the oxidation state of a substance.

replication Copying or duplication; the process by which DNA reproduces itself.

respiration "Burning" of glucose by living cells; oxygen is absorbed and carbon dioxide is given off.

restriction endonuclease Enzyme that cleaves DNA molecule at specific location of specific base sequence.

restriction fragment length polymorphin (RFLP) A pattern of segments of genetic material produced by restriction endonucleases.

ribonucleic acid (RNA) The form of nucleic acid found mainly in the cytoplasm but present in all parts of the cell.

roentgen A measure of the ability of a source of X-rays or gamma rays to ionize an air sample.

saccharide Carbohydrate; from the Latin word for sugar.

salt bridge An interaction between an acidic side chain on one amino acid residue and a basic side chain on another; the resulting charges serve as an ionic bond between peptide chains or between two parts of the same chain.

saponification Soap making; alkaline hydrolysis of fat.

saturated fat A fat composed of a large proportion of saturated fatty acids.

saturated fatty acid A fatty acid that contains no double bonds.

saturated hydrocarbon A compound of carbon and hydrogen with only single bonds.

scientific law A summary of experimental data; often expressed in the form of a mathematical equation.

second law of thermodynamics The degree of randomness in the universe, which increases in any spontaneous process.

semipermeable Pertaining to a membrane permeable to some solutes but not others.

single bond The sharing of one pair of electrons.

slow-twitch fiber Muscle fiber that is best suited for aerobic work.

solid A state of matter in which the substance maintains its shape and volume.

solute The substance that is dissolved in another substance to form a solution; usually present in a smaller amount than the solvent.

solution A homogeneous mixture of two or more substances.

solvent The substance that dissolves another substance to form a solution; usually present in a larger amount than the solute.

specific gravity The ratio of the density of a substance to that of water.

specific heat The amount of heat required to raise the temperature of 1 g of a substance by 1 °C.

starvation The withholding of nutrition from the body whether voluntary or involuntary.

state of matter A condition or stage in the physical being of matter; a solid, liquid, or gas.

steroid A lipid substance with four fused rings.

stimulant A drug that increases alertness, speeds up mental processes, and generally elevates the mood.

STP Standard temperature (273 K) and pressure (1 atm).

strong acid An acid that reacts completely, is 100% ionized in water; a powerful proton donor.

strong base A powerful proton acceptor; dissociates 100% in water.

structural formula A chemical formula that shows how the atoms of a molecule are arranged, to which other atom(s) they are bonded, and the kinds of bonds.

substrate Substance that reacts with enzymes to create a new reaction product.

substitution Replacement of one atom for another in a molecule.

sucrose Common table sugar derived principally from sugarcane or sugar beets; a disaccharide that hydrolyzes to glucose and fructose.

supersaturated Containing solute in excess of that in a solution at equilibrium.

suspension Temporary dispersion of solute in solution.

synapse The gap between nerve fibers.

synergistic effect An effect much greater than just the sum or the expected effects.

tautomers Alcohol–aldehyde or alcohol–ketone isomers.

temperature A measure of heat intensity, or how energetic the particles of a sample are.

tidal volume Normal breathing volume of the lung.

titration A technique used to detect the amount of acid or base present in a sample.

tracer Radioactive isotope used to trace movement or locate the site of radioactivity in physical, chemical, and biological systems.

transamination Formation of α-glutaric acid from an amino acid.

transcription The process by which DNA directs the synthesis of an mRNA molecule during protein synthesis.

transfer RNA (tRNA) A small RNA molecule that contains the anticodon nucleotides; the RNA molecule that bonds to an amino acid.

transition elements Metallic elements situated in the center portion of the periodic table in the B groups.

translation The process by which the information contained in the codon of an mRNA molecule is converted to a protein structure.

transmutation Change of one element into another.

triglyceride Three long-chain fatty acids bonded to glycerol.

triose A three-carbon sugar.

tripeptide A compound composed of three bonded amino acids.

triple bond The sharing of three pairs of electrons between two atoms.

triplet Three-base sequence in the tRNA that determines which amino acid will be attached to an end of tRNA.

triprotic acid An acid that can donate three protons per molecule.

tritium A rare radioactive isotope of hydrogen with two neutrons and one proton in the nucleus (a mass of 3 amu).

Tyndall effect The scattering of a beam of light as it passes through a colloid.

unsaturated hydrocarbon A hydrocarbon containing a double or a triple bond.

valence The number of covalent bonds that an atom can form.

valence shell Outermost shell of electrons of an atom.

valence-shell electron-pair repulsion theory A theory of chemical bonding: that valence-shell electrons locate themselves as far apart as possible.

vaporization The process in which a substance changes from the liquid to the gaseous (vapor) state.

vital capacity The maximum amount of air that can be forced from the lungs.

vitamins Organic compounds that the body cannot produce in the amounts required for good body health.

water-soluble vitamins Vitamins with a high proportion of oxygen and nitrogen that are able to form hydrogen bonds with water.

weak acid An acid that reacts only slightly; a poor proton donor with a low degree of ionization in solution.

weak base A base that ionizes to a small degree and produces relatively few OH^- ions in solution; a poor proton acceptor.

weight A measure of the force of attraction of the Earth on an object.

X-ray Radiation similar to visible light but of much higher energy and much more penetrating.

zwitterion A compound that contains both a positive and a negative charge; a dipolar ion.

Answers to Chapter Quizzes

Chapter Number										
1	1. b	2. b	3. c	4. c	5. a	6. b	7. a	8. c	9. a	10. c
2	1. a	2. b	3. d	4. c	5. b	6. d	7. c	8. b	9. c	10. a
3	1. b	2. b	3. a	4. c	5. c	6. b	7. c	8. c	9. d	10. d
4	1. b	2. a	3. c	4. b	5. b	6. c	7. d	8. c	9. c	10. b
5	1. b	2. a	3. d	4. c	5. d	6. b	7. b	8. d	9. b	10. b
6	1. c	2. a	3. b	4. b	5. d	6. d	7. d	8. b	9. b	10. b
7	1. c	2. b	3. a	4. c	5. d	6. c	7. c	8. d	9. a	10. b
8	1. a	2. c	3. a	4. c	5. a	6. a	7. a	8. c	9. c	10. a
9	1. c	2. c	3. b	4. b	5. a	6. b	7. b	8. c	9. b	10. c
10	1. b	2. c	3. b	4. b	5. c	6. c	7. a	8. c	9. b	10. d
11	1. b	2. c	3. b	4. b	5. a	6. c	7. a	8. b	9. a	10. c
12	1. b	2. b	3. b	4. b	5. d	6. c	7. b	8. a	9. b	10. b
13	1. b	2. d	3. b	4. a	5. b	6. c	7. b	8. b	9. b	10. c
14	1. d	2. b	3. a	4. a	5. c	6. c	7. a	8. b	9. d	10. a
15	1. b	2. c	3. a	4. a	5. b	6. b	7. b	8. d	9. a	10. d
16	1. a	2. c	3. a	4. a	5. c	6. b	7. b	8. d	9. b	10. d
17	1. a	2. d	3. c	4. d	5. a	6. b	7. c	8. a	9. b	10. a
18	1. b	2. b	3. c	4. b	5. d	6. c	7. c	8. b	9. a	10. d

Answers to Selected Problems

CHAPTER 1

3. a. physical b. chemical
4. a. chemical b. physical
5. a. physical b. chemical
7. Glucose is a pure substance.
8. pure substance: a, b; mixture: c
9. pure substance: b; mixture: a, c
10. pure substance: b; mixture: a
12. steam
14. yes
15. Both mass and weight change.
16. No; sample B weighs more on Earth than sample A weighs on the moon.
17. element: a, b; compound: c
18. element: a, c; compound: b
19. element: b; compound: a, c
21. sprinter
22. cannonball
23. automobile
24. It is impossible to choose an answer with the limited information given.
25. diver on 10-m platform
26. elevator at the 20th floor
27. roller coaster at the top of the hill
29. a. cm b. kg c. dL
30. a. L b. same
31. a. m b. lb c. gal
32. $1 \text{ cm}^3 = 1 \text{ mL}$; 15 cc = 15 mL
33. 37 mi/min
35. a. 50 000 m b. 0.25 m c. 0.076 m
36. a. 1500 mm b. 160 mm c. 2.5×10^6 mm
37. a. 10 dL b. 0.20 dL c. 0.15 dL
38. a. 2.056 L b. 47 000 L c. 0.52 L
39. a. 15 g b. 86 mg
40. a. 149 mL b. 0.047 L
41. a. 1400 mL b. 20 qt
42. a. °C b. Cal
43. a. 20°C b. −3.9°C c. −23°C
44. a. 59°F b. −24°F c. 414°F
45. a. 310 K b. 173 K c. 546 K
46. 100°C
47. coldest to hottest: 0 K < 0°F < 0°C
48. a. 2750 cal b. 740 cal
49. 1500 cal
50. 140 000 cal or 140 kcal
51. 11 g
52. 1.1 g/cm^3
53. 1.6 g/mL
54. 680 g
55. 0.73 g/mL
56. 1.02
57.
Irish mist	amber	1.02 g/mL
Triple Sec	white	1.06 g/mL
Creme de menthe	green	1.09 g/mL
Cassis	dark red	1.16 g/mL

Answers to Selected Problems

CHAPTER 2

2. atomistic: a, d, e; continuous: b, c, f
4. yes
5. no
6. no
8. See Table 2.1 in the text.
9. attract
10. The two particles should experience no electrical force.
11. protons and neutrons
12. electrons
14. a. 10
    b. The number of neutrons cannot be specified from the information given.
16. A and B are not isotopes. A and C are not isotopes. A and D are isotopes of neon (atomic number 10). B and C are isotopes of sodium (atomic number 11).
17. A and B both have masses of 21 amu. C and D have masses of 22 and 20 amu, respectively.
18. a. 2 b. 11 c. 17
    d. 8 e. 12 f. 16
19. a. 2 b. 11 c. 17
    d. 8 e. 12 f. 16
21. electrons
22. 32
23. 12
26. The atom in which an electron moves from the first energy level to the third energy level has absorbed more energy.
28. To be neutral, the atom should contain only 3 electrons, 2 in the first energy level and 1 in the second. Alternatively, the atom would be neutral if there were 4 protons in the nucleus.
30. $1s^2 2s^2$
31. He: $1s^2$
    Na: $1s^2 2s^2 2p^6 3s^1$
    Cl: $1s^2 2s^2 2p^6 3s^2 3p^5$
    O: $1s^2 2s^2 2p^4$
    Mg: $1s^2 2s^2 2p^6 3s^2$
    S: $1s^2 2s^2 2p^6 3s^2 3p^4$
32. a. Be b. N c. Al
33. a. 4   b. 7 c. 13
34. a. Before electrons are placed in the $3s$ orbital, the $2p$ orbitals should be filled.
    b. Before an electron is placed in the $3s$ orbital, the $2p$ orbitals should be filled to their capacity of 6 electrons.
    c. There are no $2d$ orbitals.
37. argon
38. neon
42. nonmetals: a and c; metal: b
43. a. IA   b. all B families
    c. VIIA   d. IIA
44. a. third period   b. sixth period
    c. first period
45. b
46. a
47. b, d, e
48. a, b
49. b, c
50. 7
51. 2
52. fourth
53. a. As   b. Cl   c. Be
54. a. Rb   b. Na   c. Zr
55. a. Cs   b. Sb   c. Ac

CHAPTER 3

5. a. 3+ b. 2− c. 1+ d. 1−
6. a. $\dot{Ba}: \rightarrow Ba^{2+} + 2\,e^-$
   b. $:\ddot{Br}\cdot + e^- \rightarrow :\ddot{Br}:^-$
   c. $\cdot \dot{Al} \cdot \rightarrow Al^{3+} + 3\,e^-$
   d. $\cdot \ddot{S} \cdot + 2e^- \rightarrow :\ddot{S}:^{2-}$
7. cations: a and c; anions: b and d
8. a. Li^+ b. I^- c. Ca^{2+}
9. a. NH_4^+ b. HCO_3^- c. PO_4^{3-}
10. a. potassium ion
    b. oxide ion
    c. aluminum ion
11. a. carbonate ion
    b. monohydrogen phosphate ion
    c. nitrate ion
    d. hydroxide ion

12. a. LiF b. CaI$_2$
    c. AlBr$_3$ d. Al$_2$S$_3$
13. a. MgSO$_4$ b. KNO$_3$
    c. NaCN d. CaC$_2$O$_4$
14. a. FeSO$_4$ b. (NH$_4$)$_3$PO$_4$
    c. Mg$_3$(PO$_4$)$_2$ d. CaHPO$_4$
15. a. sodium bromide
    b. calcium chloride
    c. aluminum oxide
16. a. potassium nitrite
    b. lithium cyanide
    c. ammonium iodide
    d. sodium nitrate
    e. calcium sulfate
    f. sodium hydrogen sulfate
    g. potassium hydrogen carbonate
    h. aluminum hydroxide
    i. sodium carbonate
17. a. magnesium acetate
    b. aluminum acetate
    c. ammonium phosphate
    d. ammonium oxalate
    e. sodium monohydrogen phosphate
    f. calcium dihydrogen phosphate
    g. magnesium hydrogen carbonate
    h. calcium hydrogen sulfate
    i. ammonium nitrite
18. NaCl, NaOH, Na$_2$S, Na$_2$SO$_3$, Na$_3$N, Na$_2$HPO$_4$, Na$_3$PO$_4$; NH$_4$Cl, NH$_4$OH, (NH$_4$)$_2$S, (NH$_4$)$_2$SO$_3$, (NH$_4$)$_3$N, (NH$_4$)$_2$HPO$_4$, (NH$_4$)$_3$PO$_4$; BaCl$_2$, Ba(OH)$_2$, BaS, BaSO$_3$, Ba$_3$N$_2$, BaHPO$_4$, Ba$_3$(PO$_4$)$_2$; CuCl$_2$, Cu(OH)$_2$, CuS, CuSO$_3$, Cu$_3$N$_2$, CuHPO$_4$, Cu$_3$(PO$_4$)$_2$; FeCl$_3$, Fe(OH)$_3$, Fe$_2$S$_3$, Fe$_2$(SO$_3$)$_3$, FeN, Fe$_2$(HPO$_4$)$_3$, FePO$_4$; AlCl$_3$, Al(OH)$_3$, Al$_2$S$_3$, Al$_2$(SO$_3$)$_3$, AlN, Al$_2$(HPO$_4$)$_3$, AlPO$_4$
19. Ti$_2$(CrO$_4$)$_3$

20.

	Element X	Element Y	Element Z
Group number	IA	IIIA	VIA
Electron-dot formula	X·	·Ÿ:	·Z̈:
Charge on ion	1+	3+	2−

21. a. VIIA, VIA, VA
    b. H:Ẍ:, :Ÿ:H, H:Z̈:H
                    H   H
    c. Na$^+$:Ẍ:$^-$; 2 Na$^+$:Ÿ:$^{2-}$
26. a. carbon disulfide
    b. carbon tetrabromide
    c. dinitrogen tetrasulfide
    d. phosphorus tribromide
27. a. N b. Cl c. F
28. a. F b. Br c. Cl
29. polar covalent
30. a. no
    b. no
    c. yes
31. a. N$_2$, CO$_2$ b. BF$_3$ c. CCl$_4$
    d. OCl$_2$ e. SCl$_6$
32. a. 109.5
    b. 120
    c. 90 and 120
    d. 109.5
33. b. (O=C=O)
    c. (Br—Be—Br)
37. a. melting
    b. vaporization
39. b and d
41. Water is the solute, and alcohol is the solvent.

CHAPTER 4

2. protium, $^1_1$H; deuterium, $^2_1$H; tritium, $^3_1$H
3. $^8_5$B
4. $^{83}_{35}$Br
5. $^{125}_{53}$I
6. a. $^{69}_{31}$Ga b. $^{98}_{42}$Mo c. $^{99}_{42}$Mo d. $^{98}_{43}$Tc
7. a. 30 protons, 32 neutrons
   b. 94 protons, 147 neutrons
   c. 43 protons, 56 neutrons
   d. 36 protons, 45 neutrons
8. not isotopes: a, c, d, e; isotopes: b
9. lithium-7
10. 10.8 amu

Answers to Selected Problems

12.

Element	Atomic No.	No. of protons	No. of Electrons	No. of Neutrons	Mass Number
Pb	82	82	82	126	208
Sr	38	38	38	50	88
N	7	7	7	7	14
Cr	24	24	24	28	52
Ag	47	47	47	60	107
As	33	33	33	42	75

14. **a.** $^{4}_{2}He$ **b.** $^{0}_{-1}e$ **c.** $^{1}_{0}n$ **d.** $^{0}_{+1}e$

16. $^{209}_{82}Pb \rightarrow \,^{0}_{-1}e + \,^{209}_{83}Bi$

18. $^{225}_{90}Th \rightarrow \,^{4}_{2}He + \,^{221}_{88}Ra$

20. $^{186m}_{79}Au \rightarrow \,^{186}_{79}Au + \text{gamma ray}$

21. $^{218}_{84}Po$, $^{214}_{82}Pb$

22. $^{82}_{36}Kr$

23. After 4.5 sec (one half-life), 1500 atoms of element 104 remain. After 9.0 sec (two half-lives), 750 atoms of the element remain.

24. 26 sec

26. $^{31}_{16}S \rightarrow \,^{0}_{+1}e + \,^{31}_{15}P$

28. $^{87}_{35}Br \rightarrow \,^{1}_{0}n + \,^{86}_{35}Br$

30. $^{21}_{12}Mg \rightarrow \,^{1}_{1}H + \,^{20}_{11}Na$

31. **a.** $^{4}_{2}He$ **b.** $^{0}_{-1}e$

32. **a.** $^{7}_{3}Li$ **b.** $^{1}_{1}H$ **c.** $^{1}_{1}H$ **d.** $^{153}_{62}Sm$

33. $^{24}_{12}Mg + \,^{1}_{0}n \rightarrow \,^{1}_{1}H + \,^{24}_{11}Na$

34. $^{215}_{85}At \rightarrow \,^{4}_{2}He + \,^{211}_{83}Bi$

36. neutrons

39. pro: far below surface, dry environment; con: can geological inactivity be extrapolated?

40. Use shielding; maintain a distance from the radioactive sample; minimize the time of exposure.

41. alpha

42. gamma

45. iodine-131

48. internal: 10 mCi; external: 500 Ci

49. treatment of malignancy: 150 mCi; imaging: 15 μCi

50. Compared to a lethal whole-body dose of about 500 rad, the diagnostic dose of 0.14 rad was about 0.028% of the lethal dose.

51. X-rays

52. gamma rays

54. ultrasonography and MRI

56. Cs^+ is chemically similar to Na^+ and K^+, which are body electrolytes. Therefore Cs^+, radioactive or otherwise, can easily find its way into body fluids.

CHAPTER 5

3. The temperature is decreasing.

4. The pressure is decreasing.

5. Assuming the temperature in both containers is the same, the pressure is the same.

6. **a.** The density of the gas in container A is higher.
   b. The density of gas in container A and B is the same.
   c. The density of the gas in container A is less.

7. **a.** 1500 mm Hg   **b.** 380 torr   **c.** 300 Pa

8. **a.** 0.10 atm   **b.** 320 mm Hg   **c.** 10 Pa

9. **a.** 150 psi   **b.** 14 cm H_2O **c.** 0.10 atm

12. 1000 mL

13. 40 m^3

14. **a.** 9000 L   **b.** 19 hr

15. 0.33 atm

17. 1.2 atm

18. **a.** 310 K   **b.** 147 °C   **c.** 191 K

19. 10 mL

20. 3.3 L

21. 40 psi

22. 0.83 atm

23. 2470 mm Hg

24. 13 L

25. 0.48 m³
26. 431 mL
27. 750 cc
28. 732 mL
29. decrease in volume: a, b; increase in volume: c
30. increase in pressure: a, b, c
32. 0.95 atm
33. 250 mm Hg
34. 708 mm Hg
35. 97 atm
36. 95%
37. about 50 mm Hg
38. 67%
39. a. Oxygen will flow from flask A to flask B.
    b. Nitrogen will flow from flask B to flask A.

CHAPTER 6

2. a. $C + O_2 \rightarrow CO_2$
   b. $CaCl_2 + H_2SO_4 \rightarrow CaSO_4 + 2\,HCl$
4. a. 4   b. 9
5. Al = 2; C = 12; H = 18; O = 12
6. balanced: a, d, e; not balanced: b, c
7. balanced: a, d, e; not balanced: b, c
8. a. not balanced   b. balanced
9. a. $4\,Al + 3\,O_2 \rightarrow 2\,Al_2O_3$
   b. $2\,C + O_2 \rightarrow 2\,CO$
   c. $N_2 + O_2 \rightarrow 2\,NO$
   d. $2\,SO_2 + O_2 \rightarrow 2\,SO_3$
   e. $2\,NO + O_2 \rightarrow 2\,NO_2$
10. a. $Zn + 2\,HCl \rightarrow ZnCl_2 + H_2$
    b. $2\,H_2S + O_2 \rightarrow 2\,H_2O + 2\,S$
    c. $Al_2(SO_4)_3 + 6\,NaOH \rightarrow 2\,Al(OH)_3 + 3\,Na_2SO_4$
    d. $Zn(OH)_2 + 2\,HNO_3 \rightarrow Zn(NO_3)_2 + 2\,H_2O$
    e. $3\,NH_4OH + H_3PO_4 \rightarrow (NH_4)_3PO_4 + 3\,H_2O$
11. a. $Cu + 2\,H_2SO_4 \rightarrow SO_2 + CuSO_4 + 2\,H_2O$
    b. $2\,NH_4Cl + CaO \rightarrow CaCl_2 + H_2O + 2\,NH_3$
12. a. 16 amu   b. 84 amu   c. 352 amu
13. a. 80 amu   b. 233 amu
    c. 98 amu   d. 123 amu
14. a. 164 amu   b. 58 amu   c. 132 amu
15. a. 2.0 mol   b. 4.0 mol   c. 0.0500 mol
16. a. 0.10 mol   b. 0.00100 mol
    c. 10.0 mol   d. 0.0300 mol
17. a. 0.20 mol N   b. 0.00400 mol O
    c. 30.0 mol H   d. 0.0300 mol K
18. a. 0.0100 mol   b. 0.10 mol
    c. 0.000100 mol
19. a. 0.016 g   b. 504 g
    c. 14 100 g or 14.1 kg
20. a. 4 g   b. 699 g
    c. 980 g   d. 2.46 g
21. a. 246 g   b. 340 g   c. 33 g
22. a. 4 mol   b. 0.6 mol   c. 0.0684 mol
23. a. 1.0 mol   b. 1.0 mol
24. a. 32 g   b. 16 g   c. 1.6 g
25. 6 (the formula is $NiCl_2 \cdot 6H_2O$)
26. a. 0.8 mol   b. 50 mol
    c. 5.0 mol   d. 0.86 mol
27. a. 0.25 mol   b. 44 g
28. a. 32 g   b. 220 g
29. a. 6 mol   b. 16 mol
30. a. 2.5 mol   b. 0.075 mol
31. a. 4.4 g   b. 264 g
32. 3.2 mol Cl_2
33. 0.087 mol
34. 39.4 g/mol; Ar
36. a. C   b. B   c. A
37.

38. exothermic
39. endothermic
42. The rate decreases.
43. The rate decreases.
44. The rate increases.
46. a. Equilibrium shifts to the left.

Answers to Selected Problems

b. Equilibrium shifts to the right.
c. Equilibrium shifts to the right.
47. Equilibrium shifts to the right.
48. Equilibrium shifts to the right.
50. Oxidation: only d
51. Oxidation: a; reduction: b, c
52. oxidized

53.

	Oxidizing Agent	Reducing Agent
a.	O_2	Al
b.	C_2H_2	H_2
c.	O_2	SO_2
d.	$AgNO_3$ (actually Ag^+)	Cu

54. a. 0 b. +6 c. −2 d. −1
55. a. +3 b. +3 c. +4
56. a. +4 b. +2

57.

	Oxidized	Reduced
a.	S	N
b.	I	Cr
c.	C	Mn

58. Indoxyl is oxidized by the oxidizing agent O_2.

CHAPTER 7

6. saturated
7. 10%; the solute is water, and the solvent is alcohol.
8. Dissolve 40 mL of acetic acid in sufficient water to give a total of 2.0 L of solution.
9. Mix 10 mL of acetone and 40 mL of water.
10. 4% by mass
11. 20.0% by mass
12. 5.0% by mass
13. Dissolve 3.0 g of $C_{12}H_{22}O_{11}$ in 97 g of water.
14. Dissolve 0.50 kg of NaCl in 4.5 kg of water.
15. Dissolve 1.0 g of Na_2CO_3 in 49 mL of water.
16. Lowest to highest concentration: 1 ppb < 1 ppm < 1 mg % < 1%
17. 100 mg %

18. a. Dissolve 30 mL of alcohol in water to give a total of 100 mL of solution.
    b. Dissolve 5.0 g of NaOH in 95 mL of water.
    c. Dissolve 1.0 g of $NaHCO_3$ in 9.0 g (9.0 mL) of water.
19. a. Dilute 0.1 mL of drug to a total volume of 100 mL.
    b. Dilute 0.0005 L (0.5 mL) of drug to a total volume of 1 L.
20. 2.0 M
21. 2.0 M
22. 0.10 M
23. 3.0 M
24. Dissolve 16 g of NaOH in water to give a total of 2.0 L of solution.
25. Dissolve 5.35 g of NH_4Cl in water to give a total of 100 mL of solution.
26. Dissolve 9.4 g of NaCl in water to give 0.40 L of solution.
27. a. Dissolve 118 g of NaCl in water to give a total of 1.00 L of solution.
    b. Dissolve 45 g of $C_6H_{12}O_6$ in water to give a total volume of 500 mL of solution.

28.

	Solute			Vol. soln
	M	mol	wt.	
Amt solute				
I	I	I	I	U
D	D	D	D	U
Amt solvent				
I	D	U	U	I
D	I	U	U	D
Amt soln				
I	U	I	I	I
D	U	D	D	D

29. a. 1 equiv b. 3 equiv c. 4 equiv
30. a. 2 equiv b. 0.2 equiv
31. 0.5 mL
33. a. 2 b. 1 c. 3 d. 2
    e. 3
34. a. 2 b. 1 c. 3 d. 2
    e. 3
35. a. 0.50 mol b. 1.0 mol c. 0.33 mol
    d. 0.50 mol e. 0.33 mol

36. 2.0 osmol/L
38. Seawater contains ions that prevent molecules of H₂O from escaping from liquid into gas. Thus seawater evaporates more slowly.
39. 5% NaCl solution
40. a. 0.1 M NaHCO₃   b. 1 M NaCl
    c. 1 M CaCl₂   d. 3 M glucose
41. a. Osmotic pressures are equal.
    b. 2 osmol/L glucose
    c. 1 osmol/L NaHCO₃
43. a. Carrots stay fresh in water since water will flow into an area of higher concentration.
    b. Water flows into the seed (place of higher concentration), causing it to swell and sprout.
    c. The bacterial cells lose their water through osmosis into the salt solution.
46. diffusion
47. more

CHAPTER 8

3. acids: b, c, d, f; bases: a, e
5. a. diprotic   b. monoprotic
   c. diprotic   d. triprotic
   e. *tetra*protic

9.
	Stronger Acid	Weaker Acid
a.	HBr	HF
b.	HF	HCN
c.	H₃PO₄	HIO₃
d.	CCl₃COOH	HCOOH
e.	HClO₃	HNO₃

10.
	Stronger Base	Weaker Base
a.	F⁻	Br⁻
b.	CN⁻	F⁻
c.	IO₃⁻	H₂PO₄⁻
d.	HCOO⁻	CCl₃COO⁻
e.	NO₃⁻	ClO₃⁻

11. a. $HCl + LiOH \rightarrow LiCl + H_2O$
     b. $Al(OH)_3 + 3\,HCl \rightarrow AlCl_3 + 3\,H_2O$
     c. $H_2SO_4 + Mg(OH)_2 \rightarrow MgSO_4 + 2\,H_2O$
12. a. $NaHCO_3 + HNO_3 \rightarrow NaNO_3 + H_2O + CO_2$
     b. $CaCO_3 + 2\,HBr \rightarrow CaBr_2 + H_2O + CO_2$
     c. $H_2SO_4 + K_2CO_3 \rightarrow K_2SO_4 + H_2O + CO_2$
13. $NH_3 + HI \rightarrow NH_4I$
15. $H_3O^+ + OH^- \rightarrow 2\,H_2O$
16. $HCO_3^- + H_3O^+ \rightarrow 2\,H_2O + CO_2$
17. a. 0.50 M   b. 0.25 M   c. 0.50 M
18. 0.125 M
19. $2\,H_3O^+ + CO_3^{2-} \rightarrow 3\,H_2O + CO_2$
20. a. 2   b. 3
21. a. 6   b. 0.5   c. 10
22. a. 1 mol   b. 0.5 mol
23. a. 2 mol   b. 0.1 mol   c. 1.5 mol
24. a. 2.00 equiv   b. 0.25 equiv
    c. 0.20 equiv
25. a. 0.0200 equiv   b. 1.5 equiv
    c. 0.25 equiv
26. yes: a, b, d; no: c
27. yes: a, c; no: b
29. $CH_3COO^- + H_2O \rightleftharpoons CH_3COOH + OH^-$
30. $NH_4^+ + H_2O \rightleftharpoons NH_3 + H_3O^+$
31. a. neutral   b. basic
    c. unable to say   d. basic
    e. acidic   f. neutral
32. basic
33. $CaCO_3 + 2\,H_3O^+ \rightarrow Ca^{2+} + CO_2 + 3\,H_2O$
37. a. 2   b. 3
38. a. 4   b. 5   c. 1
39. a. 1   b. 5
40. a. 3   b. 2
41. 37a, 12; 37b, 11; 38a, 10; 38b, 9; 38c, 13
42. 39a, 13; 39b, 9; 40a, 11; 40b, 12
43. a. basic   b. acidic
    c. neutral   d. acidic
44. a. acidic   b. basic
    c. neutral   d. basic
50. decrease
52. too high
53. higher
55. In the case of hyperventilation, a person needs more CO₂; a paper bag over the head decreases amount of O₂ available and allows the exhaled CO₂ to be inhaled again.

Answers to Selected Problems

CHAPTER 9

3. **a.** 2 **b.** 7 **c.** 4 **d.** 9
6. same compound: a, b, d; isomers: c, e
11. **a.** 3-methylhexane **b.** 4-isopropylheptane
17. **a.** 2-methyl-1-pentene
    b. 2,5-dimethyl-2-hexene
    c. cis-3-hexene
18. isomers: a, c, e; same compound: b, d
22. **a.** Cl$_2$, heat **b.** H$_2$O, H$_2$SO$_4$ **c.** Cl$_2$
24. a, b, c
25. d
31. **a.** 1-pentyne **b.** 2-butyne
33. 2 H$_2$, Ni
35. **a.** methylcyclopropane
    b. cyclobutene
39. aromatic: a, d; aliphatic: b, c
40. **a.** para **b.** ortho **c.** meta
42. **a.** ethylbenzene **b.** m-nitrotoluene
    c. 1,3,5-trinitrobenzene

CHAPTER 10

5. **a.** 2-hexanol
   b. 2-methyl-1-propanol (IUPAC), isobutyl alcohol (common)
   c. 2-propanol (IUPAC); isopropyl alcohol (common)
   d. 1-hexanol
7. **a.** ethanol **b.** methanol **c.** 2-propanol
8. lowest to highest: methanol < ethanol < 1-propanol
9. least to greatest solubility in water: 1-octanol < 1-butanol < methanol
10. oxidation: a, c; dehydration: b, e; hydration: d
18. distilled spirits
21. lowest to highest: butane < 1-propanol < ethylene glycol
22. lowest to highest: diethyl ether < 1-butanol < propylene glycol
23. least to greatest: pentane < diethyl ether < propylene glycol
25. ethanol
26. **a.** m-chlorophenol (3-chlorophenol)
    b. o-nitrophenol (2-nitrophenol)
30. **a.** dipropyl ether
    b. ethyl sec-butyl ether
39. **a.** benzaldehyde
    b. propanal (propionaldehyde)
    c. 3,3-dimethylbutanal
40. **a.** 5-methyl-3-hexanone (ethyl isobutyl ketone)
    b. cyclopentanone
42. 2-propanol
43. 1-butanol
44. acetaldehyde
47. yes: a, c, d; no: b, e
48. yes: b, c, d; no: a, e

CHAPTER 11

2. **a.** butyric acid **b.** caproic acid
   c. lauric acid **d.** stearic acid
6. **a.** formic acid (common); methanoic acid (IUPAC)
   b. propionic acid (common); propanoic acid (IUPAC)
   c. capric acid (common); decanoic acid (IUPAC)
   d. glutaric acid (common); pentanedioic acid (IUPAC)
7. butyric acid
8. butyric acid
9. butyric acid
10. lowest to highest boiling: pentane < methyl acetate < butyl alcohol < propionic acid
11. acetic acid
13. **a.** sodium benzoate
    b. calcium propionate
    c. ammonium acetate
    d. zinc butyrate
    e. calcium phthalate
14. sodium benzoate
23. **a.** methyl benzoate
    b. methyl formate (methyl methanoate)
    c. ethyl propionate (ethyl propanoate)
    d. propyl acetate (propyl ethanoate)
    e. propyl propionate (propyl propanoate)
    f. phenyl butyrate (phenyl butanoate)

25. methyl acetate
38. a. benzamide
    b. *N*-isopropylacetamide
    c. acetamide
    d. *p*-chlorobenzamide
    e. malonamide
    f. *N,N*-dimethylpropionamide
    g. butyranilide
39. butyramide
40. acetamide

CHAPTER 12

2. a. alcohol, 1°   b. amine, 1°
    c. alcohol, 2°   d. amine, 1°
    e. ether   f. amine, 2°
    g. amine, 2°   h. ether
    i. amine, 3°   j. alcohol, 3°
    k. amine, 1°   i. phenol
7. a. propylamine   b. ethylmethylamine
    c. triethylamine
8. a. *p*-nitroaniline b. *N*-ethylaniline
12. butylamine
13. butyl alcohol
14. propylamine
15. ethylamine
16. propylamine
17. 1,3,5-triaminohexane
19. diethylammonium bromide
42. The minimum lethal dose for a 70-kg person is 350 mg (0.35 g) or about 0.01 ounce.

CHAPTER 13

7. aldose: a–e; ketose: f, g
8. a. hexose   b. pentose
   c. hexose   d. triose
9. D: a; L: b–d
14. a. beta
    b. No; carbon number 1 is fixed as an acetal (not a hemiacetal).
    c. no
18. alpha: b, d; beta: a, c
19. alpha: a, b; beta: c; none: d
20. d
21. beta
22. a. glucose   b. lactose   c. glucose
    d. fructose   e. sucrose
23. glucose and galactose
24. alpha
25. Melibiose is a reducing sugar; the hemiacetal function is in the alpha arrangement.
29. glucose
30. glucose
31. glucose
32. galactose and glucose
33. fructose and glucose
34. hydrolysis
40. lactic acid
41. ATP
43. as carbon dioxide

CHAPTER 14

3. saturated: a, c, d, f; unsaturated: b, e
4. a. 6   b. 18   c. 18
   d. 16   e. 18   f. 8
5. least to most unsaturated: oleic acid < linoleic acid < linolenic acid
10. the carbon–carbon double bond
11. triolein
12. corn oil
13. liquid margarine
14. In both cases tristearin would be formed.
21. good cleaning agent: a; poor cleaning agent: b, c
26. saponifiable: a, c; nonsaponifiable: b, d
28. phospholipid: c, e
29. fatty acid ester: a, c
31. Phosphatides, soaps, and detergents have molecules with both polar and nonpolar ends.
35. steroid: b, c, f
36. cholesterol
38. cholesterol
41. lipases
45. acetyl coenzyme A

Answers to Selected Problems

CHAPTER 15

7. the L family
8. Three problems associated with a strict vegetarian diet are (1) such a diet is more likely to be lacking in some essential amino acid; (2) vitamin B_{12} cannot be obtained from plant products; and (3) calcium, iron, riboflavin, and other vitamins may be inadequately supplied.
15. amide
17. Ala-Ser-Cys-Phe-Gly-Gly
18. The secondary structure of silk is termed the *pleated sheet* conformation.
19. The secondary structure of wool has been given the designation *alpha helix*.
22. a. salt bridge
    b. hydrophobic bonding
    c. hydrogen bonding
    d. disulfide bond
26. fibrous: c, e; globular: a, b, d, f
27. globular
28. globular
30. gamma globulins or antibodies
34. a. The substrate is ethanol.
    b. The cofactor is zinc ion.
    c. The apoenzyme is the protein portion of alcohol dehydrogenase.
35. no
36. FAD is both a cofactor and a coenzyme.
39. amino acids
42. 930 cans
50. as urea

CHAPTER 16

2. Energy cannot be destroyed, but it can be transformed. The chemical energy stored in the flashlight battery is converted to heat energy and light (radiant energy).
4. The first vehicle is more efficient at making the energy stored in the gasoline do useful work.
5. The lightning-induced reaction is endergonic. The firefly reaction is exergonic.

11. a. -2.8 kcal/mol   b. -3.5 kcal/mol
    c. -3.2 kcal/mol   d. -5.1 kcal/mol
14. high-energy compounds: a, c, d; glucose 1-phosphate (e) is a borderline case.
15. a net synthesis
16. a net synthesis
17. NAD^+, FAD (Section 16.6 of text)
18. At the end of the respiratory chain oxygen (O_2) is reduced to water (H_2O).
20. pyruvic acid $\rightarrow$ FAD $\rightarrow$ cytochromes $\rightarrow$ O_2
23. aerobic oxidation of glucose
25. a net consumption
27. carbohydrates
28. fats
29. anaerobic
30. aerobic
37. pheasants: Type IIB fibers; great blue heron: Type I fibers
40. The muscle activity associated with shoveling snow could be the cause of the elevated levels of CK in the blood.

CHAPTER 17

2. DNA
5. nucleoside: b, c; nucleotide: e; neither: a, d, f
6. ribose: a, b, d; deoxyribose: c, e, f
7. in DNA: b, d; in RNA: a, c, f; in neither: e
8. pyrimidine: a; purine: b
9. purine: a, b, f; pyrimidine: c, d, e
10. a. nucleoside   b. pyrimidine   c. DNA
11. a. nucleotide   b. purine   c. RNA
13. the sequence of bases
14. a. uracil   b. cytosine
    c. adenine   d. guanine
18. hydrogen bonding
19. a. guanine   b. thymine
    c. cytosine   d. adenine
21. Each daughter DNA molecule contains one of the parent DNA strands.
25. DNA and RNA
26. RNA and RNA

27. RNA
28. RNA
29. ···TAAGC···
30. ···UAAGC···
31. ···AGGCTA···
32. a. AAA b. GUA c. UCG d. GGC
33. a. AAC b. CUU c. AGG d. GUG
34. a. Phe b. His c. Ser d. Pro
35. a. Asn b. Leu c. Arg d. Val
36. a. Leu-Pro-Arg
    b. Ala-Ser (UAA is a stop signal)
    c. Pro-Pro-Pro
37. Asn-Glu-Ser (the codons on mRNA would be AAU, GAG, AGU)

39.
DNA	mRNA	Protein
GGT	CCA	Pro
TGT	ACA	Thr
GTT	CAA	Gln
CGT	GCA	Ala

CHAPTER 18

3. a. Iodine is incorporated in the thyroid hormone thyroxine.
   b. Iron is incorporated in hemoglobin and myoglobin (proteins involved in the transport of oxygen) and in the cytochromes in the respiratory chain.
   c. Calcium plays a number of roles. Among other things, it is a major constituent of bone and teeth, it is required for blood coagulation, and it is involved in the regulation of the heartbeat.
4. Here are a few: nucleotides and nucleic acids, phosphorylated sugars, and phosphatides, and a variety of high-energy compounds like creatine phosphate.
8. All except C are B vitamins.
12. water-soluble: b, c, d; fat-soluble: a, e
13. water-soluble: b, c; fat-soluble: a, d
14. a. water-soluble   b. fat-soluble
19. hypothalamus
20. pituitary gland
24. steroids
27. The hormone binds on the outside of the cell membrane.

APPENDIX B

1. a. 100.5 cm b. 153 g
   c. 130 mL (1.3×10^2 mL)
2. a. 54.4 cm b. 436 g c. 9.3 L
3. a. 40 m^3 b. 0.196 cm^2 c. 17 cm^2
4. a. 1.1 g/mL
   b. 1500 g/mol (1.5×10^3 g/mol)
   c. 13 g/mL
5. 2.61 mL

APPENDIX C

1. a. 1×10^{-5} b. 1×10^7
   c. 3.4×10^{-3} d. 1.07×10^{-5}
   e. 4.5×10^9 f. 4.06×10^5
   g. 2×10^{-2} h. 1.24×10^5
2. a. 8.6×10^8 b. 2.5×10^{-16}
   c. 4.4×10^{-12} d. 2.2×10^{-15}
   e. 2.5×10^{16} f. 2.1×10^{-3}
   g. 5.7×10^{-30}

Index

A letter f following a page number indicates a figure; a letter t indicates a table.

Absolute temperature scale, 24, 31
Accelerator, 133
Acetal(s), 353, 412, 422
Acetaldehyde, 337
   physical properties, 351t
   structure, 347f
Acetamide, 361t, 374–375, 379t
Acetaminophen, 378
Acetanilide, 377, 379
Acetbutolol, 409
Acetic acid, 213, 256, 261, 265, 266, 269, 362, 363t
   boiling point, 364
   freezing point, 365
   odor, 365
   properties as acid, 256, 261
   structure, 361t, 362, 379
Acetoacetic acid, 455
Acetone, 335, 454–455
   physical properties, 322t
   structure, 349f
Acetyl coenzyme A (acetyl CoA), 430, 433, 452–455, 458
Acetyl phosphate, free energy of hydrolysis, 499t
Acetylene, 313
Acetylsalicylic acid, 371–372
Acid rain, 59, 61, 269, 271, 283
Acids, 254–256, 283
   aqueous solutions, 256–259
   dicarboxylic, 364t
   equivalents, 260–262, 283
   mineral, 365
   neutralization, 259–260, 283
   physiological effects, 258–259
   reactions with carbonates and bicarbonates, 268–271, 283
   strong, 257, 283
   titration, 263–264, 283
   weak, 257, 283
   See also Carboxylic acids; Nucleic acids
Acidosis, 282, 284, 286, 455
Aconitase, 334
cis-Aconitic acid, 334
Acrilan, 311t

ACTH, 551t
Actin, 508
Activation energy, 189, 192, 206, 481–482
Active site, 482, 489
Actomyosin, 508, 510, 512
Addition, significant figures, 566–567
Addition polymers, 311t
Addition reactions
   alkenes, 309–310
   alkynes, 313
Adenine, 517, 522f
Adenosine, structure, 518f
Adenosine diphosphate, *see* ADP
Adenosine monophosphate, *see* AMP
Adenosine triphosphate, *see* ATP
Adequate protein, 466
Adipic acid, structure, 364t
Adipose tissue, 431, 451, 457
ADP, 497, 499
Adrenal cortex-stimulating hormone, 551t
Adrenaline (epinephrine), 398–399, 402t, 552t
Adrenocorticotropic hormone, 551t
Adsorption, 244
Aerobic fermentation, 362
Aerobic oxidation, 430–431, 433, 503–504
   in muscle action, 508–509
Air
   composition, 141, 142t, 165
   pollution, 271
   use in medicine, 165t
Alanine, 389, 397, 463, 464t, 486f
   isoelectric point, 467t
   structure, 389
Alanylglycine, 469, 471
Albumin, isoelectric point, 469t
Alcohols, 326, 371, 412
   boiling point, 332, 355
   chemical properties, 333–335, 355, 367
   classification, 329, 355
   functional group, 326, 327t
   multifunctional, 339–340
   nomenclature, 328–331, 355
   physical properties, 332–333
   physiological properties, 335–339, 355
   protein denaturation, 479–480

   solubility in water, 332–333, 335
   structure, 326, 355
Aldehydes
   chemical properties, 351–356
   functional group, 327t, 345–346
   nomenclature, 347–348, 355
   oxidation, 351–352
   physical properties, 350–351, 355–356
   reaction with alcohols, 353–354, 416–417
Aldohexoses, 416
Aldopentoses, 416
Aldose, 413
Aldosterone, 552t
Aliphatic acids, 363t
Aliphatic compounds, 316, 320
Alkali metals, 53, 56, 64
Alkaline earth metals, 53, 57, 64
Alkaloid, 372, 391
Alkalosis, 271, 282, 284
Alkanes, 289–292, 320
   IUPAC nomenclature, 290–291, 292t, 295, 320–321
   physiological effects, 320
   prefixes of names, 291t
   properties, 300–302
Alka-Seltzer, 271
Alkenes, 289, 320
   functional group, 327t
   geometric isomerism, 305–308, 320
   IUPAC nomenclature, 303–304, 320–321
   properties, 303t, 309–312
   structure and nomenclature, 302–305
Alkoxy group, 370
Alkyl groups, 296
   structure and nomenclature, 297t
Alkynes, 289, 312–313, 320, 327t
Allotropes, 58
Alpha decay, 114, 115t
Alpha emitter, 117
Alpha form of carbohydrates, 417, 419, 432
Alpha helix, 474–475
Alpha particle, 113–114
Alpha position, amino acids, 462–463
Alpha rays, 113, 134
   penetrating power, 116–117

Aluminum, 57–58, 541
Aluminum hydroxide, 262
  as antacid, 272
Aluminum sodium dihydroxy carbonate, 272
Amides
  chemical properties, 376–377, 380
  functional group, 360–361, 374, 379
  physical properties, 376
  physiological properties, 377–379
  structure and nomenclature, 374–376
  substituted, formation, 396–398
Amines
  as bases, 392–396
  functional group, 327t
  heterocyclic, 389–391, 406, 517
  nomenclature, 387–389, 406
  odor, 392
  physical properties, 391–392
  physiological properties, 398–407
  structure and classification, 386–387, 406
Amino acids, 388
  as buffers, 467, 489
  essential, 466
  groups, 463
  isomerism, 463
  metabolism, 485–486
  physical properties, 462–463
  pool, 483–484, 486
  sequence in proteins, 469–471
  structure, 462–463, 464t–465t
L-Amino acids, 463, 489
p-Aminobenzoic acid, structure, 397f
2-Aminoethanol, 446
2-Aminopropanoic acid (α-Aminopropionic acid), structure, 369
Ammonia
  as base, 256, 258–259
  oxidative deamination, 485
  relationship to amides, 386
  structural relationship to amines, 392–393
  toxicity, 259
Ammonium benzoate, structure, 367
Ammonium chloride, 267
Ammonium hydroxide, 265
Amobarbital, 404
  toxicity, 402t
AMP, 498, 499f
  binding, 557
  free energy of hydrolysis, 499t
  structure, 499f, 519f
Amphetamine, 401–403
amu (atomic mass unit), 37, 64
Amylopectin, 424, 425f, 433
Amylose, 424, 425f, 433
Amytal, 404
Anaerobic glycolysis, 428–429, 433
  energy transformations, 500–502
  muscle action, 508
Analgesic, 371–372, 380
Anemia, 356t–457t
  screening for, 29
Anesthetics, 165t, 319
  general, 343–345, 355
  local, 396, 397f, 407

Anilide, 375
Aniline, 388, 389f
Aniline hydrochloride, 395
Anilinium chloride, structure, 395
Anilinium ion, structure, 394
Anions, 72, 103
Antacids, 271–273, 283
Anthranilic acid, structure, 389f
Antianxiety agents, 405–406
Antibodies, 481, 489
Anticodon, 530, 534
Antifreeze, 236, 339, 355
Antihistamine, 400
Antipyretic, 371, 380
Antiseptics, 204–205, 339, 341
Apoenzyme, 483, 490
Aqueous solutions, 213, 248
  acids and bases in, 256–259
  glucose, 417, 418f
  neutralization, 259–264, 283
Arginine, 465t
Argon
  in air, 55, 141
  electron configuration, 72
Aristotle, 36
Aromatic hydrocarbons, 316–320
  reactions, 317–319
Arsenic-74, 129t
Arteriosclerosis, 442
Artificial respirators, 149–150
Artificial transmutation, 121–123, 133–134
ASA (acetylsalicylic acid), 371
Ascorbic acid, see Vitamin C
Asparagine, 464t
Aspartame, 492
Aspartic acid, 465t
  isoelectric point, 467t
Aspirin, 271, 371–372, 378, 380
Atmosphere
  acidity, 269
  composition of, 141, 142t, 165
Atmosphere (unit of pressure), 144, 165
Atmospheric pressure, 143–144
Atom, 36, 64
  Bohr model 43–46, 64
  Dalton's model, 36–38, 64
  electron configuration, 48–49
  isotopes, 42–43, 64, 108–110
  nuclear model, 38–43, 64
  quantum mechanical model, 46–50, 64
Atomic mass, 134
  relation to atomic weight, 110–111
Atomic mass unit (amu), 37, 64
Atomic number, 40, 64, 108, 134
Atomic theory, Dalton's, 36–38, 64
Atomic weight, 134
  in periodic table, 51–53
  relation to atomic mass, 110–111
Atomistic theory of matter, 36, 64
ATP, 429, 433, 511
  free energy of hydrolysis, 499t
  hydrolysis, 497, 498, 499f
  structure, 499f
  synthesis, 500–502
Autoclave, 155

Automobile tires, 154
Avogadro's number, 179, 186–187, 206

Background radiation, 117–118, 134
Baking soda, see Sodium bicarbonate
Barbital
  structure, 403
  toxicity, 402t
Barbiturates, 403–405
  toxicity, 402t
Barium compounds
  light emission, 44
  use in medical imaging, 129
Barometer, 143–144, 165
Base-pairing in nucleic acids, 521–523
Bases, 254–256, 283
  amines as, 392–396
  aqueous solutions, 256–259
  equivalents, 260–262, 283
  physiological effects, 258–259
  purine, 517, 534
  pyrimidine, 405, 517, 534
  titration, 263–264, 283
Bathtub ring, 57, 444, 445
Becquerel, Antoine Henri, 38f, 40, 111–112
Beef tallow, 439t
  fatty acid distribution, 441t
  iodine number 442t
Bends, 161, 166
Benedict's reagent, 352, 356, 419–421, 433
Benzaldehyde, 346f
  physical properties, 351t
Benzamide, structure, 375
Benzedrine, 401
Benzene, 315–316
  chlorination, 318–319
  physiological properties, 319
  properties, 289t
  structure, 316
Benzocaine, structure, 397f
Benzodiazepines, 405–406
Benzoic acid, 363–364
3,4-Benzpyrene, 319–320
Benzyl acetate, structure, 368f
o-Benzyl-p-chlorophenol, 341
Benzyl ethanoate, structure, 368f
Beriberi, 542, 545t
Beryllium, 57
Beta decay, 114, 115t
Beta form of carbohydrates, 417, 419, 432
Beta particle, 113
Beta rays, 113, 134
  penetrating power, 116–117
Biacetyl, structure, 346f
Bicarbonate–carbonic acid buffer, 280–281, 283
Bicarbonate ion, physiological importance, 102
Bicarbonates, 268–271, 283
Bilayer, 446, 457
Bile salts, 448, 450–451, 457
Biotin, 546t
Birth control pills, 449, 555
Bis-GMA, 312

INDEX

Blindness, vitamin A deficiency, 548
Blood
    buffers in, 280–281, 283–284
    specific gravity, 29–30
Blood sugar, 416, 432
    See also *Glucose*
Blue-baby syndrome, 63
Body fluids, 229–232, 246, 249
    intermolecular forces and, 350t
    osmolarity, 247t
Body temperature, 190–191
Bohr model of the atom, 43–46, 64
Bohr, Niels, 44–45, 64
Boiling point, 97t
    elevation, 236–237, 249
Bonding forces, states of matter, 95–97
    See also *specific bonds*
Borax, 289
Boric acid, 57, 261
Boron, 57, 541
Boyle, Robert, 145
Boyle's law, 145–150, 165
    combined gas law, 156–158
Breathing
    mechanics, 148–149, 166
    partial pressure and, 162–166
Bromination, 309–310
Bromine, 62
    isotopes, 110
Buffers, 278–280, 283
    amino acids as, 467, 489
    in blood, 280–281, 283–284
Buret, 263
Butacaine, structure, 397f
Butanal, structure, 347f
Butane
    physical properties, 301t
    structure, 292t, 293–294
Butanedioic acid, 364t
2,3-Butanedione, structure, 346f
Butanoic acid, 363t
1-Butanol, lethal dose, 336t
Butanone, structure, 349f
1-Butene
    physical properties, 303t
    structure, 305
2-Butenes, structure, 306, 320
Butesin, structure, 397f
Butter
    fatty acid distribution, 441t
    iodine number, 442t
Butyl alcohol
    physical properties, 332t, 392t
    structure and classification, 329t
sec-Butyl alcohol, 329f
tert-Butyl alcohol, 329f, 331, 335
Butylamine, physical properties, 392t
tert-Butylamine, 388
Butyl *p*-aminobenzoate, structure, 397f
1-Butyne, structure, 313
Butyraldehyde
    physical properties, 351t
    structure, 347f
Butyranilide, structure, 376
Butyric acid, 363t 439t
    odor, 365

Cadaverine, 392
Cadmium, 63
Caffeine
    structure, 391f
    toxicity, 402t
Calciferol, 544t
    structure, 328f
Calcium, 57, 63, 541, 542
Calcium carbonate, 57, 269–270, 283
    as antacid, 272, 283
Calcium compounds, light emission, 44
Calcium hypochlorite, as disinfectant, 205
Calcium ion, physiological importance, 57, 102, 541
Calcium oxalate, 339
Calcium propanoate, structure, 367f
calorie, 25, 31
Calorie, 25, 31
    food, 25, 26t, 31
Cancer, radiation therapy, 127–128, 129t
Capric acid, 363t, 439t
Caproic acid, 363t, 439t
Carbocyclic compounds, 390, 406
Carbohydrates, 412–413, 432
    caloric content, 438
    digestion, 424, 426–427, 433
    fermentation, 336–337
    metabolism, 427–431, 433, 487f
Carbolic acid, 341
    See also *Phenol*
Carbon, 58
    electron configuration, 49–50
Carbon-12, 110
Carbon-14, 120–121
Carbon dioxide, 269
    in air, 141, 142t
    effect of pressure on solubility, 161
    electron-dot formula, 88–89
    medical use, 165t
    transport in respiration, 163–164
Carbon monoxide poisoning, 58, 162, 166
Carbon tetrachloride, 302
Carbonates, 268–271, 283
Carbonic acid, 268–269, 283
Carbonyl group, 345–347, 355, 360–361
    latent, 412
Carboxyl group, 360–361
Carboxylic acids, 352
    chemical properties, 365–367, 380
    functional group, 360, 379
    physical properties, 364–365
    structure and nomenclature, 361–364, 379
    See also *Fatty acids*
Carcinogens, 302, 319
Casein, isoelectric point, 469t
Catalysts, 191–192, 206, 310
    effect on rates of reactions, 191–192, 206
    See also *Enzymes*
Cathode ray, 39
Cation, 72
    formation, 103
Cell membranes, 102, 439, 447, 457, 556
Cellulose, 424, 425f, 433
Celsius temperature scale, 22–24
Centigrade temperature scale, 22–24

Centimeters of water (cm H_2O), 144, 165
Cephalins, 446, 457
Chadwick, James, 41
Chain reaction, nuclear, 124, 135
Changes of state, 4, 96–97
Charge, electrical, 9, 31
Charge cloud, 47, 50
Charles, J. A. C., 150
Charles's law, 150–153, 165–166
    combined gas law, 156–158
Chemical bonds, 70, 103
    See also *Covalent bond; Ionic bond*
Chemical equations, 172–176, 205, 319
Chemical pneumonia, 319
Chemical properties, 3, 31, 289
Chemical reaction(s)
    addition, 309
    defined by Dalton, 36
    energy changes, 188–189
    equilibrium, 193–195, 206, 218–219
    reversible, 190, 193, 195, 206
    spontaneous, 495–496, 512
    substitution, 309
Chemistry, 31
Chiral carbon atom, 414, 463, 489
Chirality, 414–415
Chlordiazepoxide, 405
Chloride ion
    electron configuration, 72–73
    physiological importance, 102
Chlorination, 310, 318–319
Chlorine, 61–62
    as disinfectant, 205
    electron-dot formula, 84
    isotopes, 110
Chloroacetic acid, 363
Chloroform, 302
    as anesthetic, 344
Chloromethane, 302
Chlorophyll, 57
Chlorpromazine, 400
Cholesterol, 448–449, 457
    in foods, 443t
    structure, 328f
Cholic acid, 448, 449f
Choline, 446–447
Chromium, 63
Chromium-51, 129t
Chromosomes, 524–525
Cinnamaldehyde, structure, 346f
Cis isomer, 306, 320
Cis-trans isomerism, 306–308, 320
Citric acid, 360, 365, 431
    dehydration, 334
CK (creatine kinase), 511
Cloning, 532
cm H_2O (centimeters of water), 144, 165
Cobalt, 63, 542
Cobalt-58, 129t
Cobalt-60, 129t
Cocaine, 371
    structure, 391f
    toxicity, 402t
Codeine, 372f
    toxicity, 402t
Codon, 530, 534
Coefficient(s), 173, 205, 570

Coenzyme, 483, 490
Coenzyme A, structure, 453
Cofactor, 483, 489–490
Collagen, 475, 481
Colligative properties, 236–238, 249
Collisions, elastic, 143, 165
Colloid osmotic pressure, 247
Colloidal dispersion
  properties, 242–244, 249
  types, 244t
Colloids, 242–244, 249
Combined gas law, 156–158
Combustions, 301
Compound, 5, 31
Computed tomography, 130, 132f, 134
Concentrated solutions, 220
Concentration, effect on reaction rates, 191, 206
Concentration units, *see* solutions
Condensation, 97, 160
Configuration, *see* Electron configuration
Conjugated protein, 480, 489
Conservation of matter, law, 173, 206
Continuous theory of matter, 36, 64
Conversion factors, 14–21, 31
Copper, 63, 542
Cooper (II), ion, 352, 483
Corn oil, fatty acid distribution, 441t
Corpus luteum, 554
Corrosive poisons, 258, 283
Corticosterone, 448, 449f
Cortisol, 551t
Cottonseed oil
  fatty acid distribution, 441t
  iodine number, 442t
Cough reflex, 168
Coupled reactions, 496–498, 512
Covalent bond, 83–85, 103
Covalent compounds, nomenclature, 92
Creatine, 511
Creatine kinase, 511
Creatine phosphate, 511–512
  free energy of hydrolysis, 499t
Crenation, 240, 249
Creslen, 311t
m-Cresol, structure, 341f
Crick, Francis, 521–523, 525
Crookes, Willia, 38f, 39
Cross-multiplication method, 81
Crystal, dynamic equilibrium, 218
Crystalline solid, 96–97
CT scanning, *see* Computed tomography
Curare, 344–345
Curie (unit of radiation), 118
Curie, Irene, 122
Curie, Marie Sklodowska, 38f, 40, 112
Curie, Pierre, 112
Cyanide ion, toxicity, 504
Cyanocobalamine, 547t
Cyclic AMP, 555–558
Cyclic hydrocarbons, 314–315
Cyclobutane, structure, 315
Cyclohexane, structure, 315
Cyclohexanone, structure, 349f
Cyclohexene, structure, 315
Cyclopentane, structure, 315

Cyclopropane
  as anesthetic, 165t, 319
  structure, 314, 315f
Cysteine, 464t
  isoelectric point, 467t
Cystine, 464t
Cytidine, structure, 517
Cytidine monophosphate, structure, 517
Cytochromes, 504, 505f
Cytosine, 36–38, 42, 517, 522f

Dalmane, 405
Dalton, John, 36, 37f, 158
Dalton's atomic theory, 36–38, 64
Dalton's law of partial pressures, 158–161, 166
DDT, 302
Decane, 292t
  physical properties, 301t
Decanoic acid, 363t
  solubility, 366,
Decyl alcohol, physical properties, 332t
Dehydration
  alcohols, 333–334, 355
  ester formation, 367
Denaturation, proteins, 479–480, 489
Denatured alcohol, 338
Density, 26–29, 31, 167
  mercury, 29
  water, 28
Dental sealants, 312
Deoxyadenosine, structure, 518f
Deoxycytidine, structure, 518f
Deoxyguanosine, structure, 518f
Deoxyribonucleic acid, *see* DNA
Deoxyribose, 416, 517f
  structure, 415f, 418f
Deoxythymidine, structure, 518f
Depressant, 338, 343
Dermatitis, 546t
Destructive distillation, 336, 362
Deuterium, 42, 110
Dextrose, 416
  See also Glucose
Diabetes mellitus, 421, 431–433, 454–455, 533
Diacetylmorphine, 372–373
Diagnostic scanning, 128–129
Dialysis, 244–246, 249
Dialyzing membranes, 245
1,4-Diaminobutane, 392
1,5-Diaminopentane, 392
Diazepam, 405
1,2-Dibromoethane, 309
3,5-Dibromotyrosine, 465t
Dicarboxylic acids, 364t
Dichlorodiphenyltrichloroethane, 302
Dichloromethane, 302
Diethylamine, physical properties, 392t
Diethyl ether
  as anesthetic, 343, 355
  formation, 333
  structure, 342f
Diethyl ketone
  physical properties, 351t
  structure, 349f
N,N-Diethyllysergamide, 378, 380

Diffusion
  among body fluids, 248–249
  gases, 163, 164f
Digestion, 426–427, 433
Diglycerides, 450, 451f
Diglyme, structure, 342f
Dihydrogen phosphate–monohydrogen phosphate buffer, 281, 283–284
Dihydroxyacetone, 413
3,5-Diiodotyrosine, 465t
Diisopropyl ketone, structure, 350
Dilute solutions, 220
Dilution, stock solutions, 234–235
Dilution factor, 226
Dimensional analysis, 13–21, 31
Dimer, 365
Dimethylammonium ion, 394
N,N-Dimethylaniline, structure, 388
Dimethylbenzenes, structure, 318f
Dimethyl ether, structure, 342f
5,6-Dimethyl-2-heptene, 304
Dimethylhexanes, 296, 298
Dimethyl ketone, structure, 349f
2,3-Dimethylpentane, 298
2,4-Dimethyl-3-pentanone, structure, 350
3,4-Dimethyl-2-pentene, 304–305
m-Dinitrobenzene, 317
Dioxane, structure, 342f
Dipeptide, 468
1,3-Diphosphoglyceric acid, 429f
  free energy of hydrolysis, 499t
  hydrolysis, 500, 501f
Dipole, 88
Dipole forces, 97–98, 103
Dipole interactions, 98
Diprotic acid, 256, 283
Disaccharides, 413, 422–424, 433
Discharge tube, 39
Disinfectants, 205, 341
Dispersion forces, 99–100, 103
Dissociation, 258, 283
Dissolution, 215
Distilled spirits, 337
Disulfide bond, 477, 489
Division
  exponentials, 570
  significant figures, 567–568
DNA, 70, 516, 534
  components, 520t
  replication, 524–526
  structure, 521
Dodecane, physical properties, 301t
Dodecanoic acid, 363t
Double bond, 82–84, 86, 345–346, 440–442
Double helix, 521–522
Downers, 403
Drugs, 481
  antianxiety, 405–406
  for mental problems, 399–401
  pain-relieving, 62, 371–374, 378
  psychedelic, 378
  sedative, 405
  strengths, 233–234
  toxicities, 402t
Duet of electrons, 71, 103
D5W, 241

INDEX

Dynamic equilibrium, 218–219
Dynel, 311t

Edema, 248
Efficiency, 495, 507
Eicosane, physical properties, 301t
Eijkman, Christiaan, 542
Elastic collisions, 143, 165
Electric forces, 9–10
Electrical charge, 9, 31
Electrolytes, 229, 249
    concentration in body fluids, 231t
Electron(s), 39
    discovery, 39–40
    orbitals, 47, 48t, 64
    properties, 42t, 64
    valence, 74
Electron cloud, 47, 50
Electron configuration, 48–50, 64
    elements, 49t
    energy levels, 44–49, 64, 71–73, 103
    noble gases, 71–73, 103
    order-of-filling chart, 49f
    relation to chemical properties, 52–53
Electron-dot formulas, 75–76, 103
Electronegativity, 86, 103
    elements, 87t
Elements, 5, 31
    electron-dot formulas, 75t
    periodic table, 51–54
    symbols, 5–6, 6t–7t
Emulsifying agents, 244, 249, 445
Endergonic reactions, 495, 496–497, 512
Endocrine glands, 549
Endorphins, 488, 490
Endothermic reactions, 188, 206
Endpoint, 263, 283
Energy, 7, 31
    activation, 189, 192, 206, 481–482
    conversion units, 563t
    efficiency of cellular transformations, 507
    free, 495
    kinetic, 8, 31
    light, 44–45
    potential, 7–8, 31
    units of measurement, 22–26
Energy changes, chemical reactions, 188–190, 206
Energy currency of cell, 498
Energy diagrams, 189, 206
Enflurane, 315
English system of measurement, 10–14, 16
Enkephalins, 488, 490
Enol, 419–420
Enzymes, 193, 206, 334, 336–340, 450, 477f, 481–483, 489–490
Epinephrine (adrenaline); 398–399, 402t, 552t
Equations
    chemical, 172–176, 205
    nuclear, 122–125
Equilibrium, 193–195, 206, 248
    dynamic, 193, 218–219
Equivalence point, 263, 283
Equivalent, 229–233, 249
    acid–base chemistry, 260–262, 283

Essential amino acids, 466, 489
Essential fatty acids, 466
Esterification, 367
Esters, 438–439
    chemical properties, 370, 380
    functional group, 327t, 361
    physical properties, 369
    physiological properties, 371–374
    structures and nomenclature, 367–368
Estradiol, 552t
Estrogens, 554–555
Estrone, 448, 449f
Ethanal, structure, 347f
Ethanamide, 361t, 374–375, 379
Ethananilide, structure, 379
Ethane
    physical properties, 301t
    structure, 291, 292t
Ethanoic acid, 361t, 363t, 379
Ethanol
    boiling point, 364
    dehydration, 333
    lethal dose, 336t
    physical properties, 332t
    physiological effects, 337–338
    protein denaturation, 479–480
    structure, 310, 329
    toxicity, 336, 338
Ethanolamine, 446–447
Ethene
    as anesthetic, 319
    bromination, 309
    formation, 333
    hydration, 310
    hydrogenation, 310
    physical properties, 303t
Ethers, 326, 327t, 333, 342–343, 355
Ethisterone, structure, 554f
p-Ethoxyacetanilide, 378
Ethyl acetate, 379t
Ethyl alcohol, see Ethanol
Ethyl p-aminobenzoate, structure, 397f
Ethylammonium chloride, 395
Ethylanilines, structure, 388–389
Ethylbenzene, structure, 317
Ethyl butanoate, structure, 368f
Ethyl butyrate, structure, 368f
Ethyl ethanoate, structure, 379t
Ethyl formate, structure, 367, 368f
Ethyl methanoate, structure, 367, 368f
Ethylmethylamine, structure, 387–388
Ethylmethylammonium acetate, 395
Ethyl methyl ether
    physical properties, 392t
    structure, 342f
4-Ethyl-2-methylhexane, 296
Ethyl methyl ketone
    physical properties, 351t
    structure, 349f
Ethyl n-propyl ketone, structure, 349
Ethyl valerate, structure, 369
Ethylene, structure, 303
Ethylene glycol, 339, 355
    lethal dose, 336t
    physical properties, 340t
    structure, 339
    toxicity, 339

Ethylene oxide, structure, 342f
Ethyne, 313
Exact values, 568
Excited state, 45
Exercise physiology, 510–511
Exergonic reactions, 495, 496–497, 512
Exothermic reactions, 188, 206
Expiration, change of pressure in lungs, 148–149, 166
Exponential notation, 569–573
Extracellular fluid, 246–247

Factor-label method, 13–21, 31
FAD, 503, 512
    structure, 503f
$FADH_2$, 504, 512
Fahrenheit temperature scale, 22–24, 31
Fallout, nuclear, 135
Families, in periodic table, 53, 64
Fast twitch muscle fibers, 509–510, 512
Fat(s), 440–443
    caloric content, 438
    digestion, 450–452, 457
    as energy source, 508
    metabolism, 452–457
Fat depots, 451, 457
Fat-soluble vitamins, 543, 544t, 548, 558
Fatty acid(s), 370, 439, 450–452, 457–458
Fatty acid spiral (fatty acid cycle), 452–454, 457–458
Fehling's reagent, 352, 356, 419–420, 433
Fermentation, 336–337, 355, 360, 362
Fibrous protein, 481, 489
Film badge, 118–119
Filtration, body fluids, 247–249
First law of thermodynamics, 495, 512
Fish oils, iodine number, 442t
Fission, nuclear, 124, 125f, 135
Flavin adenine dinucleotide, 503, 512
Fluorescence, 111–112
Fluorine, 62
    Bohr model, 46
    electronegativity, 86
Flurazepam, 405
Folic acid, 546t
Follicle-stimulating hormone, 551t, 552, 554
Food
    Calorie, 25, 31
    cholesterol and saturated fat levels, 443t
Force, 8, 31
    electrical, 9–10
    in solutions, 100–102
Force field, electrical, 9
Formaldehyde
    physical properties, 351t
    structure, 347f
    toxicity, 336
Formalin, 351
Formic acid, 362
    boiling point, 364
    odor, 365
    structure, 362
Formula, chemical, 103
Formula unit, 229, 237–238
Formula weights, 177–178, 206

Free energy change, (ΔG), 496–500, 512
  hydrolysis of phosphates, 499t
Freezing point, 97
  depression, 236–237, 249, 365
Fructose, 416, 496–497
  oxidation, 420–421
  structure, 415f, 418f
Fructose 1,6-diphosphate, 429f
Fructose 6 phosphate, 428, 429f
  free energy of hydrolysis, 499t
Functional groups, 326–328, 327t, 345, 355
  acids, 360–361, 379
  amides, 374
  carbonyl, 345
  esters, 361
Funk, Casimir, 542
Fusion, nuclear, 125, 135

ΔG, 496–500, 512
Galactose, 416
  structure, 415f, 418f
Gallstones, 448f
Gamma decay, 114, 115t
Gamma emitter, 117
Gamma globulins, 481
Gamma rays, 113–114, 134
  diagnostic scanning, 128
  penetrating power, 116–117
  PET scanning, 130
Gangrene, 162, 166
Gas law(s)
  Boyle's, 144–150
  Charles's, 150–153
  combined, 156–158
  Dalton's, 158–161
  Gay-Lussac's, 153–156
  ideal, 187
Gases, 140–141
  compression, 143–144
  kinetic molecular theory, 96, 142–143
  properties, 4, 31
  solubility, 216–217
Gasoline, 289, 300t
Gay-Lussac, Joseph, 153
Gay-Lussac's law, 153–156, 166
Geiger counter, 119
Gelatin, isoelectric point, 469t
General anesthetics, 343–345, 355
Genes, 526
Genetic code, 530–531
Genetic engineering, 531–533
Geometric isomerism, 305–308, 320
Germ cells, 524
Germicides, 341
Glacial acetic acid, 365
Globular proteins, 481, 489
Globulins, 481
Glucagon, 551t
Glucose, 190, 203, 415–418, 496–497
  in aqueous solution, 417, 418f
  glycolysis, 428–429, 500, 501f
  isotonic solutions, 241
  structure, 328f
  test, 352–353
  urine test, 421

Glucose 1-phosphate, 427, 429f, 497–498
  free energy of hydrolysis, 499t
Glucose 6-phosphate, 429f, 501f
  free energy of hydrolysis, 499t
Glucosides, 422
Glutamic acid, 465t, 484f–485f
  isoelectric point, 467t
Glutamine, 464t
Glutaric acid, 364t
Glyceraldehyde, structure, 347f, 413–415
L-Glyceraldehyde, 415, 463
Glyceraldehyde 3-phosphate, 429f
Glycerol, 440, 450–452, 457–458
  lethal dose, 336t
  physical properties, 340t
  structure, 340
Glycerol 3-phosphate, free energy of hydrolysis, 499t
Glycine, 463, 464t
  isoelectric point, 467t
Glycogen, 424, 427, 431, 433, 508
  in starvation, 456
Glycogenolysis, 509–510
Glycols, 339
Glycolysis, 428–429, 452, 458, 500–502
Glycylalanine, 469, 471
Glycylphenylalanine, 468
Goiter, 541
Gold foil experiment of Rutherford, 40–41
Goldstein, Eugen, 39
Grain alcohol, 336
Gram, 10–11, 31, 561
Gram formula weight, 178, 206
Gravity, 3
Greenhouse effect, 141
Ground state, 45
Groups, in periodic table, 53, 64
Guanine, 517, 522f
Guanosine, structure, 518f

Half-life, radioisotope, 119–121, 134
Hallucinogenic drug, 378
Halogenated hydrocarbons, 302
Halogenation, 310
Halogens, 53, 61–62, 64
Halothane, 344
Hard water, 57, 444–445
Heat, 24
  measurement, 24–26
Heat capacity, water, 25, 31
Heat of combustion, 301
Heating oil, 300t
Heavy metal poisoning, 480
Helium, 55
  Bohr model, 45
  electron configuration, 71
  isotopes, 113–114
  medical use, 165t
  use in deep-sea diving, 161
Heme, 484, 504
  structure, 480
Hemiacetal, 353–354, 356, 412, 417, 422, 432
Hemiketal, 356, 419, 433

Hemoglobin, 63, 102, 478, 541
  effect on specific gravity of blood, 29–30
  isoelectric point, 469t
Hemolysis, 240, 249
Hemorrhage, 544t
Henry, William, 161
Henry's law, 161–162, 166
Heptadecane, physical properties, 301t
Heptane
  physical properties, 301t
  structure, 292t
1-Heptene, physical properties, 301t
Heptose, 413, 432
Heredity, 524
Heroin, 372–374, 380
  structure, 373
  toxicity, 402t
Heterocyclic amines, 389–391, 406, 517
Heterocyclic compounds, 390, 406
Heterogeneous mixture, 5, 213, 248
Hexachlorophene, 341
Hexadecane, physical properties, 301t
Hexadecanoic acid, 363t
Hexanal, 348
Hexane
  physical properties, 301t
  structure, 292t
Hexanedioic acid, 364t
Hexanoic acid, 363t
  solubility, 365
1-Hexanol, lethal dose, 336t
3-Hexanone, structure, 349
cis-3-Hexenal, structure, 346f
1-Hexene, physical properties, 303t
Hexose, 413, 432
Hexyl alcohol, physical properties, 332t
4-Hexylresorcinol, 341f
High energy compounds, 498–499
Histidine, 465t
Homogeneous mixture, 5, 100, 103, 213, 248
Honey, as saturated solution, 219
Hopkins, F. G., 542
Hormones, 448–449, 457, 533, 549–555, 558
  cyclic AMP levels and, 556t
Human growth hormone, 533, 551t
Humidity, 160, 166
Hydration, 310
Hydrocarbons, 58, 289–290, 320
  aromatic, 316–320
  cyclic, 314–315
  physiological properties, 319–320
  saturated, 289–302, 320
  unsaturated, 302–313
Hydrochloric acid, 256–258
  as stomach acid, 271, 283
Hydrocyanic acid, 259, 261
Hydrogen
  Bohr model, 45
  electron configuration, 50
  isotopes, 42–43
Hydrogen bomb, 125, 135
Hydrogen bonding, 98–99, 103
  alcohols, 332, 339

INDEX

amides, 376
amines, 391, 406
carboxylic acids, 365, 380
involving aldehydes and ketones, 351
involving esters, 369, 380
involving ethers, 342
nucleic acids, 522–523, 525
proteins, 473–477, 489
water-soluble vitamins, 548
Hydrogen chloride, 85–86, 256–258
dipole forces, 97–98
polarity, 86–87
Hydrogen cyanide, 93, 257, 504
electron-dot formula, 93
toxicity, 259
Hydrogen fluoride, hydrogen bonding, 98–99
Hydrogen halides, 62
Hydrogen molecule, 83–84
Hydrogen peroxide, 174
as antiseptic, 204
Hydrogen sulfide, electron-dot formula, 91
Hydrogenation, 310, 313
oils, 442–443
Hydrolysis
amides, 377, 380
carbohydrates, 413, 416, 426–427, 433
esters, 370, 380
fats, 451–452, 457
phosphates, 498–499
proteins, 483
Hydrometer, 30
Hydronium ion, 257, 259, 273, 283
Hydrophobic interactions, in proteins, 476–477, 489
Hydroxide ion, 255, 259, 273, 283
p-Hydroxyacetanilide, 378
β-Hydroxybutyric acid, 454, 455f
Hydroxyl group, 326, 327, 328, 339, 353, 412
Hydroxylysine, 465t
Hydroxyproline, 464t
Hyperbaric chambers, 162, 166
Hypertension, 542
Hypertonic solution, 240, 249
Hypnotic, 338
Hypothalamus, 549–550, 558
Hypotonic solution, 240, 249

Ice, structure, 213
Imidazole, structure, 390
Imipramine, 399, 401
Immiscibility, 215
Immune response, 481, 489
Impermeable, 238
Inches of mercury, 144, 165
Indican, 435
Indicator, 263, 283
Indigo, 210, 435
Indole, structure, 390f, 391
Indoxyl, 210
Inert gases, 55
Inhibition of enzyme action, 482–483
Inorganic compounds, 288, 289t
Insolubility, 215

Inspiration, change of pressure in lungs, 148–149, 166
Insulin, 431–433, 463f, 533, 551t
isoelectric point, 469t
Interferon, 533
Interstitial fluid, 246
Intracellular fluid, 246
Iodine, 62, 541
Iodine-131, 129t, 552
Iodine chloride, 106
Iodine number, 441–442, 457
Iodized salt, 541
Ion(s), 50, 70, 103
formulas and nomenclature, 78–80
periodic relationship, 79
physiological importance, 102–103, 541–542
Ion–dipole interactions, 101
Ion product of water, 273, 283
Ionic bond, 71–74, 103
combining ratios of ions, 80–82
electron-dot formulas, 76–78
Ionic compounds, 76, 103
nomenclature and formulas, 80–82
Ionic equation, 259–260
Ionization, 257, 283
Ionization energy, 53
Ionizing radiation, 118, 134
Iron, 63, 102, 483, 541
Iron-59, 129t
Iron ions, 57, 403
in cytochromes, 504
in hemoglobin, 63, 102, 541
Iron lung, 149–150
Isobutane, structure, 293
Isobutene, structure, 305
Isobutyl alcohol
physical properties, 332t
structure and classification, 330
Isocitric acid, 486
Isoelectric point, 467, 469, 489
Isoleucine, 464t
Isomerism, 293–294, 320
geometric, 305–308, 320
mirror-image, 413–415, 432
tautomerism, 419–420, 433
Isomers, 292t, 293
Isopentane, structure, 294
Isopentyl acetate, structure, 368f
Isopentyl ethanoate, structure, 368f
N-Isopropylacetamide, structure, 375
Isopropyl, alcohol
physical properties, 332t
structure, 329
Isopropylamine, structure, 387
N-Isopropylethanamide, structure, 375
4-Isopropyl-2-methylheptane, 299–300
Isotonic solution, 241, 249
Isotopes, 42–43, 64
importance in nuclear chemistry, 108–110

IUPAC (International Union of Pure and Applied Chemistry) nomenclature, 295–300, 320–321, 375, 389
acids, 362, 363t
alcohols, 330
aldehydes, 347
alkanes, 295–296
alkenes, 303–304
esters, 368
ketones, 349

Joliot, Frédéric, 122

K_w, 273, 283
Kekulé, Friedrich August, 315
Kekulé structure of benzene, 315–316
Kelsey, Frances, 379
Kelvin, 24, 31, 151
Keratins, 481, 489
Kerosene, 300t
Ketals, 353, 356
Keto–enol tautomerism, 419–420, 432–433
α-Ketoglutaric acid, 484, 485f, 490
Ketohexose, 413, 416
Ketone bodies, 454–455
Ketones, 335, 355
chemical properties, 351–356
functional group, 327t, 346
nomenclature, 348–350, 355
physical properties, 350–351, 355–356
Ketoses, 413, 432
oxidation, 420
Ketosis, 455, 458
Kilogram, 561
Kinetic energy, 8, 31, 143, 165
Kinetic molecular theory, 95–96, 142–143, 165, 217, 248
Boyle's law and, 145
Charles's law and, 151–152
Krebs cycle, 430–431, 453, 458, 485–486, 490, 502–503, 512
Krypton, 55–56
Krypton-81m, 136
Kwashiorkor, 466

Labetalol, 409
Lactic acid, 365, 430–431, 433
muscle fatigue, 508–509, 512
Lactoglobin, isoelectric point, 469t
Lactose, 423, 433
Lard
fatty acid distribution, 441t
iodine number, 442t
Latent carbonyl group, 412
Lauric acid, 363t, 439t
Law of conservation of matter, 173, 205
Lead poisoning, 480
Least common multiple, 78
Le Châtelier, Henri Louis, 193
Le Châtelier's principle, 193–194, 206, 421
Lecithins, 446, 457
Left-handed helix, 475
Length, metric units, 10, 562t

Leucine, 464t
Leu-enkephalin, 488
Levulose, *see* Fructose
Librium, 405
Lidocaine
　structure, 397f
　toxicity, 402t
Light, as form of energy, 44–45
Limestone, acid rain effect, 269, 271, 283, 371
Limeys, 542
Lind, James, 542
Linoleic acid, 439t
Linolenic acid, 439t
Linseed oil, iodine number, 442t
Lipases, 450
Lipids, 438–439, 442, 457
　See also Fats; Fatty acids
Liquids
　kinetic molecular theory, 95–96
　properties, 4, 31
Lister, Joseph, 341
Liter, 10–11, 31, 561
Lithium, 42, 45, 56, 541
　Bohr model, 45
Lithium carbonate, 56, 399
Litmus, 254, 263, 365
Liver enzymes, 336–340
Local anesthetics, 396, 397f, 407
　toxicity, 402t
Lock-and-key model of enzyme action, 482, 489
LSD, 378, 380, 390
Lubricating oil, 30t
Lucite, 311t
Luminal, 404
Luteinizing hormone, 551t, 554
Lye, 258–259, 443
Lymph, 246
Lymphatic system, 248, 451
Lysergic acid, 378, 390
Lysine, 465t
　isoelectric point, 467t
Lysozyme, 477f
　isoelectric point, 469t

Magnesium, 57, 102, 483, 542
　oxidation, 60
Magnesium-21, half-life, 120
Magnesium carbonate, as antacid, 272
Magnesium hydroxide, as antacid, 272
Magnesium ion, physiological importance, 57, 102
Magnetic resonance imaging, 131, 132f, 135
Malonic acid, 364t
Maltose, 422–423, 433
Manganese, 63, 483, 542
Marble, acid rain effect, 269–270, 283
Margarine, 442
Marijuana, 371
Mass, 3, 31
　metric units, 10, 562t
Mass number, 108–109, 134
Matter, 1–2, 31
　classification, 2–3

physical and chemical changes, 4
properties, 3–4
states, 4, 31, 95–97, 103
Measurement, significant figures, 565–568
Measurement systems, 10–13, 561–564
Medical imaging, 129–132, 134
Melibiose, 435
Mellaril, 401
Melting, 96, 103
Melting point, 96, 97f, 103
Membranes, 439, 447, 457, 556
Mendeleev, Dmitri Ivanovich, 51–52
Menstrual cycle, 552–554
Mental illness, biochemical basis, 398–401
Meperidine, toxicity, 402t
Mercury, 63
　in barometer, 143–144
　density, 29
Mercury poisoning, 480
Mescaline, toxicity, 402t
Messenger RNA (mRNA), 527–529, 534
Mestranol, structure, 554f
meta-, 317, 321
Metals, 53
Metastable isotope, 128
Met-enkephalin, 488
Meter, 10–11, 31, 561
Methamphetamine, 403
　toxicity, 402t
Methanal, structure, 347f
Methane
　combustion, 58, 301–302
　electron-dot formula, 91–92, 290
　physical properties, 301t, 319
　structure, 290, 292t
Methanoic acid, 363t
Methanol
　classification, 330–331
　physical properties, 332t
　structure, 329
　toxicity, 335–336
Methedrine, 403
Methemoglobinemia, 63
Methionine, 464t
　isoelectric point, 467t
Methoxyflurane, 344
Methyl acetate, 361t
Methyl alcohol, *see* Methanol
Methylammonium ion, structure, 394
N-Methylaniline, structure, 388
Methylbenzene, 316–317
Methyl benzoate, hydrolysis, 370
2-Methylbutanal, 348
Methyl butanoate, structure, 368f
2-Methylbutanoic acid, structure, 362–363
3-Methyl-2-butanone, structure, 349f
Methylbutenes, structure, 308
Methyl butyrate, structure, 368f
Methyl chloride, 302
Methylcyclohexane, structure, 315
Methyl ethanoate, structure, 361t, 368
6-Methyl-3-heptanol, 331
5-Methyl-2-hexene, 303
Methyl isobutyl ketone, structure, 349f
Methyl isopropyl ketone, structure, 349f

Methylpentanes, 296
Methyl-2-pentanone, structure, 349f
4-Methyl-2-pentyne, 313
2-Methyl-2-propanol, 331, 335f
2-Methyl-1-propene, structure, 305
Methyl propyl ketone
　physical properties, 351t
　structure, 349f
Methyl salicylate, structure, 371f
Methylene chloride, 302
Metric conversions, 12t, 563t
Metric prefixes, 11t, 562t
Metric system, 10–13, 31, 561–564
Micelle, 447, 457
Milk of magnesia, 272
Milk sugar (lactose), 416, 423, 432
Milliequivalents, 231
Milligram percent concentrations, 224–225, 249
Millimeters of mercury (mm Hg), 144, 165
Mineral acids, 365
Minerals, 540–542, 558
Minipill, 555
Mirror-image isomerism, 413–415, 432
Mirror-image isomers, 463, 489
Miscibility, 215, 248
Mitochondria, 506, 612
Mixed triglycerides, 440–441, 457
Mixtures, 5, 31, 100, 103, 213, 248
mm Hg (millimeters of mercury), 144, 165
Models, importance in chemistry, 37
Molarity, 226–229, 249
Mole, 178, 206, 228
Molecule, 84
　shapes, 89–93
Molybdenum, 63, 542
Monoglycerides, 451–452
Monolayer, 446, 457
Monomers, 312
Monosaccharides, 413
　reactions, 419–422
　structure, 416–419
Morphine, 372–374, 390, 391, 488
　structure, 373
　toxicity, 402t
Morton, William, 343–344
MRI, 131, 132f, 135
mRNA, 527–529, 534
β-MSH, 472f
Multiplication
　exponentials, 570
　significant figures, 567–568
Muriatic acid, 258
Muscle fatigue, 508, 512
Muscle power, 508–511
Muscular dystrophy, 544t
Mutarotation, 417–419, 432
Myoglobin, 476, 478, 509
Myosin, 508
Myristic acid, 363t, 439t

NAD$^+$, 429f, 430, 503f, 504, 512
NADH, 429f, 430, 504, 512

INDEX

Naphthalene, 391
   structure, 318, 321
β-Naphthol, structure, 341f
Narcotic, 372
   toxicity, 402t
Natural gas, 300t
Negative charge, 9
Nembutal, 403
Neon, 55
   Bohr model, 45
   electron configuration, 72
Neopentane, structure, 294
Net ionic equation, 259–260
Neutrality, electrical, 9
Neutralization, 255, 259–264, 283
Neutron, 41–42
   discovery, 41
   properties, 42t, 64
   role in nuclear fission, 124
Niacin, 545t
Nickel, 63, 310, 541
Nicotinamide, 384f, 545t
Nicotinamide adenine dinucleotide, *see* NAD$^+$
Nicotine
   structure, 391f
   toxicity, 402t
Nicotinic acid, 545t
Night blindness, 544t
Nitric acid, 59, 261
Nitric oxide, 59
Nitro group, 317
p-Nitroaniline, structure, 389f
Nitrogen, 59, 386
   in air, 59, 142t
   energy-level diagram, 48, 49f
   medical use, 165t
   role in bends, 162, 166
Nitrogen dioxide, 59
Nitrogen molecule, 84–85
o-Nitrotoluene, structure, 318f
Nitrous oxide, as anesthetic, 165t, 344
Noble gases, 53, 54–56, 64
   chemical stability, 70–71
   electron configurations, 71, 103
Nonadecane, physical properties, 301t
Nonane
   physical properties, 301t
   structure, 292t
Nonmetals, 53
Nonpolar covalent bond, 86
Nonsaponifiable lipids, 447–450, 457
Norepinephrine (noradrenaline), 398, 403, 552t
Norlutin, structure, 554f
Normal alkanes, 295
   structures and names, 292t
Normality, 231
19-Norprogesterone, structure, 554f
Novocaine, 396
Nuclear bomb, 124–125, 135
Nuclear equations, 122–125
Nuclear fallout, 135
Nuclear fission, 134–135
Nuclear fusion, 125, 135
Nuclear medicine, 127–129

Nuclear model of the atom, 38–43, 64
Nuclear power, 108, 121, 124, 125f
Nuclear radiation
   diagnostic uses, 128–129, 134–135
   effect on living tissue, 126–127
   lethal dose, 126t
   measurement, 118–119, 134
   protection from, 127t
Nuclear reactions, 108
Nuclear reactor, 124
Nuclear symbol, 108–109
Nucleic acids, 462, 534
   polymeric backbone, 520, 534
   secondary structure, 521–523
   structure, 516–521
Nucleoproteins, 462
Nucleosides, 517, 518f, 534
Nucleotides, 516–517, 519, 534
Nucleus, 42, 64, 108

Obesity, 457
Octadecane, physical properties, 301t
Octadecanoic acid, 363t
Octane
   physical properties, 301t
   structure, 292t
Octanoic acid, 363t
1-Octene, physical properties, 303t
Octet of electrons, 71, 103
Octyl acetate, structure, 368f
Octyl alcohol, physical properties, 332t
Octyl ethanoate, structure, 368f
Oils, 441
Oleic acid, 439t, 441
Olive oil, iodine number, 442t
Oncogenes, 537
Oncotic pressure, 247
Opiates, in brain, 488
Opium, 371, 372f, 391
Orbitals, 46–50, 64
Orbits, electron, 44–45, 64
Organic chemistry, 58, 288–289, 320
Orlon, 311t
ortho-, 317, 321
Osmol, 238
Osmolarity, 238, 249
Osmosis, 238–241, 249
   sieve model, 238–239
Osmotic pressure, 240, 247, 249
Oxalic acid, 339, 364t
Oxaloacetic acid, 431
Oxidation, 60, 196, 206
   alcohols, 334–335, 355
   aldehydes, 351–352
   carbohydrates, 419–421
Oxidation numbers, 200–203, 206, 507
Oxidation–reduction reactions, 196–204, 206
Oxidative deamination, 485, 490
Oxidative phosphorylation, 502–505, 507, 512
Oxidizing agents, 198–199, 206
   as antiseptics and disinfectants, 204–205
Oxygen, 59–61, 205
   in air, 60, 141, 142t

   medical use, 160–161, 165t
   storage under pressure, 146–147
   transport in respiration, 163–164
Oxygen debt, 508–509, 512
Oxytocin, 551t
Ozone, 60–61
   as disinfectant, 205

Palladium, as catalyst, 310
Palmitic acid, 363t, 439t
   solubility, 365
Palmitoleic acid, 439t
Palm oil, fatty acid distribution, 441t
Pantothenic acid, 546t
para-, 317, 321
Paraffins, 300–301
Parathyroid hormone, 551t
Partial pressure
   Dalton's law, 158–161, 166
   respiration and, 162–166
Parts per billion concentration (ppb), 225, 249
Parts per million concentration (ppm), 225, 249
Pascal (Pa), 144, 165
Pauling, Linus, 548
Peanut oil, iodine number, 442t
Pellagra, 384, 545t
Penetrating power of nuclear radiation, 116–117, 134
Pentadecane, physical properties, 301t
Pentanal, structure, 347f
Pentane
   physical properties, 301t
   structure, 292t, 294, 296
Pentanedioic acid, 364t
Pentanoic acid, 363t
Pentanones, structure, 349f
Pentenes
   physical properties, 303t
   structure, 308
Pentobarbital, 403–404
   toxicity, 402t
Pentose, 413, 432
Pentothal, 404
Pentyl alcohol, physical properties, 332t
PEP, *see* Phosphoenolpyruvic acid
Peptide, 468, 488
Peptide (amide) bond, 467–468, 489
Percent by mass concentration, 222–224, 249
Percent by volume concentration, 220–222, 249
Percent by weight concentration, 220
Periodic table of the elements, 51–54, 64
   Mendeleev's original, 52t
Periods, in periodic table, 53–54, 64
Permeability, 238
Pernicious anemia, 63, 547t
Petroleum, 300, 320
Petroleum jelly, 320
PET scanning, *see* Positron emission tomography
pH, 273–278, 283
   blood, 280–282
   buffers, 278–280, 283

pH (*continued*)
  proteins and, 469
Phenacetin, structure, 378
Phenobarbital, toxicity, 402t
Phenol, 340, 355, 371
  as antiseptic, 341
  structure, 341f
Phenols, 340–341, 355
  antiseptic and disinfectant, 341
  functional group, 327
  structure, 341f
Phenylalanine, 464t
Phenyl benzoate, structure, 369
2-Phenylethylamine, structure, 401
Phenyl group, structure, 317
2-Phenylpentane, structure, 317
o-Phenylphenol, 341
Phenyl salicylate, 371
Phosgene, electron-dot formula, 92
Phosphate ion, physiological importance, 102
Phosphatides, 446–447, 457
Phosphoenolpyruvic acid, 429f, 499t, 500, 501f
2-Phosphoglyceric acid, 429f
3-Phosphoglyceric acid, 429f, 499t, 500, 501f
Phospholipids, 446, 457
Phosphoric acid, 260–261
  structure, 446
Phosphorus, 59, 541–542
  electron configuration, 50
Phosphorus-30, 121
Phosphorus-32, 129t
Photosynthesis, 190, 204, 206
Phthalic acid, 364t
Phylloquinone, structure, 544t
Physical properties, 4, 31
Physiological saline solution, 237–238
Picric acid, structure, 341
Piperidine, structure, 390
Pituitary gland, 549–550, 558
Plasmid, 532–533
Plasmolysis, 240, 249
Platinum, as catalyst, 310
Pleated sheet, 473–474, 489
Plexiglas, 311t
Plutonium, 137
pOH, 275, 276f, 283
Polar covalent bond, 95–98, 103
Polarity, bond, 86–87
Polio, 149
Polyacrylonitrile, 311t
Polyatomic ions, 94–95, 103
Polycyclic aromatic hydrocarbons, 318
Polyethylene, 311t
Polyhydric alcohols, properties, 340t
Polymerization, 310–312
  amino acids, 467
Polymers, 311t, 312
Poly(methyl methacrylate), 311t
Polypeptide, 468
Polypropylene, 311t
Polysaccharides, 413, 424, 425f, 427, 433
Polystyrene, 311t
Polytetrafluoroethylene, 311t

Polyunsaturated fats, 441–442
Poly(vinyl acetate) (PVA), 311t
Poly(vinyl chloride) (PVC), 311t
Poly(vinylidene chloride), 311t
Positive charge, 9
Positron, 122–123, 134
Positron emission tomography, 130, 132f, 134
Potassium chlorate, as antiseptic, 204
Potassium dichromate, 334–335
Potassium hydroxide
  as base, 255
  in soaps, 443
Potassium ion, 56, 102
Potassium permanganate, as antiseptic, 204
Potential energy, 7–8, 31
Pounds per square inch (psi), 144, 165
ppb (parts per billion), 225, 249
ppm (parts per million), 225, 249
Precipitated chalk, 272
Pregnenolone, 448, 449f
Pressure, 143, 165
  atmospheric, 143–144
  effect on solubility of gas, 161–162
  effect on temperature of gas, 153–156
  effect on volume of gas, 145–150
  units, 144, 165, 563t
Priestley, Joseph, 161, 344
Primary alcohol, 329
Primary amine, 386, 406
Primary structure of proteins, 472–473, 489
Procaine, 396
  toxicity, 402t
Procaine hydrochloride, 396
Products of reaction, 172, 205
Progesterone, 448, 449f, 552t, 555
  structure, 346f
Progestin, 555
Prolactin, 551t
Proline, 464t
Promazine, 400
Proof, alcoholic beverages, 338, 355
Propanal, structure, 347f
Propane
  physical properties, 301t
  structure, 291, 292t
Propanedioic acid, 364t
Propanoic acid, 363t
Propanols, 331
  boiling point, 364
  lethal dose, 336t
  oxidation, 334–335
Propanone, structure, 349f
Propene, 303
Properties
  chemical, 3, 31, 289
  of matter, 3–4
  physical, 4, 31
Propionaldehyde
  physical properties, 351t
  structure, 347f
Propionic acid, 363t
Propyl alcohol, physical properties, 332t
Propylamine
  physical properties, 392t

  structure, 387
Propylene, structure, 303
Propylene glycol
  physical properties, 340t
  physiological properties, 340
  structure, 339
Prostaglandin E_1, structure, 450
Prosthetic group, 480, 489
Proteins, 397, 462
  amino acid sequence, 469–471
  as blood buffers, 281, 283
  in cell membrane, 447f
  classification, 480–481, 489
  deficiency, 466
  denaturation, 479–480, 489
  digestion, 483–488, 490
  formation, 467–468
  isoelectric point, 469
  metabolism, 484–487, 490
  primary structure, 472–473, 489
  secondary structure, 473–474, 489
  synthesis, 527–531, 534
  tertiary structure, 476–477, 489
Proton, 40
  discovery, 40
  properties, 42t, 64
Proton acceptor, 254
Proton donor, 254
Proust, William, 413
psi (pounds per square inch), 144, 165
Psychedelic drug, 378
Psychotomimetic drug, 378
Pure substances, 5, 31
Purine, structure, 390f, 391
Purine bases in nucleic acids, 517, 534
Putrescine, 392
PVA [poly(vinyl acetate)], 311t
PVC [poly(vinyl chloride)], 311t
Pyridine, structure, 390–391
Pyridoxamine, structure, 545t
Pyridoxol, structure, 545t
Pyrimidine, structure, 390f, 391
Pyrimidine bases in nucleic acids, 405, 517, 534
Pyrrole, structure, 390
Pyrrolidine, structure, 390
Pyruvic acid, 340, 429f, 430–431, 500, 501f, 508

Quantum mechanical model of the atom, 46–50, 64
Quaternary ammonium salts, 393
Quaternary structure of proteins, 478, 489

R group, 463
Rad, 118, 134
Radiation adsorbed dose, 118, 134
Radiation syndrome, 126–127
Radiation therapy, 127–128, 129t
Radioactive decay, 114, 115t
Radioactivity, 40, 134
  applications, 132–133
  discovery, 40, 111–112
  half-life, 119–121, 134
  induced, 121–123
  measurement, 118–119

INDEX

penetrating power, 116–118
types, 113–115
Radioisotopes, 127
  medical use, 127–130
Radium, 113
Radium-226, 113, 129t
Radon, 55–56
Radon-86, 114
Ratio concentrations, 226, 249
Reactants, 172, 205
Reaction rates, 190–193, 206
  catalyst effects, 191–193, 206
  concentration effects, 191, 206
  temperature effects, 190–191, 206
Reactions, see Chemical reactions; Nuclear reactions
Recombinant-DNA technology, 532–534
Reconstitution of drugs, 233–234
Redox, 198, 206
Reducing agent, 198–199, 206
Reducing sugars, 420
Reduction, 196–198, 206
Relative humidity, 160, 166
Releasing factors, hormone, 549, 551t, 558
Rem, 126, 134
Replication, DNA, 524–526
Reserpine, 399–400
Respiration
  blood pH and, 282, 284
  partial pressures and, 162–166
Respiratory alkalosis, 282
Respiratory chain, 504, 505f, 512
Respiratory therapy, 149–150, 160–161
Restriction endonucleases, 532
Restriction fragment length polymorphisms, 532
Retinol, 328f, 544t
Reversible reactions, 190, 193–195, 206
RFLPs, 532
Riboflavin, 545t
Ribonuclease, isoelectric point, 469t
Ribonucleic acid, see RNA
Ribose, 354, 416, 516–517, 520
  structure, 415f, 418f
Ribosomes, 528, 534
Rickets, 543f, 544t
Right-handed helix, 474
RNA, 516, 523, 527–531, 534
  components, 520t
Roentgen, 118, 134
Roentgen, Wilhelm Konrad, 111
Roentgen equivalent in man, 126, 134
Rubbing alcohol, 339
Rutherford, Ernest, 38, 40, 112, 121

Saccharides, 413, 432–433
Safflower oil, iodine number, 442t
Salicylic acid, 371
Salol, 371f
Salt, 260, 283
Salt bridge, 476, 489
Salts
  amines, 394–396
  in aqueous solution, 264–268, 283
  carboxylic acids, 366–367, 380
  fatty acids, 370
Saponifiable lipids, 447
Saponification, 370, 380
  fats, 443–445, 457
Saturated fats, in foods, 443t
Saturated hydrocarbons, 289–302, 320
Saturated solution, 218–219, 248–249
Schlittler, Emil, 400
Schrödinger, Erwin, 46–47
Scientific notation, 569
Scurvy, 542, 546t
Secobarbital, 404
  toxicity, 402t
Seconal, 404
Second law of thermodynamics, 495, 512
Secondary alcohols, 329
Secondary amine, 386–387, 406
Secondary structure
  nucleic acids, 521–523
  proteins, 473–474, 489
Sedation, 372
Semipermeable membranes, 238–239
Serine, 463, 464t
Serotonin, 398
Serum albumin, 481
  isoelectric point, 469t
Serylalanylcysteine, 468
Shock, 248
SI, 10, 561–564
Sickle cell anemia, 471
Sieve model of osmosis, 238–239
Significant figures, 565–568
Silk, protein structure, 473–474
Silk fibroin, isoelectric point, 469t
Simple amide, 374
Simple protein, 480, 489
Simple triglycerides, 440, 457
Single bond, 103
Slow twitch muscle fibers, 509–510, 512
Soap, 370, 443–445, 451, 457
Sodium, Bohr model, 46
Sodium-24, 129t
Sodium acetate, structure, 367
Sodium bicarbonate, 269, 289
  as antacid, 271
Sodium carbonate, 268, 269
Sodium chloride, 74, 236–237
  in body fluids, 542
  crystal structure, 96
  dissolution, 101, 215
  properties, 289t
  structure, 72–74
Sodium compounds, light emission, 44
Sodium ethanoate, structure, 367
Sodium glycocholate, 450
Sodium hydroxide
  as base, 258–259, 266
  electron-dot formula, 94
  reaction with esters, 370
  in soap making, 443, 444f
Sodium hypochlorite, 205
Sodium iodide, 541
Sodium ion
  electron configuration, 71–73
  physiological importance, 56, 102
Sodium oleate, structure, 370f
Sodium palmitate, structure, 370f, 445f
Sodium salicylate, 371
Sodium stearate, structure, 370f, 444f
Solids
  ionic, 101
  kinetic molecular theory, 95–96
  properties, 4, 31
Solubility, 215, 396
  lipids, 438–439
  temperature and pressure effects, 216–217, 248–249
Solute, 100, 103–104, 213, 248
  dissolution, 215
Solutions, 5f, 100, 103–104, 213–215, 248–249
  in body, 246–249
  colligative properties, 236–238, 249
  dilution, 234–235
  forces in, 100–102
  hypertonic, 240, 249
  hypotonic, 240, 249
  isotonic, 241
  measurement of very low concentrations, 224–225, 249
  molarity, 226–229, 249
  percent concentrations, 219–224, 249
  properties, 214
  ratio concentrations, 226, 249
  saturated, 218–219, 248–249
  See also Aqueous solutions
Solvent, 100, 103, 213, 248
Sorbitol, physical properties, 340t
Sorensen, S. P. L., 275
Soybean oil
  fatty acid distribution, 441t
  iodine number, 442t
Specific gravity, 28–29, 31
  blood, 29–30
  human fat, 29
Spectator ion, 255
Speed (methamphetamine), 403
Spontaneous reactions, 495–496, 512
Sports medicine, 510–511
Standard base, 263
Standard conditions of temperature and pressure (STP), 159
Standard free energy change, 498, 499t
Starch, 425, 433
Starting materials, 172, 205
Starvation, 282
  chemistry, 456–457
States of matter, 4, 31
  bonding forces, 95–97
  changes in, 96–97, 103
Static cling, 9–10
Stearic acid, 363t, 439t, 440–441
Steroids, 447–449, 457
  synthetic, 554f, 555
Stimulants, toxicities, 402t
Stock solutions, 234–235, 249
Stone, Edward, 371
STP (standard temperature and pressure), 159
Strong acids, 257, 283
Strong solutions, 220
Structural formulas, 290

Subatomic particles, 39, 64
  properties, 42t
Substituent group, 296, 303
Substituted amide(s), 374–375
  formation, 396–398
Substitution reactions, 309
Substrate, 481–482, 489
Subtraction, significant figures, 566–567
Succinic acid, 364t
Sucrose, 423–424
  formation, 496–497
  hydrolysis, 426–427
Sugars, 412
  D and L, 415–416, 432
  reducing, 421, 433
Sulfhydryl group, 453
Sulfur, 60–61, 540
  oxidation, 60, 271
  properties, 4
Sulfur dioxide, 60–61, 271
Sulfur trioxide, 61, 271
Sulfuric acid, 61, 258, 262, 271, 333
Sulfurous acid, 60
Sunflower oil, iodine number, 442t
Supersaturated solution, 219, 249
Suspension, 242, 243t, 249
Sweat, 507
Symbols
  chemical, 5–6, 6t–7t, 31
  nuclear, 108–109
Synergism, 405
Synthetic detergents, 445
Système International (SI), 10

Table salt, 74, 96
  dissolution, 101, 215
  iodized, 541
Table sugar, 423
Tannic acid, 480
Tartaric acid, 210
Tautomerism, 433
Tautomerization, 419–420
Technetium-99m, 128, 129t, 134
Teflon, 311t
Temperature, 24–25, 31, 143, 165
  body, 190–191
  effect on pressure of gas, 153–156
  effect on reaction rates, 190–191, 206
  effect on solubility, 216–217, 248
  effect on volume of gas, 150–153
  equivalents, 564t
  scales, 22–24
Terephthalic acid, 364t
Tertiary alcohols, 329
Tertiary amine, 386, 406
Tertiary structure of proteins, 476–477, 489
Testosterone, 449, 552t
  structure, 346f
Tetrachloromethane, 302
Tetracosanol, structure, 328f
Tetradecanoic acid, 363t
Tetrahydrofuran, structure, 342f
Tetramethylammonium ion, structure, 394
Tetrose, 413, 432

Thalidomide, 379
Thermodynamics, 494–495, 512
Thermonuclear bomb, 124–125, 135
Thiamine, 545t
Thiopental, 404–405
  toxicity, 402t
Thorazine, 400
Thoridazine, 401
Three Mile Island, 117–118
Threonine, 464t
Thymidine monophosphate, structure, 519f
Thymine, 517, 522f
Thyroid (gland), 62, 128, 129t, 541, 550f, 551t, 552
Thyroid-stimulating hormone, 551t
Thyroxine, 465t, 535, 551t, 552
Tidal volume, 149
Titration, 263–264, 283
TNT, 318f
α-Tocopherol, 544t
Tollens, Bernhard, 352
Tollens' reagent, 352, 356, 419–420, 432–433
Toluene, structure, 317
o-Toluidine, structure, 389f
Torr, 144, 165
Torricelli, Evangelists, 144
Transamination, 484–485, 490
Transcription, 527, 534
Transfer RNA (tRNA), 528–530, 534
Trans isomer, 306, 320
Transition metals, 53, 62–64
Translation, 528, 534
Transmutation, artificial, 121–123, 133–134
Trichloromethane, 302
Tridecane, physical properties, 301t
Triglycerides, 440–443, 457
  digestion, 450–452
Trimethylamine
  physical properties, 392t
  reaction with nitric acid, 394
  structure, 388
Trimethylammonium ion, structure, 394
Trimethylammonium nitrate, 394
1,3,5-Trimethylbenzene, structure, 318f
2,2,4-Trimethylpentane, 296
2,4,6-Trinitrotoluene, structure, 318f
Triose, 413, 432
Tripeptide, 468
Triple bond, 85, 93, 312–313
Triple helix, 475
Triplet of bases, nucleic acids, 528, 534
Triprotic acid, 256, 283
Tristearin, 440
  saponification, 444f
Tritium, 43, 114, 129t
  half-life, 119–120
tRNA, 528–530, 534
Tryptophan, 465t
  isoelectric point, 467t
Tyndall effect, 243, 249
Type I muscle fibers, 509–510, 512

Type IIB muscle fibers, 509–510, 512
Tyrosine, 465t

UDP, 428f
Ultrasonography, 130–131, 134
Undecane, physical properties, 301t
Unit-conversion method, 13–21, 31
Unit, medication strength, 234, 249
Unsaturated fatty acids, 439–443
Unsaturated hydrocarbons, 289, 320
Unsaturated solution, 218
Uppers, 401, 403
Uracil, 517
Uranium, 112
  discovery of radioactivity, 40
Uranium-235, nuclear fission, 124
Uranium-238, half-life, 120
Urea, 484–485
Uridine diphosphate, 428f
Uridine monophosphate, structure, 519f
Uridine, structure, 518f
U.S.P. unit, 234

Valence electrons, 74
Valence energy level, 103
Valeraldehyde
  physical properties, 351t
  structure, 347f
Valeric acid, 363t
  odor, 365
Valine, 464t
Valium, 405–406
Vanadium, 63
Vanillin, structure, 346f
Vaporization, 96–97, 103
Vapor pressure, water, 159–160
Vapor pressure depression, 237
Vaseline, 320
Vasopressin, 551t
Vinegar, 213, 269, 360, 362, 365
Viruses, 462
Vital capacity, 149
Vital force, 285
Vitamin A, 544t, 548
  structure, 328f
Vitamin B_1, 545t
Vitamin B_2, 545t
Vitamin B_6, 545t
Vitamin B_{12}, 63, 547t
Vitamin C (ascorbic acid), 542, 543, 546t
  effect on bone formation, 543f
  therapy, 548–549
Vitamin D, 543, 548
Vitamin D_2, 544t
  structure, 328f
Vitamin E, 544t
Vitamin K_1, 544t
Vitamins, 542–549, 558
Volume, metric units, 10, 562t
Volume-pressure relationship of gases. *See* Boyle's law
Volume-temperature relationship of gases. *See* Charles's law

INDEX

Wastewater treatment, 205
Water
  as a dipole, 88
  in barometer, 144
  boiling point, 22
  composition, 5
  density, 28
  drinking, treatment, 205
  electron-dot formula, 87
  freezing point, 22
  heat capacity, 25
  hydrogen bonding, 99
  ion product, 273, 283
  properties, 213–214
  salts in, 264–268, 283
  as solvent, 101, 102f, 213–214
Water-soluble vitamins, 543, 545t–547t, 548, 558
Water vapor, partial pressure in gas mixtures, 159–160
Watson, James D., 521, 523, 525
Weak acids, 257, 283
Weak solutions, 220
Weight, 3, 31
Wood alcohol, 336
Wool, protein structure, 473–475

Work, 7

Xenon, 55
X rays, 111–112, 134
  use in medical imaging, 129–130
Xylenes, 318f
Xylocaine, structure, 397f

Yeast, in fermentation, 336–337

Zinc, 63, 483, 542
Zwitterion, 446, 457, 462–463, 467, 489

Symbols and Names for Simple Ions

Group	Element	Name of Ion	Symbol for Ion
IA	Lithium	Lithium ion	Li^+
	Sodium	Sodium ion	Na^+
	Potassium	Potassium ion	K^+
IIA	Magnesium	Magnesium ion	Mg^{2+}
	Calcium	Calcium ion	Ca^{2+}
IIIA	Aluminum	Aluminum ion	Al^{3+}
VA	Nitrogen	Nitride ion	N^{3-}
VIA	Oxygen	Oxide ion	O^{2-}
	Sulfur	Sulfide ion	S^{2-}
VIIA	Chlorine	Chloride ion	Cl^-
	Bromine	Bromide ion	Br^-
	Iodine	Iodide ion	I^-
IB	Copper	Copper(I) ion (cuprous ion)	Cu^+
		Copper(II) ion (cupric ion)	Cu^{2+}
	Silver	Silver ion	Ag^+
IIB	Zinc	Zinc ion	Zn^{2+}
VIIIB	Iron	Iron(II) ion (ferrous ion)	Fe^{2+}
		Iron(III) ion (ferric ion)	Fe^{3+}

Some Common Polyatomic Ions

Charge	Name	Formula
1+	Ammonium ion	NH_4^+
1−	Hydrogen carbonate (bicarbonate) ion	HCO_3^-
	Hydrogen sulfate (bisulfate) ion	HSO_4^-
	Acetate ion	$CH_3CO_2^-$ (or $C_2H_3O_2^-$)
	Nitrite ion	NO_2^-
	Nitrate ion	NO_3^-
	Cyanide ion	CN^-
	Hydroxide ion	OH^-
	Dihydrogen phosphate ion	$H_2PO_4^-$
2−	Carbonate ion	CO_3^{2-}
	Sulfate ion	SO_4^{2-}
	Monohydrogen phosphate ion	HPO_4^{2-}
	Oxalate ion	$C_2O_4^{2-}$
3−	Phosphate ion	PO_4^{3-}